Chapman & Hall/CRC Biostatistics Series

Published Titles

1. *Design and Analysis of Animal Studies in Pharmaceutical Development,* Shein-Chung Chow and Jen-pei Liu
2. *Basic Statistics and Pharmaceutical Statistical Applications,* James E. De Muth
3. *Design and Analysis of Bioavailability and Bioequivalence Studies, Second Edition, Revised and Expanded,* Shein-Chung Chow and Jen-pei Liu
4. *Meta-Analysis in Medicine and Health Policy,* Dalene K. Stangl and Donald A. Berry
5. *Generalized Linear Models: A Bayesian Perspective,* Dipak K. Dey, Sujit K. Ghosh, and Bani K. Mallick
6. *Difference Equations with Public Health Applications,* Lemuel A. Moyé and Asha Seth Kapadia
7. *Medical Biostatistics,* Abhaya Indrayan and Sanjeev B. Sarmukaddam
8. *Statistical Methods for Clinical Trials,* Mark X. Norleans
9. *Causal Analysis in Biomedicine and Epidemiology: Based on Minimal Sufficient Causation,* Mikel Aickin
10. *Statistics in Drug Research: Methodologies and Recent Developments,* Shein-Chung Chow and Jun Shao
11. *Sample Size Calculations in Clinical Research,* Shein-Chung Chow, Jun Shao, and Hansheng Wang
12. *Applied Statistical Design for the Researcher,* Daryl S. Paulson
13. *Advances in Clinical Trial Biostatistics,* Nancy L. Geller
14. *Statistics in the Pharmaceutical Industry, Third Edition,* Ralph Buncher and Jia-Yeong Tsay
15. *DNA Microarrays and Related Genomics Techniques: Design, Analysis, and Interpretation of Experiments,* David B. Allsion, Grier P. Page, T. Mark Beasley, and Jode W. Edwards
16. *Basic Statistics and Pharmaceutical Statistical Applications, Second Edition,* James E. De Muth
17. *Adaptive Design Methods in Clinical Trials,* Shein-Chung Chow and Mark Chang
18. *Handbook of Regression and Modeling: Applications for the Clinical and Pharmaceutical Industries,* Daryl S. Paulson
19. *Statistical Design and Analysis of Stability Studies,* Shein-Chung Chow
20. *Sample Size Calculations in Clinical Research, Second Edition,* Shein-Chung Chow, Jun Shao, and Hansheng Wang
21. *Elementary Bayesian Biostatistics,* Lemuel A. Moyé
22. *Adaptive Design Theory and Implementation Using SAS and R,* Mark Chang
23. *Computational Pharmacokinetics,* Anders Kallen

Chapman & Hall/CRC Biostatistics Series

Computational Methods in Biomedical Research

Edited by

Ravindra Khattree
Oakland University
Rochester, Michigan, U.S.A.

Dayanand N. Naik
Old Dominion University
Norfolk, Virginia, U.S.A.

Chapman & Hall/CRC is an imprint of the
Taylor & Francis Group, an **informa** business

Chapman & Hall/CRC
Taylor & Francis Group
6000 Broken Sound Parkway NW, Suite 300
Boca Raton, FL 33487-2742

Printed in the United States of America on acid-free paper
10 9 8 7 6 5 4 3 2 1

International Standard Book Number-13: 978-1-58488-577-1 (Hardcover)

Library of Congress Cataloging-in-Publication Data

Computational methods in biomedical research / editors, Ravindra Khattree and Dayanand Naik.
p. ; cm. -- (Biostatistics series ; 24)
"A CRC title."
Includes bibliographical references and index.
ISBN 978-1-58488-577-1 (alk. paper)
1. Medicine--Research--Data processing. 2. Biology--Research--Data processing. 3. Medicine--Research--Statistical methods. 4. Biology--Research--Statistical methods. 5. Computational biology. I. Khattree, Ravindra. II. Naik, Dayanand N. III. Series: Chapman & Hall/CRC biostatistics series ; 24.
[DNLM: 1. Computational Biology--methods. 2. Biomedical Research--methods. 3. Data Interpretation, Statistical. QU 26.5 C73756 2008]

R853.D37C63 2008
610.285--dc22 2007029936

Visit the Taylor & Francis Web site at
http://www.taylorandfrancis.com

and the CRC Press Web site at
http://www.crcpress.com

Contents

Series Introduction

The primary objectives of the Biostatistics Book Series are to provide useful reference books for researchers and scientists in academia, industry, and government, and also to offer textbooks for undergraduate and graduate courses in the area of biostatistics. This book series will provide comprehensive and unified presentations of statistical designs and analyses of important applications in biostatistics, such as those in biopharmaceuticals. A well-balanced summary will be given of current and recently developed statistical methods and interpretations for both statisticians and researchers or scientists with minimal statistical knowledge who are engaged in the field of applied biostatistics. The series is committed to providing easy-to-understand, state-of-the-art references and textbooks. In each volume, statistical concepts and methodologies will be illustrated through real-world examples.

In the last decade, it was recognized that increased spending on biomedical research does not reflect an increase in the success rate of pharmaceutical development. On March 16, 2004, the FDA released a report addressing the recent slowdown in innovative medical therapies submitted to the FDA for approval, "Innovation/Stagnation: Challenge and Opportunity on the Critical Path to New Medical Products." The report describes the urgent need to modernize the medical product development process—the critical path—to make product development more predictable and less costly. Two years later, the FDA released a Critical Path Opportunities List that outlines 76 initial projects (under six broad topic areas) to bridge the gap between the quick pace of new biomedical discoveries and the slower pace at which those discoveries are currently developed into therapies. Among the six broad topic areas, better evaluation tool (development of biomarker), streamlining clinical trial (the use of adaptive design methods), and harnessing bioinformatics (the use of computational biology) are considered the top three challenges for increasing the probability of success in pharmaceutical research and development.

This volume provides useful approaches for implementation of target clinical trials in pharmaceutical research and development. It covers statistical methods for various computational topics such as biomarker development, sequential monitoring, proportional hazard mixed-effects models, and Bayesian approach in pharmaceutical research and development. It would be beneficial to biostatisticians, medical researchers, pharmaceutical scientists, and reviewers in regulatory agencies who are engaged in the areas of pharmaceutical research and development.

Shein-Chung Chow

Preface

This edited volume is a collection of chapters covering some of the important computational topics with special reference to biomedical applications. Rapid advances in ever-changing biomedical research and methodological statistical developments that must support these advances make it imperative that from time to time a cohesive account of new computational schemes is made available for users to implement these methodologies in the particular biomedical context or problem. The present volume is an attempt to fill this need.

Realizing the vastness of the area itself, there is no pretension to be exhaustive in terms of the general field or even in terms of a topic represented by a chapter within this field; such a task, while also requiring hundreds of collaborators, would require a collection of several volumes of similar size. Hence the selection made here represents our personal view of what the most important topics are, in terms of their applicability and potential in the near future. With this in mind, the chapters are arranged accordingly, with the works of immediate applicability appearing first. These are followed by more theoretical advances and computational schemes that are yet to be developed in satisfactory forms for general applications.

Work of this magnitude could not have been accomplished without the help of many people. We wish to thank our referees for painstakingly going through the chapters as a gesture of academic goodwill. Theresa Del Forn of Taylor & Francis Group, was most helpful and patient with our repeatedly broken promises of meeting the next deadline. Our families have provided their sincere support during this project and we appreciate their understanding as well.

Ravindra Khattree, Rochester, Michigan
Dayanand N. Naik, Norfolk, Virginia

Editors

Ravindra Khattree, professor of statistics at Oakland University, Rochester, Michigan, received his initial graduate training at the Indian Statistical Institute. He received his PhD from the University of Pittsburgh in 1985. He is the author or coauthor of numerous research articles on theoretical and applied statistics in various national and international journals. His research interests include multivariate analysis, experimental designs, quality control, repeated measures, and statistical inference. In addition to teaching graduate and undergraduate courses, Dr. Khattree regularly consults with industry, hospitals, and academic researchers on various applied statistics problems. He is a chief editor of the *Journal of Statistics and Applications*, editor of *InterStat*, an online statistics journal, and an associate editor of the *Journal of Statistical Theory and Practice*. For many years, he also served as an associate editor for the *Communications in Statistics*. He is a Fellow of the American Statistical Association, an elected member of the International Statistical Institute, and a winner of the Young Statistician Award from International Indian Statistical Association. Dr. Khattree is a coauthor of two books, both with Dr. D. N. Naik, titled *Applied Multivariate Statistics with SAS Software (Second Edition)* and *Multivariate Data Reduction and Discrimination with SAS Software*, both copublished by SAS Press/Wiley. He has also coedited, with Dr. C. R. Rao, the *Handbook of Statistics 22: Statistics in Industry*, published by North Holland.

Dayanand N. Naik is professor of statistics at Old Dominion University, Norfolk, Virginia. He received his MS degree in statistics from Karnatak University, Dharwad, India, and PhD in statistics from the University of Pittsburgh in 1985. He has published extensively in several well-respected journals and advised numerous students for their PhD in statistics. His research interests include multivariate analysis, linear models, quality control, regression diagnostics, repeated measures, and growth curve models. Dr. Naik is an editor of *InterStat*, a statistics journal on the Internet, and an associate editor of *Communications in Statistics*. Dr. Naik is also actively involved in statistical consulting and collaborative research. He is an elected member of International Statistical Institute and is very active in American Statistical Association activities. Dr. Naik is coauthor of two books, both with Dr. Khattree, titled *Applied Multivariate Statistics with SAS Software (Second Edition)* and *Multivariate Data Reduction and Discrimination with SAS Software*, both copublished by SAS Press/Wiley.

Contributors

Mousumi Banerjee Department of Biostatistics, School of Public Health, University of Michigan, Ann Arbor, Michigan

N. Rao Chaganty Department of Mathematics and Statistics, Old Dominion University, Norfolk, Virginia

Kathleen A. Cronin Statistical Research and Applications Branch, National Cancer Institute, Bethesda, Maryland

Somnath Datta Department of Bioinformatics and Biostatistics, School of Public Health and Information Sciences, University of Louisville, Louisville, Kentucky

Susmita Datta Department of Bioinformatics and Biostatistics, School of Public Health and Information Sciences, University of Louisville, Louisville, Kentucky

Michael Donohue Division of Biostatistics and Bioinformatics, Department of Family and Preventive Medicine, University of California at San Diego, La Jolla, California

Eric J. Feuer Statistical Research and Applications Branch, National Cancer Institute, Bethesda, Maryland

Joseph C. Gardiner Division of Biostatistics, Department of Epidemiology, Michigan State University, East Lansing, Michigan

Kaushik Ghosh Department of Mathematical Sciences, University of Nevada at Las Vegas, Las Vegas, Nevada

Pulak Ghosh Department of Mathematics and Statistics, Georgia State University, Atlanta, Georgia

Ahmedin Jemal Department of Epidemiology and Surveillance Research, American Cancer Society, Atlanta, Georgia

Ravindra Khattree Department of Mathematics and Statistics, Oakland University, Rochester, Michigan

Lin Liu Division of Biostatistics, Department of Epidemiology, Michigan State University, East Lansing, Michigan

Lin Liu Department of Family and Preventive Medicine, University of California at San Diego, La Jolla, California

Hedibert F. Lopes Graduate School of Business, University of Chicago, Chicago, Illinois

Zhehui Luo Division of Biostatistics, Department of Epidemiology, Michigan State University, East Lansing, Michigan

Deepak Mav Constella Group, Inc., Durham, North Carolina

Jarosław Meller Division of Biomedical Informatics, Cincinnati Children's Hospital Research Foundation, Cincinnati, Ohio; Department of Informatics, Nicholas Copernicus University, Torun, Poland; and Department of Environmental Health, University of Cincinnati, Ohio

Peter Müller Department of Biostatistics, The University of Texas, M. D. Anderson Cancer Center, Houston, Texas

Dayanand N. Naik Department of Mathematics and Statistics, Old Dominion University, Norfolk, Virginia

Anne-Michelle Noone Department of Biostatistics, School of Public Health, University of Michigan, Ann Arbor, Michigan

Rudolph S. Parrish Department of Bioinformatics and Biostatistics, School of Public Health and Information Sciences, University of Louisville, Louisville, Kentucky

Nalini Ravishanker Department of Statistics, University of Connecticut, Storrs, Connecticut

William F. Rosenberger Department of Statistics, The Volgenau School of Information Technology and Engineering, George Mason University, Fairfax, Virginia

Anuradha Roy Department of Management Science and Statistics, The University of Texas at San Antonio, San Antonio, Texas

Caryn M. Thompson Department of Bioinformatics and Biostatistics, School of Public Health and Information Sciences, University of Louisville, Louisville, Kentucky

Ram C. Tiwari Statistical Research and Applications Branch, National Cancer Institute, Bethesda, Maryland

Florin Vaida Division of Biostatistics and Bioinformatics, Department of Family and Preventive Medicine, University of California at San Diego, La Jolla, California

Michael Wagner Division of Biomedical Informatics, Cincinnati Children's Hospital Research Foundation, Cincinnati, Ohio

Ronghui Xu Division of Biostatistics and Bioinformatics, Department of Family and Preventive Medicine, and Department of Mathematics, University of California at San Diego, La Jolla, California

Yanqiong Zhang Merck & Company, Rahway, New Jersey

1

Microarray Data Analysis

Susmita Datta, Somnath Datta, Rudolph S. Parrish, and Caryn M. Thompson

CONTENTS

1.1 Introduction

Microarray technology has quickly become one of the most commonly used high throughput systems in modern biological and medical experiments over the past 8 years. For most parts, a single microarray records the expression levels of several genes in a tissue sample—this number often runs in tens of thousands. At the end, a huge multivariate data set is obtained containing the gene expression profiles. A microarray experiment typically compares the expression data with two or more treatments (e.g., cell lines, experimental conditions, etc.); additionally, there is often a time component in the experiment. Owing to the relatively high production cost of microarrays, oftentimes very few replicates are available for a given set of experimental conditions that pose new challenges for the statisticians in analyzing these data sets.

Most of the early microarray experiments involved the so-called two-channel cDNA microarrays where small amounts of genetic materials (cRNA) are printed on a small glass slide with robotic print heads. The mRNA samples corresponding to two different treatments are tinted with two different fluorescent dyes (generally red and green) and allowed to hybridize (a technical term for a biological process by which an mRNA strand attaches to the complementary cDNA strand) on the same slide. At the end, the expression values of the sample under comparison are evaluated with certain specialized laser scanners. In more recent studies, the oligonucleotide arrays, also known as the Affymetrix GeneChips®, are becoming increasingly popular. These are factory-prepared arrays targeted for a particular genome (e.g., rat, humans, etc.) that contain oligonucleotide materials placed in multiple pairs—called a probe set (http://www.affymetrix.com/products/system.affx). One of each pair contains the complementary base sequences for the targeted gene; however, the other one has an incorrect base in the middle created to measure nonspecific bindings during hybridization that can be used for background correction. Expression values are computed by the relative amounts of bindings (perfect match versus perfect mismatch).

Besides the above two microarray platforms, there exist many additional choices at present including many custom arrays offered by various manufactures; in addition, serial analysis of gene expression (SAGE), which is technically not a microarray-based technique, produces gene expression data as well. Unlike microarrays, SAGE is a sequencing-based gene expression profiling technique that does not require prior knowledge of the sequences to be considered. Another important difference between the two is that, with SAGE, one does not need a normalization procedure (see Section 1.3) since it measures abundance or expression in an absolute sense.

Calculating expression itself is an issue with most, if not all, microarray platforms; in addition, there are issues of normalizations and correction for systematic biases and artifacts, some of which are discussed in Section 1.3. In addition, there have been recent studies comparing multiple microarray platforms and the amount of agreement between them. The very latest set of results (see, e.g., Irizarry et al., 2005) contradict earlier beliefs about non-reproducibility of microarray gene expression calculations and concludes that the laboratories running the experiments have more effect on the final conclusions than the platforms. In other words, two laboratories following similar strict guidelines would get similar results (that are driven by biology) even if they use different technologies. On the other hand, the "best" and the "worst" laboratories in this study used the same microarray platform but got very different answers.

In this review, we present a brief overview of various broad topics of microarray data analysis. We are particularly interested in statistical aspects of microarray-based bioinformatics. The selection of topics is, by no means, comprehensive partly because new statistical problems are emerging every day in this fast growing field. A number of monographs have come out in recent years (e.g., Causton et al., 2003; Speed, 2003; Lee, 2004; McLachlan et al., 2004; Wit and McClure, 2004), which can help an interested reader gain further familiarity and knowledge in this area.

The rest of the chapter is organized as follows. Some commonly employed statistical designs in microarray experiments are discussed in Section 1.2. Aspects of preprocessing of microarray data that are necessary for further statistical analysis are discussed in Section 1.3. Elements of statistical machine learning techniques that are useful for gaining insights into microarray data sets are discussed in Section 1.4. Hypothesis testing with microarray data is covered in Section 1.5. The chapter ends with a brief discussion of pathway analysis using microarrays as a data generation tool.

1.2 Experimental Design

In planning biological experiments, including microarray experiments, the researcher should be aware of and follow sound statistical principles. Each of these principles, as outlined below, serves a particular purpose in ensuring a valid experiment. Properties of a "good" experiment include the absence of systematic error (or bias), precision, and simplicity (Cox, 1958). The experiment should also permit calculation of uncertainty (error) and ideally have a wide range of validity.

Microarray experiments have realistic limitations that must be taken into account at the design stage. These include the cost of arrays, restrictions on the number of arrays that can be processed at one time, and the amount of material necessary to hybridize an array. In addition, there may be little or no prior information on variability available at the planning stage of a microarray experiment.

In the discussion that follows, we describe experimental design principles, primarily as applied to oligonucleotide arrays. Dual-channel arrays require specialized designs, which are discussed briefly in Section 1.2.7. While the principles described here are general, in this section we focus mainly on experiments to detect differential gene expression.

1.2.1 Data from Microarray Experiments

For most purposes, data from a microarray experiment can be described as multidimensional array of expression values Y. Usually, the first dimension (row) corresponds to genes or probe sets or ORFs. Depending on the experiment, the other dimensions may represent replicates (biological and technical), tissue types, time, and so on.

Usually, a preprocessing or normalization step is applied to the microarray readouts to calculate an accurate and "unbiased" measure of gene expression before additional statistical analyses are carried out. We describe various normalization methods in Section 1.3. In addition, certain statistical analysis may have an implicit normalization step.

1.2.2 Sources of Variation

When designing a microarray experiment, it is essential to be aware of the many sources of variation, both biological and technical, which may affect experimental outcomes. Biological variation is essentially that among subjects (i.e., natural subject-to-subject variability). Different subjects from which samples are obtained are not identical in a multitude of characteristics and thereby have gene expressions that vary as well. This type of variation occurs normally and is used as a benchmark when testing for differential expression. Biological variation is reflected in experimental error.

Technical variation includes errors or effects due to instrumentation, measurement, hybridization, sample preparation, operator, and other factors that serve to add unwanted variation to the data (Churchill, 2002; Parmigiani et al., 2003). These factors generally are uncontrolled, meaning that their presence is not only unintended but also often unavoidable. Technical variation includes systematic variation among or within arrays. It is described generally as an *array effect*. There may also be variation that is the result of how the experiment is designed or carried out; that is, variation that is due to controlled or identifiable factors. An example is variation due to batch where arrays within certain groups all share a common effect. Variation of this type usually can be accounted for through statistical analysis. This type of variation often is considered part of technical variation, and attempts can be made to eliminate it before data analysis.

Several authors, including Spruill et al. (2002), Kendziorski et al. (2005), and Zakharkin et al. (2005), have attempted to assess relative sizes of several typical sources of variation in microarray studies through experimentation. As a word of caution, it should be noted that each of these studies was conducted under a particular set of experimental conditions, and it is unclear to what extent these results may be generalized.

1.2.3 Principles of Experimental Design

The three primary principles of experimental design, namely randomization, replication, and blocking, each have a particular purpose, and are often attributed to R. A. Fisher. In discussing these principles, it is important to note that an *experimental unit* is defined to be the smallest division of experimental material such that any two units may receive different treatments in the actual experiment.

Randomization requires that experimental units be randomly allocated to treatment conditions. The random allocation may be restricted in some way, for instance, through blocking, depending on the design of the experiment. Randomization is intended to protect against bias or systematic error. As emphasized by Kerr (2003), randomization should be applied to minimize bias induced by technical artifacts, such as systematic variation in arrays within a batch according to the order in which they were printed. Also, randomization must take into account processing constraints. For example, suppose an experiment involving three treatments is run using four arrays per treatment, and it is possible to process only eight arrays per day. If the arrays from the first two treatments were processed on Day 1, and the arrays from the third treatment were processed on Day 2, it may be impossible to separate a treatment effect (whereby certain genes are differentially expressed across treatments) from a day effect (whereby expression levels for all genes tend to be higher on a particular day). That is, the treatment and day effects would be *confounded*.

Replication implies having at least one experimental unit per treatment condition, and is necessary in order to permit estimation of experimental error or variance. The microarray literature distinguishes between *biological* and *technical* replication. Biological replication refers to the number of independent biological (experimental) units assigned to each treatment condition, whereas technical replicates arise from repeated sampling of the same experimental unit and will therefore be correlated (Cui and Churchill, 2003). Increasing replication, in particular biological replication, also provides a means of increasing power and precision for comparison of treatment means. In this context, biological replicates are considered "true" replicates, whereas the technical replicates are subsamples and might be considered "pseudoreplicates." Issues related to pseudoreplication have been debated at length in the ecological literature for more than 20 years (Hurlbert,

1984, 2004; Oksanen, 2001). In particular, treating technical replicates as biological replicates in a microarray experiment tends to underestimate true experimental error, thus inflating type I error and overstating statistical significance.

Blocking involves grouping similar units together before assignment to treatment conditions, with the goal of reducing experimental error. In complete block designs, each treatment is randomly applied at least once within each block. Blocking accounts for one (or more) major source of extraneous variability, in addition to the one due to the treatments. If blocking is part of the experimental design, it should also be incorporated in the analysis.

1.2.4 Common Designs for Oligonucleotide Arrays

The simplest design is one in which two or more treatment groups are to be compared with respect to differential gene expression, at a particular time point. The design in which experimental units are assigned to treatments in an unrestricted random fashion is known as a *completely randomized design*.

Several authors, including Kerr and Churchill (2001 a,b) and Kendziorski et al. (2003) have recommended analysis of data arising from microarray experiments through linear models, provided necessary assumptions, including normality and homoscedasticity of model errors are satisfied. Data from a completely randomized design may be analyzed using a simple one-way analysis of variance (ANOVA) model for a single gene, with

$$Y_{ij} = \mu + T_i + \varepsilon_{ij}$$

where Y_{ij} is the background-corrected and normalized intensity (usually expressed on the $\log_2$ scale) of a particular gene for the jth experimental unit assigned to the ith treatment, μ is an average intensity level, T_i is the effect of the ith treatment, and the ε_{ij}'s are the model error terms, usually assumed to be independently and identically distributed with mean 0 and common variance σ^2 (i.e., $\varepsilon_{ij} \overset{iid}{\sim} N(0, \sigma^2)$).

A natural extension to mixed linear models for oligonucleotide microarray data was proposed Chu et al. (2002a), whereby analyses may be conducted on a gene-by-gene basis at the probe level, accounting for variation among experimental units within treatments, among arrays within experimental unit, and among probes for a particular gene within arrays. Following Chu et al. (2002a) a mixed linear model corresponding to a completely randomized design with one treatment factor is given by

$$Y_{ijk} = T_i + P_j + TP_{ij} + A_{k(i)} + \varepsilon_{ijk}$$

where Y_{ijk} is the background corrected and normalized measurement at the jth probe in the kth replicate for the ith treatment, T_i is the fixed effect of the

ith treatment, P^j is the fixed effect of the jth probe, and $A_{k(i)}$ is the random effect for the kth array (assuming arrays represent biological replicates) of the ith treatment. The usual model assumptions are that $A_{k(i)}^{iid} \sim N(0, \sigma_a^2)$ and $\varepsilon_{ijk}^{iid} \sim N(0, \sigma^2)$ with $A_{k(i)} + \varepsilon_{ijk} \sim N(0, \sigma_a^2 + \sigma^2)$ and

$$\text{Cov}(A_{k(i)} + \varepsilon_{ijk}, A_{k'(i')} + \varepsilon_{i',j',k'}) = \begin{cases} \sigma_a^2 + \sigma^2 & \text{if } (i,j,k) = (i',j',k'), \\ \sigma_a^2 & \text{if } i = i',\ k = k',\ j \neq j' \\ 0 & \text{otherwise,} \end{cases}$$

where σ_a^2 is the variance component associated with variation among arrays (biological replicates) within treatments.

A design in which blocks account for one extraneous source of variation, and in which experimental units are randomly applied to treatments with the restriction that each treatment appears once within each block, is known as a *randomized complete block design*. A fixed or random blocking factor may be easily incorporated into the mixed model above, through the addition of a main effect and appropriate interaction terms involving the blocks. The mixed models described here analyze each gene separately. Alternative approaches, in which all genes are included in a single large mixed linear model, have been proposed by Kerr and Churchill (2001 a,b), Wolfinger et al. (2001), and Kendziorski et al. (2003). Examples of these models are discussed in Sections 1.3.2.5 and 1.5.2.

More complex designs permit accounting for additional sources of variation. Recently, Tsai and Lee (2005) describe the use of *split plot designs* in the context of two-channel microarray experiments. Designs of this type may also find application in analysis of data arising from oligonucleotide arrays, as discussed by Chu et al. (2002a) and Li et al. (2004). For example, consider an experiment involving rat embryos with two factors: genotype (with three levels) and developmental stage (with two levels). A practical restriction of the experiment is that it will be possible to harvest material to hybridize only one array of each genotype (using pooled samples, discussed below) for a particular developmental stage at a time. Each of these sets of three arrays is replicated four times per developmental stage, using a total of 24 arrays. This is an example of a completely randomized design with a split plot arrangement, which is characterized by the two factors being applied to different experimental units. Here, the experimental unit for developmental stage (the *whole plot factor*) is a set of three arrays, although the experimental unit for genotype (*the sub plot factor*) is a single array. The model is given by

$$Y_{ijk} = D_i + R_{k(i)} + G_j + DG_{ij} + \varepsilon_{ijk}$$

where Y_{ijk} is the background corrected and normalized measurement for jth probe in the kth replicate for the ith developmental stage, D_i is the fixed effect of the ith developmental stage, G_j is the fixed effect of the jth genotype, and

$R_{k(i)}$ is the random effect for the kth replicate set of three arrays of the ith genotype. Model assumptions are similar to those for corresponding terms in the probe level mixed model presented above.

1.2.5 Power/Sample Size Considerations

An important planning issue is the question of how many biological and technical replicates to include in a microarray experiment. The number of replicates required to detect a particular size of effect with a pre-specified level of power depends not only on the several sources contributing to variation in expression levels, but also on the statistical methods used to determine differential expression, including the choice of normalization procedure. In general, biological replicates should be favored over technical replicates when the objective is to detect differentially expressed genes.

A number of authors, including Lee and Whitmore (2002), Yang et al. (2003), Gadbury et al. (2004), Jung (2005), Zhang and Gant (2005), and Tibshirani (2006) have addressed this issue. Desired minimum detectable effect sizes are often expressed in terms of *fold-changes*. For example, the objective may be to identify genes that are up or down regulated to the extent that they exhibit a twofold increase or decrease in expression levels. To answer the sample size question, prior knowledge is required with respect to the variance components associated with the various levels of the experimental structure for each gene. Estimates of these variance components may come from previously conducted similar experiments, or pilot studies. Variance components are discussed in Section 1.3.1.

In the context of an ANOVA model, Lee and Whitmore (2002) present several approaches for calculation of power and sample size, and outline applications of their methods to some standard experimental designs. Their methods account for control of the *false discovery rate* (FDR) due to multiple testing. Multiple testing issues are discussed in detail in Section 1.5. Gadbury et al. (2004) introduce the concept of *expected discovery rate* (EDR), and propose corresponding methods for power and sample size calculations adjusting for EDR. The web-based resource *Power Atlas* (Page et al., 2006), available at www.poweratlas.org, implements this methodology, currently for experiments involving just two groups, but with anticipated expansion. This resource is designed to provide researchers with access to power and sample size calculations on the basis of a number of available data sets, which may be used in lieu of conducting a pilot study to generate variance component estimates.

Recently, Tibshirani (2006) has proposed a sample size determination procedure requiring less stringent assumptions than other currently available methods. This method does not assume, for instance, equal variances or independence among genes. This approach may be implemented using the significance analysis of microarrays (SAM) software (Chu et al., 2002a) to obtain estimates of both the FDR and false negative rate (FNR) as a function of total sample size through a permutation-based analysis of pilot data.

1.2.6 Pooling

The primary motivation for pooling is to reduce the effect of biological variability and improve cost effectiveness. It is important to stress, as do Kendiorski et al. (2005) and Zhang and Gant (2005), that pooling may artificially reduce the effects of biological variability, so that statistical significance is overstated. Disadvantages of pooling include difficulty in estimating appropriate variance components, and loss of information on individuals. However, in many experiments, such as developmental studies involving extraction of tissues from mouse embryos, the amount of available RNA per individual is limiting. Therefore, an approach involving either pooling of embryo tissue or RNA amplification is necessary. A potential problem with amplification is that it tends to introduce noise, and may be nonlinear, in that all genes may not be amplified at the same rate.

One of the issues associated with the design and analysis of microarray experiments is the question of whether, and under what conditions, the pooling of RNA samples before hybridization of arrays can be beneficial. Although originally controversial (Affymetrix, 2004), it is now generally acknowledged that pooling can be useful in certain circumstances (Allison et al., 2006).

Various authors, including Kendziorski et al. (2003, 2005), Peng et al. (2003) and Zhang and Gant (2005) have discussed effects of pooling from a statistical perspective. Kendziorski et al. (2005) considered "M on n_p" pooling schemes, in which mRNA from $M = n_p \times n$ subjects was used to hybridize n_p arrays (n subjects per pool). On the basis of an experiment involving 30 female rats to compare the effects of a number of pooling schemes empirically, they concluded that there is little benefit to pooling when variability among subjects is relatively small, but suggest that pooling can improve accuracy when fewer than three arrays are available per treatment condition. In experiments with three or more arrays available per treatment condition, pooling is only likely to be beneficial if a large number of subjects contribute to each pool.

Pooling relies on the assumption of *biological averaging*, that is, that the response for each individual contributes equally to the average calculated for the pool. For example, if each array is hybridized with mRNA from n individuals, and the concentration of mRNA transcripts from kth individual for the ith gene in the jth pool is Y_{ijk}, the concentration of the ith gene's transcripts in the jth pool will be

$$\frac{1}{n}\sum_{k=1}^{n} Y_{ijk}$$

if biological averaging holds. There has been some debate as to whether this assumption is realistic. However, Kendziorski et al. (2005) observed that biological averaging occurred for most, but not all, genes involved in their experiment and argue that there is some support for the validity of this assumption.

Kendziorski et al. (2005) also introduce the notion of *equivalent designs*, and illustrate this concept through comparison of the various pooling schemes considered in their experiment. Two designs are considered to be equivalent if they lead to the same average estimation efficiency. This is an important cost-related consideration, in that it may be possible to reduce the number of arrays required to achieve a desired level of precision. However, two designs equivalent in average efficiency do not necessarily provide similar precision for individual genes.

1.2.7 Designs for Dual-Channel Arrays

Dual-channel, or spotted cDNA arrays, requires the use of specialized designs, which are discussed only briefly here. Interested readers may consult Rosa et al. (2005) for further details. The two most common classes of these designs are the *reference designs* and the *loop designs* (Kerr and Churchill, 2001a,b; Churchill, 2002; Kerr, 2003).

Despite reported advantages of loop designs, reference designs are still commonly used in practice. The term reference design refers to the fact that at least one sample is included in the experiment, to which all other samples are compared with respect to hybridization. There are several variants of this type of design. Steibel and Rosa (2005) describe three of these, and assign each a distinct name. The first of these, the *classical reference design* (Kerr and Churchill, 2001a) employs replicates of the reference sample. In contrast, the *common reference design* uses only a single reference sample as the basis for comparison. Yet another variant is the *replicated reference design*, where the reference is biologically relevant, and is considered to be a control treatment with replicates.

The basic loop design was proposed by Kerr and Churchill (2001a,b) and is essentially a balanced incomplete block design. In such designs, the number of treatments is greater than the block size. It is used when three or more treatments are to be compared using two-channel microarrays. In the microarray context, each two-channel microarray is viewed as a block of size two. Treatments are then compared with each other in a multiple pairwise fashion laid out in a daisy chain (e.g., 1 vs. 2, 2 vs. 3, 3 vs. 1). Variants of the loop design include *connected loop* design (Dobbin et al., 2003), where the same biological sample is used in two arrays connected in a daisy chain but with different dye (channel) and *interwoven loop* design. In the later, the efficiency of comparison is improved by creating additional (multiple) links amongst the treatments to be compared (e.g., 1 vs. 2, 2 vs. 3, 3 vs. 4, 4 vs. 1, 1 vs. 3, 2 vs. 4, 3 vs. 1, 4 vs. 2).

Among the reference designs, the classical reference design is generally considered to have lower statistical efficiency than other forms of reference designs. Templeman (2005) compared several design alternatives for two-channel arrays with respect to precision, power, and robustness, assuming a mixed model analysis and a single treatment factor. He found that except for cases with minimum replication (two replicates per treatment) the loop designs were superior to the reference designs in terms of smaller standard errors for treatment comparisons, higher power, and greater robustness. Although the loop designs were found to be more sensitive to missing arrays, the interwoven loop design is reported to be more robust to missing arrays than the classical loop design, and yet more statistically efficient than the common reference design. Vinciotti et al. (2005) also performed an experimental comparison of the common reference vs. the classical loop design, and found that loop designs generally have much higher precision. On the basis of further simulation studies they concluded that, for a given sample size, the classical loop design will have higher power than the common reference design to detect differentially expressed genes, while minimizing type I error.

Advances are continually being made in the area of improved designs for multichannel arrays. Recently, Wit et al. (2005) describe a procedure to find near-optimal interwoven loop designs, using the A-optimality criterion. Steibel and Rosa (2005) propose a blocked reference design, which tends to be more efficient and less expensive than common reference designs. Woo et al. (2005) present preliminary guidelines for efficient loop designs for three- and four-color microarrays.

1.3 Normalization of Microarray Data

There are different factors that contribute to variation in the observed data from microarray experiments. Typically, this variability is characterized as being of two types: biological and technical, as discussed in the previous section.

Normalization is broadly defined as any technique meant to remove or account for technical variation in the data before statistical testing or analysis. Less noise in the data, as reflected in the estimate of experimental error, translates into a stronger capability to detect differential expression (i.e., treatment effects) of smaller magnitudes (see also Parrish and Spencer, 2004). If technical variation can be removed, substantially reduced, or adjusted out, the power of statistical tests will be enhanced.

Normalization procedures mainly involve revising the observed data before statistical analysis in an effort to remove the technical variability. This relies on the assumptions that most genes should not be affected by the treatment or condition under study and that the other factors affect all or most genes in the same or similar manner. The particular assumptions depend on the method used for normalization. In the simplest case, revision of the data based on these assumptions may be the only alternative for reducing technical variation. In essence, normalization is a mechanism to "borrow" information from other variables in order to correct identifiable deficiencies in the data. The objectives of normalization also may be achieved in part by incorporating into statistical models used to analyze the data adjustment terms corresponding to known effects. This does not require assumptions about how most genes are affected by various factors and the analysis typically is conducted on a gene-by-gene basis.

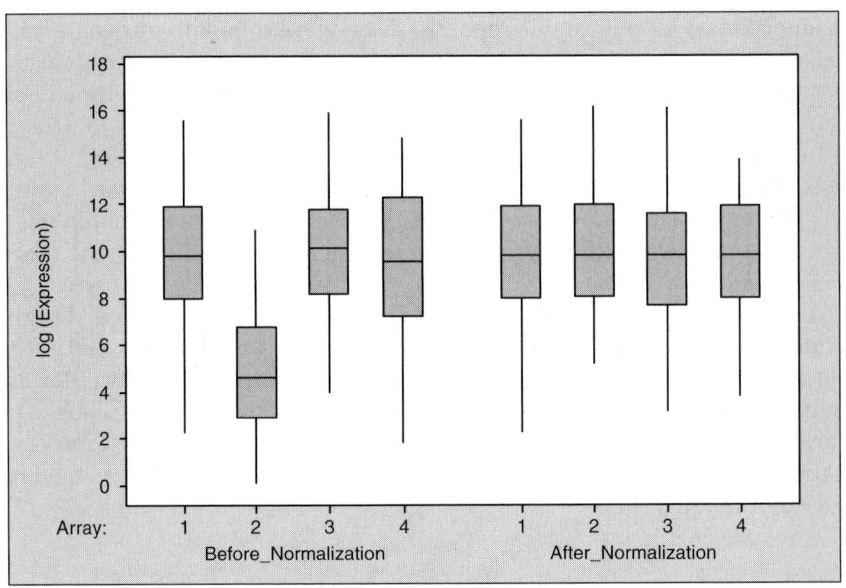

FIGURE 1.1
Panel (left) showing sample data as box plots with one of four being stochastically lower; Panel (right) showing similar box plots after normalization.

To illustrate, consider an experiment involving eight subjects, each corresponding to a unique array, who are allocated to two treatments, with four arrays per treatment. No other experimental conditions are controlled. The experimental error is based on the variation observed among arrays treated alike. The objective is to test for differential expression for all genes individually using a t-test. This can be done without normalization. If there is no systematic variation, this approach is valid and correct. Now suppose the conduct of the experiment was such that some of the arrays were affected in a manner that caused all expression readings on an affected array to be underestimated. This could be detected by comparing the collective distribution of gene expression values for each array. If one array had a distribution differing in location or dispersion characteristics compared to the majority of arrays, it might be possible to adjust all the values on the affected array so that the resulting distribution was consistent with the other arrays. Failure to make such an adjustment would result in excessive variability making its way into the error term used for testing (i.e., the denominator of the t statistic would be inflated) (see Figure 1.1).

Continuing with this example, suppose the arrays were processed in groups of two, each group being on a different day; therefore, on each day, one array from the first treatment and one from the second treatment were used. The design now has an identifiable controlled factor, DAY, which is

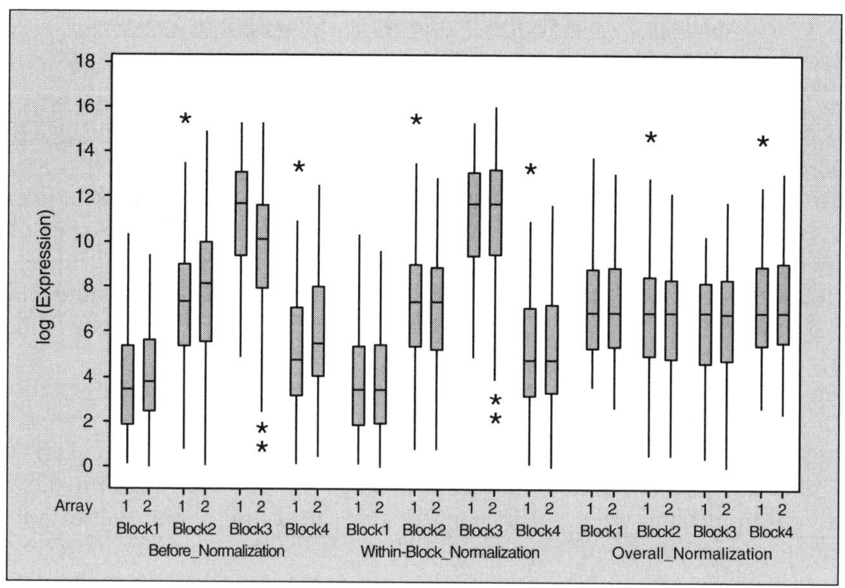

FIGURE 1.2
Panel (left) showing four groups of two box plots with some variability before normalization; Panel (middle) showing after within-group normalization; Panel (right) showing overall normalization.

a "block" effect. The experimental error now corresponds to variation due to interaction of DAY and TREATMENT (denoted DAY × TREATMENT), and the appropriate test in the ANOVA setting is an F-test. The effect of DAY can be adjusted out through the linear model used for analysis; however, any other systematic variation that impacts arrays will not be adjusted unless normalization is done. In this situation, normalization could be accomplished by comparing distributions and adjusting all expression values, as above, but doing so within each DAY grouping, followed by an ANOVA that accounts for the effect of DAY. An alternative approach is to apply a normalization method on all arrays simultaneously so as to achieve similar distributions within each array. In this case, the effect of DAY is presumably removed, and a t-test is then applied (see Figure 1.2).

Background effects may or may not be addressed by a given normalization procedure. Such techniques are directed primarily toward the objective of removing local bias, but they may also serve to reduce variability in certain situations, including those where spatial variability exists. In general, background effects may be relatively large for low expression values, and thus smaller expression values may be more significantly impacted by background correction methods. Methods such as robust multiarray average (RMA) (Bolstad et al., 2003) attempt to estimate the conditional expectation of the true gene expression level given the observed measurement.

1.3.1 Normalization and Its Implications for Estimation of Variance Components

The experimental design that is employed determines how treatment effects are tested. There are a few simple designs that are commonly used with microarray experiments, but others may be more complex. Consider the following example where there are two treatments ($k = 2$), three biological (i.e., experimental) units per treatment group ($n = 3$), and two arrays per experimental unit ($r = 2$).

It is helpful to consider the expected mean squares and the associated variance components in relation to how normalization impacts on them. These are considered for single-channel arrays.

In this design, there are two variance components: one due to variation among biological units treated alike within treatment groups (which gives rise to an estimate of "experimental error") and one due to variation among arrays within biological units. Using multiple arrays for each biological unit allows for better precision when estimating the biological unit response. Variation among arrays within units gives rise to what would normally be termed "measurement error." The correct test to use for assessing the differential effect of treatments is to form an F ratio of the mean squares for "Treatments" and "Among units (Trt)."

Strictly, from an analysis standpoint, there is no requirement to normalize the data. If many arrays could be used for each biological unit, variation among arrays would be merely a nuisance and would not impact significantly on the test for treatment effect. On the other hand, normalization could be used to try to eliminate or decrease the measurement error and thereby improve the power of the test.

As shown in Table 1.1, in the case where there is only one array per experimental unit ($r = 1$), the variance component for measurement error σ_a^2 still exists, however it is not distinguishable mathematically from the unit-to-unit variance component σ_u^2. Normalization attempts to remove unwanted variation by adjusting (typically) all the values for some arrays so that their characteristics are more similar to those of the other arrays. The net desired impact is to decrease measurement error substantially, but it may also lower the estimate of the among units variance (i.e., among biological units treated alike). When normalization is applied, it modifies an observation on an array, and hopefully this modification corrects measurement error. The correction

TABLE 1.1

Expected Mean Squares for Nested Design

Source	DF	Expected Mean Square
Treatments	$k - 1 = 1$	$Q_{\text{Trt}} + r\sigma_u^2 + \sigma_a^2$
Among Units (Trt)	$k(n - 1) = 4$	$r\sigma_u^2 + \sigma_a^2$
Arrays (Units Trt)	$kn(r - 1) = 6$	σ_a^2

will also impact on the mean of the arrays within biological units, and if it is such that the variation among the means of arrays within biological units is also reduced, then the impact shows up as a reduction in the estimate of the among units variance component, a smaller experimental error term, a larger F (or t) statistic, and a smaller p value. This is considered appropriate so long as the normalization procedure does not improperly deflate the true variance component associated with biological units treated alike. Current methods do not ensure that such super-deflation does not occur. Each particular normalization method carries with it explicit or implied assumptions.

1.3.2 Normalization Methods

Several methods for performing normalization have appeared in the microarray literature. In this section, some but not all of these are discussed. Some methods use a baseline array to establish a reference point to which other arrays are adjusted, whereas other methods use the "complete data" from all arrays simultaneously. Each method has characteristics that may be advantageous in different circumstances, but there is no agreement on which method is best for most situations. Some of the important and more relevant methods are presented here.

1.3.2.1 Method Based on Selected Invariant Genes

If it can be assumed that genes in a known set are unaffected by the treatment, subjects, or other factors present in the experiment, they can be used to make adjustments to other gene expression values (Gieser et al., 2001; Hamalainen et al., 2001). Such a list is generally specified before the experiment is conducted (Vandesompele et al., 2002) and, in fact, some microarrays are designed with so-called "housekeeping," "control," or "reference" genes. Usually, the values of all gene expression values on a given array are corrected so that the means of the reference genes for each array are all equal. This method can be simply defined as follows. Let m_i = mean of the expression values of the housekeeping genes on array i. Adjust the values from array i ($i > 1$) according to

$$y_{ij}^* = y_{ij} - (m_i - m_1).$$

The result is that all arrays then will have the same mean for the housekeeping genes.

1.3.2.2 Methods Based on Global or Local Values

Expression values can be adjusted so that every array will have the same mean or, equivalently, average signal intensity. In one approach, an array is chosen arbitrarily as a baseline reference array and the following adjustments are made:

$$y_{ij}^* = y_{ij} \times \left(\frac{y_{\text{baseline}}^{(m)}}{y_i^{(m)}} \right)$$

where $y_i^{(m)}$ is the mean of all the expression values for the *ith* array and $y_{\text{baseline}}^{(m)}$ is the mean of all the expression values for the selected baseline array (Affymetrix, 2002). The means can be ordinary or robust (e.g., trimmed means). GeneChip uses a scaling factor (SF) similar to above except that the numerator can be an arbitrary value (e.g., 500) and the means are 2% trimmed means. This approach is termed *linear scaling* and is based on an assumption of a linear relationship between corresponding values from the two arrays.

Nonlinear scaling approaches modify the linearity assumption to allow nonlinear relationships. In a method proposed by Schadt et al. (2001), a set of invariant genes is identified on the basis of rank order of the expression values from each array compared to the baseline array. They use a generalized cross-validation smoothing spline algorithm (Wahba, 1990) and produce a transformation of the form

$$y_{ij}^* = f_i(y_{ij}).$$

where f_i represents the nonlinear scaling function. This algorithm was implemented in dChip software (Schadt et al., 2001). As alternatives, a piecewise running median regression can be used (Li and Wong, 2001) and locally-weighted regression (loess) can be used on probe intensities (Bolstad et al., 2003).

A third approach involves adjusting expression values so that all arrays have the same mean. Array-specific values can be added to all expression values within the arrays that will make the array means or medians all equal. In this method, one array may be selected as a reference array and the intensity values on the other arrays are adjusted so that their means equal the mean of the reference array. This approach also can be implemented by using the mean of the array means (or the median of the array medians) as a target value and then making adjustments to values on all arrays. This process is represented as

$$y_{ij}^* = y_{ij} + (y_{.}^{(m)} - y_i^{(m)})$$

where y_{ij} is the expression value (or its logarithm) for array i and gene j, $y_i^{(m)}$ is the mean (or median) for array i, and $y_{.}^{(m)}$ is the mean of the array means (or median of array medians) computed over all arrays.

Further adjustments can be made so that all arrays have the same variability. The expression data can be scaled so that all arrays have the same or nearly the same variance, range, or other similar measure. From a distribution perspective, this method is meant to obtain the same dispersion characteristics in the data from the different arrays. This is implemented simply by applying a scaling adjustment for each array, in the form of a multiplication factor. In addition, a location adjustment can be incorporated as above. As is commonplace, the scaling constants can be based on the interquartile range (IQR) and a location adjustment can be based on the median, thereby giving the adjusted response as

$$y_{ij}^* = \left(y_{ij} - y_i^{(m)}\right) \times \left(\frac{D_{.}}{D_i}\right) + y_{.}^{(m)}$$

where D_i represents the selected measure of dispersion for the ith array and $D_.$ is the mean, median, or maximum of the D_i values over all arrays. If the IQR is used, all arrays will have approximately the same IQR after adjustment. Quasi-ranges other than the IQR (which is based on the 25th and 75th percentiles) could be used. For example, the normal range is based on the 2.5th and 97.5th percentiles.

1.3.2.3 Local Regression Methods

The M vs. A (MvA) plot (Dudoit et al., 2002) is a graphic tool to visualize variability as a function of average intensity. This plots values of M (difference of logs) vs. A (average of logs) for paired values. These are defined mathematically as

$$M_j = \log_2(y_{ij}^{(1)}) - \log_2(y_{ij}^{(2)}) \quad \text{and} \quad A_j = \frac{[\log_2(y_{ij}^{(1)}) + \log_2(y_{ij}^{(2)})]}{2}$$

for array i and gene (or probe) j. With two-channel arrays, the pairing is based on the data from the same spot corresponding to the red and green dyes. A loess smoothing function is fitted to the resulting data. Normalization is accomplished by adjusting each expression value on the basis of deviations, as follows:

$$\log_2(y_{ij}^{(1)*}) = A_j + 0.5\,M_j' \quad \text{and} \quad \log_2(y_{ij}^{(2)*}) = A_j - 0.5\,M_j'$$

where M_j' represents the deviation of the jth gene from the fitted line. This is a within-array normalization.

The MvA plot can be used in single-channel arrays by forming all pairs of arrays to generate MvA plots; this method is known as cyclic loess (Bolstad et al., 2003). The proposed algorithm produces normalized values through an iterative algorithm. This approach is not computationally attractive for experiments involving a large number of arrays.

1.3.2.4 Quantile-Based Methods

Quantile normalization forces identical distributions of probe intensities for all arrays (Bolstad et al., 2003; Irizarry et al., 2003a; Irizarry et al., 2003c; Irizarry et al., 2003b). The following steps may be used to implement this method: (1) All the probe values from the n arrays are formed into a $p \times n$ matrix (p = total number of probes on each array and n = number of arrays), (2) Each column of values is sorted from lowest to highest value, (3) The values in each row are replaced by the mean of the values in that row, (4) The elements of each column are placed back into the original ordering. The modified probe values are used to calculate normalized expression values.

1.3.2.5 Methods Based on Linear Models

Analysis of variance models have been used to accomplish normalization as an intrinsic part of the statistical analysis in two-channel arrays (Kerr et al., 2000). These can incorporate effect terms for array, dye, treatment, and gene, as in:

$$y_{ijkg} = \mu + A_i + T_j + D_k + G_g + AG_{ig} + TG_{jg} + e_{ijkg}$$

where μ is the mean log expression, A_i is the effect of the ith array effect, T_j is the effect of the jth treatment, D_k is the effect of the kth dye, G_g is the gth gene effect, and AG_{ig} and TG_{jg} represent interaction effects. Testing for differential expression is based on the TG interaction term. Normalization is achieved intrinsically by inclusion of the array, dye, and treatment terms in the model.

A mixed-effects linear model has been proposed for cDNA data (Wolfinger et al., 2001) in which the array effect is considered random, and data normalization is intrinsic to the model, as above. This model is given by

$$y_{ijg} = \mu + A_i + T_j + AT_{ij} + G_g + AG_{ig} + TG_{jg} + e_{ijg}.$$

To deal with computational issues, the authors recommended fitting the following model and then utilizing its residuals for effects analysis:

$$y_{ijg} = \mu + A_i + T_j + AT_{ij} + r_{g(ij)}.$$

The residuals are the normalized values and are treated as the dependent variables in the model

$$r_{ijg} = G_g + AG_{ig} + TG_{jg} + e_{ijg}$$

which can be written for specific genes as

$$r_{ij}^{(g)} = \mu^{(g)} + A_i^{(g)} + T_j^{(g)} + e_{ij}^{(g)}.$$

The approach also may be used with single-channel arrays. A probe-level mixed-effects model has been described (Chu et al., 2002b). Other normalization methods have been proposed that involve ANOVA and mixed-effects models (Chen et al., 2004) as well as subsets of the genes.

1.3.2.6 Probe Intensity Models

The RMA method (Bolstad et al., 2003) includes background adjustment based on convolution of gamma and normal distributions, quantile normalization of probe intensities, and a probe-level linear model fitted using robust techniques. This is implemented in Bioconductor R-based software (Bioconductor, 2003).

Wu and Irrizary (2005) described a modification of the RMA method, termed GCRMA, in which the G–C base content of probes (based on probe sequence information) is taken into account. This approach differs from RMA only in the way the background adjustment is done. Higher G–C content relates to probe affinity and is associated generally with increased binding.

Probe-level models (PLM) effectively fit models involving a probe effect and an array effect to probe intensity data on a probe-set-by-probe-set basis. This is implemented in Bioconductor R-based software known as *affyPLM*.

1.4 Clustering and Classification

Various clustering and classification methods are routinely used with microarray gene expression data. Both clustering and classification are strategies to group data (units) into collections that are similar in nature. In the machine learning literature, clustering methods are also known as unsupervised learning since no prior knowledge of the underlying grouping is utilized. In the microarray context, clustering has been primarily used as an exploratory tool to group genes into classes with the hope that the genes in a given cluster will have similar biological functions or cellular role. Classification, on the other hand, uses a training set of units whose group memberships are known. In the microarray context, this has been used mostly for classifying tissues (e.g., cancer vs. noncancer) using their gene expression profiles. In this section, we present an overview of some aspects of these techniques that are important for microarray data analysis.

1.4.1 Clustering

Microarray data involves expression on levels of thousands of genes, often recorded over a set of experiments resulting in a collection of expression profiles. A natural step in summarizing this information is to group the genes according to the similarity–dissimilarity of their expression profiles. Second, one of the central goals in microarray or expression data analysis is to identify the changing and unchanging levels of gene expression, and to correlate these changes to identify sets of genes with similar profiles. Finally, even in well-studied model systems like the yeast *Saccharomyces cerevisiae* or bacterium *Escherichia coli* (found in our waste and sewage) the functions of all genes are presently unknown. If genes of unknown function can be grouped with genes of known function, then one can find some clues as to the roles of the unknown genes. It is, therefore, desirable to exploit available tools for clustering and classifications from numerical taxonomy and statistics (Sokal and Sneath, 1963; Hartigan, 1975).

In some earlier microarray experiments, (DeRisi et al., 1997; Cho et al., 1998; Chu et al., 1998) a mainly visual analysis was performed in grouping genes into functionally relevant classes. However, this method is virtually impossible for more complicated and large-scale studies. In subsequent studies, simple sorting of expression ratios and some form of "correlation distance" were used to identify genes (Eisen et al., 1998; Roth et al., 1998; Spellman et al., 1998). Hierarchical clustering Unweighted Pair Group Method with Arithmetic mean (UPGMA) with correlation "distance" (or dissimilarity) is most often used in microarray studies after it was popularized by the influential paper by Eisen et al. (1998). One nice feature of a hierarchical clustering is that it produces a tree of clusters, also known as a dendrogram, which can be cut at various heights to see the resulting clusters. Datta (2001) introduced a novel dissimilarity measure between a pair of genes, in the presence of the remaining genes, on the basis of partial least squares (PLS) modeling gene expressions. Model-based clustering is another well-known technique that has been used for grouping microarray data (McLachlan et al., 2002). This technique is based on modeling the expression profiles by mixtures of multivariate normal distributions. The Gene Shaving algorithm (Hastie et al., 2000) allowed genes to be in more than one cluster at the same time. Sharan and Shamir (2000) introduced a novel clustering algorithm on the basis of a graph-theoretic concept. Also, there exists another noteworthy, but more complex algorithm (based on a self-organizing neural network) called SOM (self-organizing maps, Kohonen, 1997) which seems to have gained popularity in clustering microarray data.

Besides the clustering techniques mentioned above, there exist numerous clustering algorithms in statistics and machine learning literature dating back to premicroarray days. Many of them are available in statistical packages such as S-Plus and R. Thus, a microarray data analyst has many choices of clustering methods when it comes to grouping genes; for example, partition methods such as *K-means* (Hartigan and Wong, 1979), divisive clustering method *Diana* (Kaufman and Rousseeuw, 1990) and fuzzy logic–based method *Fanny* (Kaufman and Rousseeuw, 1990) are all applicable. All these clustering algorithms have their own merits and demerits. Their results might appear to be substantially different as well even when applied to the same data set of expression profiles. We take the following illustration from Datta and Arnold (2002) where a yeast data set (Chu et al., 1998) is clustered using five different clustering techniques (Figure 1.3). Thus, there is a need for careful evaluation of clustering algorithm given a particular data set. There exist a few approaches regarding the selection of a clustering algorithm and validation of the results for microarray data.

Kerr and Churchill (2001c) used a linear model (ANOVA) and residual-based resampling to access the reliability of clustering algorithms. Chen et al. (2002) compared the performances of a number of clustering algorithms by physical characteristics of the resulting clusters such as the homogeneity and separation. Yeung et al. (2001) introduced the concept of *Figure of Merit* (FOM) in selecting between competing clustering algorithms. FOM resembles the *error sum of squares* (ESS) criterion of model selection. Datta and Datta (2003) selected six clustering algorithms of various types and evaluated their performances (stability) on a well-known publicly available microarray dataset on sporulation of budding yeast, as well as on two simulated data sets. Here we provide a brief description of the stability measures introduced in

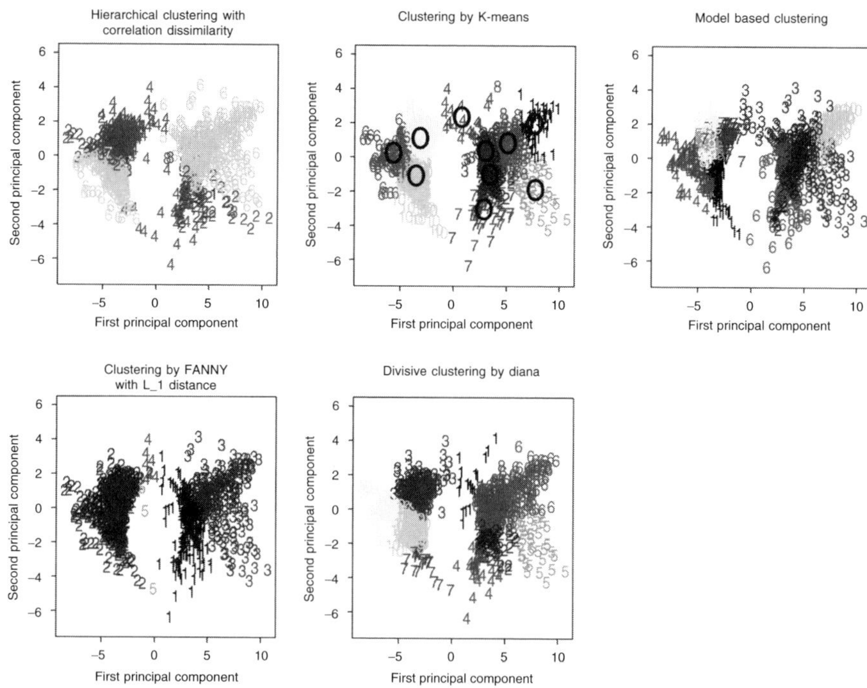

FIGURE 1.3
(See color insert following page 207.) The genes were clustered into seven groups using their expression profiles during sporulation of yeast; five different clustering algorithms were attempted (Adapted from Datta, S. and Arnold, J. (2002). In *Advances in Statistics, Combinatorics and Related Areas*, C. Gulati, Y.-X. Lin, S. Mishra, and J. Rayner, (Eds.), World Scientific, 63–74.)

Datta and Datta (2003) in order to select an algorithm that produces the most consistent results. The evaluation measures were general enough so that they can be used with any clustering algorithm.

Let K be the number of desired clusters. Datta and Datta (2003) suggested that the performance of an algorithm be investigated over an entire range of "suitable" K values. The basic idea behind their validation approach was that a clustering algorithm should be rewarded for stability (i.e., consistency of clusters it produces). Suppose expression (ratio) data are collected over all the genes under study at various experimental conditions such as time points $T_1, T_2, \ldots, T_l$. An example of such temporal data is the sporulation of yeast data of Chu et al. (1998). In that case K was around 7 (Chu et al. used $K = 7$) and number of time points $l = 7$. Thus, consider a setup where the data values are points in the l dimensional Euclidean space R^l. For each $i = 1, 2, \ldots, l$, one repeats the clustering algorithms for each of the l data set in R^{l-1} obtained by deleting the observations at experimental condition (e.g., time) T_i. For each gene g, let $C^{g,i}$ denote the cluster containing gene g in the clustering on the basis of data set with time T_i observations deleted. Let $C^{g,0}$ be the cluster in the original data containing gene g. Each of the following validation measures could be used to measure the stability of the results produced by the clustering algorithm in question. For a good clustering algorithm, one would expect these stability measures to be small.

1. The average proportion of nonoverlap measure is given by

$$V_1(K) = \frac{1}{Ml} \sum_{g=1}^{M} \sum_{i=1}^{l} \left(1 - \frac{n(C^{g,i} \cap C^{g,0})}{n(C^{g,0})}\right).$$

   This measure computes the (average) proportion of genes that are not put in the same cluster by the clustering method under consideration on the basis of the full data and the data obtained by deleting the expression levels at one experimental condition at a time.

2. The average distance between means measure is defined as

$$V_2(K) = \frac{1}{Ml} \sum_{g=1}^{M} \sum_{i=1}^{l} d(\bar{x}_{C^{g,i}}, \bar{x}_{C^{g,0}})$$

   where $\bar{x}_{C^{g,0}}$ denotes the average expression profile for genes across cluster $C^{g,0}$ and $\bar{x}_{C^{g,i}}$ denotes the average expression profile for genes across cluster $C^{g,i}$. This measure computes the (average) distance between the mean expression values (usually, log transformed expression ratios in case of cDNA microarrays) of all genes that are put in the same cluster by the clustering method under consideration on the basis of the full data and the data obtained by deleting the expression levels at one time point at a time.

3. The average distance measure is defined as

$$V_3(K) = \frac{1}{Ml} \sum_{g=1}^{M} \sum_{i=1}^{l} \left(\frac{1}{n(C^{g,0})n(C^{g,i})}\right) \times \sum_{g \in C^{g,0}, g' \in C^{g,i}} d(x_g, x_{g'})$$

   where, $d(x_g, x_{g'})$ is a distance (e.g., Euclidean, Manhattan etc.) between the expression profiles of genes g and g'. This measure computes the average distance between the expression levels of all genes that are put in the same cluster by the clustering method under consideration on the basis of the full data and the data obtained by deleting the expression levels at one experimental condition at a time.

Figure 1.5 illustrates the average proportion of nonoverlap measures for a number of existing clustering algorithms applied to two sets of simulated data. The simulated data sets were generated by adding varying degrees of random noise to a set of expression profiles (shown in Figure 1.4).

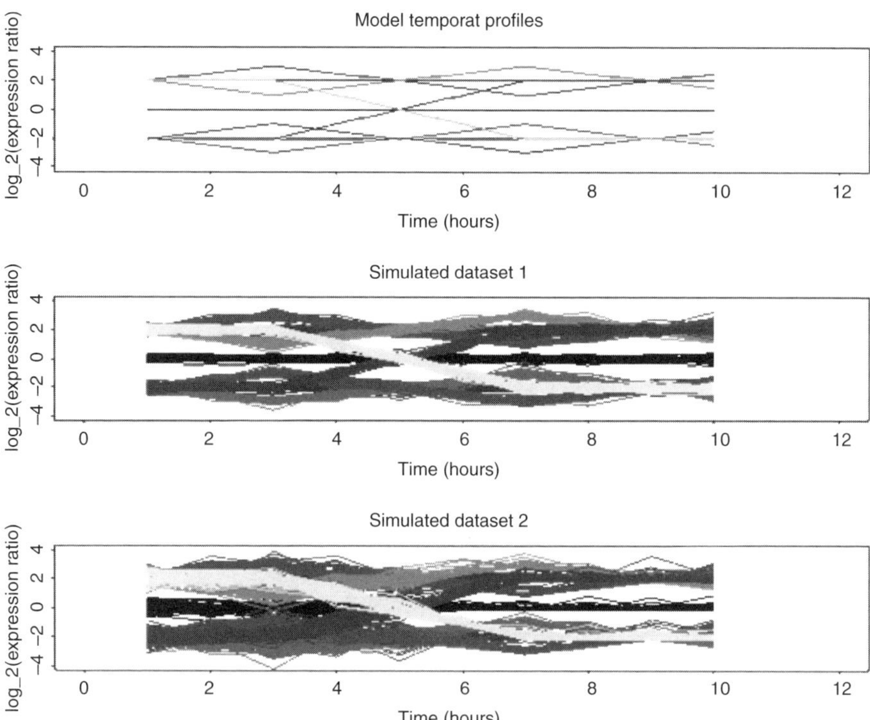

FIGURE 1.4
(See color insert following page 207.) The average proportion of nonoverlap measure for various clustering algorithms applied to simulated data sets.

S-Plus codes used to compute the above measures are available in the supplementary website of the Datta and Datta (2003) paper http://www.louisville.edu/~s0datt02/WebSupp/Clustering/SUPP/SUPP.HTM. In each plot, a profile closer to the horizontal axis indicates better performance over the usable range of *K*values.

Another approach of validating clustering results is to check whether the statistical clusters produced correspond to biologically meaningful functional classes. To this end, Datta and Datta (2003) compared the expression profiles of different statistical clustering algorithms with the model profiles of some functionally known genes. For the details, readers are referred to Datta and Datta (2003). A novel validation measure combining statistical stability and biological functional relevance was proposed in Datta and Datta (2006a). In yet another attempt (Datta and Datta, 2006b), results were validated through the gene ontology (GO) databases.

1.4.2 Classification

Unlike clustering, a classification algorithm is generally used to group tissue samples (e.g., cancer and noncancer) using their gene expression profiles

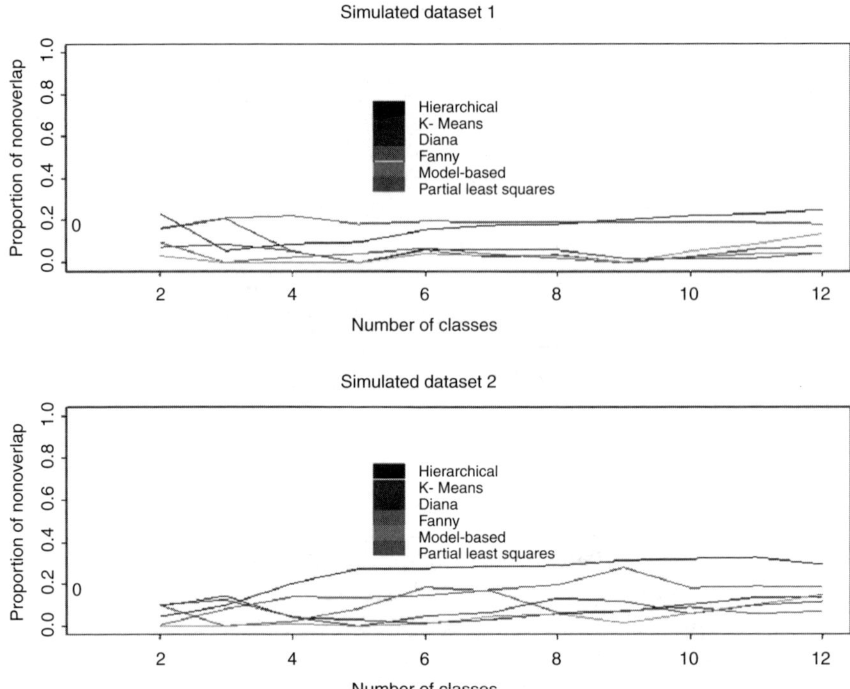

FIGURE 1.5
(See color insert following page 207.) Two simulated datasets of gene expressions were created by adding random noise to a model profile.

when partial knowledge is available about this grouping. In general, the goal of a classification technique is to predict class membership for new samples (test samples) with the knowledge of the training set (set of samples for which the classes are known).

1.4.2.1 Dimensionality Reduction

The vast amount of raw gene expression data leads to statistical and analytical challenges for using a classification algorithm to group samples into correct classes. The central difficulty in classification of microarray data is the availability of a very small number of samples in comparison with the number of genes in the sample. This violates the operating assumption of the classical statistical discriminant analysis procedures such as the linear discriminant analysis (LDA) and the quadratic discriminant analysis (QDA) or the nonparametric regression-based Neural Network procedure.

One might first attempt to reduce the number of "features" to be used in a classification algorithm from the raw gene expression profiles. For example, one might calculate the first few principal components (PCA) of the covariates (gene expressions) and then discriminate the samples on the basis of the principal components (Dudoit et al., 2000; Su et al., 2003; Datta and Delapidilla, 2006). PLS is another useful tool in constructing latent variables that can then be used for classification (Nguyen and Rocke, 2002a,b). Datta and de Padilla (2006) used PLS for feature selection, together with traditional classification algorithms such as linear discrimination and quadratic discrimination to classify multiple tumor types from proteomic data. van't Veer et al. (2002) applied a binary classification algorithm to cDNA array data with repeated measurements and classified breast cancer patients into good and poor prognosis groups. Their classification algorithm consists of the following steps. The first step is filtering, in which only genes with both small error estimates and significant regulation relative to a reference pool of samples from all patients are chosen. The second step consists of identifying a set of genes whose behavior is highly correlated with the two sample types (e.g., upregulated in one sample type but down-regulated in the other). These genes are rank-ordered so that genes with the highest magnitudes of correlation with the sample types have top ranks. In the third step, the set of relevant genes is optimized by sequentially adding genes with top-ranked correlation from the second step. However, this method involves an ad hoc filtering step and does not generalize to more than two classes. Another feature reduction technique is to consider only the genes that are deemed to be differentially expressed genes under different experimental conditions. Zhu et al. (2003), Wagner et al. (2004), Izmirlian (2004), and Datta and de Padilla (2006) used a similar approach to select the important features (mass to charge ratios) for mass spectrometry data before a classification algorithm is used.

1.4.2.2 Classification Algorithms

The literature of classification algorithms is vast. In the previous section we have mentioned about LDA (Fisher, 1936) and QDA. In addition to these two, logistic regression for two classes and log-linear models for more than two groups are also widely used. There are many more relatively modern classifiers some of which are discussed very briefly in the remainder of this section. The R-libraries (http://www.r-project.org) "class" and "MASS" contain a number of popular classifiers.

The neural network is a two-stage regression/classification model and is represented by a network diagram. Loosely speaking, it is modeled after the concept of neurons in the brain. It consists of at least three layers of nodes: the input layer, the hidden layer, and the output layer. For technical details, please refer to Hastie et al. (2001). The R function *nnet* (available in the library by the same name) fits a single hidden layer neural network.

k-nearest neighbor classification(k−NN) method is a nonparametric classifier (Devijver and Kittler, 1982; Ripley, 1996) on the basis of nonparametric estimates of the class densities or of the log posterior. The k-NN classifier finds k nearest samples in the training set and takes a majority vote among the classes of these k samples. We end this subsection by discussing three relatively new classification algorithms one of which (the Shrunken Centroid classifier) is developed primarily for classifying microarray datasets.

Support vector machine (SVM) (Vapnik, 1995; Burges, 1998; Vapnik, 1998; Cristianini and Shawe-Taylor, 2000; Hastie et al., 2001 §4.5, 12.2, 12.3) is a new generation classifier that has been reported to be very successful in a wide variety of applications. For two classes, a binary classifier constructs a hyperplane separating the two classes (e.g., cancer from noncancer samples). The basic idea of SVM is to map the data into a higher dimensional space and then find an optimal separating hyperplane. SVMs are large-margin classifiers; that is, they solve an optimization problem that finds the separating hyperplane that optimizes a weighted combination of the misclassification rate and the distance of the decision boundary to any sample vector. For further details, the reader may consult the tutorial paper by Burges (1998) or the book by Cristianini and Shawe-Taylor (2000). Brown et al. (2000) used SVM for classifying microarray data of budding yeast *S. cerevisiae* and compared the performance by other standard classification algorithms (e.g., Fisher's LDA). SVM was performed better than all the non-SVM methods compared in the paper.

Random Forest (Breiman, 2001) classification is an extension of classification trees (Breiman et al., 1984) by integrating the idea of *bagging*. Random Forest constructs many classification trees using different bootstrap samples of a fixed size m from the original data. To classify a new object from an input vector (collection of variables) it runs the input vector to each and every tree in the forest. Each tree gives a classification. The forest chooses the classification having the most votes (over all the trees in the forest). About one-third of the cases are left out of the bootstrap sample and not used in the construction of a particular tree. The samples left out of the *k-th* tree are run through the kth tree to get a classification. In this way, a test set classification is obtained for each case in about one-third of the trees that can be used to assess the accuracy of the classifier. Note that the Random Forest algorithm automatically finds the most important features/variables in order to classify the data by computing an estimate of the increase in error rate of the classifier had that variable not been used.

Finally, the nearest shrunken centroid classification introduced by Tibshirani et al. (2002) first computes a shrunken centroid for each class where each standardized centroid is shrunk towards the overall centroid. A test sample is classified into the class whose shrunken centroid is closest to, in squared distance, the gene expression profile of that sample. The shrinkage can reduce the effect of noisy genes resulting in higher accuracy of the resulting classifier. It is available under the PAM (Prediction Analysis for Microarrays, not to be confused with the clustering algorithm of the same acronym) package (http://www-stat.stanford.edu/~tibs/PAM/).

1.4.2.3 Accuracy of Classification

There are several methods for calculating the error rate of a classification algorithm (e.g., resubstitution, leave-one-out cross validation, k-fold cross validation, repeated cross validation and .632 bootstrap/bias corrected bootstrap, etc.). In microarray studies, use of a proper method of estimation of classification error is particularly important since typically one has a small sample size. Ulisses and Edward (2004) provide a comprehensive comparison of numerous estimation methods of classification error.

Resubstitution error is usually low biased for more complex algorithms. *k-fold* cross validation method is unbiased and leave-one-out is nearly unbiased for the error estimation. However, they all have high variability. The variance of the *k-fold* cross validation approach gets worse as k increases. On the other hand, the standard bootstrap resampling method (Efron, 1979) has reduced variability but high bias. However, the .632 bootstrap (Efron, 1983) method has low bias and low variability at the same time. Bias-corrected bootstrap also has low bias and low variability.

1.5 Detection of Differential Gene Expressions

One of the major goals of typical microarray studies is to determine the list of genes whose expression profiles are different among two or more sets of tissue samples usually collected under different experimental conditions. For example, in the colorectal cancer data set studied in Datta and Datta (2005), there were three types of tissues corresponding to normal, adenoma, and carcinoma cells in colon cancer patients. A major difficulty in using classical multivariate statistical methods is that the dimension of an expression data vector is huge, often exceeding tens of thousands, and at the same time, there are only a limited number of samples.

A rather impressive collection of statistical papers has emerged in this area over the past six or so years. As a result, the selective review presented here is by no means comprehensive. Although we attempt to present the development in this area in a systematic manner, it is not necessarily chronologic and reflects our own bias in selection of the highlighted papers. More comprehensive accounts and lists of references can be obtained in some recent books that have been written in the area of statistical analysis of microarray data such as McLachlan et al. (2004) and Lee (2004). The collection of methods for the detection of differential gene expression generally fall into two categories, namely those that are designed specifically for the microarray studies, including adaptation of known methods (Ideker et al., 2000; Kerr et al., 2000, 2002; Efron et al., 2001; Newton et al., 2001; Tusher et al., 2001; Dudoit et al., 2002; Efron and Tibshirani, 2002; Ge et al., 2003; Lee et al., 2003; Reiner et al., 2003; Storey and Tibshirani, 2003; Zhao and Pan, 2003, etc.) and those that are applicable to multiple testing in general such as Westfall and Young (1993), Benjamini and Hochberg (1995), Storey (2002), Efron (2004), Datta and Datta (2005).

1.5.1 Fold Change

Mostly nonstatistical in nature, this has been the most favorite method of the biologists and medical researchers where genes are ordered by the magnitude of the ratio of their gene expressions in the two samples (often referred to as fold change). The problem with this approach is that there is no provision to recognize that different genes may have different natural variability (scale) in their expression values. In addition, genes declared important or significant by an arbitrary threshold of their fold change do not control any of the various statistical error rates associated with multiple testing procedures.

1.5.2 The Two Sample t-Test and its Variants

Use of a two-sample t-test is statistically more appropriate for ranking genes in terms of their significance than the fold change approach. Usually a log-transformation is first applied to the normalized and preprocessed gene expressions to stabilize their variability and then the *t-statistic* is computed for the j-th gene as

$$t_j = \frac{\overline{Y}_{1j} - \overline{Y}_{2j}}{s_j\sqrt{\frac{1}{n_1} + \frac{1}{n_2}}} \tag{1.1}$$

where, for $i = 1, 2$, $\overline{Y}_{ij}$ is the average log-expression of the jth gene for the ith tissue type, n_i is the number of tissue samples of type i and s_j^2 is the pooled sample variance of the jth gene log-expression levels across both tissue types. Genes are ranked by absolute $|t_j|$ and all genes exceeding a threshold are declared to be significantly differentially expressed.

Sometimes, one may choose to pool sample variance across all genes as well under the assumption that the log-transformed expression values have constant variance. This leads to a gene specific statistic t as above except that an overall s is used in place of s_j.

A modified t-statistic is used in the SAM method of Tusher et al. (2001) where a constant α is added to the denominator of Equation 1.1 leading to

$$t'_j = \frac{\overline{Y}_{1j} - \overline{Y}_{2j}}{\alpha + s_j\sqrt{\frac{1}{n_1} + \frac{1}{n_2}}}.$$

This was done so that one does not get a large t-statistic simply because a gene has low sample variance (which is a likely outcome given that one deals with tens of thousands of genes in a microarray study). Of course, the null distribution of t'_j is no longer a t distribution. In SAM, one rejects for large positive or negative values of $t'_j - E_0t'_j$ where $E_0t'_j$ is an estimate of the null expectation of t'_j calculated using a null bootstrap (resample without regard to tissue type labels) or by a null permutation method (where pseudo data are generated by permuting the samples ignoring the tissue type labels). A similar "regularization" of the sample variance appears in the Bayesian methods of Baldi and Long (2001). The tuning parameter α is selected in a data-based way using a fairly complex algorithm. Then a set of MAD (median absolute difference from the median) values v of the t_j' are computed over grids of s_j and then the coefficient of variation of v is minimized.

Comparison of gene expressions for more than two tissue types is achieved by an ANOVA formulation of log-transformed gene expressions

$$Y_{ijk} = \mu + G_i + V_j + (GV)_{ij} + \varepsilon_{ijk} \tag{1.2}$$

where i denotes tissue types, j denotes genes, and k denotes replicates. We assume that the gene expressions have been normalized so that array effects, and so forth, have been taken out. Here μ stands for an overall mean log-expression and G and V denote the tissue and gene main effects. Our interest lies in testing the behaviors of the interaction terms GV. The null hypothesis of no differential expression for gene j across all tissue types corresponds to H_0: $(GV)_{ij} = 0$, for all i and the null hypothesis of no differential expression of gene j between tissue types i_1 and i_2 is given by $H_0^{i_1,i_2}$: $(GV)_{i_1j} - (GV)_{i_2j} = 0$. These are easily tested in the framework of the ANOVA model Equation 1.2 by classical methods and the genes can be ranked by the corresponding F or t-statistics. This technique has been popularized by Gary Churchill and his colleagues (Kerr et al., 2000; Kerr and Churchill, 2001 a,b; Kerr et al., 2002 etc.) in the microarray context. Note that this analysis is different from the per gene one way ANOVA model introduced in Section 1.2.4.

1.5.3 Adjustments for Multiple Testing

Let us, for simplicity, consider a balanced design so that $n_1 = n_2 = n/2$, say. Under the assumption of normality of the log-expression, the t-statistic in Equation 1.1 has a student's t- distribution with $n - 1$ degrees of freedom leading to a p value for the j-th gene comparison $p_j = 2\{1 - F_{n-2}^t(|t_j|)\}$, where F_{n-2}^t is the cumulative distribution function of a t distribution $n - 2$ degrees of freedom. If genes for which $p_j \leq 0.05$ were declared to be significantly differentially expressed, then for a typical microarray data involving 10,000 or more genes, on the average about 500 genes will be declared to be significant even if the complete null hypothesis were true (i.e., none of the genes were actually differentially expressed). In other words, the procedure would lead to too many false positives. This simple illustration demonstrates the need for the global control of error rates in microarray studies.

In general, if a gene that is not differentially expressed but it is declared to be significant by the statistical test, a type 1 error is committed. On the other hand, if the test fails to show significance for a truly differentially expressed gene, a Type II error is made. By overall or the family-wise type 1 error rate, we mean the probability of declaring at least one gene to be significant when none of the genes is truly differentially expressed. The classical Bonferonni procedure is generally not suitable for microarray studies since it would only declare genes to be significant for which $p_j \leq \alpha/g$, where α is the desired overall type 1 error probability and g is the total number of genes on the microarray. Since g is typically huge, this ratio is too small to have significance for most genes (if any) making the procedure too conservative to have any practical utility. As a middle ground, a more appropriate type 1 error control procedure can be used including the Holm (1979) procedure that first orders the p values and then declares $p_{(j)}$ to be significant if $p_{(j)} < \alpha/(g - j + 1)$. For most microarray studies, the Holm procedure will be proved to be conservative as well and the following procedure developed by Westfall and Young (1993) has been advocated by Dudoit et al. (2002). It also orders the p values and then uses resampling (or permutation) under the null to judge significance as described in the following algorithmic steps:

1. *Step 1*: Find the rank orders $r_i =$ such that $|t_{r_1}| \geq \cdots \geq |t_{r_g}|$ and let $u_i = |t_{r_i}|, 1 \leq i \leq g$ be the ordered absolute test-statistics.
2. *Step 2*: Let $\mathbf{Y}$ be the matrix of log-expression values where each column consists of the set of expression profiles for all genes corresponding to a single sample and the first n_1 columns corresponds to group 1 and the remaining columns correspond to group 2. Permute the columns of the matrix $\mathbf{Y}$ and label the first n_1 columns of the permuted matrix as group 1 and the rest as group 2. For each gene (i.e., row) j, Calculate the t test-statistics denoted t_j^* with the permuted data and reorder the absolute values as

$$u_g^* = |t_{r_g}^*|, \quad u_i^* = \max(u_{i+1}^*, |t_{r_i}^*|), \quad \text{for } g > i \geq 1.$$

3. *Step 3*: Repeat Step 2 over all possible permutations and denote the u_i^* values by

$$u_i^*(1), \cdots u_i^*(B), \text{ where } B = (n_1 + n_2)!$$

4. *Step 4*: Compute $\widetilde{P}_{r_i} = B^{-1} \sum_{l=1}^{B} I(u_i^*(l) \geq u_i)$ and monotonize them as

$$P_{r_1}^{adj} = \widetilde{P}_{r_1}, \quad P_i^{adj} = \max(\widetilde{P}_{r_1}, \widetilde{P}_{r_{i-1}}), \, for\, 1 < i \leq g.$$

Genes $r_1 \cdots r_m$ would be declared significant, given an overall type 1 error probability α where $m = \max\{k: P_{r_k}^{adj} \leq \alpha\}$.

Recently, Datta and Datta (2005) demonstrated that even the Westfall and Young procedure is too conservative for microarray applications. They propose a modification to this procedure where an empirical Bayes calculation is incorporated first to change the p values before resampling. They show that the modified procedure can enhance the overall sensitivity although maintaining the type 1 error rate. Further details of this procedure are described in Section 1.5.5.

1.5.4 False Discovery Rate

Besides controlling the overall type 1 error rate, one can consider several other performance measures in the context of microarray studies. See Dudoit et al.

(2003) for a comparative review of various error rates for several commonly used multiple testing procedures. The following error rates are often used to judge the performance of multiple tests.

Sensitivity: Expected proportion among differentially expressed genes that were declared significant

Specificity: Expected proportion amongst nondifferentially expressed genes that were not declared significant

False discovery rate (FDR): Expected proportion amongst genes declared significant that were not differentially expressed

False nondiscovery rate (FNR): Expected proportion amongst genes declared not significant that were differentially expressed

In particular, the use of FDR has become the standard in microarray studies lately largely owing to the influential paper by Benjamini and Hochberg (1995) that offers a very simple procedure for controlling the FDR on the basis of uncorrected P-values. Let r_i the rank orders of the ordered P-values so that $P_{r_i} = P_{(i)}$, $1 \leq i \leq g$. Let q be the desired upper bound on the FDR; usually q is taken to be between 5% and 10%. This Benjamini–Hochberg procedure (originally introduced by Simes, 1986) declares genes $r_1 \cdots r_m$ to be significantly differentially expressed where

$$m = \max\{k\colon P_k \leq qk/g\}.$$

In more recent works, additional variants of the FDR have been introduced such as the positive FDR (Storey, 2002, 2004) and the conditional FDR (Tsai et al., 2003) and procedures for estimating and controlling these in microarray settings have been proposed. When one declares all genes with marginal p values less than or equal to α to be significant, Storey's estimates for the FDR and the pFDR are given by

$$\text{F}\hat{\text{D}}\text{R} = \frac{\alpha\hat{\pi}}{\hat{F}(\alpha)}$$

and

$$\text{pF}\hat{\text{D}}\text{R} = \frac{\alpha\hat{\pi}}{\hat{F}(\alpha)[1-(1-\alpha)^g]},$$

respectively, where $\hat{\pi}$ estimates the number of nondifferentially expressed genes and $\hat{F}(\alpha)$ is the observed proportion of tests (genes) with marginal p values less than or equal to α.

In the context of p-FDR, Storey (2004) defined the gene specific q-value that is, roughly speaking, the p-FDR of procedures that declare genes whose statistics are as extreme as that corresponding to this particular gene to be significant.

Efron (2005) introduced the term local FDR to indicate the posterior probability that a given gene is null (i.e., nondifferentially expressed). Assuming a two-component mixture model for the Z-scores (obtained by normal transformation of the p values),

$$f(z) = \pi f_0(z) + (1 - \pi) f_1(z),$$

where f_0 and f_1 are the densities of the null (nondifferentially expressed) and non-null genes, respectively, and π is the proportion of null genes, Efron defined the local FDR(l) at $Z = z$ as

$$\text{lFDR}(z) = \pi f_0(z)/f_(z).$$

It has the property that the averages (expectation) of lFDR values for all genes whose Z values are less than or equal to z equals to the FDR of the procedure that declares all genes with Z values $\leq z$ to be significant. Ploner et al. (2006) generalized the idea of a local FDR to a tow-dimensional local FDR by considering the joint densities of a two dimensional statistic Z_1, Z_2

$$\text{2dlFDR}(z_1, z_2) = \pi f_0(z_1, z_2)/f(z_1, z_2).$$

They proposed using the t-statistic and the logarithm of its standard errors as the two components Z_1 and Z_2.

1.5.5 Procedures-Based on p Values

There are a number of recent attempts to control some global error rates on the basis of a set of uncorrected p-values. The BH procedure described above falls under this category. An advantage of such procedures is that a user can use them after obtaining the list of p values from their favorite microarray data analysis software that generally produces marginal or uncorrected p values. Allison et al. (2002) modeled the set of uncorrected p values obtained from a microarray experiment by a finite mixture of beta distributions, where the first component was taken to be uniform in order to model the null distribution. Thus, majority of the genes would correspond to the first component. They estimated the number of mixture components. In particular, in a two-component model, they ranked genes on the basis of their posterior probability of belonging to the second (i.e., non-null component).

Pounds and Morris (2003) also modeled the marginal p values by a two-component mixture of a uniform and a beta. They were able to estimate the FDR under this model scenario if one rejects all null hypotheses for which the marginal p values were below a given threshold. By inverting this relationship, their BUM procedure produces a list of significant genes corresponding to a given FDR level. See, however, Datta and Datta (2005) for a cautionary note.

More recently, Pounds and Cheng (2004) introduced the SPLOSH procedure that provides a more accurate FDR control. More importantly, since it is based on nonparametric function estimation techniques, it is expected to be more robust and work well even if the mixture of beta model is not accurate.

Datta and Datta (2005) took a different approach from the papers mentioned above. Instead of controlling the FDR, they attempted to control the overall type 1 error rate following a version of the Westfall and Young procedure that involved sampling from the uniform distribution. However, first, they transformed the set of p values by applying a normality (inverse cdf) transformation followed by computing a new set of statistics; this new set of statistics computes the empirical Bayes estimates of the location parameter θ_j in a simple normal location model $z_j = \Phi^{-1}(p_j) \sim N(\theta_j, 1)$ using nonparametric function estimation

$$\hat{\theta}_j^{EB} = z_j - h^2 \frac{\left\{\sum_{i=1}^{M}(z_j - z_i)\phi\left(\frac{z_j - z_i}{h}\right)\right\}}{\left\{\sum_{i=1}^{M}\phi\left(\frac{z_j - z_i}{h}\right)\right\}}, \tag{1.3}$$

where ϕ is the standard normal pdf and h is a user specified bandwidth.

In effect, each p value borrows strength from an overall evidence against the complete null hypothesis since the second term in the right-hand side of Equation 1.3 is an estimate of the derivative of the logarithmic marginal density $(\log f_G)'$ and its stochastic behavior under the complete null hypothesis will be different from its behavior under the alternative. We use the following real data example from Datta and Datta (2005) to illustrate this effect. Figure 1.6 shows a scatter plot of this term for a colorectal cancer data set where normal tissue gene expressions were compared with that for carcinoma tissues. We can see that for potentially informative genes (say, those corresponding to negative z_i) it tends to be below the diagonal line

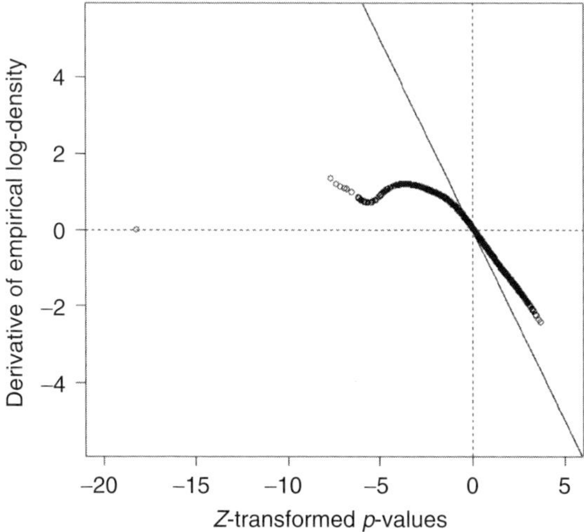

FIGURE 1.6
Scatter plot of derivative of empirical log density against transformed p-values for the "normal vs. carcinoma" comparison in a colorectal cancer data; the solid plots its theoretical values under the complete null.

(the theoretical values when the complete null is true) indicating an overall presence of "differentially expressed genes" corresponding to a given z level.

Datta and Datta (2005) showed that their empirical Bayes screening (EBS) procedure has substantially higher sensitivity than the standard step-down approach for multiple comparisons at the cost of a modest increase in the FDR. The EBS procedure also compared favorably when compared with existing FDR control procedures for multiple testing (namely BH and BUM).

1.5.6 Empirical Bayes Methods

Empirical Bayes methods have a natural place in multiple decision problems where each component decision problem can be thought of as a realization of a Bayesian decision problem, and determining differences in microarray gene expression is no exception. Besides the EBS procedure mentioned above, there have been more traditional attempts to use empirical Bayes techniques in the microarray context. Generally speaking, such approaches differ from the full Bayesian approaches in that data-based estimates are used to determine parameters at a certain stage of modeling the priors. At the end, generally, an estimate of the posterior probability that a particular gene is differentially expressed is calculated.

Newton et al. (2001) modeled the expression levels in the two channels of a cDNA microarray with gamma priors and used Bayesian and empirical Bayesian techniques to identify differentially expressed genes in two tissue types. Efron et al. (2001) formulated the distribution of normalized statistics as a two-point nonparametric mixture (as stated in Subsection 1.5.4), where one distribution corresponds to the null genes and the other to differentially expressed genes. They provided methods to nonparametrically estimate both components and the mixing proportion that in turn yielded estimated posterior probabilities for each gene being differentially expressed. Datta et al. (2004) used the empirical Bayes idea to adjust certain t-test statistics obtained from an ANOVA model to determine differential gene expressions.

1.6 Networks and Pathways

Microarray technology enables us to acquire the knowledge of genome-wide expression. Hence, this comprehensive information on gene-expression can serve as an important tool to understand the underlying biological system through genetic networks. Several attempts have been made to infer the interrelationships of the genes through unsupervised cluster analysis. A main purpose of using cluster analysis in this context is not only to group the genes but also to correlate clusters with pathways (Zien et al., 2000). Several references of cluster analysis applied to gene expression data are provided in Section 1.4 of this document. However, there are different approaches to detect the activities of all genes in consort (regulatory network). Datta (2001) used PLS method, Raychaudhuri et al. (2000) used PCA, and Spellman et al. (1998) used Fourier analysis to unravel the consorted activities of all genes. Other series of attempts to synthesize pathways using gene expression data are Boolean networks (see, e.g., Liang et al., 1998) and Bayesian networks (e.g., Friedman et al., 2000; Yamanaka et al., 2004, etc.).

Kauffman (Kauffman, 1969, 1974; Glass and Kauffman, 1973) introduced the Boolean network. However, it has recently been adapted for gene expression data by Shmulevich et al. (2002). In this model, gene expression is quantized to only two levels: ON and OFF. Also, the expression level (state) of one gene (node) is completely determined by the values of other nodes by some underlying Boolean function. Thus, a Boolean network $B(N, F)$ is defined in terms of a set of nodes (genes) $N = \{x_1, \ldots, x_n\}$ and Boolean functions $F = (f_1, \ldots, f_n)$. Each x_i represents the state (expression) of gene i, where $x_i = 1$ represents the fact that gene i is expressed and $x_i = 0$ means it is not expressed. The model is represented in the form of directed graph. Connectivity of one node to the other is updated synchronously in an iterative procedure.

Bayesian networks are another tool for connecting genes using their expression patterns. Bayesian networks (Pearl, 1988; Friedman et al., 2000), model the dependence structure between different genes using the expression levels. A Bayesian network analysis can indicate some causal relationship in the data (in particular genes in terms of their expression levels). A Bayesian network for X is a pair $BN = (D, P)$. The first component, D, denotes a directed acyclic graph (DAG) whose vertices correspond to the random variables $x_1, \ldots x_n$ (expression levels of several genes), and the edges represent direct causal dependencies between the variables. The graph D involves some conditional independence assumptions (Markov assumption) as follows: each variable x_i is independent of its nondescendants given its parents. The second component of the network pair, namely P, represents the set of parameters that quantifies the network and describes a conditional distribution for each variable (gene expression), given its parents in D. Together, these two components specify a unique distribution on $x_1, \ldots x_n$. The graph D represents conditional independence assumptions that allow the joint distribution to be decomposed, taking a minimum number of parameters. Using a Bayesian network, we might be able to answer many types of questions that involve the joint probability (e.g., what is the probability of $X = x$, the expression of a gene, given observation of some of the other gene expressions?) or independencies in the domain (e.g., are X and Y independent once we observe Z?). The Bayesian network literature contains a suite of algorithms that can be used to answer such questions efficiently by exploiting the explicit structural representation (Pearl, 1988; Jensen, 1996).

Lastly, we want to mention that it is important to evaluate these synthesized pathways with respect to already known pathways. Kurhekar et al. (2002) measured the impact of gene expression levels from a series of microarray experiments on metabolomic and regulatory pathways. They introduce the concept of a pathway scoring method. The basic idea behind it is as follows. A gene that is positively expressed at a certain time point or in a particular sample in a biological experiment indicates that the cell requires the particular protein coded by the gene. In this way, significant induction of the genes in a known pathway shows that the pathway has been used more often than say at the reference time point or in the reference cell. The significant repression of many genes in a known pathway in similar biological experiments reveals the deactivation of that specific pathway. In this way, it is possible to measure the effect of a biological process on different biochemical pathways with the help of gene expression data.

1.7 Concluding Remarks

In this chapter, we present a mostly nonmathematical overview of various statistical analyses of microarray gene expression data. Many of these techniques, notably, clustering, classification, and multiple testing can also be applied to other high-throughput data such as proteomic data obtained using mass spectrometer (Satten et al., 2004). As mentioned earlier, this review is not comprehensive—the subject is still moving in various directions. We have not reviewed papers that combine microarrays with other clinical end points, for example, survival times (Datta et al., 2007). In the future, we might see more and more studies where multiple forms of large-scale biological data (SNP, microarrays, proteomic mass-spectra) will be analyzed together and efforts in this direction are already under way (CAMDA, 2006, 2007 Conferences, http://www.camda.duke.edu).

References

Affymetrix (2002). Statistical Algorithms Description Document. *Affymetrix Inc*, Santa Clara, CA. http://www.affymetrix.com/support/ technical/ whitepapers/ sadd- whitepaper.pdf

Affymetrix (2004). Sample pooling for microarray analysis: A statistical assessment of risks and biases. Technical note, *Affymetrix Inc*, Santa Clara, CA.

Allison, D. B., Cui, X., Page, G.P., and Sabripour, M. (2006). Microarray data analysis: From disarray to consolidation and consensus. *Nat. Rev. Genet.*, **7**, 55–65.

Allison, D. B., Gadbury, G. L., Heo, M., Fernández, J. R., Les, C.-K., Prolla, J. A., and Weindruch, R. (2002). A mixture model approach for the analysis of microarray gene expression data. *Comput. Stat. Data Anal.*, **39**, 1–20.

Baldi, P. and Long, A. D. (2001). A Bayesian framework for the analysis of microarray expression data: Regularized *t*-test and statistical inferences of gene changes. *Bioinformatics*, **17**, 509–519.

Benjamini, Y. and Hochberg, Y. (1995). Controlling the false discovery rate: A practical and powerful approach to multiple testing. *J. Roy. Statist. Soc. Ser B*, **57**, 289–300.

Bioconductor. (2003). Methods for Affymetrix Oligonucleotide Arrays (affy). *The Bioconductor Project*, Version 1.2. http://www.bioconductor.org.

Bolstad, B. M., Irizarry, R. A, Astrand, M., and Speed, T. R. (2003). A comparison of normalization methods for high density oligonucleotide array data based on variance and bias. *Bioinformatics*, **19**, 185–193.

Breiman, L. (2001). Random forests. *Technical Report 567*, Statistics Department, University of California, Berkeley, CA.

Breiman, L., Friedman, J. H., Olshen, R. A., and Stone, C. (1984). *Classification and Regression Trees*. Wadsworth, Belmont, CA.

Brown, M. P. S., Grundy, W. N., Lin, D., Cristianini, N., Sugnet, C. W., Furey, T. S., Ares, M., and Haussler, D. J. (2000). Knowledge-based analysis of microarray gene expression data by using support vector machines. *Proc. Natl. Acad. Sci. USA*, **97**, 262–267.

Burges, C. J. C. (1998). A tutorial on support vector machines for pattern recognition. *Data Min. Knowl. Disc.*, **2**, 121–167.

Causton, H. C., Brazma, A., Quackenbush, J., and Kokis, S. (2003). *Microarray Gene Expression Data Analysis: A Beginner's Guide*, Blackwell Science, Malden, MA.

Chen, G., Jaradat, S. A., Banerjee, N., Tanaka, T. S., Ko, M. S. H., and Zhang, M. Q. (2002). Evaluation and comparison of clustering algorithms in analyzing ES cell gene expression data. *Stat. Sinica*, **12**, 241–262.

Chen, J. J., Delongchamp, R. R., Tsai, C. A., Hsueh, H. M., Sistare, F., Thompson, K. L., Desai, V. G., and Fuscoe, J. C. (2004). Analysis of variance components in gene expression data. *Bioinformatics*, **20**, 1436–1446.

Cho, R. J., Campbell, M., Winzeler, E., Steinmetz, L., Conway, A., Wodicka, L., Wolfsberg, T., et al. (1998). A genome-wide transcriptional analysis of the mitotic cell cycle. *Mol. Cell*, **2**, 65–73.

Chu, G., Narasimhan, B., Tibshirani, R., and Tusher, V. (2002a). Significance analysis of microarrays (SAM) software. http://www.stat.stanford.edu/~tibs/SAM/

Chu, S., DeRisi, J., Eisen, M., Mulholland, J., Botstein D., and Brown, P. O. (1998). The transcriptional program of sporulation in budding yeast. *Science*, **282**, 699–705.

Chu, T. M., Weir, B., and Wolfinger, D. (2002b). A systematic statistical linear modeling approach to oligonucleotide array experiments. *Math. Biosci.*, **176**, 35–51.

Churchill. G. A. (2002). Fundamentals of experimental design for cDNA microarrays. *Nat. Genet. Suppl.*, **32**, 490–495.

Cox, D. R. (1958). *Planning of Experiments*. Wiley, New York, NY.

Cristianini, N. and Shawe-Taylor, J. (2000). *An Introduction to Support Vector Machines*, Cambridge University Press, Cambridge.

Cui, X. and Churchill, G. A. (2003). Statistical tests for differential expression in cDNA microarray experiments. *Genome Biol.*, **4**, 210.1–210.10.

Datta, S. (2001). Exploring relationships in gene expressions: A partial least squares approach. *Gene Expression*, **9**, 257–264.

Datta, S. and Arnold, J. (2002). Some comparisons of clustering and classification techniques applied to transcriptional profiling data. In *Advances in Statistics, Combinatorics and Related Areas*, C. Gulati, Y.-X. Lin, S. Mishra, and J. Raynen (Eds.) World Scientific, 63–74.

Datta, S. and Datta, S. (2003). Comparisons and validation of statistical clustering techniques for microarray gene expression data. *Bioinformatics*, **19**, 459–466.

Datta, S. and Datta, S. (2005). Empirical Bayes Screening (EBS) of many p-values with applications to microarray studies. *Bioinformatics*, **21**, 1987–1994.

Datta, S. and Datta, S. (2006a). Validation of statistical clustering using biological information, In *Proceedings of INTERFACE 2005* (CD-ROM).

Datta, S. and Datta, S. (2006b). Methods for evaluating clustering algorithms for gene expression data using a reference set of functional classes. *BMC Bioinformatics*, **7**, 397.

Datta, S. and de Padilla, L. M. (2006). Feature selection and machine learning with mass spectrometry data for distinguishing cancer and non-cancer samples. *Statistical Methodology* (*Special Issue on Bioinformatics*), **3**, 79–92.

Datta, S., Le-Rademacher, J., and Datta, S. (2007). Predicting patient survival from microarray data by accelerated failure time modeling using partial least squares and LASSO. *Biometrics*, **63**, 259–271.

Datta, S., Satten, G. A., Benos, D. J., Xia, J., Heslin, M. J., and Datta, S. (2004). An empirical Bayes adjustment to increase the sensitivity of detecting differentially expressed genes in microarray experiments. *Bioinformatics*, **20**, 235–242.

DeRisi, J. L., Iyer, V. R., and Brown, P. O. (1997). Exploring the metabolic and genetic control of gene expression on a genomic nomic scale. *Science*, **278**, 680–686.

Devijver, P. A. and Kittler, J. (1982). *Pattern Recognition: A Statistical Approach.* Prentice-Hall, London.

Dobbin, K., Shih, J. H., and Simon, R. (2003). Statistical design of reverse dye microarrays. *Bioinformatics*, **19**, 803–810.

Dudoit, S., Fridlyand, J., and Speed, T. P. (2000). Comparison of discrimination methods for the classification of tumors using gene expression data. *Technical report 576*, Mathematical Sciences Research Institute, Berkeley, CA.

Dudoit, S., Popper, J. S., and Boldrick, J. C. (2003). Multiple hypothesis testing in microarray experiments. *Statistical Science*, **18**, 71–103.

Dudoit, S., Yang, Y. H., Callow, M. J., and Speed, T. P. (2002). Statistical methods for identifying differentially expressed genes in replicated cDNA microarray experiments. *Stat. Sinica*, **12**, 111–139.

Efron, B. (1979). Bootstrap methods: Another look at jackknife. *Ann. Statist.* **7**, 1–26.

Efron, B. (1983). Estimating the error rate of a prediction rule: Some improvements on crossvalidation. *J. Amer. Statist. Assoc.* **78**, 316–331.

Efron, B. (2004). Large-scale simultaneous hypothesis testing: The choice of a null hypothesis. *J. Amer. Statist. Assoc.*, **99**, 96–104.

Efron, B. (2005). Local false discovery rates. Preprint.

Efron, B. and Tibshirani, R. (2002). Empirical Bayes methods and false discovery rates for microarrays. *Genet. Epidemiol.*, **23**, 70–86.

Efron, B., Tibshirani, R., Storey, J. D., and Tusher, V. (2001). Empirical Bayes analysis of a microarray experiment. *J. Amer. Statist. Assoc.*, **96**, 1151–1160.

Eisen, M. B., Spellman, P. T., Brown, P. O., and Botstein, D. (1998). Cluster analysis and display of genome-wide expression patterns. *Proc. Natl. Acad. Sci. USA*, **95**, 14863–14868.

Fisher, R. A. (1936). The use of multiple measurements in taxonomic problems. *Ann. Eugenics*, **7**, 179–188.

Friedman, N., Linial, M., Nachman, I., and Pe'er, D. (2000). Using Bayesian networks to analyze expression data. *J. Comp. Biol.*, **7**, 601–620.

Gadbury, G. L., Page, G. P., Edwards, J., Kayo, T., Prolla, T. A., Weindruch, R., Permana, P.A., Mountz, J. D., and Allison, D. B. (2004). Power and sample size estimation in high dimensional biology. *Stat. Methods Med. Res.*, **13**, 325–338.

Ge, Y. Dudoit, S., and Speed, T. P. (2003). Resampling based multiple testing for microarray data analysis. *TEST*, **12**, 1–44 (with discussion pp. 44–77).

Gieser, P., Bloom, G. C., and Lazaridis, E. N. (2001). Introduction to microarray experimentation and analysis. In *Methods in Molecular Biology, Biostatistical Methods,*. S. W. Looney (Ed.), Humana Press, **184**, 29–49.

Glass, L. and Kauffman, S. A. (1973). The logical analysis of continuous non-linear biochemical control networks. *J. Theor. Biol.*, **39**, 103–129.

Hamalainen, H. K., Tubman, J. C., Vikman, S., Kyrola, T., Ylikoski, E., Warrington, J. A., and Lahesmaa, R. (2001). Identification and validation of endogenous reference genes for expression profiling of T helper cell differentiation by quantitative real-time RT-PCR. *Anal. Biochem.*, **299**, 63–70.

Hartigan, J. A. (1975). *Clustering Algorithms*, Wiley, New York, NY.

Hartigan, J. A. and Wong, M. A. (1979). A k-means clustering algorithm. *Appl. Stat.*, **28**, 100–108.

Hastie, T., Tibshirani, R., Eisen, M. B., Alizedah, A., Levy, R., Staudt, L., Chan, W. C., Botstein, D., and Brown, P. (2000). Gene shaving as a method for identifying distinct sets of genes with similar expression patterns. *Genome Biol.*, **1**(2), research0003.1–0003.21.

Hastie, T., Tishbirani, R., and Friedman, J. (2001). *The Elements of Statistical Learning*. Springer, New York, NY.

Holm, S. (1979). A simple sequentially rejective multiple test procedure. *Scand. J. Stat.*, **6**, 65–70.

Hurlbert, S. H. (1984). Pseudoreplication and the design of ecological field experiments. *Ecol. Monogr.*, **54**, 187–211.

Hurlbert, S. H. (2004). On misinterpretations of pseudoreplication and related matters: A reply to Oksanen. *OIKOS*, **104**, 591–597.

Ideker, T., Thorsson, V., Siegel, A. F., and Hood, L. (2000). Testing for differentially expressed genes by maximum-likelihood analysis of microarray data. *J. Comp. Biol.*, **7**, 805–817.

Irizarry, R. A., Bolstad, B. M., Collin, F., Cope, L. M., Hobbs, B., and Speed, T. P. (2003a). Summaries of Affymetrix GeneChip® probe level data. *Nucleic Acids Res.*, **31**, e15.

Irizarry, R. A., Gautier, L., and Cope, L. An R package for analysis of Affymetrix oligonucleotide arrays. In: Parmigiani, R. I. G., Garrett, E. S., and Ziegler, S. (Eds.). (2003b) *The Analysis of Gene Expression Data: Methods and Software*, pp. 102–119, Springer, Berlin.

Irizarry, R. A., Hobbs, B., Collin, F., Beazer-Barclay, Y. D., Antonellis, K. J., Scherf, U., and Speed, T. P. (2003c). Exploration, normalization, and summaries of high density oligonucleotide array probe level data. *Biostatistics*, **4**, 249–264.

Irizarry, R. A., Warren, D., Spencer, F., Kim, I. F., Biswal, S., Frank, B. C., Gabrielson, E., et al. (2005). Multiple laboratory comparison of microarray platforms. *Nature Met.*, **2**, 329–330.

Irizarry, R. A., Gautier, L., and Cope, L. (2003b). An R package for analysis of Affymetrix oligonucleotide arrays. In *The Analysis of Gene Expression Data: Methods and Software*, R. I. G. Parmigiani, E.S. Garrett, and S. Ziegler (Eds.), Springer, Berlin, 102–119.

Izmirlian, G. (2004). Application of the random forest classification algorithm to a SELDI-TOF proteomics study in the setting of a cancer prevention trial. *Ann. N.Y. Acad. Sci.*, **1020**, 154–174.

Jensen, F. V. (1996). *An Introduction to Bayesian Networks*, University College London Press, London.

Jung, S. H. (2005). Sample size for FDR-control in microarray data analysis. *Bioinformatics*, **21**, 3097–3104.

Kauffman, S. A. (1969). Metabolic stability and epigenesis in randomly constructed genetic nets. *J. Theor. Biol.*, **22**, 437–467.

Kauffman, S. A. (1974). Homeostasis and differentiation in random genetic control networks. *Nature*, **224**, 177–178.

Kaufman, L. and Rousseeuw, P. J. (1990) *Fitting Groups in Data. An Introduction to Cluster Analysis*. Wiley, New York, NY.

Kendziorski, C., Irizarry, R. A., Chen, K. S., Haag, J. D., and Gould, M. N. (2005). On the utility of pooling biological samples in microarray experiments. *Proc. Natl. Acad. Sci. USA*, **102**, 4252–4257.

Kendziorski, C., Zhang, Y., Lan, H., and Attie, D. (2003). The efficiency of pooling mRNA in microarray experiments. *Biostatistics*, **4**, 465–477.

Kerr, M. K. (2003). Design considerations for efficient and effective microarray studies. *Biometrics*, **59**, 822–828.

Kerr, M. K., Afshari, C. A., Bennett, L., Bushel, P., Martinez, N. W., and Churchill, G. A. (2002). Statistical analysis of a gene expression microarray experiment with replication. *Stat. Sinica*, **12**, 203–217.

Kerr, M. K. and Churchill, G. A. (2001a). Experimental design for gene expression microarrays. *Biostatistics*, **2**, 183–201.

Kerr, M. K. and Churchill, G. A. (2001b). Statistical design and the analysis of gene expression microarray data. *Genet. Res.*, **77**, 123–128.

Kerr, M. K. and Churchill, G. A. (2001c). Bootstrapping cluster analysis: Assessing the reliability of conclusions from microarray experiments. *Proc. Natl. Acad. Sci. USA*, **98**, 8961–8965.

Kerr, M. K., Martin, M., and Churchill, G. A. (2000). Analysis of variance for gene expression microarray data. *J. Comp. Biol.*, **7**, 819–838.

Kohonen, T. (1997) *Self-Organizing Maps, Second Edn.* Springer-Verlag, Berlin.

Kurhekar, M. P., Adak, S., Jhunjhunwala S., and Raghupathy, K. (2002). Genome-wide pathway analysis and visualization using gene expression data, *PSB02*, 462–473.

Lee, K. E., Sha, N., Dougherty, E. R., Vannucci, M., and Mallick, B. K. (2003). Gene selection: A Bayesian variable selection approach. *Bioinformatics*, **19**, 90–97.

Lee, M. L. T. (2004). *Analysis of Microarray Gene Expression Data*. Kluwer, Norwell, Massachusetts.

Lee, M. L. T. and Whitmore, G. A. (2002). Power and sample size for microarray studies. *Stat. Med.*, **21**, 3543–3570.

Li, C. and Wong, W. H. (2001). Model-based analysis of oligonucleotide arrays: Expression index computation and outlier detection. *Proc. Natl. Acad. Sci. USA*, **98**, 31–36.

Li, H., Wood, C. L., Getchell, T. V., Getchell, M. K., and Stromberg, A. J. (2004). Analysis of oligonucleotide array experiments with repeated measures using mixed models. *BMC Bioinformatics*, **5**, 209.

Liang, S., Fuhrman, S., and Somogyi, R. (1998). REVEAL, A general reverse engineering algorithm for inference of genetic network architectures. In *Proc. Pacific Symposium on Biocomputing*, **3**, 18–29.

McLachlan, G. J., Bean, R. W., and Peel, D. (2002). A mixture model-based approach to the clustering of microarray expression data. *Bioinformatics*, **18**, 1–10.

McLachlan, G. J., Do, K-A., and Ambroise, C. (2004). *Analyzing Microarray Gene Expression Data*, J. Wiley & Sons, Hoboken, New Jersey.

Newton, M. A., Kendziorski, C. M., Richmond, C. S., Blattner, F. R., and Tsui, K. W. (2001). On differential variability of expression ratios: Improving statistical inference about gene expression changes from microarray data. *J. Comp. Biol*, **8**, 37–52.

Nguyen, D. V. and Rocke, D. M. (2002a). Tumor classification by partial least squares using microarray gene expression data. *Bioinformatics*, **18**, 39–50.

Nguyen, D. V. and Rocke, D. M. (2002b). Multi-class cancer classification via partial least squares with gene expression profiles. *Bioinformatics*, **18**, 1216–1226.

Oksanen, L. (2001). Logic of experiments in ecology: Is pseudoreplication a pseudoissue? *OIKOS*, **94**, 27–38.

Page, G. P., Edwards, J. W., Gadbury, G. L., Yelisetti, P., Wang, J., Trivedi, P., and Allison, D. B. (2006). The PowerAtlas: A power and sample size atlas for microarray experimental design and research. *BMC Bioinformatics*, **7**, 84.

Parmigiani, G., Garrett, E. S., and Ziegler, S. (Eds.) (2003). *The Analysis of Gene Expression Data: Methods and Software*, Springer, Berlin.

Parrish, R. S. and Spencer, H. J. (2004). Effect of normalization on significance testing for oligonucleotide microarrays. *J. Biopharm. Stat.*, **14**, 575–589.

Pearl, J. (1988). *Probabilistic Reasoning in Intelligent Systems*. Morgan Kaufmann, San Francisco.

Peng, X., Wood, C. L., Blalock, E. M., Chen, K. -C., Landfield, P. W., and Stromberg, A. J. (2003). Statistical implications of pooling RNA samples for microarray experiments. *BMC Bioinformatics*, **4**, 26.

Ploner, A., Calza, S., Gusnanto, A., and Pawitan, Y. (2006). Multidimensional local false discovery rate for microarray studies. *Bioinformatics*, **22**, 556–565.

Pounds, S. and Cheng, C. (2004). Improving false discovery rate estimation, *Bioinformatics*, **20**, 1737–1745.

Pounds, S. and Morris, S. W. (2003). Estimating the occurrence of false positives and false negatives in microarray studies by approximating and partitioning the empirical distribution of p-values. *Bioinformatics*, **19**, 1236–1242.

Raychaudhuri, S., Stuart, J. M., and Altman, R. B. (2000). Principal components analysis to summarize microarray experiments: Application to sporulation time series. In *Proc. Pac. Symp. Biocomput*, 455–466.

Reiner, A., Yekutieli, D., and Benjamini, Y. (2003). Identifying differentially expressed genes using false discovery rate controlling procedures. *Bioinformatics*, **19**, 368–375.

Ripley, B. D. (1996) *Pattern Recognition and Neural Networks*, Cambridge University Press, Cambridge.

Rosa, G. J. M., Steibel, J. P., and Tempelman, R. J. (2005). Reassessing design and analysis of two-colour microarray experiments using mixed effects models. *Comp. Funct. Genom.*, **6**, 123–131.

Roth, F. P., Hughes, J. D., Estep, P. W., and Church, G. M. (1998). Finding DNA regulatory motifs within unaligned non-coding sequences clustered by whole-genome mRNA quantitation. *Natl. Biotechnol.*, **16**, 939–945.

Satten, G. A., Datta, S., Moura, H., Woolfitt, A., Carvalho, G., De, B. K., Pavlopoulos, A., Carlone, G. M., and Barr, J. (2004). Standardization and denoising algorithms for mass spectra to classify whole-organism bacterial specimens. *Bioinformatics*, **20**, 3128–3136.

Schadt, E. E., Li, C., Ellis, B., and Wong, W. H. (2001). Feature extraction and normalization algorithms for high-density oligonucleotide gene expression array data. *J. Cell. Biochem. Suppl.*, **37**, 120–125.

Sharan, R. and Shamir, R. (2000). CLICK: A clustering algorithm with applications to gene expression analysis. *Proc. Int. Syst. Mol. Biol.* **8**, 307–316.

Shmulevich, I., Dougherty, R., Kim, S., and Zhang, W. (2002). Probabilistic boolean networks: A rule-based uncertainty model for gene regulatory networks. *Bioinformatics*, **18**, 261–274.

Simes R. J. (1986). An improved Bonferroni procedure for multiple tests of significance. *Biometrika*, **73**, 751–754.

Sokal, R. R. and Sneath, P. H. A. (1963). *Principles of Numerical Taxonomy*. Freeman, New York, NY.

Speed, T. (Ed.) (2003). *Statistical Analysis of Gene Expression Microarray Data*, CRC Press, Boca Raton, FL.

Spellman, P. T., Sherlock, G., Zhang, M. Q., Iyer, V. R., Anders, K., Eisen, M. B., Brown, P. O., Botstein, D., and Futcher, B. (1998). Comprehensive identification of cell cycle-regulated genes of the yeast Saccharomyces cerevisiae by microarray hybridization. *Mol. Biol. Cell*, **12**, 3273–3297.

Spruill, S. E., Lu, J., Hardy, S., and Weir, B. (2002). Assessing sources of variability in microarray gene expression data. *Biotechniques*, **33**, 916–920; 922–923.

Steibel, J. P. and Rosa, G. J. M. (2005). On reference designs for microarray experiments. *Statistical Applications in Genetics and Molecular Biology*, **1** (4) Article 36.

Storey, J. D. (2002). A direct approach to false discovery rates. *J. Roy. Statist. Soc. Ser B*, **64**, 479–498.

Storey, J. D. (2004). The positive false discovery rate: A Bayesian interpretation and the qvalue. *Ann. Statist.*, **31**, 2013–2035.

Storey, J. D. and Tibshirani, R. (2003). Statistical significance for genome-wide studies. *Proc. Natl. Acad. Sci. USA*, **100**, 9440–9445.

Su, Y., Murali, T., Pavlovic, V., Schaffer, M., and Kasif, S. (2003). RankGene: Identification of diagnostic genes based on expression data. *Bioinformatics*, **19**, 1578–1579.

Templeman, R. J. (2005). Assessing statistical precision, power, and robustness of alternative experimental designs for two color microarray platforms based on mixed effects models. *Vet. Immunol. Immonop.*, **105**, 175–186.

Tibshirani, R. (2006). A simple method for assessing sample sizes in microarray experiments. *BMC Bioinformatics*, **7**, 106.

Tibshirani, R., Hastie, T., Narasimhan, B., and Chu, G. (2002) Diagnosis of multiple cancer types by shrunken centroids of gene expression. *Proc. Natl. Acad. Sci.* USA, **99**, 6567–6572.

Tsai, C. A., Hsueh, H. M., and Chen, J. J. (2003). Estimation of false discovery rates in multiple testing: Application to gene microarray data. *Biometrics*, **59**, 1071–1081.

Tsai, P. W. and Lee, M. L. T. (2005). Split plot microarray experiments: Issues of design, power and sample size. *Applied Bioinformatics*, **4**, 187–194.

Tusher, V. G., Tibshirani, R., and Chu, G. (2001). Significance analysis of microarrays applied to the ionizing radiation response. *Proc. Natl. Acad. Sci. USA*, **98**, 5116–5121.

Ulisses, M. B. and Edward, R. D. (2004). Is cross-validation for small-sample microarray classification? *Bioinformatics*, **20**, 374–380.

Vandesompele, J., Preter, K. D., Pattyn, F., Poppe, B., Roy, N. V., Paepe, A. D., and Speleman, P. (2002). Accurate normalization of real-time quantitative RT-PCR data by geometric averaging of multiple internal control genes. *Genome Biol.*, **3**, research0034.1–0034.11.

Vapnik, V. (1995). *The Nature of Statistical Learning Theory*. Springer-Verlag, New York.

Vapnik, V. (1998). *Statistical Learning Theory*. Wiley, New York, NY.

van't Veer, L. J., Dai, H., van de Vijver, M. J., He, Y. D., Hart A. A., Mao, M., Peterse, H. L., et al. (2002). Gene expression profiling predicts clinical outcome of breast cancer. *Nature*, **415**, 530–536.

Vinciotti, V., Khanin, R., D'Alimonte, D., Liu, X., Cattini, N., Hotchkiss, G., Bucca, G., et al. (2005). An experimental evaluation of a loop versus reference design for two-channel microarrays. *Bioinformatics*, **21**, 492–501.

Wagner, M. D., Naik, D. N., Pothen, A., Kasukurti, S., Devineni, R. R., Adam, B., Semmes, O. J., and Wright, G. L. Jr. (2004). Computational protein biomarker prediction: A case study for prostate cancer. *BMC Bioinformatics*, **5**, 26.

Wahba, G. (1990). Spline methods for observational data. *CBMS-NSF Regional Conference Series in Applied Mathematics*. SIAM, Philadelphia.

Westfall, P. H. and Young, S. S. (1993). *Resampling Based Multiple Testing: Examples and Methods for p-value Adjustment*, Wiley, New York, NY.

Wit, E. and McClure, J. (2004). *Statistics for Microarrays: Design, Analysis and Inferenc*, Wiley, New York, NY.

Wit, E., Nobile, A. and Khanin, R. (2005). Near-optimal designs for dual channel microarray studies. *Appl. Stat.*, **54**, 817–830.

Wolfinger, R. D., Gibson, G., Wolfinger, E. D., Bennett, L., Hamadeh, H., Bushel, P., Afshari, C., and Paules, R. S. (2001). Assessing gene significance from cDNA microarray expression data via mixed models. *J. Comp. Biol.*, **8**, 625–637.

Woo, Y., Krueger, W., Kaur, A., and Churchill, G. (2005). Experimental design for three-color and four-color gene expression microarrays. *Bioinformatics*, **21**, i459–i467.

Wu, Z. and Irizarry, R.A. (2005). Stochastic models inspired by hybridization theory for short oligonucleotide arrays. *J. Comp. Biol.*, **12**, 882–893.

Yamanaka, T., Toyoshiba, H., Sone, H. , Parham, F. M., and Portier, C. J. (2004). The TAO-Gen algorithm for identifying gene interaction networks with application to SOS repair in *E. coli*. *Environ. Health Persp.*, **112**, 1614–1621.

Yang, M. C. K., Yang, J. J., McIndoe, R. A., and She, J. X. (2003). Microarray experimental design: Power and sample size considerations. *Physiol. Genomics*, **16**, 24–28.

Yeung, K., Haynor, D. R., and Ruzzo, W. L. (2001). Validating clustering for gene expression data. *Bioinformatics*, **17**, 309–318.

Zakharkin, S., Kim, K., Mehta, T., Chen, L., Barnes, S., Scheirer, K. E., Parrish, R. S., Allison, D. B., and Page, G. P. (2005). Sources of variation in Affymetrix microarray experiments. *BMC Bioinformatics*, **6**, 214.

Zhang, S. D. and Gant, T. W. (2005). Effect of pooling samples on the efficiency of comparative experiments involving microarrays. *Bioinformatics*, **21**, 4378–4383.

Zhao, Y. and Pan, W. (2003). Modified nonparametric approaches to detecting differentially expressed genes in replicated microarray experiments. *Bioinformatics*, **19**, 1046–1054.

Zhu, W., Wang, X., Ma, Y., Rao, M., Glimm, J., and Kovach, J. S. (2003). Detection of cancer-specific markers amid massive mass spectral data. *Proc. Natl. Acad. Sci. USA*, **100**, 14666–14671.

Zien, A., Kueffner, R., Zimmer, R., and Lengauer, T. (2000). Analysis of gene expression data with pathway scores. In *Proc. ISMB*, 407–417.

2

Machine Learning Techniques for Bioinformatics: Fundamentals and Applications

Jarosław Meller and Michael Wagner

CONTENTS

2.1 Introduction

This chapter presents an overview of applications of machine and statistical learning techniques to problems arising in the area of molecular biology and medicine. As such, the methods and applications discussed here fall into the general area of *bioinformatics*, which is concerned with the computational analysis and interpretation of data regarding biological systems and processes. This is an active and still relatively young field of research with great potential of advancing both basic and applied biomedical research.

The Human Genome Project (Venter et al., 2001), which was completed recently, and its extensions such as the HapMap (International HapMap Consortium, 2003) project dealing with genetic variability in human populations, have triggered an enormous growth of data and research that aims at elucidating fundamental questions in medicine, biochemistry, genetics, and molecular biology. In particular, the availability of DNA sequence information has enabled the large-scale analysis of correlations between genetic variations and, for example, differences in susceptibility to diseases or other medically relevant outcomes. Machine learning-based approaches are capable of capturing complex correlations between relevant descriptors (or "features"), such as genetic mutations, and observed outcomes, such as cancer survival time. Capturing and characterizing such correlations can lead to successful prediction of various aspects of molecular systems. For a general overview of applications of machine learning in bioinformatics, see, for example, Baldi et al. (2000) and Mjolsness and DeCoste (2001).

We start this chapter with a very brief overview of central problems, data sources, and measurement techniques being used in molecular biology and genomics. This is followed by a discussion of machine learning approaches and some aspects of general importance regarding their applications to problems arising in molecular biology, such as the importance of data representation, model selection and validation, alternative learning algorithms, and their interplay with hypothesis generation and further experimental studies. We are necessarily brief and selective; rather than providing a comprehensive overview of the field, we discuss what we believe to be crucial elements of successful applications of machine learning and data mining techniques in bioinformatics.

2.2 Molecular Biology

Modern molecular biology and genomics are generating massive amounts of data that need to be transformed into knowledge and understanding of molecular systems and processes in the context of medicine and other applications. This data can be organized according to the different types of biological systems it pertains to and the experimental techniques that were used. The Central Dogma of Molecular Biology states that the information encoded by genes (coding fragments of DNA sequences) is transcribed by a complex molecular machinery into complementary messenger RNA molecules. These in turn are translated by another complex molecular machinery (the ribosome) into proteins, which form the building blocks of life. Recent technological advances now allow one to obtain high-throughput measurements and comprehensive snapshots of living systems at the level of the (whole) *genome* (DNA sequence), the *transcriptome* (messenger RNA expression levels), and the *proteome* (protein structure, protein expression, and protein interactions). Examples of these experimental advances include large-scale DNA sequencing and resequencing techniques, genome-wide microarray-based gene expression profiling, mass spectrometry-based protein characterization, high-throughput techniques to identify protein interactions as well as structural techniques such as x-ray crystallography and nuclear magnetic resonance (NMR). Although the amount of data generated by these various techniques is invariably large, their fundamental nature is very different in each case and requires significant domain insight for their analysis to result in truly meaningful conclusions.

For example, the data produced by DNA sequencers consists of strings of letters representing the four nucleotides in DNA. This type of data is essentially static in nature and is characterized by the presence of point mutations (one nucleotide in a certain position is replaced by another). The length of the sequence of interest can vary considerably and can reach billions of letters for whole-genome analyses, with point mutations observed approximately every few hundred letters. Sample sizes of resequencing projects that aim at estimating population-wide genetic variability can involve tens of thousands of individuals. Complicating factors that may need to be addressed in order to interpret sequence data include sequencing errors, the presence of two (nonidentical) copies of each chromosome in diploidal organisms, length polymorphisms such as the presence of short repetitive segments and the need for ancestral information.

Messenger RNA expression levels (the abundance of particular mRNA molecules in a sample), on the other hand, are represented by real numbers. Current technologies allow for simultaneously measuring genome-wide expression levels of tens of thousands of genes and their variants. Unlike DNA sequence, mRNA expression levels are very dynamic in nature: any given gene may be "turned off" at some point in time (or in some tissue type), or it can be highly abundant under other conditions or in other cell types.

Expression profiling at the mRNA level, and in particular observed differences in relative abundance in different sample types such as cancer versus healthy cells, are nowadays used to capture "molecular phenotypes" that result from different interpretations by the cells of the static DNA information in response to different physiological conditions. Given the large dynamic range of mRNA expression, it is especially important to consider the technical and biological variability, which requires careful experimental design in order to allow for statistically (and biologically) meaningful conclusions. Therefore, sample sizes, which presently rarely exceed one hundred, are likely to grow quickly as the cost of the technology continues to decrease.

Yet another data type is provided by structural analyses of macromolecules present in cells, and in particular protein structure data. In order to perform its physiological function, a protein typically needs to adopt a well-defined three-dimensional (3D) structure. Mutations in the genome can result in changes of the primary protein sequence, which in turn may affect the 3D structure and function of that protein. Techniques such as x-ray crystallography provide atomically detailed snapshots of macromolecular structures that can be used to elucidate functional implications of point mutations, for instance. Structural data are thus very different in nature compared with, say, gene expression data; individual atoms in a protein are characterized by their approximate coordinates in 3D space. Furthermore, some classes of proteins such as membrane proteins are difficult to resolve structurally, which limits the number of available data points.

At the same time, there are common aspects of these data sets in the context of the applications of machine learning that basically pertain to finding correlations between features in the data and other observables. In particular, and in the case of DNA sequence data, the features of interest might be point mutations in a specific gene, whereas in the case of expression data the relative abundances of the mRNA encoded by that gene would be of interest. Finally, in the case of protein structures, a feature that might be of functional consequence might be the evolutionary conservation of an individual amino acid residue in a protein sequence across related species. These "features" can then be analyzed for correlations with phenotypes such as disease states. Such plausible correlations, as identified by computational data analyses, can in turn form the basis for formulating testable hypotheses. For example, putative biomarkers of disease states can be identified for further experimental and clinical validation. We stress that conclusions obtained by analysis of any single data set are necessarily tentative; subsequent biological validation is required in order to provide ultimate verification and give credence to any computational findings. One of our goals in this chapter is to illustrate this interplay through a number of examples and specific applications.

2.3 Machine Learning

The essence of machine learning is to learn from well-characterized sample data in order to subsequently make predictions for new data instances. This definition is deliberately general and is inclusive of many techniques that arose from related fields such as data mining and statistical learning. A wide array of machine learning approaches and techniques has been developed over the past decades with applications in such varied fields as robotics, speech recognition, expert decision support systems, and bioinformatics. For a comprehensive introduction to this area of research, we refer the reader to the excellent monographs by Mitchell (1997) and Hastie et al. (2001).

A common aspect of machine learning techniques is the fact that they use a so-called training set (well-characterized sample data) to fit parameters for their different underlying models and formulations. The quality of the model can (and should) then be evaluated on independent test data sets. Machine learning techniques are also rather generic and typically not application-specific, which implies that it is *a priori* difficult to predict which technique is likely to be most successful for a particular application. On the other hand, machine learning methods are critically dependent on the input data being appropriately represented. Therefore, as mentioned before, domain knowledge is essential for making informed choices regarding representation of the problem and the choice of the machine learning technique. In this chapter, using several specific examples, we illustrate the importance of careful considerations for each of these issues.

Successful examples of machine learning applications in bioinformatics include gene prediction from primary DNA sequence (Krogh et al., 1994; Burge and Karlin, 1998) and prediction of protein secondary structures from the amino acid sequence (Rost and Sander, 1994b; Jones, 1999; Adamczak et al., 2005). In the first case, the input data consist of DNA sequence fragments and the output is the predicted location of a coding region, whereas in the latter case the input is given by amino sequence and the output is the predicted local conformation of the protein chain. Machine learning methods are also being used for prediction of protein–protein interactions, membrane domains, posttranslational modifications and other functionally relevant characteristics of proteins and other macromolecules (see, e.g., Krogh et al., 2001; Fariselli et al., 2002; Bigelow et al., 2004; Blom et al., 2004; Cao et al., 2006). Furthermore, various machine and statistical learning approaches are being used to analyze microarray gene expression data (see, e.g., Alizadeh et al., 1998; Eisen and Brown, 1999; Primig et al., 2000; Medvedovic et al., 2004), protein expression data, correlations between genotypes (patterns of variations in genomic sequences) and phenotypes, such as disease subtype, and so forth. In this review, we will focus on applications of machine learning techniques to selected problems that pertain to sequence, expression and structural data analysis.

2.4 Machine Learning in Practice

Our objective is to illustrate what we believe to be important considerations when applying learning algorithms to problems in bioinformatics. Rather than provide an exhaustive list of algorithms (we only list a few commonly used modern paradigms in the following section), we discuss "soft" issues that often have a large impact on the success or failure of an application.

2.4.1 Problem Representation

A crucial element of any machine learning application (and one that is especially important in bioinformatics) is the choice of an appropriate problem representation. Many problems allow for different data representations, and this early decision that must be made before a learning model is chosen can be critical. Many learning algorithms require numerical input data, which leads to the problem that a numerical representation for categorical data needs to be found.

A typical and common example concerns the representation of protein sequence information; this is required for, for example, various problems in the area of protein structure prediction. Proteins are linear molecules of amino acids, ranging from a few tens of residues to thousands. There are twenty amino acids in the protein "alphabet," and so one (naïve) way of encoding a protein sequence is to encode each amino acid in a 20-dimensional unit vector. The overall protein sequence of length n can thus be written as a binary vector of length $20 \times n$. A second approach is to quantify physicochemical properties of amino acid residues such as their hydrophobicity, size, and polarity. An alternative, which has proven in several instances to lead to significantly better results for protein structure prediction problems (Hansen et al., 1998; Jones, 1999; Cuff and Barton, 2000; Chen and Rost, 2002), is to take the sequence of interest and perform a multiple sequence alignment (Rost and Sander, 1994b) against a database of other known proteins. Multiple sequence alignments (such as those obtained by Psi-BLAST) find evolutionarily related proteins, and they yield an alternative way of representing a protein sequence: the position specific substitution matrix (PSSM). A PSSM consists of log-odds mutation rates for each amino acid in homologous protein. A successful option to represent protein sequence has been not only to use the (also 20-dimensional) substitution rate vector as representation for an amino acid, but to extend the number of descriptors for that amino acid by including the substitution vectors of other residues that are within a certain distance in the linear sequence. This way one captures not only the identity of the amino acid, but also a characterization of a part of its local environment. Depending on the size of the "sliding window" one chooses, this leads to a large number (hundreds) of descriptors for one amino acid, which in turn can lead to difficulties in case the number of samples is small (discussed subsequently). Finding the optimal representation, thus, requires significant insight both into the problem at hand (e.g., protein chemistry) as well as machine learning. It has been observed before that this choice often has a much more significant effect on the final prediction accuracy compared with, to say, the choice of machine learning algorithm.

2.4.2 Training and Testing Data Selection

Machine learning algorithms, as mentioned earlier, "learn" from data, and in particular, they use a "training" data set to fit parameters to models that can then be used to draw inferences about new data points. Much care must be taken in order to choose this training data, as poor choices can lead to poor performance of the predictor. In particular, there are two important criteria for choosing training data. Training instances should account for naturally occurring variability, which is the case if it is chosen as a random sample from the underlying distribution. This is often impossible to guarantee, especially when sample sizes are small as is the case when biological samples are involved. However, precautions should be taken in order to avoid unintended biases in data that can arise, for example, when samples were not treated uniformly before being assayed. The second criterion, which is somewhat related to the first, is the avoidance of redundancy in the training set. If very similar "objects" (in the most general case) are included in the training set, then the learning algorithms will likely have difficulty distinguishing them, which can lead to a phenomenon called "overfitting." In this case, a learning algorithm will be adapted too specifically for the training data and will likely not perform well in general for new data (poor generalization). Furthermore, cross-validation methods (discussed subsequently) will yield overly optimistic accuracy estimates on redundant training sets; this is a common problem with many papers that have appeared in the literature. In the particular case of protein sequences, redundancy can be avoided, for example, by ensuring that no two sequences share significant sequence similarity. Failure to ensure non-redundancy will likely negatively affect the quality of the resulting predictor.

2.4.3 Model Validation

Although the accuracy of a learning algorithm on the training set is certainly one measure of interest, a realistic performance estimate can only be obtained by checking a model's accuracy on an independent test set. Similar with the choice of training sets, care must be taken with test sets in order to avoid redundancies that can lead to biased estimates. If sufficient high-quality data are not available, as is often the case with biological data, then k-fold cross-validation can be used: the available data set is split into k sets, $k - 1$ are used for training and the remaining one for testing. This can and should be done randomly and many times over in order to obtain converged estimates of the average prediction accuracy as well as its variance (see, e.g., Wagner et al., 2004b). A special case is when $k = n$, the number of data points, which is also called *leave-one-out* cross validation, which is often used in the case that only tens of data samples are known. If sufficient data is available, then 10-fold or 5-fold cross-validation is a reasonable choice. A second, sometimes controversial, consideration in this area is the choice of accuracy measure. In the case when the prediction is a numerical value (we will discuss prediction types in more detail in a later paragraph), one can use classical measures such as the root mean squared error, the mean absolute error, and the correlation coefficient. No single one of these measures should be overinterpreted, and, in particular, two additional data points are required in order to provide context for an accuracy measurement: the standard deviation of the estimate and the "baseline accuracy," that is, the accuracy a trivial predictor would achieve. If cross-validation experiments are performed, then it is informative to report on estimates for average accuracies as well the estimated standard deviations in order to assess whether an apparent improvement in accuracy over a competing method is truly significant or not. We will illustrate this using examples in a later section. If the predictor is discrete in nature (e.g., diseased versus healthy), then the classification accuracies, sensitivity, specificity, and the Matthews correlation coefficient are natural choices. Receiver operating characteristic curves are commonly used in the field to illustrate the quality of machine learning algorithms, as they show the trade-off curve between the sensitivity and specificity of a predictor. One reasonable goal to aim for is to achieve roughly equal accuracy on both the test and training sets, as this is an indication of stability and robust performance.

2.4.4 Feature Selection and Aggregation

With billions of nucleotides in DNA, tens of thousands of genes, hundreds of thousands of protein and protein variants and order of magnitude, and more protein-protein interactions, molecular biology is very high-dimensional. High-throughput technology has allowed us to obtain various snapshots into this complexity, and these snapshots are invariably high-dimensional. When wanting to apply machine learning algorithms to this data, one is often (but not always) faced with the "large p, small n" problem; a small set of samples (data points) is characterized by a very large number of descriptors for each sample. By the way these data look in table form, they are often referred to as "short and fat." Dimension reduction (also called feature selection in machine learning) is essential in this context, as otherwise any model using this data will necessarily have more parameters than data points. A second motivation for feature selection is that one desired outcome of modeling biological data is often an increased understanding as to which features (DNA sequence features, mRNA expression levels, structural elements) are the most important ingredients that influence a measured outcome (disease versus nondisease status, secondary structure element, etc.) and which are irrelevant. In order to enable meaningful interpretation and also to prioritize the subsequent validation experiments, it is essential that the final list of features used in the model thus be reasonably small. (This is often called the principle of parsimony.) Viewed this way, feature selection can often actually be the primary goal of interpreting biological data, with the machine learning algorithm merely providing estimates as to how predictive the selected group of features might be.

The literature on feature selection is vast, as this is a challenging combinatorial problem of high practical importance. The difficulty arises fundamentally as a consequence of the fact that (1) features are rarely independent but rather have a complex dependence structure and (2) biological phenomena are rarely single factor events but rather often depend on the interaction and simultaneous occurrence of events. This introduces a combinatorial nature to the problem: whereas a single feature (e.g., DNA mutation) might not be a good predictor of, say a biological phenomenon (e.g., a complex disease), a combination of a specific subset of features might well contain valuable information. The number of possible subsets is exponential in the number of features, which implies that typically heuristics and approximations have to be used. We also note that the feature selection problem is not independent of the learning task: some feature subsets might have good predictive value with one particular learning method, although another method might not be able to extract sufficient information from the same feature set.

Care should be taken to apply feature selection techniques to the training set only and not on the entire data available; otherwise bias is introduced in the evaluation on test sets. In a k-fold cross-validation setting one is then faced with reselecting features k times. From the stability of the k feature sets one can gauge how robust the overall algorithm is; the development of principled methods to combine the k selection sets is still an open research area.

Rather than give details, we mention that there are several ways in which feature selection methods can be characterized: filter methods versus wrapper methods, and univariate versus. multivariate. Filter approaches are ones that are independent of a learning algorithm; they are typically applied to the data like a preprocessing step. As a simple, but often quite effective example, we mention the so-called F-score that was used in (Dudoit et al., 2002) microarray data classification. Wrapper approaches to feature selection are integrated with the learning method; for example, the learning algorithm iteratively determines the relative importance of different features by discarding variables that do not contribute significantly to the model (e.g., the support vector machine [SVM] method discussed later in this chapter); as such they tend to be more computationally involved.

Finally, we mention another strategy to deal with the high-dimensionality of biological data: feature aggregation techniques. Traditional statistical techniques such as principal component analysis (PCA), for example, can be used to determine the mutually orthogonal directions that account for the largest proportions of the overall variance in the data (Garrido et al., 1995; Romero et al., 1997; Hastie et al., 2000; Hastie et al., 2001; Bair and Tibshirani, 2004; Du et al., 2006). The resulting principal component vectors (that correspond to weighted linear combinations of features) can subsequently be used as input for machine learning methods. As with feature selection methods, it is crucial to compute the PCA decomposition on training data only, as otherwise the evaluation on the test set would necessarily be biased.

2.4.5 Supervised versus Unsupervised Learning

There are two fundamentally different paradigms in machine learning, both of which find ample applications in bioinformatics. In the first case, also called unsupervised learning or clustering, the task is to discover structures in the primary data by grouping similar data points into clusters. Approaches differ mainly in their choices of similarity measure (i.e., the underlying metric) and the actual clustering algorithm. Well-known approaches include k-means and hierarchical clustering. By clustering genes from microarray expression data under different conditions, for example, one can hope to discover coregulated genes, that is, genes that require the same transcription factors. Coregulation of genes is often also taken as an indication of functional similarity or potential interaction of protein product, which motivate the need for robust clustering algorithms that can prioritize further experimental validation. One difficulty with clustering is that it is generally very difficult to evaluate the quality of one solution versus another. Clustering is a discovery algorithm (structure in data is revealed), and it is unclear how clusters can be validated without resorting to experimental (wetlab) techniques. Furthermore, the most commonly used clustering procedures are, however, rather ad hoc in nature and incapable of separating statistically significant patterns from artifacts of random fluctuations and uncertainty in the data. Moreover, clustering approaches based on statistical modeling of the data often require the number of patterns to be specified in advance. One promising approach that circumvents this problem is the Bayesian infinite mixtures model (IMM) based clustering. IMM applies "model-averaging" and offers credible assessments of statistical significance of detected patterns (Medvedovic and Sivaganesian, 2002; Medvedovic et al., 2004). This "model-averaging" approach also allows one to circumvent effectively the problem of identifying the "correct" number of patterns. Furthermore, these models are capable of directly modeling various sources of experimental variability as well as accounting for noninformative features (i.e., context specificity of different patterns).

The other class of learning algorithms are so-called supervised learning methods. As implied by the name, these algorithms use a training set (a "gold standard") where the "answer" (e.g., in the form on a response variable such as a phenotype, a protein fold, or a gene sequence) is given. The learning algorithm's task can thus be viewed as trying to approximate the function that takes the primary data (e.g., protein sequence) as input and yields the output of interest (e.g., the type of secondary structure adapted by the protein in its native state). There are essentially two types of response variables that machine learning methods can handle. If the variable is continuous is nature (e.g., the degree to which an amino acid residue is exposed to solvent in a native protein structure), then regression-type methods are generally used. This includes simple ordinary least-squares regression, constrained least-squares regression as well as more involved methods such as support vector regression (SVR) and neural network (NN) regression. We will elaborate on some of these in later subsections. On the other hand, if the response variable is discrete or categorical (e.g., the type of secondary structure conformation a residue adopts in a native fold, α-helix, β-strand, or coil), then the problem is also called a classification problem and a different arsenal of methods is applicable. For classification problems, a further distinction can be made between binary problems (those with only two classes) and multiclass problems. Some classification methods are only suitable for two-class problems; heuristics such as majority voting algorithms need to be applied in the multiclass case. As mentioned earlier, the validation measures will also necessarily change depending on the nature of the response variable.

2.4.6 Model Complexity

Finally, we want to bring up a last consideration for machine learning methods: the complexity of the model used. Model complexity can be roughly defined as the number of free internal parameters that are optimized when the model is trained. Another view is split complexity into the degree of nonlinearity of the model and the number of parameters (features) by which the input data is described. Although nonlinear models are generally more powerful in the sense that they can capture more complex relationships between the input data and the response variable, they also suffer from pitfalls and dangers that need to be weighed carefully. If the number of internal parameters (including the number of features of the input data) is significantly greater than the number of samples in the training set, then any method is likely to suffer from overfitting: the method may have good accuracy on the training set but will not generalize well. This situation is all too frequent with high-dimensional biological data, with possible remedies including careful feature selection (as mentioned in a previous subsection) and the use of the simplest model that achieves balanced accuracy in the sense that the accuracy measures on the training and test sets are roughly equal.

2.5 Examples of Modern Machine and Statistical Learning Techniques

The field of machine learning has produced a myriad of different algorithms that is impossible to completely survey in a short chapter. Instead, we rather focus on briefly introducing a few of the most successful methods that are commonly applied to problems in bioinformatics (for a comprehensive introduction to these and other statistical learning approaches, see, for example, Hastie et al., 2001). As emphasized in the previous chapter, there are numerous other very important issues besides the choice of algorithms that influence the quality of a machine learning approach to real-world problems. Nevertheless, the choice of learning algorithm is certainly a critical one, especially in the all-too-frequent cases where the signal in the data is weak.

2.5.1 Linear Discriminant Analysis and Support Vector Machines

Linear models are commonly used to solve classification problems (as well as regression problems discussed later). These models are attractive because of their simplicity and ease of interpretation of their results. In particular, they involve roughly as many parameters as there are features, thus facilitating extrapolation from a limited number of sample data points. The essence of these methods is that they find a hyperplane in the vector space X that separates vectors of one group (for example, samples from individuals affected by a disease) from another group (healthy samples). A hyperplane $w^T x + \beta = 0$ for samples $x \in X$ may be found using many algorithms so that samples from the two groups lie on two different sides of the plane:

$$\begin{aligned} w^T x + \beta &< 0 \quad \text{for } x \in \text{Group 1} \\ w^T x + \beta &> 0 \quad \text{for } x \in \text{Group 2} \end{aligned} \tag{2.1}$$

One particularly well-known linear discriminant method is Fisher's linear discriminant analysis (LDA) (Hastie et al., 2001). A second example that has enjoyed tremendous popularity and success is the (linear) SVM, which may be regarded as a generalization of LDA classifiers and is based on a learning algorithm that uses only vectors close to the decision boundary to "support" the orientation of the hyperplane (including misclassified vectors). As a consequence SVMs are less sensitive to outliers. SVMs have emerged in the last 10 years as powerful machine learning tools and have shown excellent performance in the context of bioinformatics for classification problems (e.g., Brown et al., 2000; Furey et al., 2000) The motivation for the SVM separating hyperplane stems from the desire to achieve maximum separation between the two groups. SVMs solve an optimization problem that maximizes the margin, that is, the distance between the separating hyperplane and the closest correctly classified data vector. This margin is inversely proportional to the norm of the normal vector $\|w\|$, which means that this term should be minimal. At the same time, SVMs optimize the training accuracy by minimizing the sum of violations of the constraints (Equation 2.1), resulting in the following overall formulation:

$$\begin{aligned} &\min \|w\| + C\|\xi\| \\ &\text{s.t. } w^{\mathrm{T}} x_i + \beta - \xi_i < 0 \quad \text{for } x_i \in \text{Group 1} \\ &w^{\mathrm{T}} x_i + \beta + \xi_i > 0 \quad \text{for } x_i \in \text{Group 2} \end{aligned} \tag{2.2}$$

Here C is a trade-off parameter between the generalization term $\|w\|$ and the training misclassification term $\|\xi\|$ that must be chosen *a priori*. The SVM problem (Equation 2.2) is, in general, an optimization problem that can be solved with standard methods. Using (also standard) optimization duality theory one can show that the number of constraints that are satisfied with equality at the optimal solution is small, which implies that only a few data points actually influence the hyperplane defined by (w, β). Linear SVMs are suitable for solving very large-scale problems since they can be solved very efficiently on parallel computers (Wagner et al., 2004a; 2005) and thus allow for the kinds of extensive cross-validation experiments necessary to obtain a reliable classifier.

Finally, we mention briefly that SVMs can be extended using nonlinear kernel functions (e.g., Schoelkopf and Smola, 2002) that compute inner products $\langle w, x\rangle$ in high-dimensional spaces. Kernel SVMs thus still compute maximum margin linear separating hyperplanes, but these now can correspond to nonlinear separation boundaries in the original feature space. A number of kernels are commonly used, including polynomial kernels $k(w,x) = (w^T x + \beta)^p$ and radial basis function (RBF) kernels for which $k(w,x) = \exp(-\|w - x\|^2/\sigma)$. More sophisticated kernels, reflecting some prior knowledge about the specific problem analyzed, have been designed (Leslie et al., 2003), and, in particular, recent years have seen the development of specialized string kernels for particular bioinformatics applications.

There are a number of SVM software packages available, both in commercial packages and as open-source software. One particular package that can be recommended for its numerical stability and computational efficiency is libsvm (Fan et al., 2005), which is freely available for researchers over the Internet. Other packages and a wealth of background information on SVMs is available at http://www.support-vector.net.

2.5.2 Linear Regression and Support Vector Regression

If the response variable is quantitative as opposed to categorical then regression approaches are needed. If the data are given by vectors x_i and the corresponding response variable is y_i, then the straightforward ordinary least squares (LS) approach computes weights w and a scalar β such that the squared prediction error is minimized:

$$\min_{w,\beta} \sum_i (w^T x_i + \beta - y_i)^2$$

The SVR approach, which is closely related to SVMs for classification, offers a more flexible solution to the regression problem, overall numerical efficiency, and applicability to large-scale problems. In particular, the so-called ε-insensitive SVR models allow for the error measure to be defined in a flexible way, for example, reflecting the expected level of errors by varying error bars, ε, for different types of training examples that may differ in their characteristics (Wagner et al., 2005). SVR models can be seen as extensions of LS models where an ε-insensitive penalty function is minimized instead of the sum of squared errors (see, e.g., Hastie et al., 2001). For the purposes of these considerations, we restrict ourselves to stating the overall optimization problem that is solved by SVRs, which reads

$$\begin{aligned} &\min \|w\| + C\|\xi\| \\ &\text{s.t. } |w^T x_i + \beta - y_i| - \xi_i \le \varepsilon \quad \text{for all } i \end{aligned} \tag{2.3}$$

Here C is again an *a priori* penalty parameter that balances the regression error term $\|\xi\|$ and the regularization term $\|w\|$ (that corresponds to the margin maximization term for SVMs). If $(\tilde{w}, \tilde{\beta})$ denotes the optimal SVR solution to Equation 2.3, then the predicted outcome for a new data point characterized by x_k is given by $\hat{y}_k = \tilde{w}^T x_k + \tilde{\beta}$. To compare and contrast with the standard LS approach, note that the term $w^T x_i + \beta - y_i$ in the constraints corresponds to the objective function for LS. SVRs will penalize this deviation through the slack variable ξ_i if it exceeds ε, which is an error insensitivity parameter that must be set by the user *a priori*. In addition, SVRs generally use the 1-norm to penalize the regression error $\|\xi\|_1$ and are thus less outlier sensitive, which again is especially advantageous in many bioinformatics applications with noisy data. Finally, we also mention that although C and ε are typically chosen as fixed parameters, there is no real reason not to allow them to be functions of y_i, choosing different error insensitivities $\varepsilon_i = \varepsilon(y_i)$, depending on the expected (or naturally occurring) error for the response variable y_i. These ideas were successfully applied in a number of protein structure prediction contexts; see, for example, Wagner et al., 2005.

2.5.3 Neural Networks for Classification and Regression

Another solution that goes beyond LDA is provided by NNs, which can generate arbitrary nonlinear decision boundaries by addition of many simple functions. This is typically achieved by a multistage transformation, which may be represented graphically as a network (directed graph) of interconnected layers of "computing" nodes that integrate input signals from the previous layer. In particular, the input features (attributes) are represented by individual nodes in the input layer and are subsequently transformed into a new set of features using several hyperplanes, w_k, corresponding to the hidden layer nodes (here, for simplicity, we assume that only one hidden layer is used). In other words, the inputs for the hidden layer nodes are linear combinations of the original N features with the coefficients of the linear combination, w_{ik}, associated with connections between input node i and hidden layer node k. The hidden layer nodes transform such defined inputs using some functions $h_k(x) = s(w_k^T x + w_{ik})$, where the scalar functions $s(\cdot)$ are usually chosen to be logistic functions. Thus, they have a sigmoidal shape with output bounded by maximum and minimum values. As a result, the outputs are in general nonlinear functions of inputs.

There is a distant analogy between the activity of biological neurons that sum input signals weighted by the strength of synaptic connections and send output signals that are bounded by some maximum values. For this reason, such nodes are called artificial neurons or perceptrons. A number of these nodes connected to the same input form a layer that transforms the input vector to the vector of hidden layer activities H. Connecting input and hidden and output nodes form a network representation of such function, and this is called a *neural network*. If the sigmoidal-shape functions are used for all layers, then this network is called a "multilayered perceptron" (MLP). If the hidden layer functions are of Gaussian or similar type, then the network is called RBF network (Duda and Hart, 1973; Ripley, 1994; Hassoun 1995). In general, MLP and RBF neural networks are basically function-mapping systems for classification and regression that can learn how to associate numerical inputs with arbitrary outputs, changing their internal parameters.

The training of NNs typically involves minimizing the following error function:

$$\mathrm{SSE}(w) = \sum_i \alpha(y_i)(\hat{y}_i(w) - y_i)^2$$

where $\hat{y}_i$ denotes the predicted value of the variable of interest, given the parameters w of the network (weights and biases), and y_i represents the target value imposed in the training. In analogy to the error insensitivities ε_i, the weights $\alpha(y_i)$ can be chosen to reflect uncertainties associated with the target values. Many training algorithms have been devised to find parameters w that fit inputs to the desired outputs. One standard example of a commonly used training algorithm is the so-called back propagation algorithm. The important point is that the training of NNs involves an attempt to solve a global nonlinear optimization problem, which is inherently difficult and requires heuristics such as different starting points and exit strategies from local minima. It is also difficult to assess the quality of a NN configuration, that is, to estimate how far from the global optimum one is. This is in stark contrast to SVMs that solve a convex optimization problem with a unique minimum and are thus computationally more efficient and less prone to overfitting.

On the other hand, NNs are very flexible and in general accurate if trained properly. Furthermore, they can be used for either classification or regression problems. For classification problems, NNs can be designed to have several output nodes, each one indicating the strength of prediction for a class. For regression problems, NNs typically have a single output node that emits a real-valued predictor for the quantity in question. We note that the number of free parameters that need to be chosen during the training phase for NNs can typically be large, depending on the number of activation nodes in the hidden layers and the topology of the network, thus requiring a comparatively large number of training samples in order to avoid overparametrization. This is in contrast to linear models that in general only have as many model parameters (weights) as there are features in the data representation. However, if a sufficient number of training instances is available and the models are carefully trained, then NNs are often the methods of choice. (see, e.g., the upcoming section on protein secondary structure prediction).

2.5.4 Hidden Markov Models

Hidden Markov Models (HMMs) are widely used probabilistic machine learning techniques based on a finite state machine representation of the structure ("grammar") of a "language" that is to be modeled. In the case of bioinformatics the language typically relates to genomic DNA sequence or protein sequences, with the corresponding "grammars" encoding, for example, the exon/intron/intergenic region constitution of genomic sequences from a given genome or, say, the membership of an amino acid sequence in a family of evolutionarily related proteins. There is a vast literature covering methodology and numerous applications of HMM models; for an excellent review of HMM developments and applications we refer the reader to Durbin et al. (1998).

HMMs are characterized by a set of "states" as well as the transition probabilities between these states. Furthermore, HMMs can emit letters in states (from which the words of the "language" are assembled), and the emission probabilities for each letter depend on the underlying state the HMM is in. Finally, the Markov property dictates that the subsequent state the HMM transitions to depends on its current state alone. An HMM may thus be viewed as a machine that generates sequences—each trajectory that follows a number of states through transition arrows generates a sequence.

Given transition and emission probabilities for an HMM, one can compute a total probability that a particular sequence was emitted by that HMM. Conversely, given a number of well-characterized training instances (e.g., conserved protein sequences belonging to the same family) one can design an HMM by choosing a topology as well as transition and emission probabilities that maximize the likelihood that these sequences were generated by the HMM. There are several algorithms to perform this "training" (choice of model parameters based on well-characterized data); the most successful ones are based on variants of the well-known expectation-maximization (EM) algorithm. We refer the reader to Durbin et al. (1998) for technical details.

We would like to comment that the availability of various HMM packages for development and training of such models contributed significantly to the widespread use of HMM and similar statistical learning techniques in bioinformatics. For example, the HMMer package (Durbin et al., 1998; http://hmmer.janelia.org), a general-purpose HMM simulator for biological sequence analysis, which is available freely for academic use, can be used, among other applications, in gene finding and discovery of new patterns in genomic sequences.

In this part of the chapter, we discuss several applications of machine learning approaches in bioinformatics. We start the discussion with applications to structural bioinfomatics, including canonical examples of secondary structure and solvent accessibility prediction for amino acid residues in proteins. Next, we present the problem of gene prediction and applications of HMMs to that problem. Finally, we briefly discuss issues related to drug design and applications of machine learning techniques to prediction of quantitative structure–activity relationship (QSAR) in this context.

2.6 Applications of Machine Learning to Structural Bioinformatics

2.6.1 Secondary Structure Prediction

In order to perform their function, proteins typically adopt a specific 3D structure that remains stable under a range of physiological conditions. This is known as the folding process, in which an extended or unfolded protein conformation undergoes a series of conformational transitions into the folded (compact) structure. As part of this process, regular local conformations known as secondary structures emerge. These ordered local structures include α-helices and β-strands that are largely defined by local propensities of amino acid residues with nonlocal contacts between residues additionally stabilizing (or destabilizing) such local conformations. One of the first successful application of machine learning techniques in the field of protein structure prediction was the prediction of secondary structures in proteins (Qian and Seinowski, 1988; Rost and Sander, 1994a). (See also Figure 2.1 for an illustrative example of secondary structure prediction.)

In their pioneering work Rost and Sander (1994a) demonstrated the importance of the multiple alignment representation and used NNs to train a successful classifier capable of assigning each residue to one of the three classes (helix, β-strand or coil) with over 70% classification accuracy. In related work, Rost and Sander (1994b) proposed to extend this approach to predict relative solvent accessibility (RSA), which quantifies relative solvent exposure compared with an extended conformation. Since then many different machine and statistical learning techniques have been devised and used to improve secondary structure prediction (for a recent review see, e.g., Przybylski and Rost, 2002). In fact, increasingly accurate prediction methods, which achieve about 80% accuracy for classification into three states (Eyrich et al., 2001; Przybylski and Rost, 2002; Adamczak et al., 2005) have already and significantly contributed to the improved performance of fold recognition and de novo protein structure prediction methods (Fischer et al., 2001; Venclovas et al., 2001; Schonbrun et al., 2002).

2.6.2 Solvent Accessibility Prediction

The solvent accessible surface area of an amino acid residue in a protein structure is another important attribute that, if known, can be used to facilitate and enhance the overall structure prediction in fold recognition or de novo folding simulations (Fischer et al., 2001; Venclovas, 2001). For example, globular proteins in aqueous solution are characterized by the formation of a hydrophobic core, which is shielded from the solvent. Therefore, estimates of the solvent accessibility can be compared with the solvent accessible surface areas observed in known protein structures and can thus help identify

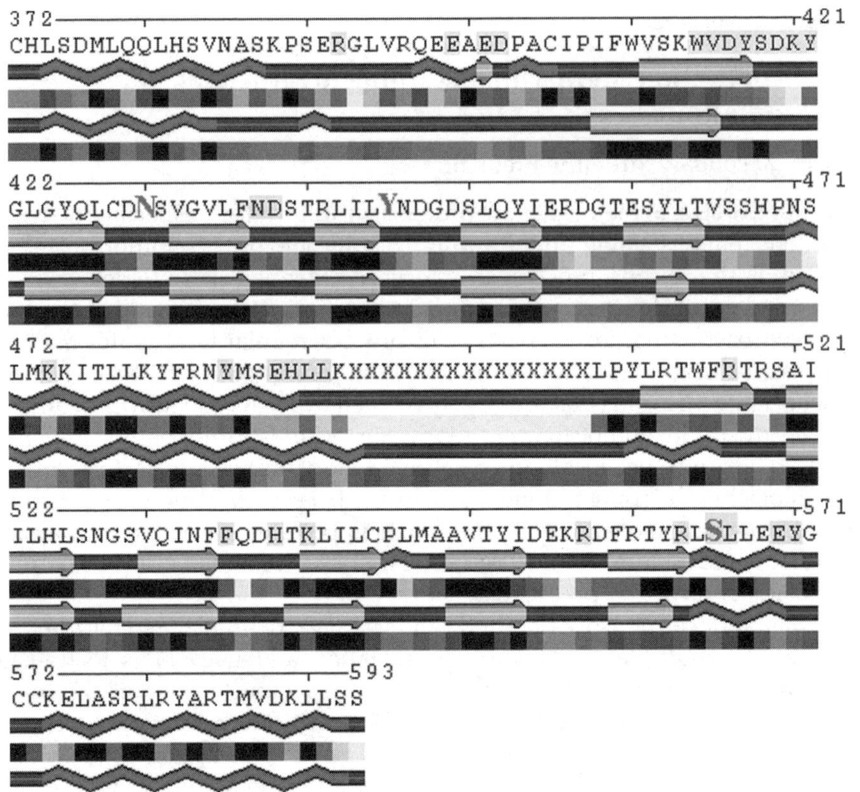

FIGURE 2.1
(See color insert following page 207.) Comparison of experimentally observed (PDB structure lq4k, chain A, two upper rows) and predicted (using the SABLE (From R. Adamczak, A. Porollo and J. Meller, *Proteins*, 56, 753–767, 2004.) server, lower rows) structures of polo kinase PIki. Helices are indicated using red braids, beta-strands are indicated using green arrows and loops are shown in blue. The relative solvent accessibility is represented by shaded boxes, with black boxes corresponding to fully buried residues. Sites located in known protein–protein interaction interfaces are highlighted using yellow, whereas residues corresponding to polymorphic sites are highlighted in red and Xs represent fragments unresolved in the crystal structure. Figure generated using the POLYVIEW server (http://polyview.cchmc.org).

the most compatible structural template in fold recognition. Similarly, in folding simulations the search through the space of all possible conformations can be biased toward those conformations that are consistent with the predicted pattern of solvent accessibility (Adamczak et al., 2004). In addition, identifying surface exposed residues is an important step in recognition of protein–protein interaction interfaces and may help classify functional effects of mutations (Glaser et al., 2003). The problem of predicting RSA appears to be more difficult than prediction of secondary structures (Rost and Sander, 1994b; Adamczak et al., 2004). The reasons RSA prediction methods are comparatively less successful lie primarily in the nature of the problem, but to some extent they are also rooted in the way the problem is typically solved.

The RSA of an amino acid residue in a protein structure is a real number between 0% and 100%, with 0% RSA corresponding to fully buried and 100% RSA to fully exposed residues, respectively, that represents the solvent exposed surface area of this residue in relative terms. The level of solvent exposure is weakly conserved in families of homologous structures (Rost and Sander, 1994b; Adamczak et al., 2004). Using PFAM multiple alignments for protein families, it has been estimated that the average correlation of the RSAs for pairs of equivalent residues in homologous structures is equal to 0.57 (Adamczak et al., 2004). Thus, contrary to the prediction of secondary structures, the highly variable real valued RSA does not support the notion of clearly defined distinct classes of residues and suggests that a regression-based approach is appropriate for this problem. In light of these difficulties, it is striking that most existing RSA prediction methods cast this problem within the classification framework, attempting to predict whether a given amino acid exceeds some (arbitrary) RSA threshold and would thus be predicted to be "exposed," as opposed to "buried." Recent examples of such attempts to further improve the RSA prediction include both feed forward (Ahmad and Gromiha, 2002) and recurrent NN (Pollastri et al., 2002), as well as SVM-based (Kim and Park, 2004) approaches.

In an effort to go beyond the classification paradigm and provide real valued RSA predictions we recently developed several alternative regression based models, including standard LS, linear SVR and NNs-based nonlinear models. This allowed us to investigate the prediction limits of the simplest kinds of regression models (linear models) on the same dataset, perform extensive cross-validation and feature selection with these simple models and use the results to assess the relative benefits of the more involved nonlinear NNs (Adamczak et al., 2004; Wagner et al., 2005). In order to represent an amino acid residue we used evolutionary information as encoded by PSSMs and we trained our methods using a large set of representative and nonredundant protein structures. In rigorous tests, following an evaluation of automatic protein structure prediction (EVA)-like methodology (Eyrich et al., 2001) for evaluation of the accuracy of secondary structure prediction methods (see http://cubic.bioc.columbia.edu/eva for details.), the new NN-based methods achieved significantly higher accuracy than previous methods from the literature, with mean absolute errors between 15.3% and 15.8% RSA and correlation coefficients between observed and predicted RSAs of about 0.64–0.67 on different control sets. In two state projections (e.g., using 25% RSA as a threshold between buried and exposed residues), the new method outperformed current state-of-the-art classification-based approaches, achieving an overall classification accuracy of about 77% (for details see Adamczak et al., 2004). These estimates of the accuracy have since been confirmed by independent studies (see, e.g., Garg et al., 2005).

2.6.3 Structural Predictions for Membrane Proteins

Although high-resolution structural data for soluble proteins and their interactions are relatively abundant (and quickly growing), it is not the case for membrane proteins. Owing to difficulties in applying experimental techniques, such as x-ray crystallography or NMR, the number of high resolution structures of membrane proteins that have been solved to date is still very limited. From January 2006, there were only 129 unique integral membrane proteins with known 3D structures in the Protein Data Bank, of which 101 were α-helical membrane protein and 28 were β-barrel and other membrane proteins (http://blanco.biomol.uci.edu/mpex/). On the other hand, it is estimated that integral membrane proteins constitute about 20%–30% of all proteins in the sequenced genomes (Wallin and von Heijne, 1998). The computational prediction of membrane proteins has therefore become an important alternative and complementary tool for genomic annotations and membrane protein studies (for a comprehensive review of the state of the art in membrane protein prediction see, e.g., Chen and Rost, 2002).

Many successful prediction protocols, especially for the prediction of transmembrane helices, have been devised and employed since the pioneering efforts of Kyte and Doolittle more than 20 years ago (Kyte and Doolittle, 1982). The first class of methods for transmembrane helix prediction relies on hydropathy scales and other physicochemical properties of amino acids. Successful examples of such methods include SOSUI (Hirokawa et al., 1998) and TopPred (Claros and von Heijne, 1994). The second class includes statistical methods based on the observed preferences of amino acids for membrane proteins. An example of a successful method of this type is TMpred (Hofmann and Stoffel, 1993). Finally, the third group of methods is based on evolutionary information, as encoded by the multiple alignment (MA). Examples of such highly successful methods include the NN-based PHDhtm (Rost et al., 1994; Rost et al., 1995; Rost et al., 1996), and a HMM-based TMHMM (Krogh et al., 2001) and HMMTOP (Tusnady and Simon, 1998; Tusnady and Simon, 2001) methods. Whereas similar concepts apply to prediction of β-barrel membrane proteins as well, this problem appears to be even more difficult owing to the weakly hydrophobic nature of membrane spanning β-strands in these proteins (Bigelow et al., 2004).

The accuracy of membrane protein prediction has recently been reevaluated by several groups, suggesting that, despite some optimistic estimates in the literature, existing methods are still rather limited in their predictive power. For example, according to an independent assessment performed in the Rost Lab (Chen et al., 2002; Chen and Rost, 2002; Kernytsky and Rost, 2003), the top performing methods for transmembrane helix prediction, such as PHDhtm, HMMTOP, or TMHMM, achieved (Chen and Rost, 2002; Kernytsky and Rost, 2003) per-segment accuracy of up to 84% and per-residue accuracy of up to 80% in the TMH Benchmark evaluation, with an estimated 1–6% false positive rate among globular proteins and between 20% and 30% false positive matches among signal peptides. These rates of confusion with globular proteins and especially signal peptides make proteome-wide annotation of membrane proteins likely to result in large number of false positives (Moeller et al., 2000, 2001; Chen et al., 2002). However, recent improvements in that regard are likely to further increase the accuracy of membrane domain prediction (Viklund and Elofsson, 2004; Cao et al., 2006).

2.6.4 Computational Protocols for the Recognition of Protein–Protein Interaction Sites

Elucidating complex networks of molecular interactions and the resulting dynamic states of cells and higher order systems has become one of the major challenges in systems biology. In particular, the importance of protein–protein interactions continues to stimulate the development of both experimental and computational protocols that aim at elucidating protein function and the underlying physical interactions. One important stepping stone toward these bigger goals is the prediction of protein–protein interaction interfaces, which is an active field of research, as summarized below.

In general, the problem of recognizing protein–protein interaction sites can be cast as a classification problem, that is, each amino acid residue is assigned to one of the two classes: interacting (interfacial) or noninteracting (noninterfacial) residues. Consequently, the problem may be solved using statistical and machine learning techniques, such as NNs (Zhou and Shan, 2001; Fariselli et al., 2002; Ofran and Rost, 2003) or SVMs (Bock and Gough, 2001; Zhang et al., 2003; Koike and Takagi, 2004). From the point of view of a representation (feature space) used to capture characteristic signatures (or fingerprints) of interaction interfaces, one may distinguish two main groups of approaches. The first group of methods attempts to predict interaction sites using just sequence information (Gallet et al., 2000; Bock and Gough, 2001; Ofran and Rost, 2003), whereas the second group takes available structural information into account as well (Fariselli et al., 2002; Neuvirth et al., 2004; Bordner and Abagyan, 2005).

In the latter case, the problem typically involves the identification of specific patches on the surface of a monomeric protein structure with residues that are either evolutionarily conserved or have a propensity for interaction interfaces (see, e.g., Jones and Thornton, 1997; Armon et al., 2001; Valdar and Thornton, 2001; Glaser et al., 2003; Yao et al., 2003; Caffrey et al., 2004; Bordner and Abagyan, 2005). In addition, various statistical techniques were used to analyze amino acid distributions at interaction interfaces and to identify distinct types of complexes with different amino acid biases (see, e.g., Gallet et al., 2000; Ofran and Rost, 2003) For example, amino acid conservation patterns were utilized to map binding sites in the SH2, PTB, and ATP domains (Armon et al., 2001).

The advantage of using structural information, for example, in the form of a resolved monomeric protein structure, is that the surface exposed residues and spatial neighbors (residues in contact in 3D) can be identified, defining potential interacting patches on the surface of a protein. Furthermore, geometric characteristics and the topology of potential interacting patches can be taken into account. On the other hand, it has been suggested that the predicted solvent accessibility can be used to address some of the limitations of sequence-based methods, in order to enable a more reliable prediction of interaction sites even when structural information is not available (Ofran and Rost, 2003; Berezin et al., 2004). However, owing to uncertain information about structural neighbors and the additional uncertainty introduced by RSA predictions, it remains to be seen if significant progress can be made without incorporating experimentally derived (or obtained using reliable modeling techniques) high-resolution structural information.

2.6.5 Phosphorylation as a Crucial Signal Transduction Mechanism

Phosphorylation constitutes one of the most important posttranslational modification of proteins by virtue of covalently linking phosphate groups to side chains of serine, threonine, and tyrosine amino acids by the respective protein kinases (with phosphatases catalyzing the opposite enzymatic reaction). Relatively large, negatively charged phosphate groups change the chemical nature of these amino acids, modulating protein structure and function by affecting protein conformation, interactions with cofactors, the overall protein stability, and other functionally relevant characteristics. Examples of phosphorylation and dephosphorylation events controlling important cellular and physiological processes include cell cycle regulation, various signal transduction cascades, activation of transcription elongation complex, or cytoskeleton reorganization.

The computational prediction of phosphorylation and other posttranslational modification sites, both from the structure and the primary amino acid sequence is an active field of research (Kreegipuu et al., 1998; Berry et al., 2004; Blom et al., 2004; Iakoucheva et al., 2004; Rychlewski et al., 2004; Zhou et al., 2004). Examples of methods for phosphorylation site prediction are NetPhos (Blom et al., 1999), a NN-based predictor, Scansite (Obenauer et al., 2003), a sequence-motif based predictor, and disorder ehhanced phosphorylation sites predictor (DISPHOS) (Iakoucheva et al., 2004), which uses indicators of intrinsic disorder, such as secondary structure predictions in addition to sequence information. The latter method, together with previous reports that most of the sites of phosphorylation lay in coil regions and close to the surface (Kreegipuu et al., 1998; Nakai and Kanehisa, 1988), suggest that accurate prediction-based profiles proposed here may help in improving the accuracy of the recognition of phosphorylation sites. It should be also noted that some "negative" sites might include those that just have not (yet) been reported as phosphorylated. Therefore, phosphorylation site prediction is also likely to be improved by one-class machine learning protocols briefly discussed in an earlier section of this review.

2.6.6 Assessment of Functional and Structural Consequences of Mutations

The analysis of human genetic variation can shed light on the problem of human phenotypic variation in general and the genetic basis of complex disorders, in particular. Single nucleotide polymorphisms in coding regions (cSNPs) and regulatory regions are most likely to affect gene function (Chakravarti, 1998; Collins et al., 1998; Syvaenen and Taylor, 2004). According to various estimates (which are to some extent tentative and will be most likely refined with the rapid progress of large-scale SNP sequencing and genotyping efforts) half of cSNPs cause missense mutations in the corresponding proteins, whereas the other half are silent (synonymous SNPs) and do not cause any change in the amino acid (Cargill et al., 1999; Halushka et al., 1999).

Nonsynonymous SNPs (nsSNPs), which occur in evolutionarily conserved regions or domains of proteins, can imply functional significance. Currently available methods for assessing the significance of the SNPs such as PolyPhen (Ramensky et al., 2002), which classifies a SNP as "possibly damaging" or "probably damaging" and so forth, are primarily based on analysis of conservation and frequency of substitutions observed in families of homologous proteins. However, a thorough understanding of the structural and functional properties of the normal and the "altered" proteins owing to nsSNPs is dependent on improved understanding and prediction of protein structure, stability, and interactions. In particular, improved annotations pertaining to protein–protein interactions and recognition of sites of posttranslational modifications would help classify nsSNPs occurring in protein coding regions. Such refined classification will help focus experimental research by prioritizing targets for mutagenesis and other studies.

2.7 Computational Gene Identification

An accurate recognition of all gene components is hindered by the limitations of our knowledge of complex biological processes and signals regulating gene expression. Therefore, gene prediction from genomic DNA sequences is an important and challenging problem, especially in the case of eukaryotic organisms (including the human genome), which are characterized by the presence of noncoding fragments (called introns) that separate noncontiguous coding fragments (called exons) within genes. In the human genome, there are roughly ten introns per gene (in some cases the number of introns is as high as 40 and more). Vast combinatorial possibilities of potential exon/intron assemblies and relatively high content of noncoding sequences make the problem of prediction the full coding sequence of a gene computationally very difficult. Nevertheless, motivated by the practical importance of the gene prediction problem, computational methodology for finding genes and other functional sites in genomic DNA has evolved significantly over the past 10 years.

Among the types of informative attributes (features) that correspond to sequence motifs and functional sites in genomic DNA that researchers have sought to recognize are splice sites (exon/intron boundaries that are recognized by the spliceosome machinery that removes introns before the mature gene without noncoding fragments can be translated into protein), start and stop codons (these correspond to the first and last amino acid residues in the resulting protein sequence), branch points (also recognized by the spliceosome), promoters and other regulatory elements of transcription, length distributions and compositional features of exons, introns, and intergenic regions. Even though existing methods achieved only moderate accuracy (Durbin et al., 1998), computational gene prediction has become a widely used strategy in genomic research, facilitating greatly experimental analysis, annotation, and biological discovery in newly sequenced genomes (Lander et al., 2001; Venter et al., 2001). These approaches have been described in depth previously (Snyder and Stormo, 1993; Krogh et al., 1994; Solovyev et al., 1994; Gelfand et al., 1996; Burge and Karlin, 1997; Henderson et al., 1997; Burge and Karlin, 1998; Durbin et al., 1998; Salzberg et al., 1998) and only those aspects of gene recognition methods that are most relevant for the illustration of some important aspects of applications of machine learning in this context are briefly discussed herein.

From the methodological standpoint, two approaches to gene recognition may be distinguished: *ab initio* gene prediction and similarity-based spliced alignments methods. In both approaches, gene identification proceeds formally through genomic sequence inspection and relies on prior knowledge of gene structure. The *ab initio* methods, such as HMM models for exon and intron sequences, incorporate the knowledge of known genes and their structure into a probabilistic model whose parameters are optimized to accurately recognize representative genes included in a training set. In contrast, splice alignment methods rely on similarities to other genes or gene products facilitated by alignment of genomic DNA sequence to known protein or cDNA sequences to reveal coding regions. In practice, various hybrid methods are often used that incorporate knowledge of predicted splice junctions to optimize splicing alignments (Gelfand et al., 1996). Although *ab initio* methods are biased to succeed for genes similar to those in the training set, the success of splicing alignments depends on the suitability of the chosen target sequence of an established gene or its product.

An important consideration of *ab initio* gene finding methods is the requirement for training by known examples when general principles governing the transcription process and mRNA-genomic relationships are not known. This training serves to optimize a classification model, which distinguishes coding from noncoding and intergenic DNA sequences. Thus, since learning from labeled examples forms the basis for *ab initio* gene prediction, essentially all methods for supervised learning are applicable in this context (although certainly not all are equally well suited for the task). We focus, however, on HMMs, which, as introduced before, have traditionally been considered as part of the statistical learning field.

Figure 2.2 (adopted from Henderson et al., 1997) illustrates the HMM model employed in the VEIL program (Henderson et al., 1997) for gene identification in eukaryotic genomes. Training sequences (with exons and introns) are treated as if they were generated by the HMM model (with its topology), and the goal is to adjust the emission and transition probabilities until

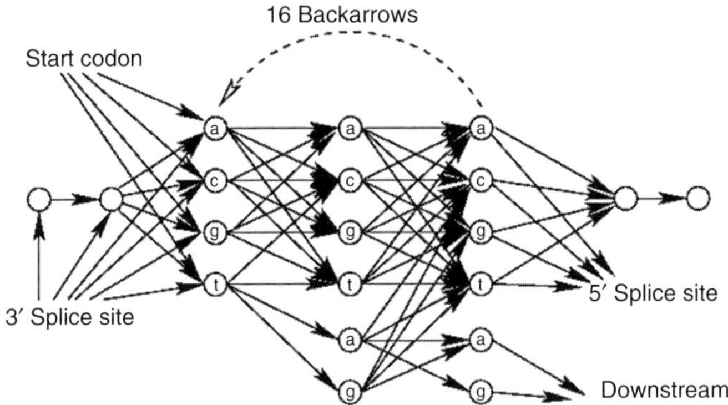

FIGURE 2.2
A typical HMM model of an exon. (Adopted from J. Henderson, S. Salzberg and K. H. Fasman, *J. Comput. Biol.*, 4, 127–141, 1997.)

the total probability of the training sequences is maximized. This is usually achieved using the expectation-maximization algorithm (Durbin et al., 1998). Once the model is trained, prediction for new sequences may be obtained by "threading" the query sequence through the model, to find an optimal trajectory generating the query sequence (using the Viterbi algorithm Durbin et al., 1998). Aligning each base in the query sequence to a particular state optimally in the model defines location of exons, introns, and their boundaries in the sequence. For example, the bases that are aligned to states representing exons are predicted to belong to an exon.

One HMM approach that proved to be especially successful, although more complex, is the Generalized Hidden Markov Model (GHMM) (Kulp et al., 1996). The GHMM approach is employed in the widely used and very successful HMMgene (Krogh, 2000) and GenScan (Burge and Karlin, 1997) programs, for example.

2.8 Biomarkers, Drug Design, and QSAR Studies

In the past, identification of drugs and their design was done largely using trial and error approach that was clearly not very effective. Moreover, the mechanism of action of a successful drug would typically remain obscured. As an alternative approach to drug design the QSAR is being widely used. The idea behind these approaches was to use the known responses (activities) of simple compounds (structures) to predict the responses of complex compounds, made from different combinations of the simple ones. Only those compounds that were predicted to have desired potential were then tested in the laboratory. Thus, QSAR approaches lend themselves naturally to applications of machine learning.

Ligands may be represented by multiple structural and other descriptors. Thus, selection of key descriptors is an important step in any QSAR study. Another important step is the identification of patterns that correlate with activity. Furthermore, compounds exhibiting promising properties may be compared with other candidates in order to identify other potential drugs that share critical features. Therefore, it is evident that the machine learning approaches to feature-selection, pattern-recognition, classification, and clustering can be applied to this problem. In fact, various clustering methods (e.g., hierarchical divisive clustering, hierarchical agglomerative clustering, nonhierarchical clustering, and multidomain clustering) have been applied to such problems (Kitchen et al., 2004). Self-organized NNs, SVMs, binary kernel discrimination, and other genetic algorithm-based classification techniques have also been discussed in the review (Walters et al., 2005) with regard to compound classification.

Application of clustering techniques and genetic algorithms towards predicting molecular interactions has been reviewed in Dror et al. (2004) The role of feature selection in QSAR has also been reviewed recently (Walters et al., 2005). For example, the clustering of receptor proteins on the basis of their structural similarity has been shown to be of great significance to drug design in Koch and Waldmann's (2005) review. Many other studies have made use of machine learning techniques to address similar problems. In particular, NNs have been widely used to solve many problems in drug design. A comprehensive review of the applications of NNs in variety of QSAR problems has been presented by Winkler (2004).

In recent years, SVMs have become relatively widely used in this context as well. For example, Zhao et al. made use of SVM for predicting toxicity suggesting that SVM outperforms multiple linear regression and radial basis function NNs for this application (Zhao et al., 2006). A novel method called least squares support vector machine (LSSVM) was employed to screen calcium channel antagonists in a QSAR study (Yao et al., 2005). SVM was also used to predict oral absorption in humans involving molecular structure descriptors (Liu et al., 2005) and to calculate the activity of certain enzyme inhibitors and were found to perform better than classical QSAR methods (Zernov et al., 2003).

2.9 Conclusions

Machine learning has clearly become a tool of great importance in bioinformatics, as evidenced by a number of promising applications in diverse areas such as protein structure prediction, gene finding, and QSAR studies. We have focused on a number of "soft" issues surrounding machine learning that in our view are essential ingredients for its successful application: significant domain insight leading to appropriate representations and descriptions of the data, an appropriate, unbiased choice of training (or gold standard) sets, stringent, independent evaluation on test sets, dimension reduction through feature selection and careful consideration and evaluation of the most appropriate level of model complexity. A number of machine learning algorithms have arisen in the last decade that have been particularly successful, and while we could not possibly provide a complete survey we hope to have led the reader to a number of relevant references for more information. Finally, we surveyed a (necessarily biased) sample of applications to illustrate the keys to success in practice and to show the wide range of exciting applications in bioinformatics that all have potential of providing significant new insights through computational methods.

References

Adamczak R., Porollo A., and Meller J. (2004). Accurate prediction of solvent accessibility using neural networks-based regression. *Proteins*, 56, 753–767.

Adamczak R., Porollo A., and Meller J. (2005). Combining prediction of secondary structure and solvent accessibility in proteins. *Proteins*, 59, 467–475.

Ahmad S. and Gromiha M. M. (2002). NETASA: Neural network based prediction of solvent accessibility. *Bioinformatics*, 18, 819–824.

Alizadeh A., Eisen M., Botstein D., Brown P. O., and Staudt L. M. (1998). Probing lymphocyte biology by genomic-scale gene expression analysis. *J. Clin. Immunol.*, 18, 373–379.

Armon A., Graur D., and Ben-Tal. N. (2001). ConSurf: An algorithmic tool for the identification of functional regions in proteins by surface mapping of phylogenetic information. *J. Mol. Biol.*, 307, 447–463.

Bair E. and Tibshirani R. (2004). Semi-supervised methods to predict patient survival from gene expression data. *PLoS Biol.*, 2, E108.

Baldi P., Brunak S., Chauvin Y., Andersen C. A., and Nielsen H. (2000). Assessing the accuracy of prediction algorithms for classification: An overview. *Bioinformatics*, 16, 412–424.

Berezin C., Glaser F., Rosenberg J., Paz I., Pupko T., Fariselli P., Casadio R., and Ben-Tal N. (2004). ConSeq: The identification of functionally and structurally important residues in protein sequences. *Bioinformatics*, 20, 1322–1324.

Berry E. A., Dalby A. R., and Yang Z. R. (2004). Reduced bio basis function neural network for identification of protein phosphorylation sites: Comparison with pattern recognition algorithms. *Comput. Biol. Chem.*, 28, 75–85.

Bigelow H. R., Petrey D. S., Liu J., Przybylski D., and Rost B. (2004). Predicting transmembrane beta-barrels in proteomes. *Nucleic Acids Res.*, 32, 2566–2577.

Blom N., Gammeltoft S., and Brunak S. (1999). Sequence and structure-based prediction of eukaryotic protein phosphorylation sites. *J. Mol. Biol.*, 294, 1351–1362.

Blom N., Sicheritz-Pontén T., Gupta R., Gammeltoft S., and Brunak S. (2004). Prediction of post-translational glycosylation and phosphorylation of proteins from the amino acid sequence. *Proteomics*, 4, 1633–1649.

Bock J. R. and Gough D. A. (2001). Predicting protein–protein interactions from primary structure. *Bioinformatics*, 17, 455–460.

Bordner A. J. and Abagyan R. (2005). Statistical analysis and prediction of protein–protein interfaces. *Proteins*, 60, 353–366.

Brown M. P., Grundy W. N., Lin D., Cristianini N., Sugnet C. W., Furey T. S., Ares M., and Haussler D. (2000). Knowledge-based analysis of microarray gene expression data by using support vector machines. *Proc. Natl. Acad. Sci. USA.*, 97, 262–267.

Burge C. and Karlin S. (1997). Prediction of complete gene structures in human genomic DNA. *J. Mol. Biol.*, 268, 78–94.

Burge C. B. and Karlin S. (1998). Finding the genes in genomic DNA. *Curr. Opin. Struct. Biol.*, 8, 346–354.

Caffrey D. R., Somaroo S., Hughes J. D., Mintseris J., and Huang E. S. (2004). Are protein–protein interfaces more conserved in sequence than the rest of the protein surface? *Protein Sci.*, 13, 190–202.

Cao B., Porollo A., Adamczak R., Jarrell M., and Meller J. (2006). Enhanced recognition of protein transmembrane domains with prediction-based structural profiles. *Bioinformatics*, 22, 303–309.

Cargill M., Altshuler D., Ireland J. et al. (1999). Characterization of single-nucleotide polymorphisms in coding regions of human genes. *Nat. Genet.*, 22, 231–238.

Chakravarti A. (1998). It's raining SNPs, hallelujah? *Nat. Genet.*, 19, 216–217.

Chen C. P., Kernytsky A., and Rost B. (2002). Transmembrane helix predictions revisited. *Protein Sci.*, 11, 2774–2791.

Chen C. P. and Rost B. (2002). State-of-the-art in membrane protein prediction. *Appl. Bioinformatics*, 1, 21–35.

Claros M. G. and von Heijne G. (1994). TopPred II: An improved software for membrane protein structure predictions. *Comput. Appl. Biosci.*, 10, 685–686.

Collins F. S., Brooks L. D., and Chakravarti A. (1998). A DNA polymorphism discovery resource for research on human genetic variation. *Genome Res.*, 8, 1229–1231.

Cuff J. A. and Barton G. J. (2000). Application of multiple sequence alignment profiles to improve protein secondary structure prediction. *Proteins*, 40, 502–511.

Dror O., Shulman-Peleg A., Nussinov R., and Wolfson H. J. (2004). Predicting molecular interactions in silico: I. A guide to pharmacophore identification and its applications to drug design. *Curr. Med. Chem.*, 11, 71–90.

Du Q.-S., Jiang Z.-Q., He W.-Z., Li D.-P., and Chou K.-C. (2006). Amino Acid Principal Component Analysis (AAPCA) and its applications in protein structural class prediction. *J. Biomol. Struct. Dyn.*, 23, 635–640.

Duda R. O. and Hart P. E. (1973). *Pattern Classification and Scene Analysis*. John Wiley & Sons, New York.

Dudoit S., Fridlyand J., and Speed T. P. (2002). Comparison of discrimination methods for the classification of tumors using gene expression data. *J. Am. Stat. Assoc.*, 97, 77–87.

Durbin R., Eddy S., Krogh A., and Mitchison J. (1998). *Biological Sequence Analysis*. Cambridge University Press, New York, NY.

Eisen M. B. and Brown P. O. (1999). DNA arrays for analysis of gene expression. *Methods Enzymol.*, 303, 179–205.

Eyrich V. A., Martí-Renom M. A., Przybylski D., Madhusudhan M. S., Fiser A., Pazos F., Valencia A., Sali A., and Rost B. (2001). EVA: Continuous automatic evaluation of protein structure prediction servers. *Bioinformatics*, 17, 1242–1243.

Fan R.-E., Chen P.-H., and Lin C.-J. (2005). Working set selection using the second order information for training SVM. *J. Mach. Learn. Res.*, 6, 1889–1918.

Fariselli P., Pazos F., Valencia A., and Casadio R. (2002). Prediction of protein–protein interaction sites in heterocomplexes with neural networks. *Eur. J. Biochem.*, 269, 1356–1361.

Fischer D., Elofsson A., Rychlewski L. et al. (2001). CAFASP2: The second critical assessment of fully automated structure prediction methods. *Proteins*, Suppl. 5, 171–183.

Furey T. S., Cristianini N., Duffy N., Bednarski D. W., Schummer M., and Haussler D. (2000). Support vector machine classification and validation of cancer tissue samples using microarray expression data. *Bioinformatics*, 16, 906–914.

Gallet X., Charloteaux B., Thomas A., and Brasseur R. (2000). A fast method to predict protein interaction sites from sequences. *J. Mol. Biol.*, 302, 917–926.

Garg A., Kaur H., and Raghava G. P. S. (2005). Real value prediction of solvent accessibility in proteins using multiple sequence alignment and secondary structure. *Proteins*, 61, 318–324.

Garrido L., Gaitan V., Serra-Ricart M., and Calbet X. (1995). Use of multilayer feedforward neural nets as a display method for multidimensional distributions. *Int. J. Neural. Syst.*, 6, 273–282.

Gelfand M. S., Mironov A. A., and Pevzner P. A. (1996). Gene recognition via spliced sequence alignment. *Proc. Natl. Acad. Sci. USA.*, 93, 9061–9066.

Glaser F., Pupko T., Paz I., Bell R. E., Bechor-Shental D., Martz E., and Ben-Tal N. (2003). ConSurf: Identification of functional regions in proteins by surface-mapping of phylogenetic information. *Bioinformatics*, 19, 163–164.

Halushka M. K., Fan J. B., Bentley K., Hsie L., Shen N., Weder A., Cooper R., Lipshutz R., and Chakravarti A. (1999). Patterns of single-nucleotide polymorphisms in candidate genes for blood-pressure homeostasis. *Nat. Genet.*, 22, 239–247.

Hansen J. E., Lund O., Tolstrup N., Gooley A. A., Williams K. L., and Brunak S. (1998). NetOglyc: Prediction of mucin type O-glycosylation sites based on sequence context and surface accessibility. *Glycoconj. J.*, 15, 115–130.

Hassoun M. (1995). *Fundamentals of Artificial Neural Networks*. MIT Press, Cambridge.

Hastie T., Tibshirani R., Eisen M. B., Alizadeh A., Levy R., Staudt L., Chan W. C., Botstein D., and Brown P. (2000). "Gene shaving" as a method for identifying distinct sets of genes with similar expression patterns. *Genome Biol.*, 1, RESEARCH0003.

Hastie T., Tibshirani R., and Friedman J. (2001). The Elements of Statistical Learning. Springer Verlag, New York, NY.

Henderson J., Salzberg S., and Fasman K. H. (1997). Finding genes in DNA with a Hidden Markov Model. *J. Comput. Biol.*, 4, 127–141.

Hirokawa T., Boon-Chieng S., and Mitaku S. (1998). SOSUI: Classification and secondary structure prediction system for membrane proteins. *Bioinformatics*, 14, 378–379.

Hofmann K. and Stoffel W. (1993). Tmbase—A database of membrane spanning proteins segments. *Biol. Chem.*, 374, 166.

Iakoucheva L. M., Radivojac P., Brown C. J., O'Connor T. R., Sikes J. G., Obradovic Z., and Keith Dunker A. (2004). The importance of intrinsic disorder for protein phosphorylation. *Nucleic Acids Res.*, 32, 1037–1049.

Jones D. T. (1999). Protein secondary structure prediction based on position-specific scoring matrices. *J. Mol. Biol.*, 292, 195–202.

Jones S. and Thornton J. M. (1997). Prediction of protein–protein interaction sites using patch analysis. *J. Mol. Biol.*, 272, 133–143.

Kernytsky A. and Rost B. (2003). Static benchmarking of membrane helix predictions. *Nucleic Acids Res.*, 31, 3642–3644.

Kim H. and Park H. (2004). Prediction of protein relative solvent accessibility with support vector machines and long-range interaction 3D local descriptor. *Proteins*, 54, 557–562.

Kitchen D. B., Stahura F. L., and Bajorath J. (2004). Computational techniques for diversity analysis and compound classification. *Mini. Rev. Med. Chem.*, 4, 1029–1039.

Koch M. A. and Waldmann H. (2005). Protein structure similarity clustering and natural product structure as guiding principles in drug discovery. *Drug Discov. Today*, 10, 471–483.

Koike A. and Takagi T. (2004). Prediction of protein–protein interaction sites using support vector machines. *Protein Eng. Des. Sel.*, 17, 165–173.

Kreegipuu A., Blom N., Brunak S., and Järv J. (1998). Statistical analysis of protein kinase specificity determinants. *FEBS Lett.*, 430, 45–50.

Krogh A. (2000). Using database matches with for HMMGene for automated gene detection in Drosophila. *Genome Res.*, 10, 523–528.

Krogh A., Larsson B., von Heijne G., and Sonnhammer E. L. (2001). Predicting transmembrane protein topology with a Hidden Markov Model: Application to complete genomes. *J. Mol. Biol.*, 305, 567–580.

Krogh A., Mian I. S., and Haussler D. (1994). A Hidden Markov Model that finds genes in E. coli DNA. *Nucleic Acids Res.*, 22, 4768–4778.

Kulp D., Haussler D., Reese M. G., and Eeckman F. H. (1996). A generalized hidden Markov model for the recognition of human genes in DNA. *Proc. Int. Conf. Intell. Syst. Mol. Biol.*, 4, 134–142.

Kyte J. and Doolittle R. F. (1982). A simple method for displaying the hydropathic character of a protein. *J. Mol. Biol.*, 157, 105–132.

Lander E. S. and International Human Genome Sequencing Consortium (2001). Initial sequencing and analysis of the human genome. *Nature*, 409, 860–921.

Leslie C., Eskin E., Weston J., and Noble W. S. (2003). Mismatch string kernels for SVM protein. *Neuroinformation Processing Systems*, 15, 1441–1448.

Liu H. X., Hu R. J., Zhang R. S., Yao X. J., Liu M. C., Hu Z. D., and Fan B. T. (2005). The prediction of human oral absorption for diffusion rate-limited drugs based on heuristic method and support vector machine. *J. Comput. Aided. Mol. Des.*, 19, 33–46.

Medvedovic M. and Sivaganesan S. (2002). Bayesian infinite mixture model based clustering of gene expression profiles. *Bioinformatics*, 18, 1194–1206.

Medvedovic M., Yeung K. Y., and Bumgarner R. E. (2004). Bayesian mixture model based clustering of replicated microarray data. *Bioinformatics*, 20, 1222–1232.

Mitchell T. (1997). *Machine Learning*. McGraw Hill, New York.

Mjolsness E. and DeCoste D. (2001). Machine learning for science: State of the art and future prospects. *Science*, 293, 2051–2055.

Möller S., Croning M. D., and Apweiler R. (2001). Evaluation of methods for the prediction of membrane spanning regions. *Bioinformatics*, 17, 646–653.

Möller S., Kriventseva E. V., and Apweiler R. (2000). A collection of well characterised integral membrane proteins. *Bioinformatics*, 16, 1159–1160.

Nakai K. and Kanehisa M. (1988). Prediction of *in-vivo* modification sites of proteins from their primary structures. *J. Biochem. (Tokyo)*, 104, 693–699.

Neuvirth H., Raz R., and Schreiber G. (2004). ProMate: A structure based prediction program to identify the location of protein-protein binding sites. *J. Mol. Biol.*, 338, 181–199.

Obenauer J. C., Cantley L. C., and Yaffe M. B. (2003). Scansite 2.0: Proteome-wide prediction of cell signaling interactions using short sequence motifs. *Nucleic Acids Res.*, 31, 3635–3641.

Ofran Y. and Rost B. (2003). Predicted protein–protein interaction sites from local sequence information. *FEBS Lett.*, 544, 236–239.

Pollastri G., Baldi P., Fariselli P., and Casadio R. (2002). Prediction of coordination number and relative solvent accessibility in proteins. *Proteins*, 47, 142–153.

Primig M., Williams R. M., Winzeler E. A., Tevzadze G. G., Conway A. R., Hwang S. Y., Davis R. W., and Esposito R. E. (2000). The core meiotic transcriptome in budding yeasts. *Nat. Genet.*, 26, 415–423.

Przybylski D. and Rost B. (2002). Alignments grow, secondary structure prediction improves. *Proteins*, 46, 197–205.

Qian N. and Sejnowski T. J. (1988). Predicting the secondary structure of globular proteins using neural network models. *J. Mol. Biol.*, 202, 865–884.

Ramensky V., Bork P., and Sunyaev S. (2002). Human non-synonymous SNPs: Server and survey. *Nucleic Acids Res.*, 30, 3894–3900.

Ripley B. D. (1994). Neural networks and related methods for classification. *J. R. Stat. Soc. [Ser B]*, 56, 409–456.

Romero P., Obradovic Z., and Dunker A. K. (1997). Sequence data analysis for long disordered regions prediction in the calcineurin family. *Genome Inform. Ser. Workshop Genome Inform.*, 8, 110–124.

Rost B., Casadio R., Fariselli P., and Sander C. (1995). Transmembrane helices predicted at 95% accuracy. *Protein Sci.*, 4, 521–533.

Rost B., Fariselli P., and Casadio R. (1996). Topology prediction for helical transmembrane proteins at 86% accuracy. *Protein Sci.*, 5, 1704–1718.

Rost B. and Sander C. (1994a). Combining evolutionary information and neural networks to predict protein secondary structure. *Proteins*, 19, 55–72.

Rost B. and Sander C. (1994b). Conservation and prediction of solvent accessibility in protein families. *Proteins*, 20, 216–226.

Rost B., Sander C., and Schneider R. (1994). PHD—an automatic mail server for protein secondary structure prediction. *Comput. Appl. Biosci.*, 10, 53–60.

Rychlewski L., Kschischo M., Dong L., Schutkowski M., and Reimer U. (2004). Target specificity analysis of the Abl kinase using peptide microarray data. *J. Mol. Biol.*, 336, 307–311.

Salzberg S., Delcher A. L., Fasman K. H., and Henderson J. (1998). A decision tree system for finding genes in DNA. *J. Comput. Biol.*, 5, 667–680.

Schoelkopf B. and Smola A. (2002). *Learning with Kernels.* MIT Press,Cambridge.

Schonbrun J., Wedemeyer W. J., and Baker D. (2002). Protein structure prediction in 2002. *Curr. Opin. Struct. Biol.*, 12, 348–354.

Snyder E. E. and Stormo G. D. (1993). Identification of coding regions in genomic DNA sequences: An application of dynamic programming and neural networks. *Nucleic Acids Res.*, 21, 607–613.

Solovyev V. V., Salamov A. A., and Lawrence C. B. (1994). Predicting internal exons by oligonucleotide composition and discriminant analysis of spliceable open reading frames. *Nucleic Acids Res.*, 22, 5156–5163.

Syvänen A.-C. and Taylor G. R. (2004). Approaches for analyzing human mutations and nucleotide sequence variation: A report from the Seventh International Mutation Detection meeting, 2003. *Hum. Mutat.*, 23, 401–405.

The International HapMap Consortium. (2003). The International HapMap Project. *Nature*, 426, 789–796.

Tusnády G. E. and Simon I. (1998). Principles governing amino acid composition of integral membrane proteins: Application to topology prediction. *J. Mol. Biol.*, 283, 489–506.

Tusnády G. E. and Simon I. (2001). The HMMTOP transmembrane topology prediction server. *Bioinformatics*, 17, 849–850.

Valdar W. S. and Thornton J. M. (2001). Conservation helps to identify biologically relevant crystal contacts. *J. Mol. Biol.*, 313, 399–416.

Venclovas C. (2001). Comparative modeling of CASP4 target proteins: Combining results of sequence search with three-dimensional structure assessment. *Proteins*, Suppl. 5, 47–54.

Venclovas C., Zemla A., Fidelis K., and Moult J. (2001). Comparison of performance in successive CASP experiments. *Proteins*, Suppl. 5, 163–170.

Venter J. C., Adams M. D., Myers E. W. et al. (2001). The sequence of the human genome. *Science*, 291, 1304–1351.

Viklund H. and Elofsson A. (2004). Best alpha-helical transmembrane protein topology predictions are achieved using hidden Markov models and evolutionary information. *Protein Sci.*, 13, 1908–1917.

Wagner M., Adamczak R., Porollo A., and Meller J. (2005). Linear regression models for solvent accessibility prediction in proteins. *J. Comput. Biol.*, 12, 355–369.

Wagner M., Meller J., and Elber R. (2004a). Large-scale linear programming techniques for the design of protein folding potentials. *Mathematical Programming*, V101, 301–318.

Wagner M., Naik D. N., Pothen A. et al. (2004b). Computational protein biomarker prediction: A case study for prostate cancer. *BMC Bioinformatics*, V5, 26.

Wallin E. and von Heijne G. (1998). Genome-wide analysis of integral membrane proteins from eubacterial, archaean, and eukaryotic organisms. *Protein Sci.*, 7, 1029–1038.

Walters W. P. and Goldman B. (2005). Feature selection in quantitative structure-activity relationships. *Curr. Opin. Drug. Discov. Devel.*, 8, 329–333.

Winkler D. A. (2004). Neural networks as robust tools in drug lead discovery and development. *Mol. Biotechnol.*, 27, 139–168.

Yao H., Kristensen D. M., Mihalek I., Sowa M. E., Shaw C., Kimmel M., Kavraki L., and Lichtarge O. (2003). An accurate, sensitive, and scalable method to identify functional sites in protein structures. *J. Mol. Biol.*, 326, 255–261.

Yao X., Liu H., Zhang R., Liu M., Hu Z., Doucet J. P., and Fan B. (2005). QSAR and classification study of 1,4-dihydropyridine calcium channel antagonists based on least squares support vector machines. *Mol. Pharm.*, 2, 348–356.

Zernov V. V. , Balakin K. V., Ivaschenko A. A., Savchuk N. P., and Pletnev I. V. (2003). Drug discovery using support vector machines. The case studies of drug-likeness, agrochemical-likeness, and enzyme inhibition predictions. *J. Chem. Inf. Comput. Sci.*, 43, 2048–2056.

Zhang Y., Yin Y., Chen Y., Gao G., Yu Y., Luo J., and Jiang Y. (2003). PCAS—A precomputed proteome annotation database resource. *BMC Genomics*, 4, 42, (http://www.biomedcentral.com/1471–2164/4/42), Open Access.

Zhao C. Y., Zhang H. X., Zhang X. Y., Liu M. C., Hu Z. D., and Fan B. T. (2006). Application of support vector machine (SVM) for prediction toxic activity of different data sets. *Toxicology*, 217, 105–119.

Zhou H. X. and Shan Y. (2001). Prediction of protein interaction sites from sequence profile and residue neighbor list. *Proteins*, 44, 336–343.

Zhou F.-F., Xue Y., Chen G.-L., and Yao X. (2004). GPS: A novel group-based phosphorylation predicting and scoring method. *Biochem. Biophys. Res. Commun.*, 325, 1443–1448.

3

Machine Learning Methods for Cancer Diagnosis and Prognostication

Anne-Michelle Noone and Mousumi Banerjee

CONTENTS

3.1 Introduction

Machine learning methods have become increasingly popular as powerful analytic tools for exploring complex data structures. The applications of these methods are far reaching. The best documented, and arguably most popular uses of machine learning methods are in biomedical research where classification is a central issue. For example, a clinician may be interested in the following question: Does this patient with an enlarged prostate gland have prostate cancer, or does he simply have a benign disease of the prostate? To answer this question, various clinical information on the patient must be collected, and a good diagnostic test utilizing such information must be in place. The goal of machine learning methods is to provide a solution for constructing such diagnostic tests. For applications of machine learning methods in molecular biology and genomics, see Meller and Wagner (2007) in this volume.

In statistical nomenclature, *learning* or training refers to the process of finding the values of unknown parameters. This process can be broadly classified into two categories: supervised and unsupervised learning. Supervised learning, also called learning with a teacher, occurs when there is a known target value associated with each input in the training set. Unsupervised learning is needed when the training data lack target output values corresponding to input patterns. The algorithm must learn to group or cluster the input patterns on the basis of some common features, similar to factor analysis and principal components. This type of training is also called learning without a teacher because there is no source of feedback in the training process.

In this chapter, we discuss methodological and practical aspects of four supervised learning techniques, namely, classification and regression trees, artificial neural networks, random forests, and logic regression, in the context of cancer diagnosis and prognostication. Tree-based methods were originally introduced by Morgan and Sonquist (1963) and later popularized by Breiman et al. (1984) through practical and theoretical advances described in their monograph on Classification and Regression Trees. Generally, tree-based methods recursively partition the covariate space into disjoint regions and the corresponding data into groups (nodes). For each node to be split, some measure of separation in the response distribution between the two daughter nodes resulting from a split is calculated. All possible splits for each of the covariates are evaluated, and the variable and corresponding split point that best separates the daughter nodes is chosen. The same procedure is applied recursively to increase the number of nodes until each contains only a few subjects. The resulting model can be represented as a binary tree. After a large tree is grown, there are rules for pruning and for readjusting the size of the tree. Extensions of original tree-based methods for survival data have been studied by several authors (Gordon and Olshen, 1985; Ciampi et al., 1987, 1988; Segal, 1988, 1995; Davis and Anderson, 1989; LeBlanc and Crowley, 1992; LeBlanc and Crowley, 1993; Intrator and Kooperberg, 1995; Zhang, 1995; LeBlanc, 2001). Some applications of tree-based methods for cancer prognostication are given by Albain et al. (1990, 1992), Banerjee et al. (2000, 2004), Zhang et al. (2001), Freedman et al. (2002), and Katz et al. (2001).

Artificial neural networks were originally developed as models for the human brain. It is generally believed that the human brain learns by adjusting the connection strengths between individual nerve cells (neurons) and thus altering its patterns of electrical activity. It is this ability to change the patterns of interconnections and their relative strengths that is credited with the brain's ability to learn, and to store and use knowledge. Artificial neural networks are loosely based on such concepts in human synaptic physiology.

A neural network is a set of simple computational units that are highly interconnected. The units are also called nodes, and loosely represent the biological neurons. The connections model the synaptic connections in the brain. Each connection has a weight associated with it, called the synaptic weight. The neurons fire when the total signal passed to that unit is activated according to a certain activation function. Several monographs on neural networks have been written during the last decade; two excellent ones are by Ripley (1996) and Bishop (1995). Ripley (1993), and Warner and Misra (1996) discuss neural networks from a statistical perspective. Some applications of artificial neural networks in oncology are given by Ravdin and Clark (1992), Burke (1994), Neiderberger (1995), Errejon et al. (2001), Khan et al. (2001), and Takahashi et al. (2004).

Random forest is an ensemble of trees, originally developed by Breiman (2001) with the goal of improving predictive performance and addressing the problem of instability that is often inherent in a single tree. Applications of random forest are given by Zhang et al. (2003) and Segal et al. (2004). In this chapter, we present an adaptation of Breiman's (2001) random forest methodology to the survival data setting. The strategy involves substituting null martingale residuals for the survival endpoint and enabling inheritance of the random forest algorithm applicable to continuous outcomes, thereby bypassing difficulties that result from censoring. Hothorn et al. (2006) proposed a unified and flexible framework for ensemble learning in the presence of censoring.

Logic regression is an adaptive regression methodology that attempts to construct predictors as Boolean combinations of binary covariates (Ruczinski et al., 2003). The method is particularly well suited to situations in which interactions between many variables result in large differences in response. Indeed, logic regression has been applied to studies of single nucleotide polymorphism (SNP) where high-order interactions between SNPs may define genetic pathways to disease. Applications to SNP data can be found in Kooperberg et al. (2001, 2007), and Kooperberg and Ruczinski (2005). Other applications of logic regression in the cancer literature include combining biomarkers to detect prostate cancer (Etzioni et al., 2003) and identifying populations for colorectal cancer screening (Janes et al., 2005).

The chapter is organized as follows. In Section 3.2, we discuss classification and regression trees (CARTs), with algorithms for the survival data setting described in Section 3.2.1. In Section 3.3 we discuss artificial neural networks for classification and censored regression. Sections 3.4 and 3.5 present random forest and logic regression, respectively. Analyses of the breast cancer data are presented in Section 3.6. Finally, Section 3.7 contains concluding remarks.

3.2 Tree-Based Methods

The literature on tree-based methods dates from work in the social sciences by Morgan and Sonquist (1963), and Morgan and Messenger (1973). In statistics, Breiman et al. (1984) had a seminal influence both in bringing the work to the attention of statisticians and in proposing new algorithms for constructing trees. At around the same time decision tree induction was beginning to be used in the field of machine learning and in engineering.

The terminology of trees is graphic: a tree T has a *root* which is the top node, and observations are passed down the tree, with decisions being made at each *node* (also called daughters) until a *terminal node* or *leaf* is reached. Each nonterminal node (also called *internal node*) contains a question on which a split is based. The terminal nodes of a tree T are collectively denoted by $\tilde{T}$, and the number of terminal nodes is denoted by $|\tilde{T}|$. Each terminal node contains the class label (for a classification problem) or an average response (for a least-squares regression problem). The branch T_t that stems from node t includes t itself and all its daughters. A *subtree* of T is a tree with root a node of T; it is a *rooted subtree* if its root is the root of T.

In the CART paradigm, the covariate space is partitioned recursively in a binary fashion. The partitioning is intended to increase within-node homogeneity, where homogeneity is determined by the dependent variable in the problem. There are three basic elements for constructing a tree under the CART paradigm. These are (1) tree growing, (2) finding the "right-sized" tree, and (3) testing. The first element is aimed at addressing the question *how and why a parent node is split into daughter nodes.* CART uses binary splits, phrased in terms of the covariates, that partition the predictor space. Each split depends upon the value of a single covariate. For ordered (continuous or categorical) covariates, X_j, only splits resulting from questions of the form "Is $X_j \leq c$?" for $c \in \text{domain}(X_j)$ are considered, thereby allowing at most $n-1$ splits for a sample of size n. For nominal covariates no constraints on possible subdivisions are imposed. Thus, for a nominal covariate with M categories, there are $2^{M-1} - 1$ splits to examine.

The natural question that comes next is, how do we select one or several preferred splits from the pool of allowable splits? Before selecting the best split, one must define the goodness of split. The objective of splitting is to make the two daughter nodes as homogeneous as possible. Therefore, the goodness of a split must weigh the homogeneities in the two daughter nodes. Extent of node homogeneity is measured quantitatively using an impurity function. Potential splits are evaluated for each of the covariates, and the covariate and split value resulting in the greatest reduction in impurity is chosen.

Corresponding to a split s at node t into left and right daughter nodes t_L and t_R, the reduction in impurity is given by

$$\Delta I(s,t) = i(t) - P(t_L)i(t_L) - P(t_R)i(t_R),$$

where $i(t)$ is the impurity in node t, and $P(t_L)$ and $P(t_R)$ are the probabilities that a subject falls in nodes t_L and t_R, respectively. For classification problems, $i(t)$ is measured in terms of entropy or Gini impurity. For regression problems, $i(t)$ is typically the mean residual sum of squares. The probabilities $P(t_L)$ and $P(t_R)$ are estimated through corresponding sample proportions. The splitting rule that maximizes $\Delta I(s,t)$ over the set S of all possible splits is chosen as the best splitter for node t.

A useful feature of CART is that of growing a large tree and then *pruning* it back to find the "right-sized tree." During the early development of recursive partitioning, stopping rules were proposed to quit the partitioning process before the tree becomes too large (Morgan and Sonquist, 1963). Breiman et al. (1984) argued that depending on the stopping threshold, the partitioning tends to end too soon or too late. Accordingly, they made a fundamental shift by introducing a second step, called pruning. Instead of attempting to stop the partitioning, they propose to let the partitioning continue until it is saturated or nearly so. Beginning with this generally large tree, they prune it from the bottom up. The point is to find a subtree of the saturated tree that is most "predictive" of the outcome and least vulnerable to the noise in the data.

Let $c(t)$ be the misclassification cost of a node t. Now define $C(T)$ to be the misclassification cost of the entire tree T: $C(T) = \sum_{t \in \tilde{T}} P(t)c(t)$. Note that $C(T)$ is a measure of the quality of the tree T. The purpose of pruning is to select the best subtree of an initially overgrown (or saturated) tree, such that $C(T)$ is minimized. In this context, an important concept introduced by Breiman et al. (1984) is the concept of tree cost-complexity. It is defined as

$$C_\alpha(T) = C(T) + \alpha|\tilde{T}|,$$

where α (≥ 0) is a penalty parameter for the complexity of the tree. The total number of terminal nodes, $|\tilde{T}|$, is used as a measure of tree complexity. Note that the total number of nodes in a tree T (i.e., its size) is twice the number of its terminal nodes minus 1. Thus, tree complexity is really another term for the size of the tree. The difference between $C_\alpha(T)$ and $C(T)$ as a measure of tree quality resides in that $C_\alpha(T)$ penalizes a large tree.

For any tree, there are many subtrees, and therefore many ways to prune. The challenge is how to prune, that is, which subtrees to cut first. Breiman et al. (1984) showed that (1) for any value of the penalty parameter α, there is a unique smallest subtree of T that minimizes the cost-complexity, and (2) if $\alpha_1 > \alpha_2$, the optimal subtree corresponding to α_1 is a subtree of the optimal subtree corresponding to α_2. The use of tree cost-complexity therefore allows one to construct a sequence of nested optimal subtrees from any given tree T. This is done by recursively pruning the branch(es) with the weakest link; that is, the node t with the smallest value of α such that $C_\alpha(t) \leq C_\alpha(T_t)$. Having obtained a nested sequence of pruned optimal subtrees, one is left with the problem of selecting a *best* tree from this sequence. Using the learning sample (resubstitution) estimate of misclassification cost results in selecting the largest tree. Breiman et al. (1984) suggest using a test sample or cross-validation to obtain honest estimates of $C(T)$. The subtree with the smallest estimate of misclassification cost is chosen as the final tree. Details of the cross-validation method are described in Breiman et al. (1984) and Zhang and Singer (1999).

3.2.1 Tree-Based Methods for Survival Data

Interest in tree-based methods for survival data naturally came from the need of clinical researchers to define interpretable prognostic classification rules both for understanding the prognostic structure of data (by forming a small number of groups of patients with differing prognoses) and for designing future clinical trials. Several authors have studied extensions of original tree-based methods in the setting of censored survival data (Gordon and Olshen, 1985; Ciampi et al., 1987, 1988; Segal, 1988; Davis and Anderson, 1989; LeBlanc and Crowley, 1992; LeBlanc and Crowley, 1993; Intrator and Kooperberg, 1995; Zhang, 1995). Some applications of tree-based survival analyses are given by Albain et al. (1990, 1992), Banerjee et al. (2000, 2004), Freedman et al. (2002), and Katz et al. (2001).

Consider the usual setting for censored survival data, which includes a measurement of time under observation and covariates that are potentially associated with the survival time. Specifically, an observation from a sample of size n consists of the triple $(y_i, \delta_i, \mathbf{X}_i), i = 1, \ldots, n$ where y_i is the time under observation for individual i, δ_i is the event indicator for individual i (i.e., $\delta_i = 1$ if the i-th observation corresponds to an event ("failure"), and =0 if the i-th observation is censored), and $\mathbf{X}_i = (X_{i1}, \ldots, X_{ip})$ is the vector of p covariates for the i-th individual. For simplicity, we will assume that there are no tied events.

Algorithms for growing trees for survival data are broadly classified under two general approaches. One approach is to measure the within-node homogeneity with a statistic that measures how similar the subjects in each node are and choose splits that minimize the within-node error. The alternative is to summarize the dissimilarity in survival experiences between two groups induced by a split and choose splits that maximize this difference.

Tree growing and pruning based on measures of within-node homogeneity adopt the CART algorithm directly, since the measures defined are all subadditive, allowing comparisons between subtrees. Gordon and Olshen (1985) presented the first extension of CART to censored survival data, which involved a distance measure (the Wasserstein metric) between Kaplan-Meier curves and certain point masses. A likelihood based splitting criterion was proposed by Davis and Anderson (1989) in which they assumed that the survival function in a node is exponential with a constant hazard. The measure for within-node homogeneity is based on the negative log-likelihood of the exponential model at a node, that is, for node h, this is given by

$$R(h) = D_h \left[1 - \log \left(\frac{D_h}{y_h} \right) \right]$$

where $D_h = \sum_{i \in h} \delta_i$ is the total number of events and $y_h = \sum_{i \in h} y_i$ is the sum of observation times for all subjects in node h. LeBlanc and Crowley (1992) developed a splitting method based on the popular semiparametric proportional hazards model, using the deviance residual as the measure of within-node homogeneity. Therneau et al. (1990) proposed using the null martingale residuals from a proportional hazards model as the outcome variable in a regression tree. In the absence of time-dependent covariates these residuals are given by

$$\hat{M}_i = \delta_i - \hat{\Lambda}_0(y_i),$$

where $\hat{\Lambda}_0(\cdot)$ is the Breslow estimator (1972) of the baseline cumulative hazard. Since this transforms the censored data into uncensored values in the form of the martingale residuals, they can be used directly as continuous outcome in CART without modification to the regression tree algorithm.

A different approach to splitting is to recursively partition the data by maximizing the dissimilarity of the two daughter nodes resulting from a split (Segal, 1988). One such algorithm was proposed by LeBlanc and Crowley (1993) who use the two-sample log-rank statistic to measure the separation in survival times between two daughter nodes. The two-sample log-rank statistic was chosen because of its extensive use in the survival analysis setting, and also because it is an appropriate measure of dissimilarity in survival between two groups. The numerator of the log-rank statistic can be expressed as a weighted difference between estimated hazard functions

$$G = \int_0^\infty w(u) \frac{n_1(u)n_2(u)}{n_1(u) + n_2(u)} (d\hat{\Lambda}_1(u) - d\hat{\Lambda}_2(u)),$$

where $w(\cdot) = 1$, $n_1(u)$ and $n_2(u)$ are the number of subjects at risk in each group at time u, and $\hat{\Lambda}_1$ and $\hat{\Lambda}_2$ are the Nelson cumulative hazard estimators for each group. In general, other weights could be chosen to have greater sensitivity to early or late differences in the hazards of the two groups. LeBlanc and Crowley (1993) propose using the ratio of G-squared divided by an estimate of its variance as the splitting statistic. Partitioning at node h, involves finding the split s, among all variables that maximize the standardized two-sample log-rank statistic. Pruning is done using split complexity (LeBlanc and Crowley, 1993) and the final tree is selected using a bias-corrected version of the split complexity and bootstrap to estimate the bias.

3.3 Artificial Neural Networks

In statistical nomenclature, a neural network is a two-stage regression or classification model, typically represented by a network diagram as in Figure 3.1. The network in Figure 3.1 is a feed-forward neural network, named such because units in one layer are connected only to units in the next layer, and not to units in a preceding layer or units in the same layer. This network applies to both regression and classification. For regression, typically $K = 1$ and there is only one output unit Y_1 at the top. For K-class classification, there are K units at the top, with the kth unit modeling the probability of class k. There are K target measurements Y_k, $k = 1, \ldots, K$, each being coded as a 0-1

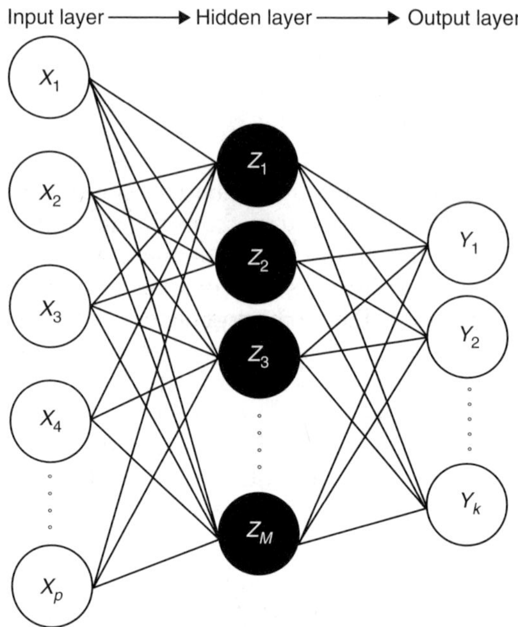

FIGURE 3.1
Schematic representation of a feed-forward neural network with one hidden layer.

variable for the k-th class. The number of input measurements is p, and these are denoted by $X_1, X_2, \ldots, X_p$.

Derived features Z_j are created from linear combinations of the inputs, and then the target Y_k is modeled as a function of linear combinations of the Z_j,

$$Z_j = \sigma(\alpha_{0j} + \alpha_j^T X), \quad j = 1, \ldots, M,$$

$$T_k = \beta_{0k} + \beta_k^T Z, \quad k = 1, \ldots, K,$$

$$f_k(X) = g_k(T), \quad k = 1, \ldots, K,$$

where $X = (X_1, X_2, \ldots, X_p)$, $Z = (Z_1, Z_2, \ldots, Z_M)$, $T = (T_1, T_2, \ldots, T_K)$, and the α and β coefficients represent the synaptic weights. The units Z_j in the middle layer of the network are called *hidden units*, since the Z_j are not directly observed. In general, there can be more than one hidden layer, with a variable number of hidden units per layer. The intercept terms α_{0j} and β_{0k} can be thought of as weights introduced by adding a new unit (the *bias unit*) that is permanently at $+1$ and connected to every unit in the hidden and output layers. This is the same idea as incorporating the constant term in the design matrix of a regression by including a column of 1's.

The function $\sigma(\cdot)$ is referred to as the activation function in the neural network literature. The only practical requirement for $\sigma(\cdot)$ is that it be differentiable. Although several choices of $\sigma(\cdot)$ exist in the literature, the one that is used most commonly is the sigmoid function $\sigma(v) = 1/(1 + e^{-v})$.

The output function $g_k(T)$ allows a final transformation of the vector of outputs T. For regression, we typically choose the identity function $g_k(T) = T_k$. For K-class classification, the function $g_k(T) = e^{T_k}/\sum_{l=1}^{K} e^{T_l}$ is used, which is exactly the transformation used in the multilogit model, and produces positive estimates that sum to one.

The brain learns by adapting the strength of the synaptic connections. Similarly, the (synaptic) weights in neural networks, similar to coefficients in regression models, are iteratively adjusted to make the model fit the training data well. Assume that the training data consists of N observations on the output and input units. We denote the complete set of weights by θ, which consists of $\{\alpha_{0j}, \alpha_j;\ j = 1, \ldots, M\}$ and $\{\beta_{0k}, \beta_k;\ k = 1, \ldots, K\}$, for a total of $M(p+1) + K(M+1)$ weights. For regression, we use sum-of-squared errors as our measure of fit

$$R(\theta) = \sum_{i=1}^{N} \sum_{k=1}^{K} (y_{ik} - f_k(x_i))^2.$$

For classification, we use either squared error or deviance

$$R(\theta) = -\sum_{i=1}^{N} \sum_{k=1}^{K} y_{ik} \log f_k(x_i).$$

The generic approach to minimizing $R(\theta)$ is by gradient descent, called *back propagation* in this setting. Because of the compositional form of the model, the gradient can be derived using the chain rule for differentiation. This can be computed by a forward and backward sweep over the network, keeping track only of quantities local to each unit. In the back propagation algorithm, each hidden unit passes and receives information only to and from units that share a connection. Hence, it can be implemented efficiently on a parallel architecture computer. For derivations of the back propagation equations in the setting of a multilayered feedforward network, the reader is referred to Ripley (1996) and Hastie et al. (2001).

For the iterative back propagation algorithm, starting values for weights are usually chosen to be random values near zero. This makes the model start out nearly linear, and progressively nonlinear as the weights increase. Some care is needed that the starting values are not taken to be too large, for if all the $\alpha_j^T X$ are initially large, the hidden units start in a "saturated state" (with outputs very near zero or one).

Often neural networks will overfit the data at the global minimum of R. One way to avoid overfitting is to discourage large weights and hence large inputs to units. *Weight decay* modifies the classic algorithm by adding a penalty to the error function $R(\theta) + \lambda J(\theta)$ where $J(\theta) = \sum_{k=1}^{K} \sum_{j=1}^{M} \beta_{kj}^2 + \sum_{j=1}^{M} \sum_{l=1}^{p} \alpha_{jl}^2$, and $\lambda \geq 0$ is a tuning parameter. This is analogous to ridge regression used for linear models. Larger values of λ will shrink the weights toward zero; typically cross-validation is used to estimate λ.

Generally, in fitting neural network models, it is better to have too many hidden units than too few. With too few hidden units, the model might not have enough flexibility to capture the nonlinearities in the data; with too many hidden units, the extra weights can be shrunk toward zero if weight decay is used. Typically the number of hidden units is somewhere in the range of 5–100, with the number increasing with the number of inputs and number of training cases. It is most common to put down a reasonably large number of units and train them with weight decay.

3.3.1 Neural Networks for Survival Data

Several methods have been proposed in the literature to adapt neural networks to survival analysis. The simplest approach involves the application of so-called "single time-point models" (De Laurentiis and Ravdin, 1994). Since they are identical to a logistic perceptron or a feed-forward neural network with a hidden layer, they correspond to fitting of logistic regression models or their generalizations to survival data. In practice, a single time point t is fixed and the network is trained to predict the t-year survival probabilities. This approach was used by Burke (1994) and McGuire et al. (1992). Of course, such a procedure can be repeatedly applied for the prediction of survival probabilities at fixed time points $t_1 < t_2 < \cdots < t_k$. For example, Kappen and Neijt (1993) trained several ($k = 6$) neural networks to predict survival of patients with ovarian cancer after $1, 2, \ldots, 6$ years. However, note that without restriction on the parameters, such an approach does not guarantee monotonicity of the estimated survival curves. Ravdin et al. (1992), and Ravdin and Clark (1992) developed a method that incorporates the censoring by creating multiple data records for each subject that span the follow-up time. Since these authors use number of the time interval as input unit, the estimated survival probabilities do not depend of the length of the time intervals. Schwarzer et al. (2000) caution against such naive applications of neural networks to survival data.

Standard requirements for the analysis of survival data are incorporated in the approaches of Liestol et al. (1994), Faraggi and Simon (1995), and Biganzoli et al. (1998) among others. For continuous time data Liestol et al. (1994) proposed a piecewise constant hazard approach, whereas Faraggi and Simon (1995) extended the proportional hazard Cox model with a neural network predictor. Specifically, Faraggi and Simon (1995) suggested replacing the functional $\boldsymbol{\beta}'\mathbf{X}_i$ of the Cox model by the output function of a single hidden layer feed-forward neural network. However, the maximization of the resultant partial likelihood is complex and does not have a straightforward implementation using standard software. Biganzoli et al. (1998) proposed a flexible neural network approach, in a discrete survival time context, which provides smoothed hazard function estimation and allows for nonlinear covariate effects. The authors demonstrated that by treating the time interval as an input variable in a standard feed forward network with logistic activation and entropy error function, it is possible to estimate smoothed discrete hazards as conditional probabilities of failure. Biganzoli et al. (2002) also introduced specific error functions and data representations for multilayer perceptron and radial basis function extensions of generalized linear models for survival data. Following by Therneau et al. (1990), original observation, Ripley and Ripley (2001) suggested using the null martingale residuls from a proportional hazards model as the output to a feed-forward neural network. Kattan et al. (1998) and Kattan (2003) explored the above approach and found that it performs competitively with the Cox model for prediction.

3.4 Random Forest

Random forest (Breiman, 2001) was developed as a second-generation CART method. It is an ensemble (collection) of trees whose predictions are combined to make the overall prediction for the forest. The mechanism of selecting a best split in CART and the recursive partitioning of data leads to smaller and smaller data sets. This can lead to instability (Breiman, 1996) in the tree structure, whereby small changes in the data and/or algorithm inputs can have dramatic effects on the nature of the solution (variables and splits selected). Another major shortcoming of tree-based methods is their modest prediction performance, attributable to algorithm greediness and constraints which, although enhancing interpretability, reduce flexibility of the fitted functional forms. Growing an ensemble of trees and aggregating is a way to fix these problems. The advantage in growing many trees and using an aggregated estimate is that it is a way to reduce variance (Breiman, 2001). It also leads to classifiers and predictors that are drawn from a richer class of models (Hastie et al., 2001). Ensemble methods like bagging (Breiman, 1996; Quinlan, 1996), boosting (Freund and Schapire, 1996; Quinlan, 1996), and random forest (Breiman, 2001) yield substantial performance improvement over a single tree and are known to be stable.

Bagging involves random manipulation of the training data through bootstrap. A large number of pseudo datasets are generated by resampling the original observations with replacement, and a tree grown on each pseudo dataset. This results in an ensemble of trees, some of which may be close to a global or local maxima. In boosting, the data are iteratively reweighted instead of random resampling. The algorithm alternates between fitting a tree and reweighting the data. The weights are adaptively chosen, with more weight given to observations that the tree models poorly. Again, an ensemble of trees result. The simple mechanism whereby bagging and boosting reduce prediction error, is well understood in terms of variance reduction resulting from averaging (Hastie et al., 2001). Such variance gains can be enhanced by reducing the correlation between the quantities being averaged. It is this principle that motivates random forest.

Random forest (Breiman, 2001) is an ensemble of unpruned classification or regression trees, induced from bootstrap samples of the training data, using random feature selection in the tree induction process. Correlation reduction is achieved by the random feature selection. Instead of determining the optimal split of a given node of a tree by evaluating all allowable splits on all covariates, as is done with growing a single tree, a subset of the covariates drawn at random is employed. Prediction is made by aggregating (majority vote for classification or averaging for regression) the predictions of the ensemble. Random forests demonstrate exceptional prediction accuracy (Breiman, 2001) comparable to artificial neural networks and support vector machines.

3.4.1 Random Forest for Survival Data

The published literature on ensemble techniques for survival data is sparse owing to the difficulties induced by censoring. Hothorn et al. (2004) studied an aggregation scheme for bagging survival trees. Breiman (2002), introduced a software implementation of a random forest variant for survival data; however, it does not come with a formal description of the methodology. Ishwaran et al. (2004) proposed a method that combines random forest methodology with survival trees grown using Poisson likelihoods. In a very recent article, Hothorn et al. (2006) proposed a unified and flexible framework for ensemble learning in the presence of censoring.

In this section, we present an adaptation of Breiman's (2001) random forest methodology to the survival data setting. The strategy involves substituting suitably chosen residuals for the survival endpoint and enabling inheritance of the random forest algorithm applicable to continuous outcomes, thereby bypassing difficulties that result from censoring. This general strategy has been employed to adapt additive (Cox) models (Grambsch et al. 1995; Segal et al. 1995), multivariate adaptive regression splines (MARS) (LeBlanc and Crowley, 1999), regression trees (LeBlanc and Crowley, 1992; Keles and Segal, 2002), artificial neural networks (Kattan et al. 1998; Ripley and Ripley, 2001) and least angle regression-lasso (Segal, 2006) to censored survival outcomes.

For growing random forest in the survival data setting, we propose using the null martingale residuals from a Cox proportional hazards model as the outcome variable in the random forest algorithm. This approach is easy to implement and circumvents the complexity induced by censoring. Ensemble predictions are computed by aggregating across different trees in the forest. This reduces variance and avoids the instability of working with a single tree. We illustrate this approach using data from the breast cancer prognostic study.

Following Breiman (2001), the idea is to grow trees by injecting two types of randomness into the process. To grow the trees in the forest: (1) Bootstrap the training data. Grow each tree on an independent bootstrap sample using null martingale residuals from a Cox proportional hazards model as the outcome variable. (2) At each node, randomly select m covariates out of all M possible covariates. Find the best split on the selected m covariates. (3) Grow the tree to maximal depth under the restriction of minimum nodesize $= 5$

(i.e., splitting is stopped when a node has fewer than five subjects). No pruning is performed. (4) Repeat for each bootstrap sample. (5) Average the trees to get predictions.

Steps 1 and 2 introduce randomness. To ensure that random forests have good prediction properties, it is important to check that the correct amount of randomization has been introduced. This means that we need to determine an appropriate number of randomly selected covariates, m, to be used in step 2 of the procedure. If we select too few covariates, the trees might be too sparse, and the ensemble estimator will have suboptimal properties. Choosing too many covariates will make the trees highly correlated, which can also degrade performance. As discussed in Breiman (2001), one method for assessing the accuracy of a forest is through its generalization error. As m increases, the strength of a tree increases, which contributes to a lower forest generalization error; at the same time, however, the correlation between residuals increases, which increases error.

An estimate of the prediction error rate is obtained, based on the training data, as follows: (1) At each bootstrap iteration, predict the data not in the bootstrap sample (Breiman calls this "out-of-bag" data) using the tree grown with the bootstrap sample. (2) Average the out-of-bag predictions. Calculate the error rate, and call it the out-of-bag estimate of error rate. Given that enough trees have been grown, the "out-of-bag" estimate of error rate is an accurate estimate of test set prediction error rate (Breiman, 2001).

In addition to excellent prediction performance, random forests possess a number of advantages. These include the distinction of forests from so-called black-box methods (e.g., neural nets), and accurate, internal estimates of test set prediction error. Furthermore, a by-product of forest is a collection of variables that are frequently used in the forests, and the frequent uses are indicative of the importance of these variables. Zhang et al. (2003) examined the frequencies of the variables in a forest and used them to rank the variables. We illustrate these in our analysis of the breast cancer data.

3.5 Logic Regression

In the logic regression framework, given a set of binary covariates X, the goal is to try to create new, better predictors for the response by considering Boolean combinations of the binary covariates. For example, if the response is binary, the goal is to find decision rules such as "if X_1, X_2, X_3, and X_4 are true," or "X_5 or X_6 but not X_7 are true," then the response is more likely to be in class 0. Boolean combinations of the covariates, called logic trees, are represented graphically as a set of and/or rules. An example of such a tree is shown in Figure 3.2, representing the Boolean expression $(X_1 \wedge X_2^c) \wedge [(X_3^c \wedge X_4^c) \vee (X_5 \wedge (X_3^c \vee X_6))]$. Note that a logic tree is similar to CART in the sense that any tree produced by CART can be written as a Boolean combination of covariates. However, there are some Boolean expressions that can be very

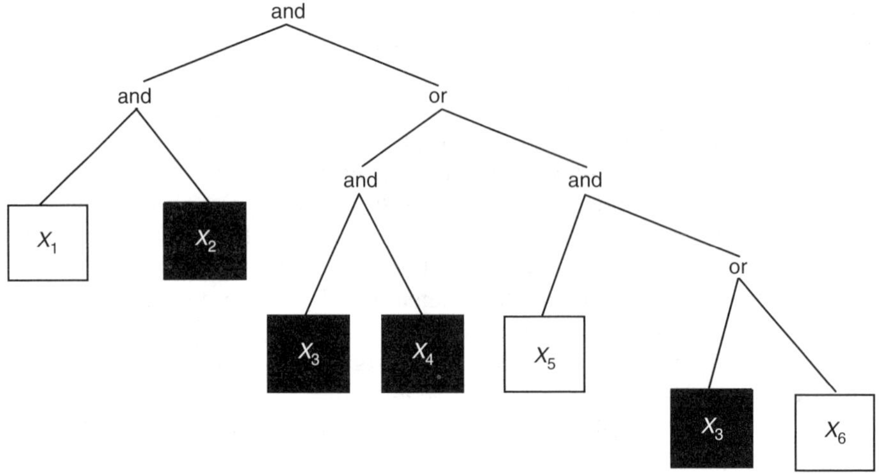

FIGURE 3.2
Schematic diagram of a logic tree representing the Boolean expression $(X_1 \wedge X_2^c) \wedge [(X_3^c \wedge X_4^c) \vee (X_5 \wedge (X_3^c \vee X_6))]$. White letters on black background denote the conjugate of the variable.

simply represented as logic trees, but which require fairly complicated CART rules (Ruczinski et al., 2003).

Let $X_1, X_2, \ldots, X_k$ be binary (0/1) predictors and let Y be the response. The general logic regression model is written as

$$g(E[Y]) = \beta_0 + \beta_1 L_1 + \cdots + \beta_p L_p, \tag{3.1}$$

where L_j is a Boolean combination of the predictors X_i. The above model can be applied to any type of regression outcome by choosing the correct scoring and link functions. The score function can be thought of as a measure of quality of the model under consideration. For example, for linear regression, $g(E[Y]) = E[Y]$, and the residual sum of squares could be used for the score function. For logistic regression, $g(E[Y]) = \log(E[Y]/(1 - E[Y]))$, and the score could be the binomial deviance. The goal is to find Boolean expressions in Equation 3.1 that minimize the scoring function associated with the model type, estimating the parameters β_j simultaneously with the search for the Boolean expressions L_j. The output from logic regression is represented as a series of trees, one for each Boolean predictor, L_j, and the associated regression coefficient.

Ruczinski et al. (2003) provide a detailed description of logic regression and the algorithm used to fit it. Finding good candidates for the logic trees can be challenging since the total number of possible trees is huge. Since an exhaustive search of all logic trees is infeasible, the trees are selected by a simulated annealing algorithm (van Laarhoven and Aarts, 1987), in which a modification to the current logic tree is proposed at random. The proposed change is chosen from a set of permissible moves that include replacing a covariate with another, changing an and/or operator, adding or removing a branch, or splitting a node. The proposed modification to a tree is always accepted if the score of the new tree is better than the score of the previous tree. If the score of the new tree is not better, then the new tree is accepted with a certain probability. This acceptance probability depends on the difference between the two scores and the stage of the algorithm.

In addition to the specification of the scoring function, the fitting algorithm also requires that the maximum number of trees (p) be specified. This is computationally necessary, since all trees in the model are fit simultaneously. Furthermore, in order to avoid overfitting, model size, measured by the maximum number of variables, or leaves, that make up a tree, is fixed. During model fitting, modifications to the tree are not proposed if they result in a tree exceeding the fixed size. The model size can be selected using external test sets, cross-validation, or randomization tests (Ruczinski et al., 2003). For interpretability the model size can also be chosen a priori, as implemented by Janes et al. (2005).

Logic regression models are not restricted to classification or linear regression. Any other regression model can be considered as long as a scoring function can be determined. Specifically, for survival data, Ruczinski et al. (2003) considered the Cox proportional hazards model using partial likelihood as score. Software for fitting logic regression models is available as R or Splus packages at http://bear.fhcrc.org/ingor/logic/.

3.6 Detroit Breast Cancer Study

As an illustrative example, we present analyses of data from a cohort study of breast cancer patients. Women eligible for this study were newly diagnosed patients with stage I, II, or III breast cancer, diagnosed between January 1990 and December 1996 at Harper Hospital in Detroit, Michigan. Detailed demographic, clinical, pathological, treatment, and follow-up information were obtained from the Surveillance, Epidemiology, and End Results (SEER) database, hospital, and clinic records. Recurrence-free survival (RFS) was the primary endpoint of the study, defined as the interval between diagnosis and documented regional/local or distant recurrence. The primary goal of the study was to identify patient subgroups with homogeneous RFS within a group but also different RFS between groups (i.e., prognostic grouping of patients).

The analysis cohort consisted of 764 patients. A total of ten covariates were considered for the analysis. These included sociodemographic variables (age, race, marital status, and socioeconomic status), factors characterizing tumor (tumor size, number of positive lymph nodes, tumor differentiation, estrogen receptor (ER), and progesterone receptor (PR) status), and body mass index (BMI) as a comorbid factor. Patients were classified as obese if their BMI was >30, per the standard guideline recommended by the World Health

Organization (WHO, 1998). Number of positive lymph nodes was categorized as: 0, 1–3, 4–9, and >10 positive nodes. Tumor differentiation was categorized as: well, moderate, and poor. Estrogen and progesterone receptors are binary categorical variables (positive/negative).

Data analysis was performed in R (R Development Core Team, 2005). We performed tree, neural network, random forest, and logic regression analyses of the breast cancer data using null martingale residuals from a proportional hazards model as the outcome variable. Null martingale residuals are obtained simply in R by specifying residual type and zero iterations in the call to function *coxph*(). The 764 patients were split randomly into two-thirds for training ($N = 524$), and one-third for testing ($N = 240$). The primary motivation for this analysis was to compare and contrast the four methods in yielding simple, interpretable prognostic rules, as well as compare their predictive performances on the testing set.

For constructing the survival tree, we used the `RPART` package (Therneau and Atkinson, 2005) in R. For constructing the neural network, we used the `nnet` package in R (Venables and Ripley, 2002). We set the number of hidden units to be 20 a priori. The decay parameter was varied between 0 and 1 and the value that gave the lowest error on the training set was chosen. After the final model was chosen (decay = 0.0001), the model was trained and prediction error was reported on the basis of the testing set. For growing random forest, we used the `randomforest` software, available as an R interface (Liaw and Wiener, 2002). We grew 500 trees in a forest. The primary tuning parameter m (i.e., the number of covariates selected randomly at each node, among all possible covariates) was varied from 2 to 5. The smallest prediction error (based on "out-of-bag" estimates of prediction error variance) was achieved by $m = 2$ among the range of forests examined. Also, the size of the individual trees constituting the forest is controlled by a tuning parameter, which specifies the number of cases in a node below which the tree will not split. This was set to the default value of 5, which is claimed to give generally good results. The logic regression analyses were done using the R code available from http://bear.fhcrc.org/ingor/logic/. The scoring function used was residual sum of squares. In our analyses, we chose model sizes a priori; for interpretability we fit models with four leaves per tree.

Figure 3.3 shows the survival tree based on using null martingale residuals. At each level of the tree, we show the best splitter (covariate with cutpoint). Circles denote terminal nodes in the trees. Within each terminal node, n denotes the number of patients, R denotes the (crude) number of recurrences, and 5 Yr is the 5-year RFS rate. The root node was split by tumor size $\leq$2 cm vs. >2 cm. Patients with tumors smaller than 2 cm had significantly better RFS than patients with larger tumors (>2 cm). The former formed a terminal node in the tree; notably, this group had the best prognosis, with a 5-year RFS rate of 85%. The subgroup with tumors >2 cm was next split by number of positive lymph nodes (<4 vs. $\geq$4 positive nodes). None of the resulting subgroups had any further split and formed terminal nodes, thereby resulting in a tree with three terminal nodes. Patients with tumor size >2 cm,

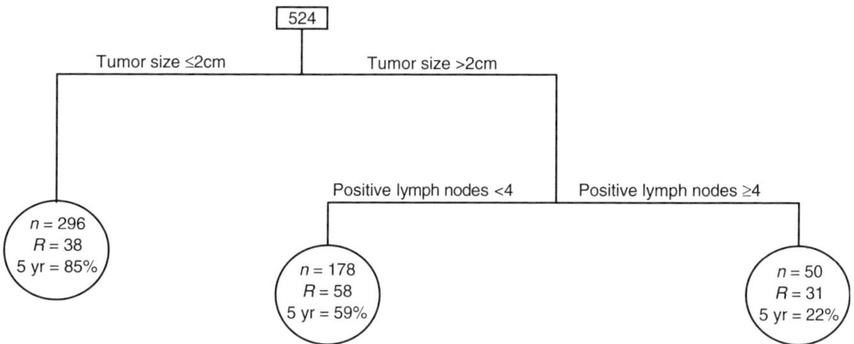

FIGURE 3.3
Survival tree based on martingale residuals.

and more than four positive nodes had the worst prognosis, with a 5-year RFS rate of 22%. In comparison, patients with tumor size >2 cm, but fewer than four positive nodes did better, with an intermediate prognosis (5-year RFS rate = 59%).

Figure 3.4 shows a plot of variable importance from the random forest analysis. The variable importances correspond to the forest with minimal prediction error ($m = 2$). Note consistency with the tree results in terms of the prominence of tumor size, and number of positive lymph nodes.

For the logic regression analysis, we first fitted a model with a single Boolean tree predictor, that is, $p = 1$. The tree is shown in Figure 3.5. The estimate and the 95% confidence interval associated with the coefficient for the tree are 0.43 and (0.34, 0.52) respectively, with p-value <0.001. Patients with poorly differentiated and large (≥2 cm) tumors, who have positive nodes or whose age is ≤50 years have a worse outcome. Note that for prognostic grouping, with only one tree, there is only one distinct nondegenerate positivity criterion to consider, namely, whether or not the tree is satisfied ($L_1 = 1$). For patients satisfying the tree (i.e., those with poorly differentiated and large (≥2 cm) tumors, who have positive nodes or whose age is ≤50 years), the 5-year RFS rate was 39%, compared with 81% for patients who do not satisfy the tree. We also fit logic regression models with two trees, that is, $p = 2$. The model was fit six times, resulting in three unique two-tree models. Since the simulated annealing algorithm used to fit the logic regression models is not guaranteed to find the "best" model, this variation is to be expected. However, the prediction errors based on the test set for all three two-tree models were larger than the prediction error for the single Boolean tree model, therefore, we do not present the results of the analyses from the two-tree models.

Table 3.1 shows the prediction errors from the tree, neural network, random forest, and logic regression (single Boolean tree) analyses. The entries are prediction error variances based on the testing set ($N = 240$). The best prediction error was achieved by the random forest. However, this is only a marginal improvement from the prediction error attained from a single

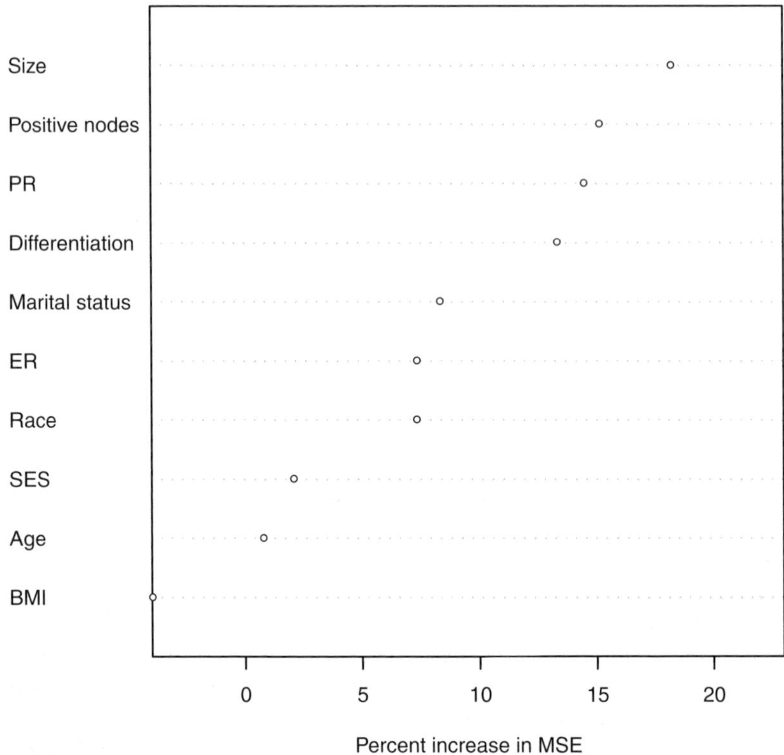

FIGURE 3.4
Plot of variable importance in the random forest with minimal prediction error ($m = 2$).

pruned survival tree. A possible reason for this is the strong correlation between the covariates in this study. Roughly 65% of all possible pairwise correlations between covariates were significant. This could have potentially hindered the effectiveness of the random forest variance reduction strategy. The single logic tree (Figure 3.5) was very comparable to the survival tree (Figure 3.3) based on predictive performance, indicating that for this problem there are several models that perform equally well. The neural network had the largest prediction error variance. Furthermore, the so-called black-box nature of the method makes it unattractive as a tool for prognostication. In contrast to neural network, the other three methods allow us to describe and evaluate the influence of individual covariates. This information is often of equal importance as the decision rule, since it allows the clinician to better understand the underlying process. Indeed, all the other three methods (survival tree, random forest, and logic regression) give consistent evidence about the importance of tumor size and positive lymph nodes, concurring with previous reports in the literature. In addition, the survival tree and logic tree both provide simple characterizations for prognostic grouping of patients.

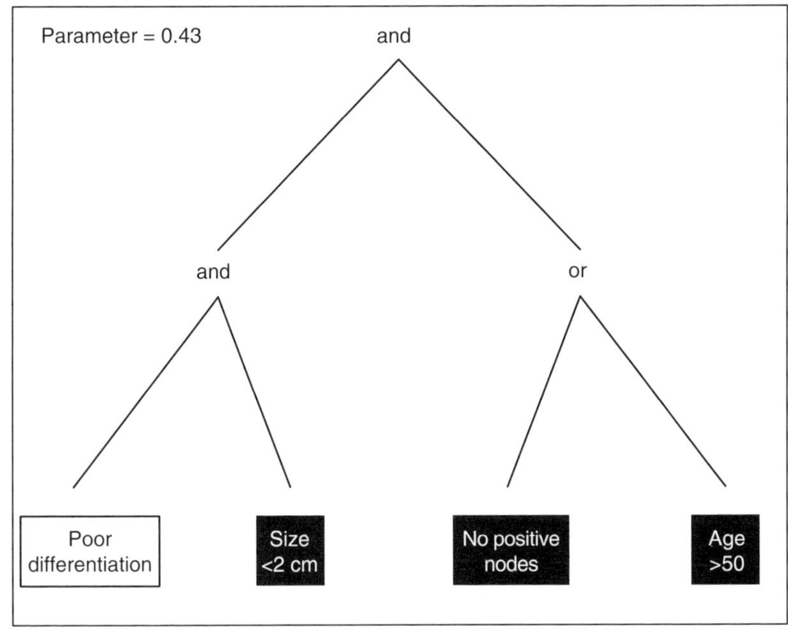

FIGURE 3.5
The single logic tree, L_1, fitted to the breast cancer data. Variables that are printed white on a black background are the complements of those indicated.

TABLE 3.1
Prediction Error Variances from Survival Tree, Neural Network, Random Forest, and Logic Regression for Breast Cancer Data

Method	Prediction Error Variance
Tree	59.84
Neural net	74.80
Random forest	55.77
Logic regression	59.50

3.7 Conclusions

In this chapter, we discussed methodological and practical aspects of classification and regression tree, artificial neural network, random forest, and logic regression for cancer diagnosis and prognostication. Using data on breast cancer patients, we compared and contrasted these methods in terms of their predictive performance, and their capability to yield simple, interpretable prognostic rules. We found that a random forest constructed using null martingale residuals from a proportional hazards model had the smallest prediction error. Results from the logic regression and survival tree analyses were very comparable in terms of predictive performance. Both methods also provide a natural way to identify patient subgroups based on covariate profiles. On the other hand, a by-product of the random forest analysis is a collection of variables that are frequently used in the forest, and the frequent uses are indicative of the importance of these variables. These variable importance summaries can be used to assess the relative importance of the covariates in a forest. In contrast, neural networks fail to provide such information, and therefore their utility in assessing the effect of individual covariates is limited. In our breast cancer data example, the neural network constructed using null martingale residuals also had considerably worse predictive performance compared with the tree models and random forest.

Applications of neural networks in the clinical literature are often accompanied by grossly overstated claims, such as neural networks' "ability to learn … make them formidable tools in the fight against cancer" (Burke, 1994), and "neural computation may be as beneficial to medicine and urology in the twenty-first century as molecular biology has been in the twentieth" (Neiderberger, 1995). There are many other instances in the clinical literature praising neural networks as the ultimate solution to the problem of diagnosis and prognosis. However, as pointed out by Schwarzer et al. (2000), there is no evidence that artificial neural networks have provided real progress in the field of diagnosis and prognosis in oncology. Feed-forward neural networks are nothing more than regression models, the only difference being that feed-forward neural networks (with hidden layers) provide a larger class of regression functions. This is often referred to as the greater flexibility of neural networks. However, greater flexibility is only of value if the true regression function is far away from that of a linear or logistic regression model. Small deviations from a linear or logistic model do not matter, because owing to the small sample sizes of a few hundred typical in oncological applications, such a difference may be small relative to random errors. Large deviations, especially functions with many jumps, are not very plausible, because biological relationships tend to be smooth. Hence one cannot expect that the greater flexibility of neural networks helps them to outperform regression models, especially if the latter are combined with careful model building, allowing use of quadratic or higher interaction terms, for example. Schwarzer et al. (2000) also discuss methodological deficiencies often associated with applications of artificial neural networks in oncology.

Support vector machine (SVM) is another supervised machine learning technique, that has been shown to perform well in multiple areas of biological analysis. SVMs (Burges, 1998; Cristianini and Shawe-Taylor, 2000) were originally introduced by Vapnik and coworkers (Boser et al., 1992; Vapnik, 1998) and successively extended by a number of other researchers. SVMs differ from other linear discriminant methods in that they produce nonlinear boundaries by constructing a linear boundary in a large transformed version of the predictor space. Owing to their robustness to sparse and noisy data, SVMs have been used for a wide range of classification problems, especially in cancer genomic and proteomic studies. Some recent applications of SVMs to genomic data are found in Furey et al. (2000), Listgarten et al. (2004), Man et al. (2005), Hayashida et al. (2005), and Ehlers and Harbour (2005).

There are many versions of free-wares implementing the methods described in this chapter. Most of these are available through Statlib (http://lib.stat.cmu.edu) and CRAN (http://cran.r-project.org/). In particular, the RPART program (Therneau and Atkinson, 2005) could be used to implement the classification and regression tree methods, including the methods of LeBlanc and Crowley (1992), and Therneau et al. (1990) for censored data. Free, open-source code for random forests is available from http://www.stat.berkeley.edu/users/breiman/RandomForests. There is also an R implementation of random forests (Liaw and Wiener, 2002). Software for artificial neural network is available as `nnet` package in R (Venables and Ripley, 2002). Finally, R code for logic regression analysis is available from http://bear.fhcrc.org/ingor/logic/.

References

Albain, K.S., Crowley, J.J., LeBlanc, M., and Livingston, R.B. (1990). "Determinants of improved outcome in small-cell lung cancer: An analysis of the 2,580-patient Southwest Oncology Group data base." *Journal of Clinical Oncology*, 8, 1563–1574.

Albain, K.S., Green, S., LeBlanc, M., Rivkin, S., O'Sullivan, J., and Osborne, C.K. (1992). "Proportional hazards and recursive partitioning and amalgamation analyses of the Southwest Oncology Group node-positive adjuvant CMFVP breast cancer data base: a pilot study." *Breast Cancer Research and Treatment*, 22, 273–284.

Banerjee, M., Biswas, D., Sakr, W., and Wood, D.P., Jr. (2000). "Recursive partitioning for prognostic grouping of patients with clinically localized prostate carcinoma." *Cancer*, 89, 404–411.

Banerjee, M., George, J., Song, E.Y., Roy, A., and Hryniuk, W. (2004). "Tree-based model for breast cancer prognostication." *Journal of Clinical Oncology*, 22, 2567–2575.

Biganzoli, E., Boracchi, P., Mariani, L., and Marubini, E. (1998). "Feed forward neural networks for the analysis of censored survival data: A partial logistic regression approach." *Statistics in Medicine*, 17, 1169–1186.

Biganzoli, E., Boracchi, P., and Marubini, E. (2002). "A general framework for neural network models on censored survival data." *Neural Networks*, 15, 209–218.

Bishop, C.M. (1995). *Neural Networks for Pattern Recognition*. New York: Oxford University Press.

Boser, B.E., Guyon, I.M., and Vapnik, V.N. (1992). "A training algorithm for optimal margin classifiers." In *Proceedings of the 5th Annual ACM Workshop on Computational Learning Theory*, ACM Press, Pittsburgh, PA.

Breiman, L. (1996). "Bagging predictors." *Machine Learning*, 24, 123–140.

Breiman, L. (2001). "Random forests." *Machine Learning*, 45, 5–32.

Breiman, L. (2002). *"How to use survival forests."* URL http://www.stat.berkeley.edu/users/breiman/.

Breiman, L., Friedman, J.H., Olshen, R.A., and Stone, C.J. (1984). *Classification and Regression Trees*. Wadsworth: Belmont, California.

Breslow, N. (1972). "Contribution to the discussion of the paper by D.R. Cox." *Journal of the Royal Statistical Society Series B*, 34, 216–217.

Burges, C.J.C. (1998). "A tutorial on support vector machines for pattern recognition." *Data Mining and Knowledge Discovery*, 2, 121–167.

Burke, H.B. (1994). "Artificial neural networks for cancer research: Outcome prediction." *Seminars in Surgical Oncology*, 10, 73–79.

Ciampi, A., Chang, C.-H., Hogg, S., and McKinney, S. (1987). "Recursive partitioning: A versatile method for exploratory data analysis in biostatistics." In: MacNeil, I.B. and Umphrey, G.J., eds. *Biostatistics*. Dordrecht: Reidel, 23–50.

Ciampi, A., Hogg, S., McKinney, S., and Thiffault, J. (1988). "RECPAM: A computer program for recursive partitioning and amalgamation for censored survival data." *Computer Methods and Programs in Biomedicine*, 26, 239–256.

Cristianini, N. and Shawe-Taylor, J. (2000). *An Introduction to Support Vector Machines*. Cambridge: Cambridge University Press.

Davis, R.B. and Anderson, J.R. (1989). "Exponential survival trees." *Statistics in Medicine*, 8, 947–961.

De Laurentiis, M. and Ravdin, P. (1994). "A technique for using neural network analysis to perform survival analysis of censored data." *Cancer Letters*, 77, 127–138.

Ehlers, J.P. and Harbour, J.W. (2005). "NBS1 expression as a prognostic marker in uveal melanoma." *Clinical Cancer Research*, 11, 1849–1853.

Errejon, A., Crawford, E.D., Dayhoff, J., O'Donell, C., Tewari, A., Finkelstein, J., and Gamito, E.J. (2001). "Use of artificial neural networks in prostate cancer." *Molecular Urology*, 5, 153–158.

Etzioni, R., Kooperberg, C., Pepe, M., and Smith, R. (2003). "Combining biomarkers to detect disease with applications to prostate cancer." *Biostatistics*, 4, 523–538.

Faraggi, D. and Simon, R. (1995). "A neural network model for survival data." *Statistics in Medicine*, 14, 73–82.

Freedman, G.M., Hanlon, A.L., Fowble, B.L., Anderson, P.R., and Nicoloau, N. (2002). "Recursive partitioning identifies patients at high and low risk for ipsilateral tumor recurrence after breast-conserving surgery and radiation." *Journal of Clinical Oncology*, 20, 4015–4021.

Freund, Y. and Schapire, R.E. (1996). "Experiments with a new boosting algorithm." In *Proceedings of the Thirteenth International Conference on Machine Learning*, Bari, Italy, 148–156.

Furey, T.S., Cristianini, N., Duffy, N., Bednarski, D.W., Schummer, M., and Haussler, D. (2000). "Support vector machine classification and validation of cancer tissue samples using microarray expression data." *Bioinformatics*, 16, 906–914.

Gordon, L. and Olshen, R. (1985). "Tree-structured survival analysis." *Cancer Treatment Reports*, 69, 1065–1069.

Grambsch, P.M., Therneau, T.M., and Fleming, T.R. (1995). "Diagnostic plots to reveal functional form for covariates in multiplicative intensity models." *Biometrics*, 51, 1469–1482.

Hastie T., Tibshirani, R., and Friedman, J. (2001). *The Elements of Statistical Learning*. New York: Springer.

Hayashida, Y., Honda, K., Osaka, Y., et al. (2005). "Possible prediction of chemoradiosensitivity of esophageal cancer by serum protein profiling." *Clinical Cancer Research*, 11, 8042–8047.

Hothorn, T., Buhlmann, P., Dudoit, S., Molinaro, A., and Laan, M.J. van der (2006). "Survival ensembles." *Biostatistics*, 7, 355–373.

Hothorn, T., Lausen, B., Benner, A., and Radespiel-Tröger, M. (2004). "Bagging survival trees." *Statistics in Medicine*, 23, 77–94.

Intrator, O. and Kooperberg, C. (1995). "Trees and splines in survival analysis." *Statistical Methods in Medical Research*, 4, 237–261.

Ishwaran, H., Blackstone, E.H., Pothier, C.E., and Lauer, M.S. (2004). "Relative risk forests for exercise heart rate recovery as a predictor of mortality." *Journal of the American Statistical Association*, 99, 591–600.

Janes, H., Pepe, M., Kooperberg, C., and Newcomb, P. (2005). "Identifying target populations for screening or not screening using logic regression." *Statistics in Medicine*, 24, 1321–1338.

Kappen, H.J. and Neijt, J.P. (1993). "Neural network analysis to predict treatment outcome." *The Annals of Oncology*, 4, S31–S34.

Kattan, M.W. (2003). "Comparison of Cox regression with other methods for determining prediction models and nomograms." *Journal of Urology*, 170, S6–S10.

Kattan, M.W., Hess, K.R., and Beck, J.R. (1998). "Experiments to determine whether recursive partitioning (CART) or an artificial neural network overcomes theoretical limitations of Cox proportional hazards regression." *Computers and Biomedical Research*, 1, 363–373.

Katz, A., Buchholz, T.A., Thames, H., Smith, C.D., McNeese, M.D., Theriault, R., Singletary, S.E., and Strom, E.A. (2001). "Recursive partitioning analysis of locoregional recurrence patterns following mastectomy: Implications for adjuvant irradiation." *International Journal of Radiation Oncology, Biology, Physics*, 50, 397–403.

Keles, S. and Segal, M.R. (2002). "Residual-based tree-structured survival analysis." *Statistics in Medicine*, 21, 313–326.

Khan, J., Wei, J.S., Ringner, M., Saal, L.H., Ladanyi, M., Westermann, F., Berthold, F., Schwab, M., Antonescu, C.R., Peterson, C., and Meltzer, P.S. (2001). "Classification and diagnostic prediction of cancers using gene expression profiling and artificial neural networks." *Nature Medicine*, 7, 673–679.

Kooperberg, C., Bis, J.C., Marciante, K.D., et al. (2007). "Logic regression for analysis of the association between genetic variation in the renin-angiotensin system and myocardial infarction or stroke." *American Journal of Epidemiology*, 165, 334–343.

Kooperberg, C. and Ruczinski, I. (2005). "Identifying interacting SNPs using Monte Carlo logic regression." *Genetic Epidemiology*, 28, 157–170.

Kooperberg, C., Ruczinski, I., LeBlanc, M.L., and Hsu, L. (2001). "Sequence analysis using logic regression." *Genetic Epidemiology*, 21, S626–S631.

Laarhoven, P.J.M. van and Aarts, E.H.L. (1987). *Simulated annealing: Theory and applications*. Norwell: Kluwer Academic Publishers.

LeBlanc, M. (2001). "Tree-based methods for prognostic stratification." In: Crowley, J., ed. *Handbook of Statistics in Clinical Oncology*. New York: Marcel Dekker, Inc., 457–472.

LeBlanc, M. and Crowley, J. (1992). "Relative risk trees for censored survival data." *Biometrics*, 48, 411–425.

LeBlanc, M. and Crowley, J. (1993). "Survival trees by goodness of split." *Journal of the American Statistical Association*, 88, 457–467.

LeBlanc, M. and Crowley, J. (1999). "Adaptive regression splines in the Cox model." *Biometrics*, 55, 204–213.

Liaw, A. and Wiener, M. (2002). "Classification and regression by randomForest." *R News*, 2, 18–22.

Liestol, K., Andersen, P.K., and Andersen, U. (1994). "Survival analysis and neural nets." *Statistics in Medicine*, 13, 1189–1200.

Listgarten, J., Damaraju, S., Poulin, B., et al. (2004). "Predictive models for breast cancer susceptibility from multiple single nucleotide polymorphisms." *Clinical Cancer Research*, 10, 2725–2737.

Man, T.-K., Chintagumpala, M., Visvanathan, J., et al. (2005). "Expression profiles of osteosarcoma that can predict response to chemotherapy." *Cancer Research*, 65, 8142–8150.

McGuire, W.L., Tandon, A.K., Allred, D.C., et al. (1992). "Treatment decisions in axillary node-negative breast cancer patients." *Journal of the National Cancer Institute Monographs*, 11, 173–180.

Meller, J. and Wagner, M. (2007). "Machine learning techniques for bioinformatics: Fundamentals and applications." In: Khattree, R. and Naik, D.N., eds. *Computational Methods in Biomedical Research*, CRC Press, Boca Raton, FL, 45–76.

Morgan, J.N. and Messenger, R.C. (1973). "THAID: A sequential search program for the analysis of nominal scale dependent variables." Technical Report, Institute for Social Research, University of Michigan, Ann Arbor, MI.

Morgan, J.N. and Sonquist, J.A. (1963). "Problems in the analysis of survey data and a proposal." *Journal of the American Statistical Association*, 58, 415–434.

Neiderberger, C.S. (1995). "Commentary on the use of neural networks in clinical urology." *Journal of Urology*, 153, 1362.

Quinlan, J. (1996). "Bagging, boosting, and C4.5." In *Proceedings of the Thirteenth American Association for Artificial Intelligence National Conference on Artificial Intelligence*, AAAI Press, Menlo Park, CA, 725–730.

R Development Core Team (2005). R: A language and environment for statistical computing. R Foundation for Statistical Computing, Vienna, Austria. ISBN 3-900051-07-0, URL http://www.R-project.org.

Ravdin, P.M. and Clark, G.M. (1992). "A practical application of neural network analysis for predicting outcome of individual breast cancer patients." *Breast Cancer Research and Treatment*, 22, 285–293.

Ravdin, P.M., Clark, G.M., Hilsenbeck, S.G., et al. (1992). "A demonstration that breast cancer recurrence can be predicted by neural network analysis." *Breast Cancer Research and Treatment*, 21, 47–53.

Ripley, B.D. (1993). "Statistical aspects of neural networks." In Barndorff-Nielsen, O., Jensen, J., and Kendall, W., eds. *Networks and Chaos—Statistical and Probabilistic Aspects*. London: Chapman & Hall, 40–123.

Ripley, B.D. (1996). *Pattern Recognition and Neural Networks*. New York: Cambridge University Press.

Ripley, B.D. and Ripley, R.M. (2001). "Neural networks as statistical methods in survival analysis." In: Dybowski, R. and Gant, V., eds. *Artificial Neural Networks: Prospects for Medicine*. Cambridge: Cambridge University Press, 237–255.

Ruczinski, I., Kooperberg, C., and LeBlanc, M. (2003). "Logic regression." *Journal of Computational and Graphical Statistics*, 12, 475–511.

Schwarzer, G., Vach, W., and Schumacher, M. (2000). "On the misuses of artificial neural networks for prognostic and diagnostic classification in oncology." *Statistics in Medicine*, 19, 541–551.

Segal, M. (1988). "Regression trees for censored data." *Biometrics*, 44, 35–48.

Segal, M. (1995). "Extending the elements of tree-structured regression." *Statistical Methods in Medical Research*, 4, 219–236.

Segal, M. (2006). "Microarray gene expression data with linked survival phenotypes: Diffuse large B-cell lymphoma revisited." *Biostatistics*, 7, 268–285.

Segal, M., Barbour, J.D., and Grant, R.M. (2004). "Relating HIV-1 sequence variation to replication capacity via trees and forests." *Statistical Applications in Genetics and Molecular Biology*, 3(1), Article 2.

Segal, M., James, I.R., French, M.A.H., and Mallal, S. (1995). "Statistical issues in the evaluation of markers of HIV progression." *International Statistical Review*, 63, 179–197.

Takahashi, H., Masuda, K., Ando, T., Kobayashi, T., and Honda, H. (2004). "Prognostic predictor with multiple fuzzy neural models using expression profiles from DNA microarray for metastases of breast cancer." *Journal of Bioscience and Bioengineering*, 98, 193–199.

Therneau, T.M. and Atkinson B. (2005). R port by Brian Ripley <ripley@stats.ox.ac.uk>. rpart: Recursive Partitioning. R package version 3.1-27.

Therneau, T.M., Grambsch, P.M., and Fleming, T.R. (1990). "Martingale-based residuals for survival models." *Biometrika*, 77, 147–160.

Vapnik, V. (1998). *Statistical Learning Theory*. New York: Wiley.

Venables, W.N. and Ripley, B.D. (2002). *Modern Applied Statistics with S. Fourth Edition*. New York: Springer.

Warner, B. and Misra, M. (1996). "Understanding neural networks as statistical tools." *The American Statistician*, 50, 284–293.

World Health Organization. (1998). *Obesity: Preventing and managing the global epidemic—Report of a WHO consultation presented at the World Health Organization, June 3–5, 1997*. Geneva, Switzerland, WHO, 1998.

Zhang, H. (1995). "Splitting criteria in survival trees." In: Seeber, G.U.H., Francis, B.J., Hatzinger, R., and Steckel-Berger, G., eds. *Statistical Modelling. Proceedings of the 10th International Workshop on Statistical Modelling, Innsbruck, Austria, July 1995*. New York: Springer, 305–314.

Zhang, H. and Singer, B. (1999). *Recursive Partitioning in the Health Sciences*, New York: Springer.

Zhang, H., Yu, C.Y., and Singer, B. (2003). "Cell and tumor classification using gene expression data: Construction of forests." *Proceedings of the National Academy of Sciences*, 100, 4168–4172.

Zhang, H., Yu, C.Y., Singer, B., and Xiong, M.M. (2001). "Recursive partitioning for tumor classification with gene expression microarray data." *Proceedings of the National Academy of Sciences*, 98, 6730–6735.

4

Protein Profiling for Disease Proteomics with Mass Spectrometry: Computational Challenges

Dayanand N. Naik and Michael Wagner

CONTENTS

4.1 Introduction

Proteins are the building blocks of life, as they constitute the basis for much of the molecular machinery that enables cells to function and replicate efficiently. Understanding protein function and interactions, and, in particular, loss of function and/or interactions in case of disease, are thus important goals in biomedical science in general. Although many factors can contribute to the breakdown of parts of this molecular machinery, one important question when studying a particular disease often concerns which proteins are at the root of disease manifestation. Answers to this questions can lead to fundamental insights into protein function and, possibly, to new diagnostic techniques and/or promising therapeutic targets. The prospect of having technology that can reliably identify proteins that are effectors of disease (or at least affected by disease) is thus understandably very exciting for biomedical research in general.

Proteomic approaches to this problem attempt to capture a global snapshot of protein expression in a biological sample. The great difficulty that needs to be overcome lies fundamentally in the tremendous diversity of the physical properties of protein molecules, making them very difficult targets to capture and purify experimentally. Furthermore, proteins are very dynamic in nature, exhibiting flexibility and variability in their conformational states, their expression levels, and their cellular localization. This is in stark contrast to, say, genetic sequence that remains essentially intact and static over the entire lifetime of an organism. Last but not least, a typical cell contains tens of thousands of different protein species that undergo a number of post-translational modifications, further complicating their identification and analysis. And so, although proteomics has tremendous appeal and would invariably have revolutionary impact on biomedicine in general, many technical challenges have yet to be overcome for it to translate into the clinic.

One particular proteomic approach that has been investigated in the last several years and has generated heated and sometimes controversial discussion is that of mass spectrometry-based *protein profiling*. The idea here is to use mass spectrometry (for details see Section 4.2) to generate a one-dimensional mass profile of a complex biological sample. Peaks in this profile are presumed to indicate the presence of one (or more) proteins of the corresponding mass in that biological sample, and by comparing mass profiles from, say, populations of diseased specimens with those of healthy controls one hopes to find clues as to which mass regions would be interesting to examine in greater detail. Early publications on this general approach suggested that the analysis of the mass spectra might lead directly to diagnostic tools (i.e., without identification of the actual molecular origins of the observed peaks in the spectra), but more careful studies have shown this to probably be overly ambitious and unrealistic. And so, although it is still unclear to date whether mass spectrometry-based protein profiling as a technique will finally have the impact on disease diagnosis and lead to the generation of insights into molecular disease mechanisms, that were promised in early publications, it is worth discussing computational and statistical aspects associated with the kind of data generated by the technique. In this Chapter we outline some of the steps that have been proposed to handle the high-dimensional nature of mass spectra, and critically discuss problems such as reproducibility, lack of test data as well as validation.

4.2 Mass Spectrometry Profiling of Complex Biological Samples

We first outline the experimental techniques that generate the primary data that are later subjected to the analytical techniques that are of particular interest here. Mass spectrometry-based protein profiling techniques have relied primarily on so-called MALDI-TOF (matrix assisted laser desorption/ionization-time of flight)-based technology. Briefly, the protein content is extracted from clinical samples (typically from fluids, such as blood serum, urine, spinal fluid, or saliva), mixed with a so-called *matrix* to cocrystallize with the protein sample, and the mixture is spotted and immobilized on a target plate. A laser shot ionizes the proteins in the sample and transfers them into gas phase, in general without breaking their structural integrity. The ions are then accelerated by an electric field and fly to a detector, with their flight time being a function of their mass to charge ratio. The time an ionized protein takes to fly through the vacuum tube and into the detector is a function of its mass, so that intensities in the resulting mass spectrum above a certain signal-to-noise-threshold can be inferred to indicate the presence of a protein at the corresponding mass-to-charge ratio. It should be stressed at this point that MALDI-TOF-based profiling of undigested samples does *not* actually allow for the *identification* of proteins that are expressed in a given biological sample. It is rather only a somewhat rough screening step that, if performed properly, aims to answer the question whether statistically significant differences in protein profiles can be detected. If answered in the affirmative, then the masses corresponding to differentially expressed peaks can be used to narrow the search for differentially expressed proteins using complementary technology. This brief outline should also make it clear that there are numerous parameters at the experimental level that need to be chosen appropriately, such as the matrix to be used for sample crystallization, the number of spectra to sum per acquisition, the instrument calibration procedure, the laser intensity, and so forth. The quality and shape of the resulting mass spectra depend on the particular choices of these to various degrees, and good choices will also depend on the nature of the biological sample being profiled, but one should certainly be aware of these choices that, if chosen inconsistently, make a comparative analysis between experiments difficult.

Biological fluids such as serum, plasma, or urine, which are often targeted for profiling, are exceedingly complex in their proteomic content. In order to reduce complexity and thus increase the sensitivity of mass spectrometry-based profiling several steps can be taken. Abundant common proteins (such as albumin in the case of serum), which can result in large peaks that mask smaller ones, can (and should) be depleted. Furthermore, samples are commonly fractionated using chromatographic methods, either in the form of magnetic beads or, in the case of the so-called surface-enhanced laser desorption/ionization (SELDI) technology, by spotting samples onto target chips that have affinity to certain subpopulations of proteins (e.g., proteins with a certain degree of hydrophobicity). The details of these wetlab sample preparation steps are beyond the scope of this paper. However, interested reader may refer to Hutchens and Yip (1993), Merchant and Weinberger (2000), Srinivas et al. (2001), and Srinivas et al. (2002) for a detailed discussion of these technologies. It suffices, however, to say that appropriate choices for a given experimental setup are crucially important and will often be the determining factor if the resulting data will be useful (e.g., contain signature peaks that discriminate between the two sample populations) or not. Finally, we point out that mass spectrometry exists in many guises and is used for many purposes other than profiling (e.g., protein identification, to name another very important application). In particular, another application of MALDI-TOF mass spectrometry is the so-called peptide mass fingerprinting (PMF) approach to protein identification. In this context, MALDI is used to obtain a mass spectrum of a tryptic *digest* of a *purified* protein in order to identify it by comparing the spectrum with that of a theoretical digest. We stress that our application is quite distinct from PMF as it profiles *complex* and *undigested* protein samples and over a larger mass range. Other mass spectrometry techniques that are quite relevant for the field of proteomics include liquid chromatography mass spectrometry (LC/MS-MS) and FTQ mass spectrometry. The nature of the data generated by these instruments is very different from the profiles generated by MALDI-TOF, making these topics beyond the scope of this present paper. Interested reader may refer to Aebersold and Mann (2003) and Listgarten and Emili (2005).

We continue in the following sections with an overview of various computational and statistical techniques that have been proposed in the literature to deal with MALDI-TOF-based mass spectrometry profiling data. Recently, several articles, for example, by Fung and Enderwick (2002), Yasui et al. (2003), Wu et al. (2003), and White et al. (2004) have appeared providing steps for these type of data analysis. Although a set of strategic steps of protein biomarker discovery in prostate cancer data are provided in Yasui et al. (2003), the article by Wu et al. (2003) provides a comparison of several classification methods for analyzing ovarian cancer data. Also, see Aebersold and Mann (2003) and Listgarten and Emili (2005) for review of various steps in the analysis of LC-MS data.

4.3 Data Description

A mass spectrum $m = (x, y)$ consists of a (typically very high-dimensional) real valued vector x of mass-to-charge values x_i and a real valued vector y of equal dimension representing the measured intensities y_i at the corresponding x_i. The x_is have unit Dalton/charge (m/z) and are a function of the (measured) time-of-flight of the ionized molecule in the vacuum tube. The units of the intensity values y_i are arbitrary, the measured value depends on the signal measured by the detector at the corresponding time of flight as well as the number of acquisitions summed for the measurement. This number can vary from spectrum to spectrum, depending on instrument settings that determine when an acquisition is discarded owing to lack of signal strength.

Early publications in the field reported each sample being profiled just once, resulting in a single mass spectrum representing the information about each sample. See Figure 4.1 for such a spectrum for a sample. Samples are often fractionated, however, and thus it is possible that each clinical sample be represented by multiple complementary spectra. Furthermore, in order to assess reproducibility one would want to collect technical replicate mass

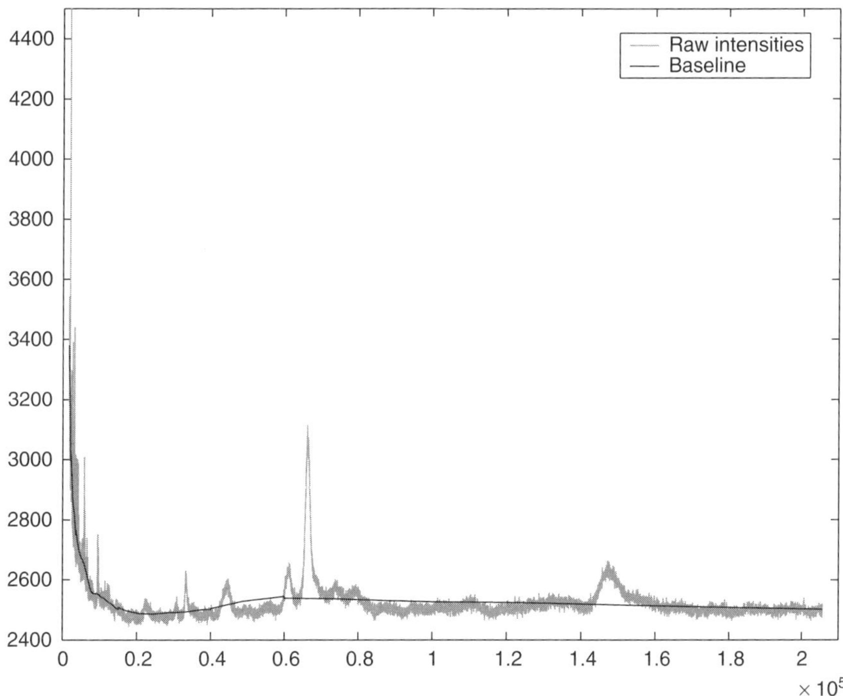

FIGURE 4.1
A mass spectrum. Masses are shown on the horizontal axis and intensities on the vertical axis.

spectra and integrate this information. The literature on dealing with technical replicates is thin, however, we will revisit this topic later and for now focus on the case where every sample is represented by a single mass spectrum.

One of the first tasks is to prepare this information (data) produced by mass spectrometry into a matrix of data, containing variables (say, in columns) and observations (in rows). The variables here are the different m/z ratios and the observations are the samples. The response variable is the intensity measured at a particular m/z ratio. The initial data will have samples from different groups, and each sample (spectrum) will have a very large number of measurements corresponding to it.

Suppose we have g groups and the ith group contains $n_i, i = 1, \ldots, g$ subjects. Our data constitute $n_1 + \cdots + n_g$ mass spectra on all the subjects from different groups. For example, in the context of a lung cancer study, $g = 2$ groups may be "normal healthy" group and "lung cancer group"; in the context of a prostate cancer study, $g = 4$ groups may be "normal group" containing n_1 subjects with healthy controls, "benign prostatic hyperplasia" (BPH) group containing n_2 subjects with benign prostatic hyperplasia, "early stages of cancer" group containing n_3 subjects and "late stages cancer" group containing n_4 subjects. Data items collected from each spectrum can be denoted by $(x_{ij1}, y_{ij1}), \ldots, (x_{ijp_j}, y_{ijp_j})$, where x_{ijk} and y_{ijk} respectively are the kth observed ($k = 1, \ldots, p_j$) values of m/z ratio and the intensity value at that m/z ratio, for the jth subject $j = 1, \ldots, n_i$, from the ith group $i = 1, \ldots, g$. Note that $y_{ij1}, \ldots, y_{ijp_j}$, represent measurements on the same subject, these are usually correlated and p_j in general will be very large. This leads to a very high dimensional correlated data.

At first instance the data may appear to be a multivariate data set with different number of variables observed for different samples. However, the fact that the samples have intensities measured at different masses (m/z ratios) makes these data very different from the usual multivariate data. Various stages are involved in preparing and analyzing these data. In the following text we overview data analysis strategies that have been adopted for the analysis of these high-dimensional data.

The following steps are generally used in the literature and in various commercial packages that are available (e.g., ClinProTools by Bruker, or CiphergenExpress by Ciphergen) to process mass spectrometry profiling data:

- Baseline subtraction
- Preprocessing of the data
- Peak identification
- Normalization of intensities
- Peak alignment
- Peak (feature) selection, and
- Classification methods with cross-validation analysis

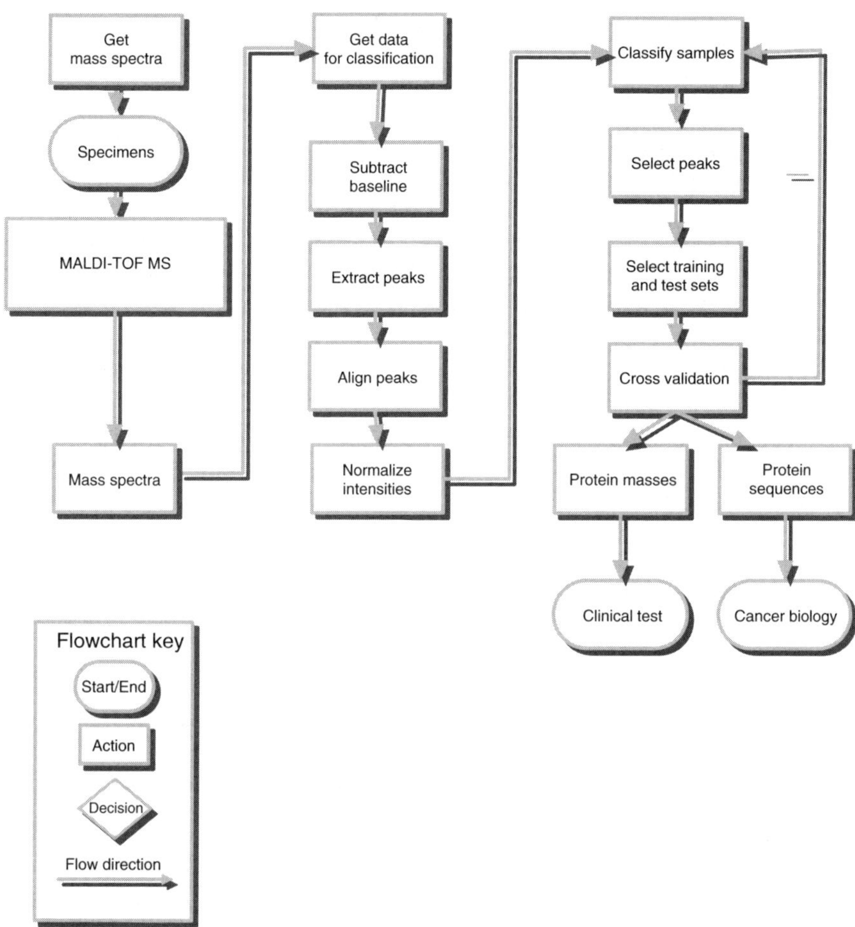

FIGURE 4.2
Biomarker discovery process.

A slightly modified data analysis strategy diagram (from one of our earlier publications) that summarizes various steps mentioned above is provided in Figure 4.2. We will proceed by discussing each of them in a separate section.

4.4 Baseline Subtraction Methods

Each mass spectrum exhibits its own base intensity level (*a baseline*) that varies from sample to sample and hence is to be identified and subtracted. Two approaches to handle this are (1) subtracting a fitted model and (2) applying a filtering method to smooth the signals. In some cases filtering as well as subtraction of the fitted model can be performed (Listgarten and Emili, 2005).

Various smoothing methods are available; however, we have found local linear regression to be most useful. Given the data $(x_i, y_i), i = 1, \ldots, n$, the idea behind local regression is that a regression function, say $g(x)$, can be locally approximated by the value of a smooth function at a predictor variable, say x. Such a local approximation is obtained by fitting a regression surface to the data points within a selected neighborhood of the point x. The percentage of the data points used in each local neighborhood for fitting is the smoothing parameter here. Weighted least squares can be used for fitting by taking a smooth decreasing function of the distance from the data point to the center of the neighborhood as weights. Statistical softwares, like SAS' LOESS method can easily implement this procedure.

Listgarten and Emili (2005) have suggested various other filters for data smoothing including moving average, median, and moving geometric mean filters. The "top hat" filter suggested in Sauve and Speed (2004) is another choice. Recently, Williams et al. (2005) as an improvement over Wagner et al. (2003)'s local linear regression technique, suggested a robust algorithm for computing the baseline correction of MALDI mass spectra that seems to work well.

4.5 Peak Identification and Extraction: Preprocessing Techniques

The peaks in the spectrum (local maxima in the intensity values) can be located by assigning each peak to a prespecified subinterval of masses, or by iteratively merging peaks that are closer than a threshold. For implementing this, we may suggest the following approaches:

1. Divide the whole range of mass into equal subintervals of width, say 8, 12, or 16 Da (mass units), and for each sample, determine the maximum intensity value in these subintervals. If there are no peaks found at a certain subinterval assign a value of zero to the intensity.
2. Alternatively, peaks may be extracted first (peaks are the ones having intensities greater than a certain signal-to-noise threshold value) from each sample. Once all the peaks for all the samples are collected, a data file can be created by taking the union of all the mass values and providing the observed intensities of the peaks at those mass values. Once again the intensity is taken as zero if a peak did not occur for any sample at a mass value. This method has some drawbacks in that, if new samples have to be included in the analysis the whole exercise of processing the data has to be done once again.

4.6 Normalization of Intensities

The experimental setup is such that the absolute peak intensities are not comparable across different samples. MALDI-type assays are generally not quantitative, in the sense that one cannot estimate protein concentrations from peak heights directly. However, individual peak intensities have been shown to correlate highly with the amount of the particular underlying protein (Bakhtiar and Nelson, 2001), and so it is reasonable to use changes in relative intensities as indicators of changes of protein expression levels.

This motivates the need for normalization of the intensities. One could think of a number of choices of how to normalize, for example, with respect to the maximum intensity in a sample, using the sum of all peak intensities, or, possibly, using the total area under the peaks as reference value. None of these is an obvious choice, and all have severe defects in the presence of pathological examples. Suppose we choose to normalize with respect to the sum of the intensities. This can be implemented by dividing peak intensities of a sample by the sum of all peak intensities in that sample and multiplying by 100, so that the processed intensities could be interpreted as percentages of total intensity in the sample (see Baggerly et al., 2003; Wagner et al., 2003). If area under the curve (AUC) is the choice, then normalize each of the mass spectra by the dividing coefficient AUC of spectrum divided by average AUC over all spectra. For implementation of this method see Fung and Enderwick (2002) and Sauve and Speed (2004).

Satten et al. (2004) provide a systematic way to normalize the intensities. If x_i is the m/z ratio at which y_i is the intensity then y_i is normalized as

$$y_i^* = \frac{y_i - Q_{0.5}(x_i)}{Q_{0.75}(x_i) - Q_{0.25}(x_i)},$$

where $Q_\alpha(x)$ is an estimate of the α-th quantile of spectral intensities at m/z ratio x. That is, the spectra is centered using a (local) estimate of the median spectral intensity, and scaled by dividing by a (local) estimate of the interquartile range. See Satten et al. (2004) for details on how to implement this standardization.

4.7 Peak Alignment

Finally, in order to make the peak profiles comparable across different samples, we need to align them, that is, to find one common set of peak locations across all samples that will work as coordinates for the vectors we will use for each sample in the classification schemes to follow. One way we propose for this step is, if two peaks are within a certain small percentage (depending on the resolution and mass accuracy of the instrument) of each other, say 0.4% of each other, then they should be considered identical and their masses are to be reassigned. One can also use an algorithm and R routine provided by Jeffries (2005) for this purpose.

4.8 Dimension Reduction and Peak Selection

The processed data at this stage are generally of a reasonable size with about 800–1600 variables (masses) and the intensities at those masses. Unfortunately, the number of samples available in each group is usually small, sometimes as small as five. Thus, we are faced with the problem of high-dimensional data even after the preprocessing. Generally, in the processed data a large percentage (many times 60% or higher) of the peaks appear in only very few samples and are thus not likely to be helpful in classifying the majority of the samples. Hence, one can ignore any peaks that occurs in fewer than a certain prespecified number of samples. This step usually reduces the dimension of the identifying vectors down to 200–300. However, this number is still much larger than the number of samples available. There is no hope of getting statistically meaningful results with lack of degrees of freedom, so we need to further reduce the number of peaks used in the classification.

Feature selection, that is, the reduction of the number of input variables (or, in this case, peaks), is a crucially important step. Many classification methods are known to perform poorly when "irrelevant" features or ones without information content are added. Second, computational biologists are frequently faced with the problem of having only few (tens) samples but many (thousands) descriptors, as is the case with microarray analysis. This presents the challenge of designing models that are not "overfitted" to the data. One approach to prevent this is to try to decrease the feature dimensionality by performing feature selection.

Here we are interested in finding a reasonably small set of peaks in order to then enable the identification of the underlying proteins and, eventually, understand the biological function they have in the disease pathway. In this sense the classification methods used can be viewed as validation methods for the feature selection algorithms.

Unfortunately, finding the "best" set of features to build a predictive model is a hard combinatorial problem, and so one must live with heuristic approaches. The literature on this subject is vast, and one generally distinguishes between filtering methods (those which rank individual features according to some criterion) and more involved wrapper algorithms, which use classification methods directly to evaluate a particular set of features. Here we suggest only simple filtering methods, since they seem to do reasonably well for our purposes.

4.8.1 Principal Component Method

Principal component analysis is a very general method of dimension reduction with minimal assumptions about the probability distribution of the data. The idea behind principal component analysis is to find a best fitting q-dimensional subspace for p-dimensional vectors $\mathbf{y}_1, \ldots, \mathbf{y}_n$, which minimizes the sum of squares of the perpendicular distance from the vectors to the subspace. This subspace is spanned by the q-dimensional vectors $\mathbf{u}_i = \mathbf{L}'(\mathbf{y}_i - \bar{\mathbf{y}})$, $i = 1, \ldots, n$, where $\mathbf{L} = (\boldsymbol{\ell}_1, \ldots, \boldsymbol{\ell}_q)$ is $p \times q$ matrix of q eigenvectors corresponding to the largest q eigenvalues of the variance covariance matrix of $\mathbf{y}_i$. Here, $\bar{\mathbf{y}} = 1/n \sum_{i=1}^{n} \mathbf{y}_i$ and $\mathbf{L}'$ denotes the transpose of matrix $\mathbf{L}$.

In fact, $\mathbf{u}_i$, $i = 1, \ldots, n$ are the principal component scores and the elements of the vector are the principal components. These scores can be used as the data for further statistical analysis.

In the present context, this method can be adopted to create smaller dimensional data. Suppose $\mathbf{y}_{ik}$ is the kth sample vector (of dimension p) from the ith group, where $k = 1, \ldots, n_i$ and $i = 1, \ldots, g$. We compute the pooled sample variance covariance matrix of all the data as follows:

$$\mathbf{S} = \frac{1}{n-g} \sum_{i=1}^{g} \sum_{k=1}^{n_i} (\mathbf{y}_{ik} - \bar{\mathbf{y}}_i)(\mathbf{y}_{ik} - \bar{\mathbf{y}}_i)'.$$

We note that $\mathbf{S}$ is a symmetric positive semidefinite matrix.

Suppose the eigenvalue (spectral) decomposition of $\mathbf{S}$ is written as $\mathbf{S} = \boldsymbol{\Gamma}\boldsymbol{\Delta}\boldsymbol{\Gamma}' = \sum_{i=1}^{p} \lambda_i \boldsymbol{\ell}_i \boldsymbol{\ell}_i'$ with $\boldsymbol{\Delta}$, a diagonal matrix containing the eigenvalues λ_i of $\mathbf{S}$ in an increasing order at the diagonals and $\boldsymbol{\Gamma}$ containing the corresponding eigenvectors $\boldsymbol{\ell}_i$ in the columns. An approximation to $\mathbf{S}$ is $\mathbf{S} \approx \sum_{i=1}^{q} \lambda_i \boldsymbol{\ell}_i \boldsymbol{\ell}_i'$, where $q < p$ is selected so that at least about 90% of the variation in the data, measured by the total variance, is accounted for by the q principal components. In the present context q will be generally much smaller than p. Using these eigenvectors we generate the data on principal component scores as: $\mathbf{u}_{ik} = \mathbf{L}'(\mathbf{y}_{ik} - \bar{\mathbf{y}}_i)$, $k = 1, \ldots, n_i$, $i = 1, \ldots, g$. For all the future statistical analysis, one can use $\{\mathbf{u}_{ik}\}$ as the data.

Although in the present context we are using the principal component method mainly for reducing the data, Xiong et al. (2000) found recently that higher accuracy in discrimination can be achieved when the principal components are used for discriminant analysis in classifying gene expression data. This work also supports our contention that principal components can be used here both for dimension reduction and classification.

One disadvantage of using principal components for reducing the dimension, however, is that in this process of creating principal component scores, we lose the identity of the original variables. However, using the magnitudes of the elements of the eigenvectors, a selection of original variables representing each of the principal components can be made. See Khattree and Naik

(2000) for an illustration of this approach and Lilien et al. (2003) for further comments on this issue.

Wavelet transformation is another approach that can be used for dimension reduction instead of principal component method. Recently Qu et al. (2003) used wavelet transformations for dimension reduction in the context of their prostate cancer spectrometry data analysis.

4.8.2 Peak Selection Method

In the following we propose a protein (variable) selection method that is similar to gene selection method adopted in Dudoit et al. (2002), which directly identifies most discriminating variables. Also see Golub et al. (1999) for an alternative standardization for selecting genes.

The ratio of between-group sum of squares and within-group sum of squares (B/W ratio) can be used, for feature selection. Suppose, that y_{ikj} is the observed intensity of the jth feature of the kth sample belonging to the ith group, that the number of groups is denoted by g, n_i is the number of samples in the ith group, and that

$$\bar{y}_{ij} = \frac{1}{n_i}\sum_{k=1}^{n_i} y_{ikj}, \quad \bar{y}_j = \frac{1}{\sum_{i=1}^{g} n_i}\sum_{i=1}^{g}\sum_{k=1}^{n_i} y_{ikj}.$$

Then the between-group sum of squares for the jth feature is

$$B_j = \sum_{i=1}^{g}(\bar{y}_{ij} - \bar{y}_j)^2$$

and the within group sum of squares is

$$W_j = \sum_{i=1}^{g}\sum_{k=1}^{n_i}(y_{ikj} - \bar{y}_{ij})^2.$$

For every feature $j = 1, \ldots, p$ we compute B_j/W_j or equivalently (for ordering purposes), the ANOVA F-statistic

$$F_j = \frac{B_j/\nu_1}{W_j/\nu_2},$$

where $\nu_1 = g - 1$ and $\nu_2 = \sum n_i - g$ are the degrees of freedoms of B_j and W_j respectively. Then the reduced data set will be the data corresponding to the q largest F_j values. An advantage of this method over the other data reduction techniques, like principal component analysis, where the data on few linear combinations of the original variables (features) are used, is that we can obtain clues (their molecular weight or mass-to-charge ratio) as to the identity of the important proteins that are used as classifiers.

A slightly more general ordering criterion for feature selection is the Wilks' likelihood ratio (Λ), which is the ratio of the maximized likelihood function under the assumption that there is no difference between the groups and the maximized likelihood function without any assumption. Smaller values of Λ indicate more significant differences between the groups. If the probability density functions (*pdf*) are normal then this method reduces to the B/W ratio (F-statistic) method described above. The advantage of this method is that it allows us to use a different *pdf* for this process.

4.9 Classification/Discrimination Methods

The next task is to perform a discriminant analysis to construct discriminant functions so that the classification of the new unknown samples obtained from MS can be performed. Various classical and modern methods are available for this purpose. Classical statistical methods (parametric as well as nonparametric) have stood the test of time and proved to be very useful. However, two modern classification methods have emerged recently. One set of methods is bagging with boosting of classification trees, and the other set is based on support vector machines. Boosting methods have been utilized by Qu et al. (2002), and several classical statistical methods and support vector machines are adopted by Wagner et al. (2004) in their analyses.

In the following sections, we will only briefly describe these methods and more details can be found in Khattree and Naik (2000) and Hastie et al. (2001).

4.9.1 Parametric Discriminant Procedures

In general, in discriminant analysis, the decision rule to classify a new (or test) sample into one of the several groups by taking the prior probabilities and the cost of misclassifications into consideration, is as follows. Classify the sample with an observation vector $\mathbf{y}$ into the ith group if the expected cost of misclassification, $\sum_{s=1}^{g} \pi_s f_s(\mathbf{y}) c(i \mid s), s \neq i$, is smallest for $i = 1, \ldots, g$. Here π_s and $f_s(\mathbf{y})$, $s = 1, \ldots, g$, are respectively the prior probability and the probability density function for the sth group and $c(i \mid s)$ is the cost of misclassification when the sample is classified into the ith group when it actually comes from the sth group. Of course, for $s = i$, $c(i \mid i) = 0$. If all the costs of misclassification are assumed to be equal then the classification rule is based on minimizing the expected total probability of misclassification and we classify the sample into the ith group if $\sum_{s=1}^{g} \pi_s f_s(\mathbf{y})$, $s \neq i$, is smallest for $i = 1, \ldots, g$. This rule further reduces to simply checking whether or not $\pi_i f_i(\mathbf{y}) > \pi_j f_j(\mathbf{y})$, for all $j = 1, \ldots, g, j \neq i$ (see Anderson, 1984). In practice, different known multivariate probability densities can be used for $f_i(\cdot)$, but the most common density used is the multivariate normal density. If the form of the density is not assumed to be known then the nonparametric methods are used for estimating the density using the data.

If the probability density function for the ith group is assumed to be multivariate normal with mean vector μ_i and variance covariance matrix Σ_i for $i = 1, \ldots, g$ then the above classification rule simplifies to classifying an observation $\mathbf{y}$ into the ith group, if $D_i^2(\mathbf{y}) > D_j^2(\mathbf{y})$, for all $j = 1, \ldots, g, j \neq i$. Here $D_i^2(\mathbf{y})$ are defined as

$$D_i^2(\mathbf{y}) = (\mathbf{y} - \boldsymbol{\mu}_i)'\Sigma_i^{-1}(\mathbf{y} - \boldsymbol{\mu}_i) + ln|\Sigma_i| - 2\,ln\,\pi_i, \quad i = 1, \ldots, g.$$

Here and at later occurrences $|\Sigma|$ denote the determinant of the matrix Σ. Also note that the quantity $(\mathbf{y} - \boldsymbol{\mu}_j)'\Sigma_j^{-1}(\mathbf{y} - \boldsymbol{\mu}_j)$ is the Mahalanobis distance between the observation vector $\mathbf{y}$ and the mean of the jth population. This is one of the most popular statistical distance that measures the distances between two vectors by taking into account the covariance matrix of the random vector. The above rule of classification is called *quadratic discrimination rule* owing to the presence of quadratic terms in $\mathbf{y}$.

With $D_i^2(\mathbf{y}), i = 1, \ldots, g$ defined as above, the posterior probability that given $\mathbf{y}$, the sample will be classified into the ith group, is given in terms of $D_i^2(\mathbf{y})$ as

$$P(i|\mathbf{y}) = e^{-\frac{1}{2}D_i^2(\mathbf{y})} \Big/ \sum_{j=1}^{g} e^{-\frac{1}{2}D_j^2(\mathbf{y})}, \quad i = 1, \ldots, g.$$

The estimated posterior probabilities are obtained by replacing $\boldsymbol{\mu}_i'$s and Σ_i's in $D_i^2(\mathbf{y})$ by their maximum likelihood estimates $\bar{\mathbf{y}}_i = 1/n_i \sum_{k=1}^{n_i} \mathbf{y}_{ik}$ and $\mathbf{S}_i$, the sample variance covariance matrix computed using the data from the ith group, $i = 1, \ldots, g$ respectively. The criterion for classifying an observation to the closest group is equivalent to classifying it to the group with maximum posterior probability given the observation. The quadratic discrimination procedure described above reduces to *linear discrimination rule* if the variance covariance matrices Σ_i's are all equal for the g groups.

The quadratic and linear discrimination rules described here require that the variables used for classification are continuous and multivariate normally distributed. If some or all the variables are categorical then predictors based on *logistic regression* can be developed.

4.9.2 Nonparametric Discriminant Procedures

In the parametric discriminant analysis the density functions $f_i(\mathbf{y})$ were assumed to be multivariate normal. However, often, the functional form of these densities are unknown (or suggested to be nonnormal by the data in hand). In many cases we are able to use some suitable transformations of the variables to achieve multivariate normality. This in turn enables us to use one of the linear or quadratic discriminant analysis. However, sometimes such attempts may fail. This necessitates a search for alternative approaches where the density functions themselves have to be estimated from the data available as training sets. Since in this case no parametric forms for densities are assumed, this approach is termed as the nonparametric approach.

We now describe a nonparametric approach based on *kernel method*. Suppose $\mathbf{y}_{i1}, \ldots, \mathbf{y}_{in_i}$, is a random sample from the ith group, and $\mathbf{y}$ is an additional observation from this group that has a (unknown) probability density function $f_i(\mathbf{y})$. The unknown density $f_i(\mathbf{y})$ is estimated (for using in the discrimination rule) by $\hat{f}_i(\mathbf{y}) = \frac{1}{n_i}\sum_{k=1}^{n_i} K_i(\mathbf{y} - \mathbf{y}_{ik})$, where the function $K_i(\mathbf{z})$ is a kernel function defined for the p-dimensional vector $\mathbf{z}$, normalized such that $\int_{\mathbf{R}^p} K_i(\mathbf{z})d\mathbf{z} = 1$. We often assume that $K_i(\mathbf{z})$ is also nonnegative. Thus, any multivariate density can be a prospective choice for the kernel function. One popular choice for the kernel is the normal kernel $K_i(\mathbf{z}) = (1/c_0(i))\exp(-(1/2)\mathbf{z}'\mathbf{V}_i^{-1}\mathbf{z}/r^2)$, where $c_0(i) = (2\pi)^{p/2}r^p \mid \mathbf{V}_i \mid^{1/2}$ and recall that $\mid \mathbf{V}_i \mid$ is the determinant of the matrix $\mathbf{V}_i$. In the above expressions, the matrix $\mathbf{V}_i$ is used to assign an appropriate metric in the computation of distances and densities. In particular, some of the choices for $\mathbf{V}_i$ are, $\mathbf{V}_i = \mathbf{S}_i$, the estimated variance covariance matrix for the ith group, $\mathbf{V}_i = \mathbf{S}$, the pooled estimated variance covariance matrix, $\mathbf{V}_i = diag\,(\mathbf{S})$, the diagonal matrix of the pooled variance covariance matrix, and so on. The value of r in the selected kernel is chosen to get the degree of smoothness we want for the density that is being estimated. Further details may be found in Khattree and Naik (2000).

4.9.3 Fisher's Canonical Discriminant Analysis

Fisher's linear discriminant analysis, which is also known as *canonical discriminant analysis*, is performed using only a few (less than or equal to $(g-1)$, where g is the number of groups) canonical variables that are certain linear combinations of the original variables. The canonical variables have a better capacity to discriminate between the groups than any individual variable because these are created such that the between-group sum of squares for these variables is large relative to the within-group sum of squares.

Suppose $\{\mathbf{y}_{11}, \ldots, \mathbf{y}_{1n_1}\}, \ldots, \{\mathbf{y}_{g1}, \ldots, \mathbf{y}_{gn_g}\}$ of sizes $n_1, \ldots, n_g$ are independent samples from the respective populations with different mean vectors and common variance covariance matrix. Then the sample mean vectors $\bar{\mathbf{y}}_i = \frac{1}{n_i}\sum_{k=1}^{n_i}\mathbf{y}_{ik}$ $i = 1, \ldots, g$, estimate the corresponding population means and the average of the population means, is estimated by $\bar{\mathbf{y}} = (1/\sum_{i=1}^{g} n_i)\sum_{i=1}^{g}\sum_{k=1}^{n_i}\mathbf{y}_{ik} = (1/\sum_{i=1}^{g} n_i)\sum_{i=1}^{g} n_i\bar{\mathbf{y}}_i$. The common variance covariance matrix is estimate by the pooled sample variance covariance matrix $\mathbf{S} = \mathbf{E}/(\sum_{i=1}^{g} n_i - g)$, where $\mathbf{E}$ is the pooled within-group sums of squares matrix and is given by $\mathbf{E} = \sum_{i=1}^{g}\sum_{k=1}^{n_i}(\mathbf{y}_{ik} - \bar{\mathbf{y}}_i)(\mathbf{y}_{ik} - \bar{\mathbf{y}}_i)'$. Also, $\mathbf{D}$, the between-group sums of squares matrix is given by $\mathbf{D} = \sum_{i=1}^{g} \sum_{k=1}^{n_i} (\bar{\mathbf{y}}_i - \bar{\mathbf{y}})(\bar{\mathbf{y}}_i - \bar{\mathbf{y}})' = \sum_{i=1}^{g} n_i(\bar{\mathbf{y}}_i - \bar{\mathbf{y}})(\bar{\mathbf{y}}_i - \bar{\mathbf{y}})'$.

Now, the canonical variables are obtained as the linear combinations $\mathbf{v}_l'\mathbf{y}, 1 \leq l \leq (g-1)$, where the vectors $\mathbf{v}_1, \mathbf{v}_2, \ldots,$ respectively, solve the optimization problems, $\max_{\mathbf{b} \neq \mathbf{0}}(\mathbf{b}'\mathbf{D}\mathbf{b}/\mathbf{b}'\mathbf{S}\mathbf{b})$, subject to the restrictions that $\mathbf{v}_1, \mathbf{v}_2, \ldots$ are all such that $\mathbf{v}_l'\mathbf{S}\mathbf{v}_l = 1$ and $\mathbf{v}_l'\mathbf{S}\mathbf{v}_m = 0$ for $l \neq m$. Note that for $\mathbf{b} = \mathbf{v}_1$ the above ratio of between-group sums of squares to within-group sums of squares is maximized. It may be pointed out that the vectors $\mathbf{v}_1, \mathbf{v}_2, \ldots$ are nothing but the eigenvectors of the nonsymmetric matrix $\mathbf{S}^{-1}\mathbf{D}$. The corresponding eigenvalues say w_l are such that $w_1 \geq w_2 \geq \ldots$.

In Fisher's discrimination method, canonical variables corresponding to all the nonzero eigenvalues or corresponding to only the first few can be used. Suppose only the first r canonical variables are used. Then the classification procedure will classify an observation vector $\mathbf{y}$ to the ith group if

$$\sum_{l=1}^{r} [\mathbf{v}_l'(\mathbf{y} - \bar{\mathbf{y}}_i)]^2 \leq \sum_{l=1}^{r} [\mathbf{v}_l'(\mathbf{y} - \bar{\mathbf{y}}_j)]^2, \quad \text{for } j = 1, \ldots, g, \quad j \neq i.$$

More details can be found in Khattree and Naik (2000) and Johnson and Wichern (2007). For only two groups and under multivariate normal distribution assumption, this procedure is same as the one derived earlier on the basis of density functions.

4.9.4 Nearest-Neighbor Methods

Another approach that is also nonparametric in nature is the *nearest-neighbor method*. This approach is based on a criterion involving distances from "immediate neighbors" and hence, bypasses the need for a density altogether. An affinity measure to determine the nearest neighbors is selected first. Two common measures are the Mahalanobis distance or Euclidean distance and one minus the correlation coefficient between the two samples. In the nearest-neighbor method we first compute the affinity measure between the unknown sample and all the other samples. There will be as many values of the affinity measure as there are the number of samples. For example, for an observation $\mathbf{y}$, which is to be classified into one of the g groups, there will be $\sum_{i=1}^{g} n_i$ such measures. For $k = 1, \ldots, n_i$, and $i = 1, \ldots, g$, the Mahalanobis distances d_{ik} between $\mathbf{y}$ and $\mathbf{y}_{ik}$, where $d_{ik}^2 = (\mathbf{y} - \mathbf{y}_{ik})'\mathbf{S}^{-1}(\mathbf{y} - \mathbf{y}_{ik})$ or one minus the absolute value of correlation coefficients, $1 - |r_{ik}|$ between $\mathbf{y}$ and $\mathbf{y}_{ik}$, where

$$r_{ik} = \frac{(\mathbf{y} - \bar{y}\mathbf{1})'(\mathbf{y}_{ik} - \bar{y}_{ik}\mathbf{1})}{\sqrt{(\mathbf{y} - \bar{y}\mathbf{1})'(\mathbf{y} - \bar{y}\mathbf{1})(\mathbf{y}_{ik} - \bar{y}_{ik}\mathbf{1})'(\mathbf{y}_{ik} - \bar{y}_{ik}\mathbf{1})}},$$

with $\mathbf{1}$ as the vector of all ones, $\bar{y} = (1/p)\sum_{j=1}^{p} y_j$, and $\bar{y}_{ik} = (1/p)\sum_{j=1}^{p} y_{ikj}$, assuming that y_j and y_{ikj} respectively are the jth components of $\mathbf{y}$ and $\mathbf{y}_{ik}$, can be calculated.

Next, we find the samples corresponding to the k smallest values of the selected affinity measure, where k is prespecified number. The unknown sample is classified as belonging to the group to which the maximum number of the k closest samples belong. We may have undecided cases due to tied number of samples belonging to each group. If this happens a new group (category) "undecided" is created and the proportion of samples classified to this group is also computed.

4.9.5 Support Vector Machines

Support vector machines (SVMs) are powerful classification tools that arose out of the machine learning and optimization communities in the 1960s (e.g., Mangasarian, 1965). They have recently found immense popularity, as evidenced in the large numbers of books and research articles (see, e.g., Vapnik, 1995, Cristianini and Shawe-Taylor, 2000, and Hastie et al., 2001, and references therein).*

We will confine ourselves here to introducing the main ideas underlying the modeling approach and algorithms as well as our approach to the multiclass case. Instead of going into details our goal here is merely to give a brief and intuitive introduction.

We start with the simple case of two classes. SVMs take a list of features (in our case peak intensities) and associated class labels (such as healthy or cancerous) as input and attempt to find a hyperplane that cleanly separates the two classes, that is, one that has all members of one class lying on one side and all members of the other class on the other. In case one such hyperplane exists, infinitely many other feasible hyperplanes will provide clean separation, and SVMs choose the (unique) hyperplane with *maximum margin*, that is one that maximizes the distance to *any* data point. This is in contrast to, say, Fisher's discriminant method, which can be interpreted as finding a linear hyperplane, which maximizes the distance to the class medians.

If the data vectors are denoted by $\mathbf{y}_i$ and the classes by index sets C_1 and C_0, then the basic classification problem can be written as: Find vector $\mathbf{w}$ and scalar b such that

$$\begin{cases} \mathbf{y}_i'\mathbf{w} + b > 0 & \text{for all } i \in C_1 \\ \mathbf{y}_i'\mathbf{w} + b < 0 & \text{for all } i \in C_0. \end{cases} \tag{4.1}$$

Here $\mathbf{w}$ is the normal vector that together with the displacement b defines the separating hyperplane. If we set $a_i = 1$ for $i \in C_1$ and $a_i = -1$ for $i \in C_0$ then Equation 4.1 can be rewritten more compactly: Find vector $\mathbf{w}$ and scalar b such that

$$a_i(\mathbf{y}_i'\mathbf{w} + b) > 0 \quad \text{for all } i \in C_1 \cup C_0. \tag{4.2}$$

*The interested reader should check the web site www.kernel-machines.org for an updated and comprehensive bibliography.

Finding the hyperplane with maximum margin can be shown (see, e.g., Proposition 6.1 in Cristianini and Shawe-Taylor (2000)) to be equivalent to minimizing the norm of **w**, which gives rise to an optimization problem (known as the hard-margin SVM):

$$\begin{aligned} \min_{\mathbf{w},b} \quad & \|\mathbf{w}\| \\ \text{such that} \quad & a_i(\mathbf{y}_i'\mathbf{w} + b) \geq 1 \quad \text{for all } i \in C_1 \cup C_0. \end{aligned} \tag{4.3}$$

Given an optimal solution $(\mathbf{w}^*, b^*)$, a new sample $\mathbf{y}$ will be classified as being in C_1 if $\mathbf{y}'\mathbf{w}^* + b^* > 0$, and in C_0 if $\mathbf{y}'\mathbf{w}^* + b^* < 0$. In practice, cases where $|\mathbf{y}'\mathbf{w}^* + b^*|$ is very small are often interpreted as ambiguous and no prediction is made. The margin of the optimal hyperplane will be given by $1/\|\mathbf{w}\|$. If the Euclidean norm is chosen, this is a (convex) quadratic optimization problem that can be solved very efficiently with standard software.

Of course it can happen that the constraints in Equation 4.3 are inconsistent. This will occur when the classification points are not linearly separable, and, as is usually the case with real problems, happens very often when dealing with real data. The typical SVM approach to deal with this case is to allow for *misclassification* by introducing slack variables ξ_i and relaxing the constraints. A secondary goal now is to minimize the total misclassification (along with maximizing the margin). Clearly, there is a tradeoff between these two objectives, and their relative importance is reflected in a trade-off parameter C that must be chosen a priori by the modeler. The SVM formulation with misclassification (also known as soft-margin SVM) is now

$$\begin{aligned} \min_{\mathbf{w},b,\xi_i} \quad & \|\mathbf{w}\| + C\|\xi\| \\ \text{such that} \quad & a_i(\mathbf{y}_i'\mathbf{w} + b) + \xi_i \geq 1 \quad \text{for all } i \in C_0 \cup C_1, \end{aligned} \tag{4.4}$$

where ξ is the vector of ξ_i's.

The idea can be extended to nonlinear decision boundaries by introducing nonlinear *kernels*. By choosing an appropriate kernel functions $h(\mathbf{y}_i)$ that transform the data point, one can, for example, define SVMs that handle polynomial or Gaussian kernels. The reader is referred to Cristianini and Shawe-Taylor (2000) for details.

The extension of SVMs to the case with multiple classes is still an active research topic. One can use a simple pairwise approach that constructs all $g(g-1)/2$ pairwise discriminators for g classes (groups). The final classifier is taken to be the one that dominates all others, provided it exists. Otherwise, the result is considered to be inconclusive, an event that occurs in only a very small percentage of cases. This approach works reasonably well for small values of g. We want to stress here that even in the inconclusive cases it is sometimes possible to rule out certain classes (in case they are dominated by all others), which is an outcome that might still be of some medical relevance. Lee et al. (2004) have found some natural and theoretically satisfying extensions.

4.9.6 Classification Trees

The classification and regression trees (CART) introduced by Breiman et al. (1984) have played a major role in improving classification and machine learning methods. The classification trees are constructed by recursively splitting the samples into two subsets starting with all the samples. Each terminal node is assigned a group or class label and the resulting partition provides a classifier. See Hastie et al. (2001) for an easy exposition of these and other machine learning algorithms. Adam et al. (2002) have used CART in their prostate cancer spectrometry data analysis.

4.9.7 Bagging and Boosting for Improving Prediction

Breiman (1996, 1998) in a series of papers suggested methods to improve the performance of CART and other classifiers by aggregating them. Aggregating (unstable) classifiers can greatly improve predictive accuracy. Bootstrap aggregating, termed by Breiman (1998) as bagging, is a popular aggregating method and it has been successfully used in mass spectrometry data analysis by Wu et al. (2003). In bagging algorithm, a large number, say B, of bootstrap samples are selected with replacement from the original training data set and tree classifiers are computed using each of the B bootstrap samples. Classification of a sample is made using all these classifiers and the final classification is based on a simple majority rule. Computer routines written in R-language are available to implement bagging for classification.

Freund and Schapire (1997) proposed an algorithm called boosting, which is especially popular among machine-learning community. Whereas bagging algorithm works by taking bootstrap samples from the training data set, the boosting algorithm works by changing the weights on the training set. Construction of the predictor is such that it can incorporate weights on the samples. Depending on the training data set classification errors, the weights are changed and a new predictor is constructed. Variations such as Adaboost and other algorithms to implement boosting are available in R routines.

Boosting algorithm with regression and classification trees has been used by Qu et al. (2002), and it is used along with predictor based on logistic regression discussed by Yasui et al. (2003) in their prostate cancer mass spectrometry data analysis.

4.9.8 Random Forest

Random forests proposed by Breiman (2001) are a combination of tree predictors such that each tree depends on the values of an independently sampled vector having the same distribution for all trees in the forest. This method combines bagging and random feature selection and this leads to improvement in predictive accuracy of the predictors. An algorithm provided by Breiman (2003) for implementing random forest is as follows: (1) Draw B bootstrap samples with replacement from the training data set. (2) Use the ith bootstrap sample and construct a classification tree and use it to predict those samples that are not in the bootstrap sample. These samples are termed as out-of-bag samples and the predictions are called out-of-bag estimators. (3) When constructing the prediction tree, at each node splitting first randomly select a smaller number, say *m*, variables then choose one best split from these variables. (4) The final prediction is the average of out-of-bag estimators over all bootstrap samples.

Using the Random forest package in R it is easy to apply this method and Wu et al. (2003) have successfully used this for their analysis of ovarian cancer mass spectrometry data.

4.10 Validation Techniques

Providing methods and tools for assessing and validating the predictors is another important data analysis step. Using an appropriate measure of prediction error estimation is very important for the comparisons, especially where both smoothed and unsmoothed predictors are included.

4.10.1 Cross-Validation Method

Leave-one-out or leave-many-out cross-validation method has been a very popular tool for validating a predictor by computing prediction error estimates. In order to assess the generalization power of the classification methods and to estimate their prediction capabilities for unknown samples, the data are split into training and test sets. The predictor that is constructed using a training sample (e.g., all but one) is used to classify the sample that is left out and this process is repeated on all samples. Finally, prediction error estimate is obtained by computing the proportion of misclassified samples. This cross-validation procedure is easy to implement and provides an excellent estimate of the prediction error if the predictors are smooth like linear discriminants. We stress that feature selection also needs to be performed in every experiment on the training set only (unlike what is often seen in the literature) in order not to bias the feature selection procedure unfairly. Several papers (e.g., Wagner et al., 2003) have shown that performing feature selection on the entire data set often grossly underestimates the prediction error.

4.10.2 Bootstrap Method

Although for smooth predictors cross-validation method will provide an excellent estimate of prediction error, for unsmooth predictors, like nearest neighbor predictors, cross-validation method is found to provide unstable estimates of prediction error. See Efron and Tibshirani (1997) for details. In general, bootstrap provides an alternative method for estimating the prediction error. However, Efron and Tibshirani (1997) proposed a specific bootstrap, *the 632+ bootstrap*, which is an improvement on cross-validation method.

4.11 Examples

Herein we provide two examples that we have dealt with in the past. The first example is Duke's lung cancer data obtained through MALDI-TOF technology (Wagner et al., 2003) and the second example is Eastern Virginia Medical College prostate cancer data obtained through SELDI-TOF technology (Wagner et al., 2004). Although we have concentrated on MALDI-TOF here, the indicated steps for data analysis are general and can be applied to any mass spectrometry data. We have implemented the entire scheme from Figure 4.2 using a combination of languages and tools such as Perl, SAS, and Matlab. The SVM^{light} software (Joachims et al., 1999) was used for the SVM.

4.11.1 Duke Lung Cancer Data

Herein we provide a summary of the analysis of these data that is detailed in Wagner et al. (2003). A total of 820 protein mass spectra obtained from serum samples of 24 lung cancer patients and 17 healthy patients with each sample split into 20 fractions were available for the analysis. Each mass spectrum consists of 60,831 intensity measurements at discrete mass/charge (m/z) values. Both the raw data sets as well as a processed set, which contained the locations and raw intensities of peaks as identified by the software that comes with the MALDI-TOF instruments, were available for the analysis. The problem of interest was to find patterns among the protein mass spectra of these samples that characterize and distinguish healthy individuals from diseased ones.

In order to make the data amenable to classification, we needed to convert the mass spectra of each fraction into lower dimension vectors, that characterize the samples (the *peak profiles*). Each raw spectrum consists of 60,831 intensity measurements at discrete mass/charge (m/z) values. Given the small sample size of 41, our first goal is to reduce this data to, say, less than 20 peaks that discriminate between healthy and diseased states. However, first we go through a series of preprocessing steps as indicated below.

Baseline identification and subtraction: It can be seen that each mass spectrum (an example is given in Figure 4.1) exhibits a base intensity level (a *baseline*) that varies from sample to sample (fraction to fraction in this example) and consequently needs to be identified and subtracted. We see in the figure a near-exponential decay in the noise at the beginning, after which the noise level appears to be a linear function of the mass/charge ratio. This noise varies across the m/z-axis, and it generally varies across different fractions, so that a one-value-fits-all strategy cannot be applied. Using local linear regression (as implemented in the software package SAS by the LOESS Procedure) we get a rough approximation of the baseline iteratively in order to smooth over the peaks. To deal with the exponential decay at the beginning we used varying degrees of smoothness, depending on the mass/charge interval being considered. A second iteration of smoothing was applied by identifying intensity values that deviated from the baseline by more than one standard deviation. Those values were (temporarily) replaced by their corresponding baseline values, and the smoothing technique was reapplied. As can be seen in the example in Figure 4.1, the resulting baseline appears to be satisfactory.

Peak identification and extraction: Here we took the masses of the peaks from the processed data that were identified by the software provided with the mass spectrometry instrument, coupled with some human processing.

Intensity normalization: Details of the experimental setup are such that the absolute peak intensities are not comparable across different fractions, let alone samples. One could think of a number of choices of how to normalize, but we chose to normalize with respect to the sum of the intensities. Each peak intensity was divided by the sum of all peak intensities in that fraction and multiplied by 1000, so that the processed intensities could be interpreted uniformly over different fractions and samples.

Merger of fraction data into sample data: There are numerous cases where several fractions display peaks at very similar mass points, and so the question arises how to decide when two peaks in different fractions are to be considered to stem from the same protein and when they represent different proteins. In order to extract a single peak profile per sample one needs to merge the normalized peak profiles from the 20 fractions. The mass accuracy of the instrument was given to be approximately 0.1%. We chose the following heuristic when merging peaks. If the masses of two peaks are within 0.2%, we merge them and assign the new peak to have a mass of the average of the two and its intensity to be the maximum of the two peaks. The tolerance of 0.2% was intentionally chosen to be larger than the instrument accuracy to additionally smoothen the data. This scheme was applied iteratively, with subsequent new peak masses to be chosen as the weighted average of the previous peaks.

Peak alignment across samples: In order to make the peak profiles comparable across different samples, we used the same to the one used when merging fractions into samples, that is, if two peaks are within 0.2% of each other then they were considered identical and their masses were reassigned.

Peak selection: The preceding steps resulted in vectors of length 603 for each sample. Of the 603 peaks in this reduced data set, over 60% appear in only very few samples and are thus not likely to be helpful in classifying the majority of the samples. Hence, we chose to ignore any peaks that occurred in fewer than 8 samples, a step that reduced the dimension of the identifying vectors down to 229. Next, using B/W ratio (or the F-statistic as described earlier) we ordered the peaks in order to select the most important peaks for discrimination.

Classification methods: We experimented with the classification algorithms using between 3 and 15 peaks as ordered by the B/W criterion. The methods used are the linear and quadratic discrimination methods, nonparametric discrimination method using a kernel, k-nearest neighbor classification (kNN) method using the Mahalanobis distance, and linear SVMs. We chose $k = 6$ for kNN, $r = 0.5$ for the kernel method and $C = 1$, where C is the tradeoff parameter between margin maximization and misclassification error in the SVM.

Results and discussion: Leave-one-out cross-validation error rates using top four peaks for classification, when the selection of peaks was based on the entire data set, averaged around 13% indicating that most of the methods performed similar. It is encouraging that just based on four peaks these methods were able to correctly discriminate 87% of the time. When these classification procedures were applied on a set of randomized data sets the average error rate was about 50% as it ought to be. This indicates that these methods have significant discriminant power and the results obtained are not just the artifacts of the choices made in the preprocessing stages.

We were able to increase the success rate considerably when the top thirteen peaks were used. For SVM and kNN the error rate dropped down, close to 2%. However, for the quadratic discriminant method the error rate jumped to the highest level at 24%. A reason for such a high rate for the quadratic discrimination method, which requires estimation of 13×13 variance covariance matrix within each group, is due to very small sample size (only 17) in one of the two groups. When cross-validation studies are performed, care must be taken to ensure that the test set does not influence the choice of the peaks used in the classification. When peak selection was done at each iteration (on every training set), the error rates for all the methods except SVM increased considerably. It was found that SVM is least affected either by how the peak selection was made or by the small sample size.

The four most significant peaks that we used in order to generate the results occur at m/z values 28,088.9, 11,695.2, 9,481.7, and 8,712.4. The first, third, and fourth of these peaks are down-regulated in lung cancer, whereas the second is upregulated; indeed, the second peak appears only in one of the healthy samples, at a low intensity. Identifying these proteins or protein fragments, and understanding their role in lung cancer would provide credence to the data processing techniques and classification algorithms that we have employed.

4.11.2 Prostate Cancer Data

Serum samples were obtained from the Virginia prostate center tissue and body fluid bank. SELDI-TOF mass spectrometry protein profiles of serum from 82 unaffected healthy men, 77 patients diagnosed with BPH, 83 patients with organ-confined prostate cancer (PCA), and 82 patients with nonorgan-confined PCA were available, in duplicate, for the analysis. For details on sample preparation and the particular kind of chromatographic affinity chip used, see Adam et al. (2002).

Peak detection and alignment were performed with Ciphergen ProteinChip Software 3.0 with some modifications. Compilation of 326 samples providing 779 peaks in the mass range from 2 to 40 kDa were selected by the ProteinChip software for analysis. For details on sample preparation, the particular kind of chromatographic affinity chip used, and various steps involved in preprocessing of the data see Adam et al. (2002) and Qu et al. (2002).

We first chose to disregard any peaks appearing in 30 or fewer samples (i.e., $\leq 10\%$), thus preventing the classification methods from taking advantage of what are likely to be spurious peaks or data artifacts, possibly contaminants. This resulted in a reduction from 779 peaks in the original data set to 220. In order to further reduce the number of features to, say, under 25, we used the ratio of *between group sum of squares* and *within group sum of squares* (B/W ratio), for feature selection.

For classification/discrimination, as in the previous example, we used the quadratic discriminant method, nonparametric (kernel) method, Fisher's canonical (linear) discriminant method, the kNN method, and the SVMs.

We used a standard cross-validation technique and split the data randomly and repeatedly into training and test sets. The training sets consisted of randomly chosen subsets containing 90% of each class (for a total of 294 per run); the remaining 10% of the samples from each class (a total of 32) were left as test sets. Feature (peak) selection was performed in every experiment on the training set only. Repeated cross-validation runs were used to estimate the average classification accuracy as well as the standard deviation. The following conclusions are made on the basis of our experiments. Details can be found in Wagner et al. (2004).

On the basis of 100 repetitions, all the methods achieved rather comparable prediction accuracies, which ranged from 75% to 84%, with SVM performing the best. The standard deviations were rather high, which indicated that there was a wide range of observed classification accuracies over the 100 runs performed. Our results also indicated that all methods are rather sensitive to noise. Increasing the number of peaks at times deteriorates the classification accuracy, underscoring the need for high-quality feature selection procedures.

These results should be viewed in the context of what one would expect to see if the peaks considered contained no information with regard to the various phenotypes. Since there are four classes, a random classifier would be expected to achieve about 25% accuracy. In order to get a sense of the significance of these results and to attempt to rule out data artifacts, we checked the performance of the classifiers on the same data but with randomized group assignments. We generated 1000 randomized data sets (the labels of the entire data set were permuted at random) and averaged the performance of the linear SVM using 15 peaks on 10 random choices of test and training set (so that in fact 10,000 random runs were performed). The best classification accuracy average out of those 1000 runs was 34.4%, although the median classification accuracy was 24.1%. This is significantly below the 79.3% observed for SVM method using 15 peaks, and is an indication that these results are not merely due to some spurious structure in the data.

We also want to mention that when all 5 classification methods are trained using the entire data set and 15 peaks, 74% of all samples are correctly classified by all methods simultaneously. We take this high level of concordance of the classification methods as a strong indication and additional evidence that a large majority of samples are indeed well separated in this low-dimensional space, and that there is significant information content in this data set that can be used to discriminate between the four classes.

Finally, we want to mention the top masses that repeatedly appear in the peak selection list of various classifiers: 9720.0, 9655.7, 5074.2, 3896.6, 3963.2, 7819.8, 7844.0, 6949.2, 8943.1, 4079.5. Some of these masses, for example, 7819.8 and 9655.7, had also been used in previous studies (Adam et al., 2002) as being important discriminators. Note, however, that all masses are in the range of those of typical proteins. Although the identification of the underlying proteins and our understanding of their biological significance is still outstanding, we believe that the results we provide here do indeed indicate that they are good candidates for biomarkers, and that their identification can provide new insights of clinical relevance.

4.12 Remarks about Quality Control Issues

The quality control issues of the data generated by the MALDI, SELDI, and other technologies have to be addressed before the data are used for drawing conclusions. The reproducibility of the spectra can be checked by studying the effects of time, machines, operators, chips, spots, and other factors using the analysis of variance methods on the data for the same specimen sample, but produced under these different values of the factors. See Coombes et al. (2003) for a discussion of quality control issues with the spectra produced by SELDI technology in the context of samples of nipple aspirate fluid. They used control samples on two spots on each of the three ProteinChip arrays (Ciphergen Biosystems, Inc.) on four successive days to generate twenty-four SELDI spectra. Using these data the effects of spots, chips, and time can be studied on the spectra by taking the intensities at various m/z ratios as response variable. Coombes et al. (2003), however, used Mahalanobis distance in principal component space as a measure to assess the reproducibility of proteomics spectra. Also, see Reese (2006) for a detailed study on quality assessment of SELDI-TOF mass spectrometry data.

Computer programs and routines to carry out most of the analysis that we have proposed here are easily available in public (e.g., R) and professional (e.g., SAS, S+, MATLAB) softwares. In our previous works the entire scheme from Figure 4.1 was implemented using a combination of languages and tools including Perl, SAS, and Matlab. The linear support vector machine was implemented using a freely available software package SvmFu (available at: http://five-percent-nation.mit.edu/SvmFu/) and SVM^{light} software.

References

Adam, B.-L., Qu, Y., Davis, J. W., Ward, M. D. et al. (2002). Serum protein fingerprinting coupled with a pattern-matching algorithm distinguishes prostate cancer from benign prostate hyperplasia and healthy men. *Cancer Research*, 62, 3609–3614.

Aebersold, R. and Mann, M. (2003). Mass spectrometry-based proteomics. *Nature*, 422, 198–207.

Anderson, T. W. (1984). *An Introduction to Multivariate Statistical Analysis*, Second ed. John Wiley & Sons, New York.

Baggerly, K. A., Morris, J. S., Wang, J., Gold, D., Xiao, L. C., and Coombes, K. R. (2003). A comprehensive approach to the analysis of MALDI-TOF proteomics spectra from serum samples. *Proteomics*, 3, 1667–1672.

Bakhtiar, R. and Nelson, R. W. (2001). Mass spectrometry of the proteome. *Molecular Pharmacology*, 60, 405–415.

Breiman, L. (1996). Bagging predictors. *Machine Learning*, 26, 123–140.

Breiman, L. (1998). Arcing classifiers. *Annals of Statistics*, 26, 801–824.

Breiman, L. (2001). Random Forests. *Machine Learning*, 45, 5–32.

Breiman, L. (2003). Manual: Setting up, using and understanding random forests. *Technical Report, UC Berkeley Statistics Dept.*

Breiman, L., Friedman, J. H., Olshen, R. A., and Stone, C. (1984). *Classification and Regression Trees.* Chapman & Hall.

Coombes, K. R., Fritsche, Jr., H. A., Clarke, C., Chen, J.-N., Baggerly, K. A., Morris, J. S., Xiao, L.-C., Hung, M.-C., and Kuerer, H. M. (2003). Quality control and peak finding for proteomics data collected from nipple aspirate fluid by surface-enhanced laser desorption and ionization. *Clinical Chemistry*, 49, 1615–1623.

Cristianini, N. and Shawe-Taylor, J. (2000). *An Introduction to Support Vector Machines.* Cambridge University Press, Cambridge, UK.

Dudoit, S., Fridlyand, J., and Speed, T. P. (2002). Comparison of discrimination methods for the classification of tumors using gene expression data. *Journal of American Statistical Association*, 97, 77–87.

Efron, B. and Tibshirani, R. (1997). Improvements on cross-validation: The .632+ bootstrap method. *Journal of American Statistical Association*, 92, 542–560.

Freund, Y. and Schapire, R. E. (1997). A decision-theoretic generalization of on-line learning and an application to boosting. *Journal of Computer and System Sciences*, 55, 119–139.

Fung, E. T. and Enderwick, C. (2002). ProteinChip clinical proteomics: Computational challenges and solutions. *Computational Proteomics Supplement*, 32, 34–41.

Golub, T. R., Slonim, D. K., Tamayo, P., Huard, C. et al. (1999). Molecular classification of cancer: Class discovery and class prediction by gene expression monitoring. *Science*, 286, 531–537.

Hastie, T., Tibshirani, R., and Friedman, J. (2001). *The Elements of Statistical Learning: Data Mining, Inference, and Prediction*. Springer Series in Statistics, Springer Verlag, New York, NY.

Hutchens, T. W. and Yip, T. T. (1993). Techniques for producing high throughput spectometry data. *Rapid Communications in Mass Spectrometry*, 7, 576–580.

Jeffries, N. (2005). Algorithms for alignment of mass spectrometry proteomic data. *Bioinformatics*, 21, 1–8.

Joachims, T. (1999). Making large-scale SVM learning practical. In *Advances in Kernel Methods—Support Vector Learning*, B. Schölkopf, C. Burges and A. Smola (ed.), 169–172, MIT Press, Cambridge, MA.

Johnson, R. A. and Wichern, D. W. (2007). *Applied Multivariate Statistical Analysis*, Sixth edn. Prentice Hall, New Jersey.

Khattree, R. and Naik, D. N. (2000). *Multivariate Data Reduction and Discrimination with SAS Software*. SAS Press, Cary, NC and John Wiley & Sons, New York.

Lee, Y., Lin, Y., and Wahba, G. (2004). Multicategory Support Vector Machines. *Journal of the American Statistical Association*, 99, 67–81.

Lilien, R., Farid, H., and Donald, B. (2003). Probabilistic disease classification of expression-dependent proteomic data from mass spectrometry of human serum. *Journal of Computational Biology*, 10, 925–946.

Listgarten, J. and Emili, A. (2005). Statistical and computational methods for comparative proteomic profiling using liquid chromatography-tandem mass spectrometry. *Molecular and Cellular Proteomics*, 4, 419–434.

Mangasarian, O. L. (1965). Linear and nonlinear separation of patterns by linear programming. *Operations Research*, 13, 444–452.

Merchant, M. and Weinberger, S. R. (2000). Recent advancements in surface-enhanced laser desorption/ionization-time of flight (SELDI-TOF) mass spectrometry. *Electrophoresis*, 21, 1164–1167.

Qu, Y., Adam, B-L., Thornquist, M., Potter, J. D. et al. (2003). Data reduction using a discrete wavelet transform in discriminant analysis of very high dimensionality data. *Biometrics*, 59, 143–151.

Qu, Y., Adam, B-L., Yasui, Y., Ward, M. et al. (2002). Boosted decision tree analysis of SELDI mass spectral serum profiles discriminates prostate cancer from noncancer patients. *Clinical Chemistry*, 48, 1835–1843.

Reese, M. (2006). Quality assessment of SELDI mass spectrometry data. *Old Dominion University, Department of Mathematics and Statistics, MS project*.

Satten, G. A., Datta, S., Moura, H., Woolfitt, A. et al. (2004). Standardization and denoising algorithms for mass spectra to classify whole-organism bacterial specimens. *Bioinformatics*, 20, 3128–3136.

Sauve, A. C. and Speed, T. P. (2004). Normalization, baseline correction and alignment of high-throughput mass spectrometry data. *Proceedings Gensips*, 4 pages.

Srinivas, P. R., Srivastava, S., Hanash, S., and Wright, Jr., G. L. (2001). Proteomics in early detection of cancer. *Clinical Chemistry*, 47, 1901–1911.

Srinivas, P. R., Verma, M., Zhao, Y., and Srivastava, S. (2002). Proteomics for cancer biomarker discovery. *Clinical Chemistry*, 48, 1160–1169.

Vapnik, V. (1995). *The Nature of Statistical Learning Theory*. Springer, New York.

Wagner, M., Naik, D., and Pothen, A. (2003). Protocols for disease classification from mass spectrometry data. *Proteomics*, 3, 1692–1698.

Wagner, M., Naik, D., Pothen, A., Kasukurti, S., Devineni, R. R., Adam, B.-L., Semmes, O. J., and Wright, Jr., G. L. (2004). Computational protein biomarker prediction: A case study for prostate cancer. *BMC Bioinformatics*, 5, 26, (http://www.biomedcentral.com/1471–2105/5/26) Open Access.

White, C. N., Chan, D. W., and Zhang, Z. (2004). Bioinformatics strategies for proteomic profiling. *Clinical Biochemistry*, 37, 236–241.

Williams, B., Cornett, S., Crecelius, A., Caprioli, R., Dawant, B., and Bodenheimer, B. (2005). An algorithm for baseline correction of MALDI mass spectra. *Proceedings of the 43rd Annual Southeast Regional Conference*, 1, 137–142.

Wu, B., Abbott, T., Fishman, D., McMurray, W. et al. (2003). Comparison of statistical methods for classification of ovarian cancer using mass spectrometry data. *Bioinformatics*, 19, 1636–1643.

Xiong, M., Jin, W. L., Li, W., and Boerwinkle, W. (2000). Computational methods for gene expression-based tumor classification. *BioTechniques*, 29, 1264–1270.

Yasui, Y., Pepe, M., Thompson, M., Adam, B-L. et al. (2003). A data-analytic strategy for protein biomarker discovery: profiling of high-dimensional proteomic data for cancer detection. *Biostatistics*, 4, 449–463.

5

Predicting US Cancer Mortality Counts Using State Space Models

Kaushik Ghosh, Ram C. Tiwari, Eric J. Feuer, Kathleen A. Cronin, and Ahmedin Jemal

CONTENTS

5.1 Introduction

Despite the decrease in overall cancer mortality rates in the United States since the early 1990s, cancer remains a major public health problem. The total number of recorded cancer deaths in the United States continues to increase every year, due to an aging and expanding population. Cancer is currently the second leading cause of death in the United States after heart disease, and is estimated to cause approximately 563,700 fatalities in 2004—which is more than 1500 people a day (see American Cancer Society, 2004). The National Institutes of Health (NIH) predicted the overall annual costs of cancer for 2001 at $156.7 billion: $56.4 billion for direct medical costs, $15.6 billion for indirect morbidity costs, and $84.7 billion for indirect mortality costs.

With so much at stake, many private and public agencies depend on reliable and accurate prediction of future mortality and incidence counts for program planning and resource management. Examples include research investigators, health care providers, the media, and the government, among others.

For more than 40 years, the American Cancer Society (ACS) has been producing such figures annually. These predictions are based on models fitted to past data and appear in the ACS' publications *Cancer Facts & Figures* (CFF) and *CA-A Cancer Journal for Clinicians* every year and are also available from the ACS website `http://www.cancer.org`. Past data are in the form of causes of death filed by the certifying physicians and are available from the National Center for Health Statistics (NCHS). Owing to administrative and procedural delays, the latest available data are always 3 years old, necessitating 3-year-ahead projections to get an idea of the current year's numbers.

Before 1995, the ACS model was based on linear projections. Since 1995, the ACS has used a time series model with a quadratic trend and an autoregressive error component. This model was fitted on the basis of the mortality counts from 1979 through the most recently available year and then projected 3 years ahead to get predictions for the current calendar year (see Wingo et al., 1998). Since *SAS*™ `PROC FORECAST` was used to do the fitting and prediction, we shall henceforth refer to this model as the PF model. Even though the PF model was a substantial improvement over the ones in place prior to 1995, it was slow in capturing sudden and rapid changes in the trend and performed well only in cases where the trend changed gradually. To compensate for this slow adjustment, ACS made subjective modifications to the forecasts before publishing them in *Cancer Facts & Figures*. The published number for a particular year was one of five possible choices (see Wingo et al., 1998): the point prediction from `PROC FORECAST`, the 95% prediction limits and the midpoints between the forecast and the prediction limits. These choices varied among different sites and also from year to year for the same site.

Recently, the National Cancer Institute (NCI), in collaboration with ACS, developed and tested a new method of obtaining 3-year-ahead predictions of the mortality counts. After extensive testing and validation, the new method was found to be a significant improvement. It was used to obtain the prediction counts published in Cancer Statistics, 2004 (see Jemal et al., 2004), and has been decided that the method be used in all future predictions of *Cancer Facts & Figures* and *Cancer Statistics*. In this chapter, we present theoretical details and motivation for the new method. Detailed validation results based on runs on various cancer sites are published in Tiwari et al. (2004).

The proposed method uses a special form of state space model (SSM) (see Harvey, 1989, p. 100), (Harvey, 1993, p. 82) known as the local quadratic trend model (see Harvey, 1989, pp. 294–295). This model assumes that locally (i.e., at each point of time), the mortality counts have a quadratic trend, and that the coefficients randomly vary over time. The *local* nature of the new model makes it quite flexible and it is quick to adjust to rapidly changing trends in the data, when compared to models where coefficients do not vary over time. The introduction of this extra randomness, although giving rise to improved predictions, can, however, result in sharp year-to-year variations in the predicted series, especially when the observed series is not smooth. To further improve these predictions to be closer to their observed values and to control the excess variability, we introduce "tuning parameters" in the error variances. These parameters are estimated by minimizing the sum of squares of prediction errors. The resulting predictions are sometimes smoother than the "untuned" model, and are still an improvement over the fixed coefficient method.

The layout of this chapter is as follows: In Section 5.2, we describe the NCHS cancer mortality data. In Section 5.3, we review the previous methods used for cancer mortality prediction. In Sections 5.4 and 5.5, we present the SSM and examples of some applications. In Sections 5.6 and 5.7, we present a variant of the SSM and examples of its use. Finally, in Section 5.8, we discuss future research. Derivations of some relevant results are presented in the Appendix.

Throughout this chapter, we will use the notation d_t for the number of (cancer) deaths. Furthermore, we will use the notations Δ, Δ^2, and Δ^3 to denote first, second, and third differences respectively of the appropriate time series. For example, $\Delta d_t = d_t - d_{t-1}$, $\Delta^2 d_t = \Delta d_t - \Delta d_{t-1}$, and $\Delta^3 d_t = \Delta^2 d_t - \Delta^2 d_{t-1}$ respectively. The notation $\gamma(k)$ will be used to denote the autocovariance at lag k and finally, $\rho(k)$ will be used to denote the corresponding autocorrelation.

5.2 NCHS Cancer Mortality Data

Information on cancer deaths is based on causes-of-death reported by the certifying physicians on patients' death certificates filed by the states. Such data are cleaned and compiled each year by the NCHS of the Centers for Disease Control (CDC) (`http://www.cdc.gov/nchs`). Owing to the complex process of data collection, tabulation, and publication, and the sheer number of records involved, the latest data are about 3 years old when they become available to the public. For example, in January 2004, the latest available actual cancer mortality figures were from 2001.

Even though data from earlier years were readily available, ACS used data going back only upto 1979 to obtain the predicted national and state level mortality counts in *Cancer Facts & Figures*. For comparison of the proposed method with its competitors, we initially based our predictions on the same mortality data, that is, 1979 through the currently available year. However, since the SSM was found to result in improved predictions with larger data sets, we based all subsequent analyses on 1969 through the most recently available data (see Section 5.4), ensuring International Classification of Disease (ICD) compatibility. Our final data are thus of the form $(d_t)_{t=1}^{33}$, where

$t = 1$ corresponds to 1969 and $t = 33$ corresponds to 2001. We will use the notation T to denote the length of the observed data series (here, $T = 33$).

In our analyses, we have used data for several sites including most common cancers such as breast cancer, lung cancer, and prostate and rare cancers such as testicular cancer. These data were broken down by gender (if appropriate), by state, and the observed number of cases in 1998 and separate analyses were run for each such group. The data were obtained from the NCI's Surveillance, Epidemiology, and End Results (SEER) program using the SEER*Stat software (see National Cancer Institute, 2004).

5.3 Review of Previous Method

As mentioned earlier, prior to 1995, the predictions from ACS were based on linear projections. Since 1995 until 2003, ACS used the PF model consisting of a quadratic time trend to capture the long-term behavior, and an autoregressive process for the errors to capture the short-term fluctuations in cancer deaths (see Wingo et al., 1998). More precisely, the PF model fitted by ACS was of the form

$$d_t = b_0 + b_1 t + b_2 t^2 + u_t, \tag{5.1}$$

$$u_t = a_1 u_{t-1} + \cdots + a_p u_{t-p} + \epsilon_t. \tag{5.2}$$

Here $\{\epsilon_t\}$ is assumed to be an independent sequence of zero-mean, random errors with constant variance. Note also that d_t and t are the only observed values in this model.

The PF model was fitted in two sequential steps. First, the quadratic time trend model

$$d_t = b_0 + b_1 t + b_2 t^2 \tag{5.3}$$

was fitted using ordinary least squares to detrend the series.

Once the parameters b_0, b_1, and b_2 were estimated, the residuals

$$\hat{u}_t = d_t - (\hat{b}_0 + \hat{b}_1 t + \hat{b}_2 t^2) \tag{5.4}$$

were obtained. Next, the autoregressive model (5.2) was fitted to the residuals $\{\hat{u}_t\}$ to generate the overall forecasting model.

Standard Box–Jenkins-type of techniques (Box and Jenkins, 1976) were used for model fitting using SAS procedure `PROC FORECAST` in the SAS/ETS library (see SAS Institute, 1993, Chapter 9). At least seven observations are necessary to fit the previous model.

Once a model was fitted, `PROC FORECAST` was further used to obtain a 3-year-ahead prediction and the corresponding 95% prediction interval.

The point estimates of the mortality predictions obtained directly from PROC FORECAST will henceforth be denoted by PF estimates, or simply by PF. The numbers published in *Cancer Facts & Figures*, however, were not necessarily the same as PF and will henceforth be denoted by CFF. As mentioned earlier, a priori, there were five possible choices for CFF: the PF estimate itself, the upper and lower 95% prediction limits, and the midpoint between these limits and the PF estimate. CFF was a *subjective* choice among these five possibilities, made to compensate for the effects of recently changing mortality rates or large year-to-year variations in the number of cancer deaths that could not be captured through the model. For example, for 1998 estimates, lower 95% limit was selected for prostate, upper midpoint was selected for colon cancer in females and the lower midpoint was selected for stomach and cervix uteri in females (see Wingo et al., 1998).

Figure 5.1 shows the outcome of this 3-year-ahead prediction procedure using data on all cancer deaths (1979–2000). Since at least 7 years' data are necessary to fit the PF model, PF prediction starts from 1988 onwards. Note that each one of the PF estimates in Figure 5.1 are 3-year-ahead predictions and hence are based on potentially different model fits. For example, the PF estimate for 1995 is obtained by fitting the PF model to data from 1979 to 1992 and extrapolating 3 years ahead. The PF estimates in the figure are possibly different from the published CFF estimates.

In practice, the PF model was applied to individual site- and gender-specific data and the predictions were summed up to obtain an overall national level

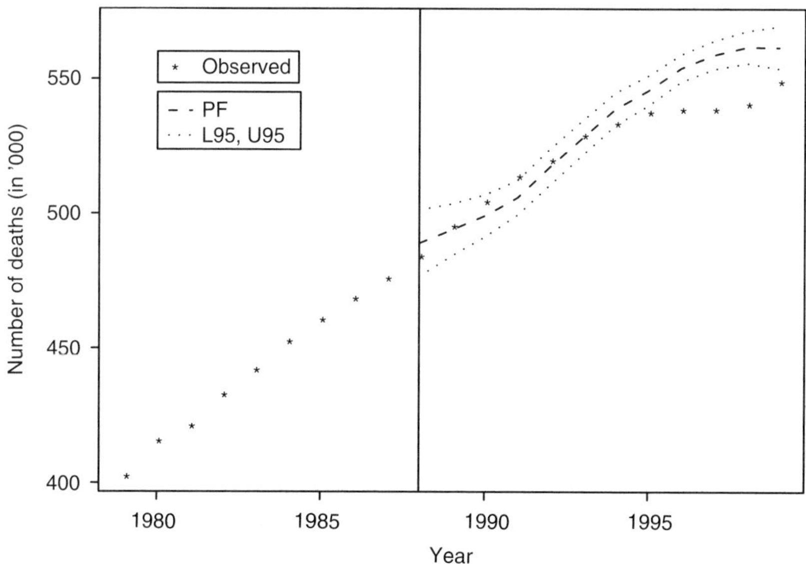

FIGURE 5.1
Three year-ahead predictions of all US cancer deaths, 1988–1999, using the current method. L95 and U95 denote the lower and upper 95% prediction limits.

prediction. State-level predictions were also obtained using the same methodology; however, the estimates for individual states were proportionally adjusted to make sure that the state totals equal the corresponding national PF estimate.

5.4 The State Space Model

We model the observed number of deaths d_t by

$$d_t = \alpha_t + \epsilon_t \quad (t = 1, 2, \ldots), \tag{5.5}$$

where α_t is the unobserved trend and ϵ_t is the (measurement) error at time t. ϵ_ts are assumed to be serially uncorrelated with mean 0 and constant variance σ_ϵ^2.

Since our goal is to predict mortality counts at the national and state levels, and the observed number of mortality counts for most of the cancers at the national and state levels are large, the errors ϵ_t are assumed to be normally distributed, similar to the assumption in Equation 5.1. One may also model the mortality counts using a Poisson or a general exponential family of distributions. However, as discussed in Section 5.8, these models require the knowledge of the population at risk and their estimates. The resulting estimates of d_t are hence biased.

Instead of using a deterministic function to model the trend (e.g., as in Equation 5.1), we propose using a trend that changes with time. This allows the model to quickly make adjustments and get closer to the observed series. Possible forms of the time-varying trend are

1. *Local level*

$$\alpha_t = \alpha_{t-1} + \eta_{1t} \quad (t = 1, 2, \ldots). \tag{5.6}$$

2. *Local linear*

$$\left.\begin{aligned} \alpha_t &= \alpha_{t-1} + \beta_{t-1} + \eta_{1t}, \\ \beta_t &= \beta_{t-1} + \eta_{2t}, \end{aligned}\right\} \quad (t = 1, 2, \ldots). \tag{5.7}$$

3. *Local quadratic*

$$\left.\begin{aligned} \alpha_t &= \alpha_{t-1} + \beta_{t-1} + \gamma_{t-1} + \eta_{1t}, \\ \beta_t &= \beta_{t-1} + 2\gamma_{t-1} + \eta_{2t}, \\ \gamma_t &= \gamma_{t-1} + \eta_{3t}, \end{aligned}\right\} \quad (t = 1, 2, \ldots). \tag{5.8}$$

Similar higher-order generalizations are also possible. In all these formulations, the errors η_{it} are assumed to be serially uncorrelated with mean 0 and variance σ_i^2. Furthermore, they are assumed to be uncorrelated with each other and with the ϵ_ts.

As the names suggest, Equation 5.6 generalizes a trend with fixed level, Equation 5.7 generalizes a trend that is linear in time, and Equation 5.8 generalizes one that is quadratic in time. Assuming $\sigma_i^2 \equiv 0$ $(i = 1, 2, 3)$ in all the above cases give the corresponding constant, linear, and quadratic trends with deterministic coefficients. In particular, Equation 5.8 becomes $d_t = \alpha_0 + \beta_0 t + \gamma_0 t^2 + \epsilon_t$ and hence is similar to the PF model, except that it uses independent errors.

Models (5.6), (5.7) or (5.8), when combined with Equation 5.5 can be written in the general state-space formulation

$$d_t = F_t \boldsymbol{\theta}_t + \epsilon_t, \tag{5.9}$$

$$\boldsymbol{\theta}_t = G_t \boldsymbol{\theta}_{t-1} + \boldsymbol{\eta}_t. \tag{5.10}$$

For example, Equation 5.5 combined with Equation 5.8 can be written as Equations 5.9 and 5.10 where

$$F_t \equiv F = (1, 0, 0), \quad \boldsymbol{\theta}_t = (\alpha_t, \beta_t, \gamma_t)', \quad G_t \equiv G = \begin{pmatrix} 1 & 1 & 1 \\ 0 & 1 & 2 \\ 0 & 0 & 1 \end{pmatrix}$$

and $\boldsymbol{\eta}_t = (\eta_{1t}, \eta_{2t}, \eta_{3t})'$. In Equation 5.8, β_t can be interpreted as the local slope and γ_t as the local curvature of the trend at time t.

Equations 5.9 and 5.10 are called the measurement and transition equations respectively of a SSM. $\boldsymbol{\theta}_t$ is a random quantity called the state vector and contains all the relevant information about the process generating the observed series at time t. ϵ_t and $\boldsymbol{\eta}_t$ are called the measurement and transition errors respectively and are assumed to have zero mean with covariance matrices given by V_t and W_t respectively. F_t, G_t, V_t, and W_t are called system matrices and may depend on unknown parameters ψ. The specification of a SSM is completed with two further assumptions: (1) the initial state vector $\boldsymbol{\theta}_0$ has mean $\boldsymbol{a}_0$ and covariance matrix C_0 and (2) the disturbances ϵ_t and $\boldsymbol{\eta}_t$ are uncorrelated with each other and with the initial state vector $\boldsymbol{\theta}_0$. For more details on SSMs, see Harvey (1989, 1993), West and Harrison (1997).

When the system matrices along with $\boldsymbol{a}_0$ and C_0 are completely known, the Kalman filter (KF) algorithm (Kalman, 1960; Kalman and Bucy, 1961) can be applied to recursively calculate the optimal estimator of the state vector at time t on the basis of all the information at time t (i.e., $d_1, \ldots, d_t$). Once the end of the series is reached, further application of the KF allows one to obtain optimal predictions of future observations. See Meinhold and Singpurwalla (1983), Harvey (1989, 1993) for more details on the KF. Various software packages are available that readily implement the filtering algorithm.

To keep our proposed model comparable to the fixed-coefficient model, we will henceforth work only with the local-quadratic model. For simplicity, we make the model time-invariant by assuming $V_t \equiv V$ and $W_t \equiv W$. Note that the transition equation in this simplified model is not stationary, since the eigenvalues of the transition matrix G are all equal to 1 (see Harvey, 1989). In the absence of any prior information on $\boldsymbol{\theta}_0$, we settle for a diffuse prior by setting $\boldsymbol{a}_0$ to be a solution of

$$\begin{pmatrix} d_1 \\ d_2 \\ d_3 \end{pmatrix} = \begin{pmatrix} 1 & 1 & 1 \\ 1 & 2 & 4 \\ 1 & 3 & 9 \end{pmatrix} \begin{pmatrix} \alpha_0 \\ \beta_0 \\ \gamma_0 \end{pmatrix}$$

and $C_0 = \kappa I$ where $\kappa = 10{,}000$ (see Harvey, 1993).

Since V and W are not known, they are estimated from the data before application of the KF. Assuming that the measurement and transition errors as well as the initial state $\boldsymbol{\theta}_0$ are normally distributed, one can write the log-likelihood of the observations as

$$\log L(\psi) = -\frac{T}{2}\log 2\pi - \frac{1}{2}\sum_{t=1}^{T}\log K_t - \frac{1}{2}\sum_{t=1}^{T}\nu_t' K_t^{-1}\nu_t,$$

where ν_t is the innovation vector, K_t is its mean squared error (MSE) (see Appendix Equations 5.13 and 5.14), and ψ is the vector of model parameters (called hyperparameters) determining V and W. The hyperparameters can then be estimated by maximizing the likelihood. Plugging in the estimated hyperparameter values give us the maximum likelihood estimators (MLEs) of V and W. This V and W can be used in conjunction with the previous information to run the KF for prediction. In addition to a point prediction, one can get a 95% prediction interval calculated from the MSE. Note that one of the byproducts of the maximum likelihood (ML) estimation is Akaike's information criterion (AIC) (see Akaike, 1974), which gives a measure of goodness of fit of the Gaussian model to the observed series. Various packages such as SsfPack2.2 (see Koopman et al., 1999) allow one to do the model estimation and prediction along the lines described above.

Note that the normal-error SSM may be inappropriate for modeling data on rare cancers or data at the state level, owing to the nonnormality of such data. In such cases, one may use Dynamic Generalized Linear Models (DGLM) described in Ferreira and Gamerman (2000), West and Harrison (1997). For our problem, we have only worked with national-level data. It was also desirable that we have a single and easily explainable method of modeling for all cancer sites. Hence, we did not explore the DGLM aspect further.

5.5 Application

Figures 5.2 and 5.3 give the 3-year-ahead predictions of mortality counts using PF model and SSM for lung cancer, in males, breast cancer in females, prostate cancer, and testicular cancer. The predictions are shown for the years

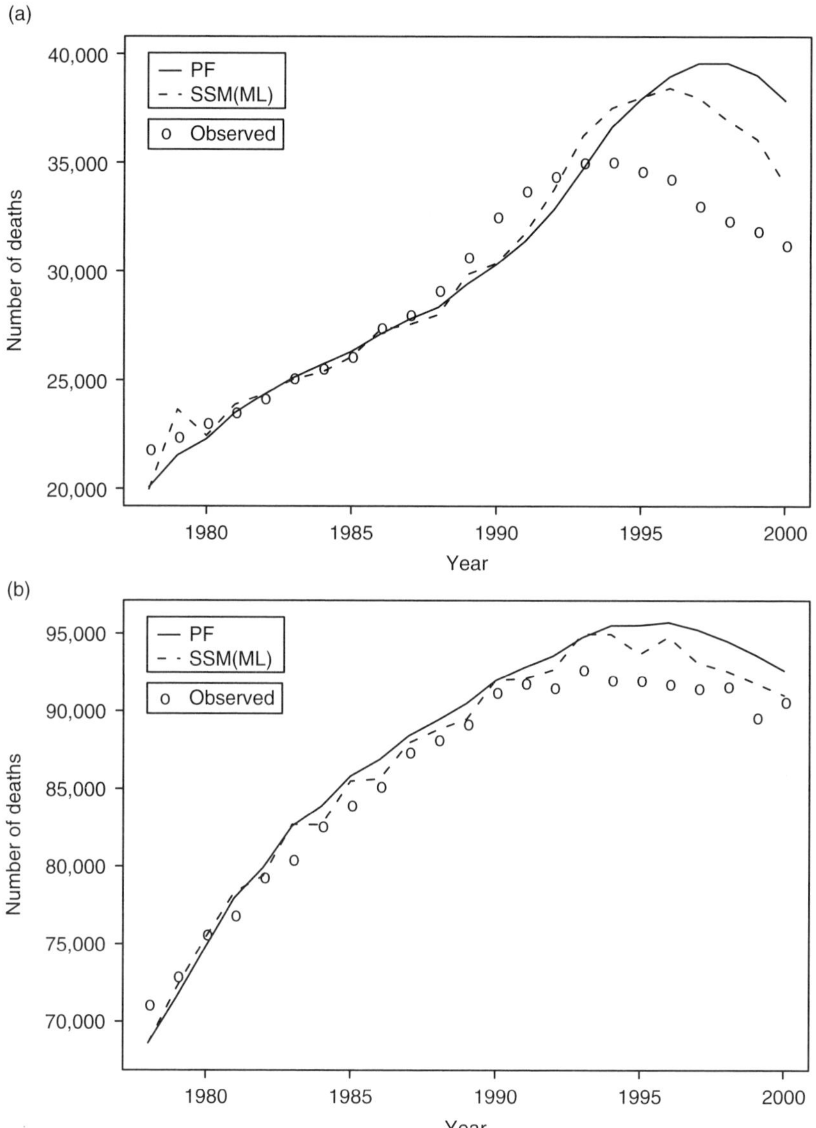

FIGURE 5.2
Three year-ahead predictions using PF and SSM (ML) methods. (a) Prostate cancer and (b) lung cancer, males.

1978–2000 along with the corresponding observed values. It is important to note that the prediction for any year is based on data from 1969 to 3 years prior to that prediction year. For example, the prediction for 1995 is based on data from 1969 to 1992, and so on. Thus, even when using the same method (say PF), the input data for the model increases as one moves along in time.

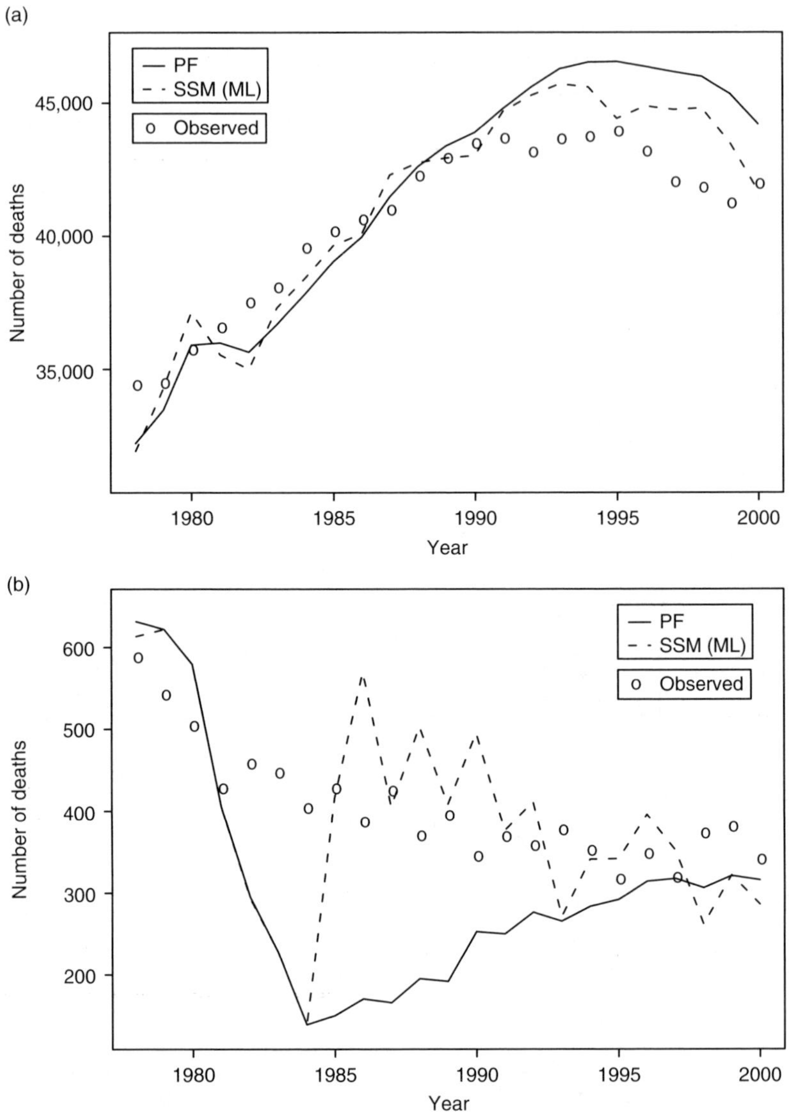

FIGURE 5.3
Three year-ahead predictions using PF and SSM (ML) methods. (a) Breast cancer, females and (b) testicular cancer.

Note that the predicted series obtained using SSM is able to adapt to changing trends in the observed series faster than that obtained using PF model. To be able to adjust fast enough, the predicted series from SSM exhibits larger jumps from year-to-year compared to that from PF model. In contrast, the fixed time trend in the PF model is more robust and the model adjusts very little in the presence of additional data. Although the resulting SSM series stays close to the observed data on the average owing to the flexibility of the model, the year-to-year fluctuations are somewhat of a disadvantage from a practitioner's point of view. The uncertainty generated by these variations make the results unusable, especially if the variations are very pronounced (e.g., as in testicular cancer). When the observed data are inherently oscillatory (e.g., in rare cancers like testicular cancer, see Figure 5.3), these oscillations get magnified by the SSM prediction process.

5.6 "Tuned" SSM

The SSM predictions are further improved (resulting in predicted values being closer to their observed values) by introducing "tuning parameters." These tuning parameters may also control excessive variability in the prediction curve.

Let V_7^* and W_7^* be the time-invariant covariance matrices used to fit $(d_t)_{t=1}^7$, using which d_{10} is predicted. Note that the suffix 7 in V_7^* and W_7^* denotes that error covariances V and W relate to the data $\{d_t\}_{t=1}^7$ but otherwise, they are time invariant. As V_7^* and W_7^* are not known, they are estimated from the data (say, using method of moments or ML approach). We denote the resulting predicted value by $\hat{d}_{10}$. Similarly, let V_8^* and W_8^* be used in fitting $(d_t)_{t=1}^8$ and then getting $\hat{d}_{11}$. Proceeding like this, we will have V_{29}^* and W_{29}^*, which are used to obtain $\hat{d}_{32}$. The proposed method works in several steps. First, it estimates $V_7^*, \ldots, V_{29}^*$ and $W_7^*, \ldots, W_{29}^*$. Then, it replaces V_t^* by $\kappa_V V_t^*$ and W_t^* by $\kappa_W W_t^*$ where κ_V and κ_W are unknown constants in the interval $(0, 1)$ and are called "tuning parameters." For a known value of (κ_V, κ_W), the "tuned" covariance matrices $\kappa_V V_t^*$ and $\kappa_W W_t^*$ can be used to refit the corresponding model and get the corresponding predicted value $\hat{d}_{t+3}$. The resulting prediction error is given by $e_{t+3}(\kappa_V, \kappa_W) = \hat{d}_{t+3} - d_{t+3}$ and is obviously a function of (κ_V, κ_W). Repeating this process for $t = 7, \ldots, 29$, we can get the corresponding prediction errors for $t = 10, \ldots, 32$. Defining the sum of squares of prediction errors by

$$\text{SSPE}(\kappa_V, \kappa_W) = \sum_{t=7}^{29} e_{t+3}^2(\kappa_V, \kappa_W),$$

the tuning parameters are estimated by minimizing $\text{SSPE}(\kappa_V, \kappa_W)$. Hence,

$$(\hat{\kappa}_V, \hat{\kappa}_W) = \arg \min_{(\kappa_V, \kappa_W)} \text{SSPE}(\kappa_V, \kappa_W).$$

In the second step, the predictions are recalculated using the "tuned" covariance matrices based on the estimated tuning parameters, that is, with $\hat{\kappa}_V V_t^*$ and $\hat{\kappa}_W W_t^*$.

As indicated earlier, the V_t and W_t values were estimated using the ML method—in this case, we used SsfPack 2.2 for this purpose. We then used these estimated values to estimate the tuning parameters κ_V and κ_W using the Nelder–Mead algorithm (Nelder and Mead, 1965) of optimization. For programming convenience, the "tuning" part of this method was written in "R" (Ihaka and Gentleman, 1996) and the routine `optim` was used for this purpose.

5.7 Results

Figures 5.4 and 5.5 compare the "tuned" and "untuned" versions of the SSM to the PF and the observed values. Note the success of the tuning parameters in bringing the predictions closer to the observed values and in some cases, dampening the oscillations. Table 5.1 gives the square root of the mean of sum of squares of relative prediction errors (RMSPE) for the years 1978–2000. Note that for prostate cancer, SSM reduces the RMSPE by 29% and introduction of tuning reduces it further by 30%. The corresponding figures for testicular cancer are 23% and 11% respectively. The smaller decrease for testicular cancer can be attributed to the fact that the observed data are more oscillatory compared to that for prostate cancer. The proposed 2-step method was extensively studied and compared to (1) PF and (2) the published predictions in *Cancer Facts & Figures*. Analysis was done for mortality data both at the state level and at the national level. In each case, separate analyses were run for each gender/site combination, wherever applicable. At the time this analysis was done, data were available only upto 1999. For the last 3 years (1997–1999), we calculated the prediction errors for each of the site/gender combinations. These prediction errors were averaged over different years for the same site, averaged over all sites for the same gender or averaged over different sites which fall into the same rarity category. The results are summarized in Tables 5.2 and 5.3. The runs on state-specific data were, however, not as favorable to SSM as their national counterpart, although, overall SSM was once again better on the average. For a more detailed description of the findings, see Tiwari et al. (2004).

As pointed out by the referee, predictions from the tuned model may still suffer from considerable jaggedness. This is due to the fact that from one year to the next, the proposed method fits a separate model and the tuning procedure only adjusts on the basis of the overall prediction error, not looking at year-to-year variation. One may, however, use smoothing techniques to reduce the amount of jaggedness, but the results may come at the cost of decreased overall accuracy. We have not investigated this aspect in our research.

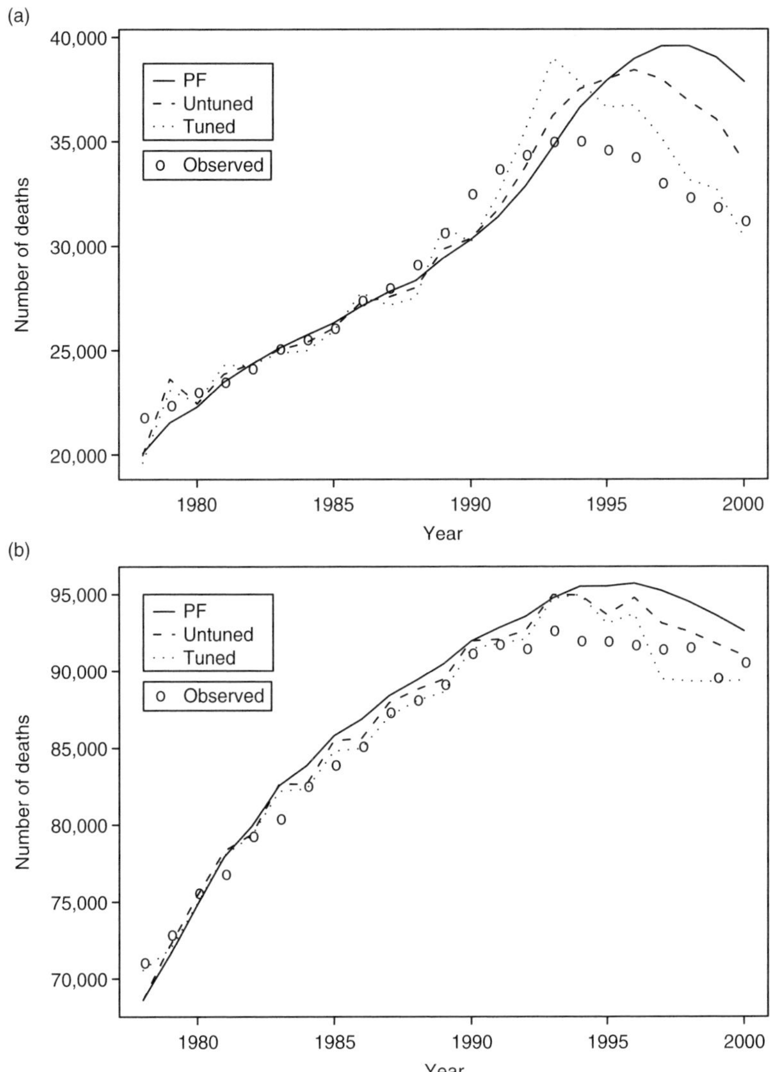

FIGURE 5.4
Comparison of 3-year-ahead predictions using PF and SSM (tuned and untuned). (a) Prostate and (b) lung cancer, males.

5.8 Conclusion

We have developed a new method for short-term projection of cancer mortality counts on the basis of SSM. This method combines the flexibility of a

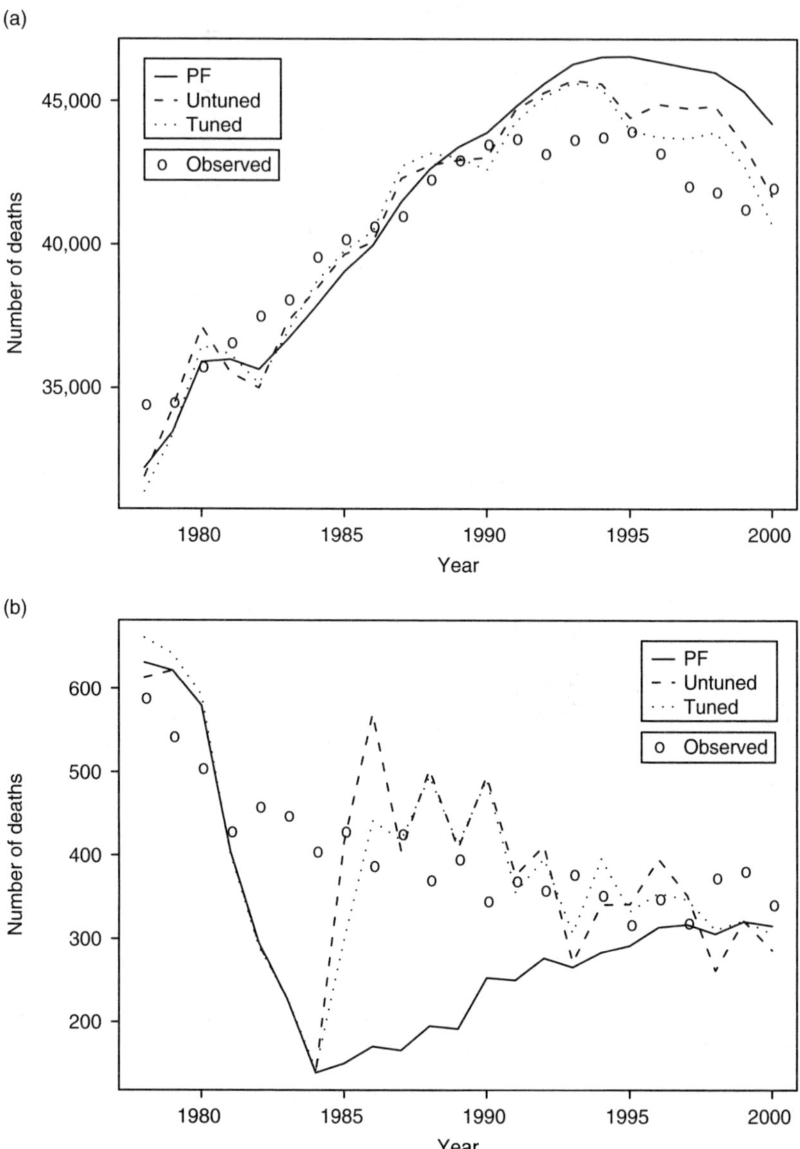

FIGURE 5.5
Comparison 3-year-ahead predictions using PF and SSM (tuned and untuned) methods. (a) Breast cancer, females and (b) testicular cancer.

local quadratic model with the smoothness of a fixed-trend autoregressive model. The resulting predictions are seen to quickly adjust to changes in the observed trends and are an improvement over the previously used fixed-trend PF model with subjective choice at the end. On the basis of extensive studies, we have found that SSM performs better on the average, but the superiority

TABLE 5.1

Root Mean Square of Relative Prediction Errors for the Years 1978–2000 for Selected Cancer Sites

	RMSPE		
Site	**PF**	**SSM (untuned)**	**SSM (tuned)**
Lung, males	0.0276	0.0192	0.0147
Prostate	0.1035	0.0741	0.0519
Breast, females	0.0542	0.0401	0.0362
Testis	0.3504	0.2714	0.2492

TABLE 5.2

PMSE Using the Two Models at Various Sites

	PMSE	
Site	**PF**	**SSM**
All malignant cancers	12.90×10^3	$\mathbf{7.38 \times 10^3}$
Colon and rectum	$\mathbf{1.37 \times 10^3}$	1.61×10^3
Liver and intrahepatic bile duct	$\mathbf{1.86 \times 10^3}$	2.68×10^3
Breast (females)	2.25×10^3	$\mathbf{1.46 \times 10^3}$
Prostate	3.07×10^3	$\mathbf{1.66 \times 10^3}$
Lung and bronchus	5.15×10^3	$\mathbf{2.06 \times 10^3}$
Cervix uteri	3.60×10^2	$\mathbf{2.69 \times 10^2}$
Non–Hodgkin lymphoma	1.04×10^3	$\mathbf{0.88 \times 10^3}$
Leukemia	$\mathbf{6.22 \times 10^2}$	6.88×10^2
Stomach	5.89×10^2	$\mathbf{5.36 \times 10^2}$

Bold numbers denote **smaller** values.

is not uniform. Starting January 2004, ACS has employed the new procedure for obtaining mortality predictions. The tables and figures appearing in Jemal et al. (2004) were also obtained using this method.

One valid criticism of the method is the use of normal measurement errors for count data. This is, however, not a practical problem when we are dealing with common cancers at the national level, but can potentially lead to erroneous predictions when dealing with rare cancers. In such cases, one may use measurement equations from the exponential family, such as a Poisson. Such DGLMs are discussed in West and Harrison (1997), Ferreira and Gamerman (2000).

Another set of improvements may be obtained by using different local polynomials for different cancer sites. For example, instead of using a local quadratic, a local linear may be sufficient for testicular cancer. Addition of the quadratic term possibly increases the variability of the predictions, as evident in Figure 5.5. Addressing this problem would require one to manually go through each cancer site and decide on the optimal model individually.

TABLE 5.3

Observed and 3-year-ahead Predictions for Selected Cancer Site/Sex Combinations for 1999, Using Data from 1969 to 1996

		Three Year Predicted Values		
Cancer Sites/Sex Combinations	**Observed**	**PF**[a]	**CFF**[b]	**SSM**[c]
Colon and rectum (Females)	28,909	27,957	**28,800**	26,589
Lung and bronchus (Females)	62,662	69,421	68,000	**62,658**
Melanoma of the skin (Females)	2,686	2,821	**2,700**	2,807
Breast (Females)	41,144	45,345	43,300	**42,746**
Ovary	13,627	14,317	14,500	**13,294**
Hodgkin lymphoma (Females)	618	715	**600**	656
Non–Hodgkin lymphoma (Females)	11,008	12,472	12,300	**12,126**
Acute lymphocytic leukemia (Females)	582	707	600	**591**
Colon and rectum (Males)	28,313	**27,831**	27,800	27,793
Lung and bronchus (Males)	89,399	93,619	90,900	**89,257**
Melanoma of the skin (Males)	4,529	4,745	**4,600**	4,760
Prostate	31,728	39,027	37,000	**32,733**
Testis	378	**321**	300	**321**
Hodgkin lymphoma (Males)	785	892	**700**	623
Non–Hodgkin lymphoma (Males)	11,794	13,692	13,400	**13,163**
Acute lymphocytic leukemia (Males)	779	866	800	**789**

Bold numbers indicate best prediction.

[a] Three-year-ahead predicted value from `PROC FORECAST`.

[b] *Cancer Facts & Figures, 1999,* using data from 1979 to 1996.

[c] Three-year-ahead predicted value from the State space Method.

Although we explored the above two issues by trying different models for different cancer sites, it was eventually decided that one method (local quadratic with normal errors) be used for the prediction of all cancer mortality counts. Such a uniform method is not only easy to implement, but is also easy to explain to the broad audience of nonstatisticians. Moreover, an advantage of the normal model is that it allows one to estimate the national counts from a joint model of the state counts.

For completeness, we would like to mention that we analyzed several sites using estimates of V and W on the basis of method of moments, and the predictions obtained were quite similar. The method of moments estimates V and W were obtained by equating the sample autocovariances of $\Delta^3 d_t$ to the corresponding population versions (see Equation 5.15). Unlike the ML method, the method of moments is easier to implement as the estimates are explicitly obtained.

Although the SSM (with tuning) has been shown to perform better in most cases, theoretical properties of the tuning parameter estimators and hence of the tuned predictors are very difficult to obtain. Hence, we only present results from various runs instead of doing a theoretical study. An alternate approach using full Bayesian analysis was also investigated by the authors. This is fully model based and is computationally more involved. The results, however, were comparable to those obtained in this paper, with a slight edge for the SSM. See Ghosh and Tiwari (2007) for details on the Bayesian method and the results.

Improved predictions can also possibly be obtained by incorporating covariates in the model being used. For example, the model can be modified as

$$\begin{aligned} d_t &= \alpha_t + \boldsymbol{\delta}'\boldsymbol{x}_t + \epsilon_t, \\ \alpha_t &= \alpha_{t-1} + \beta_{t-1} + \gamma_{t-1} + \eta_{1,t}, \\ \beta_t &= \beta_{t-1} + 2\gamma_{t-1} + \eta_{2,t}, \\ \gamma_t &= \gamma_{t-1} + \eta_{3,t}, \end{aligned}$$

where $\boldsymbol{x}_t$ is the vector of covariates. Variables such as age-group and gender are natural choices for covariates and incorporation of them is expected to improve predictions. For lung cancer, we could include the percentage of smokers in the population as one of the components of $\boldsymbol{x}_t$ (note that such modifications are not possible with the PF method). Smoking figures are not available for all the years 1979–1998 and hence would possibly need to be interpolated/extrapolated to be used in this case. Moreover, the effect of smoking on cancer is not immediate but is with a pronounced lag. It is an open question how we would incorporate such lagged effect in our model.

By using preliminary mortality data (which become available earlier than the final numbers), the number of steps to predict ahead can be reduced to two. This would result in improved predictions and definitely narrower intervals.

The noisy prediction in rare cancer sites (such as testicular cancer) can be corrected by using a non-normal model—for example, a Poisson model in the set-up of West et al. (1985), West and Harrison (1997), Ferreira and Gamerman (2000). A model of the form

$$\begin{aligned} d_t &\sim \text{Poisson}(n_t g(\lambda_t)), \\ g(\lambda_t) &= \log\left(\frac{\lambda_t}{1-\lambda_t}\right) = x_t'\beta_t \end{aligned}$$

was tried in particular. This, however, required knowledge of n_t for the past years and also for the future year for which prediction is desired. As population census is conducted every 10 years, most of the values of n_t are based on interpolation and extrapolation. The variability in the n_t values translated to increased uncertainty in the estimated mortality counts. We thus found this method to be impractical for public use.

It is worth noting that the proposed SSM can be used to generate predictions for either counts, crude rates, or age-adjusted rates of mortality. For example, suppose one denotes by r_t the age-adjusted mortality rate at time t, w_i to be the adjustment factor for age-group i, d_{it} to be the mortality count in age-group i at time t and n_{it} to be the population at risk (of death) in age-group i at time t.

Then, we have

$$r_t = \sum_{i=1}^{I} w_i \frac{d_{it}}{n_{it}}, \quad t = 1, \ldots, T.$$

One may use the r_t values as input to the SSM routine to get the desired prediction of r_{T+3}. Extensive runs of the SSM procedure on age-adjusted rates for various cancer sites have been recently conducted and the results look quite promising.

Finally, another approach would be to incorporate cancer incidence information (available from SEER) along with the cancer survival rates and use a back-calculation method for mortality prediction (see Brookmeyer and Gail, 1988; Bacchetti et al., 1993; Mezzetti and Robertson, 1999). This is the subject of ongoing research on incidence prediction and results will be published in a future article.

5.A Appendix

5.A.1 The Kalman Filter

Let $\hat{\boldsymbol{\theta}}_t$ be the optimal estimator of the state vector $\boldsymbol{\theta}_t$ based on $d_1, \ldots, d_t$ and C_t be its $m \times m$ MSE matrix. We then have

$$C_t = E[(\boldsymbol{\theta}_t - \hat{\boldsymbol{\theta}}_t)(\boldsymbol{\theta}_t - \hat{\boldsymbol{\theta}}_t)'].$$

Suppose we are at time t and $\hat{\boldsymbol{\theta}}_t$ and C_t are available. Then, based on the data upto and including time t, the optimal estimator of $\boldsymbol{\theta}_{t+1}$ is

$$\hat{\boldsymbol{\theta}}_{t+1|t} = G_{t+1}\hat{\boldsymbol{\theta}}_t, \tag{5.11}$$

and the updated MSE matrix for $\hat{\boldsymbol{\theta}}_{t+1|t}$ is

$$C_{t+1|t} = G_{t+1}C_tG'_{t+1} + W_{t+1}. \tag{5.12}$$

Equations 5.11 and 5.12 are called the *prediction equations*. The corresponding estimator of d_{t+1}, called the predicted value, $\tilde{d}_{t+1|t}$ is then

$$\tilde{d}_{t+1|t} = F_{t+1}\hat{\boldsymbol{\theta}}_{t+1|t}.$$

Let the prediction error of d_{t+1} based on data upto t (also called the *innovation* vector), be denoted by ν_{t+1}. Then,

$$\nu_{t+1} = d_{t+1} - \tilde{d}_{t+1|t} = F_{t+1}(\boldsymbol{\theta}_{t+1} - \hat{\boldsymbol{\theta}}_{t+1|t}) + \epsilon_{t+1}, \tag{5.13}$$

and its MSE is given by

$$K_{t+1} = F_{t+1}C_{t+1|t}F'_{t+1} + V_{t+1}. \tag{5.14}$$

Once a new observation d_{t+1} becomes available, the estimator $\hat{\boldsymbol{\theta}}_{t+1|t}$ of the state vector $\boldsymbol{\theta}_{t+1}$ and its corresponding MSE can be updated. The *updating equations*, known as KF updating equations, are given by

$$\hat{\boldsymbol{\theta}}_{t+1} = \hat{\boldsymbol{\theta}}_{t+1|t} + C_{t+1|t}F'_{t+1}K_{t+1}^{-1}(d_{t+1} - F_{t+1}\hat{\boldsymbol{\theta}}_{t+1|t})$$

and

$$C_{t+1} = C_{t+1|t} - C_{t+1|t}F'_{t+1}K_{t+1}^{-1}F_{t+1}C_{t+1|t}.$$

Starting with initial conditions $\hat{\boldsymbol{\theta}}_0$ and C_0, the above equations are used recursively for $t = 0, 1, \ldots, T-1$ to finally get $\hat{\boldsymbol{\theta}}_T$, which contains all the information for predicting future values of $d_t, t > T$. The l-step ahead estimator of $\boldsymbol{\theta}_{T+l}$ given information upto T is then

$$\hat{\boldsymbol{\theta}}_{T+l|T} = G_{T+l}\hat{\boldsymbol{\theta}}_{T+l-1|T}, \quad l = 1, 2, \ldots$$

with $\hat{\boldsymbol{\theta}}_{T|T} = \hat{\boldsymbol{\theta}}_T$. The associated MSE matrix is given by

$$C_{T+l|T} = G_{T+l}C_{T+l-1|T}G'_{T+l} + W_{T+l}, \quad l = 1, 2, \ldots$$

with $C_{T|T} = C_T$.

The l-step predictor of d_{T+l} given $d_1, \ldots, d_T$ is

$$\tilde{d}_{T+l|T} = F_{T+l}\hat{\boldsymbol{\theta}}_{T+l|T}$$

with its prediction MSE being

$$\text{MSE}(\tilde{d}_{T+l|T}) = F_{T+l}C_{T+l|T}F'_{T+l|T} + V_{T+l}.$$

See Harvey (1993) for derivations of the results in this section and further details.

5.A.2 Autocovariances of the Local Quadratic Trend Model

$$\begin{aligned}
\Delta d_t &= \beta_{t-1} + \gamma_{t-1} + \eta_{1t} + \Delta\epsilon_t, \\
\Delta^2 d_t &= 2\gamma_{t-2} + \eta_{2,t-1} + \eta_{3,t-1} + \Delta\eta_{1,t} + \Delta^2\epsilon_t, \\
\Delta^3 d_t &= 2\eta_{3,t-2} + \Delta\eta_{2,t-1} + \Delta\eta_{3,t-1} + \Delta^2\eta_{1,t} + \Delta^3\epsilon_t.
\end{aligned}$$

Hence, autocovariances of $\Delta^3 d_t$ are given by

$$\left.\begin{aligned}\gamma(0) &= 20\sigma_\epsilon^2 + 6\sigma_1^2 + 2\sigma_2^2 + 2\sigma_3^2,\\ \gamma(1) &= -15\sigma_\epsilon^2 - 4\sigma_1^2 - \sigma_2^2 + \sigma_3^2,\\ \gamma(2) &= 6\sigma_\epsilon^2 + \sigma_1^2,\\ \gamma(3) &= -\sigma_\epsilon^2,\\ \gamma(k) &= 0, \quad k \geq 4.\end{aligned}\right\} \tag{5.15}$$

Acknowledgments

This work has been partly supported by NIH contracts 263-MQ-109670 and 263-MQ-116978.

References

Akaike, H. (1974), "A New Look at the Statistical Model Identification," *IEEE Transactions on Automatic Control*, AC-19, 716–723.

American Cancer Society. (2004), *Cancer Facts & Figures, 2004*, American Cancer Society, Atlanta, GA.

Bacchetti, P., Segal, M. R., and Jewell, N. P. (1993), "Backcalculation of HIV Infection Rates," *Statistical Science*, 8, 82–119.

Box, G. E. P. and Jenkins, G. M. (1976), *Time Series Analysis: Forecasting and Control*, 2nd ed. Holden-Day, San Francisco, CA.

Brookmeyer, R. and Gail, M. H. (1988), "A Method for Obtaining Short-term Projections and Lower Bounds on the AIDS Epidemic," *Journal of the American Statistical Association*, 83, 301–308.

Ferreira, M. A. R. and Gamerman, D. (2000), "Dynamic Generalized Linear Models," in *Generalized Linear Models: A Bayesian Perspective*, Dey, D. K., Ghosh, S. K., and Mallick, B. K. (eds.), Marcel Dekker, New York, pp. 57–72.

Ghosh, K. and Tiwari, R. C. (2007), "Prediction of U.S. Cancer Mortality Counts Using Semiparametric Bayesian Techniques," *Journal of the American Statistical Association*, 102, 7–15.

Harvey, A. C. (1989), *Forecasting, Structural Time Series Models and the Kalman Filter*, Cambridge, UK: Cambridge University Press.

——. (1993), *Time Series Models*, 2nd ed. Cambridge, MA: The MIT Press.

Ihaka, R. and Gentleman, R. (1996), "R: A Language for Data Analysis and Graphics," *Journal of Computational and Graphical Statistics*, 5, 299–314.

Jemal, A., Tiwari, R. C., Murray, T., Ghafoor, A., Samuels, A., Ward, E., Feuer, E. J., and Thun, M. J. (2004), "Cancer Statistics, 2004," *CA: A Cancer Journal for Clinicians*, 54, 8–29.

Kalman, R. E. (1960), "A New Approach to Linear Filtering and Prediction Problems," *Journal of Basic Engineering, Transactions ASME, Series D*, 82, 35–45.

Kalman, R. E. and Bucy, R. S. (1961), "New Results in Linear Filtering and Prediction Theory," *Journal of Basic Engineering, Transactions ASME, Series D*, 83, 95–108.

Koopman, S. J., Shephard, N., and Doornik, J. A. (1999), "Statistical Algorithms for Models in State Space Using Ssfpack2.2," *Econometrics Journal*, 2, 113–166.

Meinhold, R. J. and Singpurwalla, N. D. (1983), "Understanding the Kalman Filter," *American Statistician*, 37, 123–127.

Mezzetti, M. and Robertson, C. (1999), "A Hierarchical Bayesian Approach to Age-specific Back-calculation of Cancer Incidence Rates," *Statistics in Medicine*, 18, 919–933.

National Cancer Institute. (2004), *Seer*Stat software, version 5.1.14*, Surveillance Research Program, National Cancer Institute, Bethesda, MD, `http://seer.cancer.gov/seerstat`.

Nelder, J. A. and Mead, R. (1965), "A Simplex Algorithm for Function Minimization," *Computer Journal*, 7, 308–313.

SAS Institute. (1993), *SAS/ETS User's Guide*, 2nd ed., SAS Press, Cary, NC.

Tiwari, R. C., Ghosh, K., Jemal, A., Hachey, M., Ward, E., Thun, M. J., and Feuer, E. J. (2004), "A New Method of Predicting US and State-level Cancer Mortality Counts for the Current Calendar Year," *CA: A Cancer Journal for Clinicians*, 54, 30–40.

West, M. and Harrison, J. (1997), *Bayesian Forecasting and Dynamic Models*, 2nd ed., Springer-Verlag, New York.

West, M., Harrison, J., and Migon, H. S. (1985), "Dynamic Generalized Linear Models and Bayesian Forecasting (with discussion)," *Journal of the American Statistical Association*, 80, 73–96.

Wingo, P., Landis, S., Parker, S., Bolden, S., and Heath, C. (1998), "Using Cancer Registry and Vital Statistics Data to Estimate the Number of New Cancer Cases and Deaths in the United States for the Upcoming Year," *Journal of Registry Management*, 25, 43–51.

6

Analyzing Multiple Failure Time Data Using SAS® Software

Joseph C. Gardiner, Lin Liu, and Zhehui Luo

CONTENTS

6.1 Introduction

Survival models have found numerous applications in many disciplines. Beginning with their use in analyses of *survival times* in clinical and health related studies and failure times of machine components in industrial engineering, these models have now found application in several other fields, including demography (e.g., time intervals between successive child births, duration of marriages), criminology (e.g., studies of recidivism), and labor economics (e.g., spells of unemployment, duration of strikes, time to retirement). The terms *duration analysis, event-history analysis, failure time analysis, reliability analysis,* and *transition analysis* refer essentially to the same collection of techniques although the emphases in certain modeling aspects would differ across disciplines.

The main outcome in a survival model is a time T to some well-defined event (such as death) measured from some time origin ($t = 0$). Often interest lies in assessing the impact of measured covariates $\mathbf{z}$ on the survival distribution $S(t \mid \mathbf{z}) = P[T > t \mid \mathbf{z}]$, $t \geq 0$, on derivative measures such the mean survival $E(T \mid \mathbf{z})$ or the median survival and other percentiles. Several regression-type models, such as the *accelerated failure time model* and the *proportional hazards model* have been extensively developed for the single event survival data. However, in many applications in epidemiology, medicine, sociology, and econometrics, individuals can experience several events over time. For example, in cancer studies patients often undergo periods of remission and relapse before succumbing to death. The events "remission" and "relapse" have different clinical characteristics and the event history path from initiation of cancer treatment to death could involve several periods of remission and relapse. In labor economics, workers may experience several spells of unemployment interspersed with periods of employment. Here the events "employed," "unemployed" are repeated events. Studies of recidivism in criminology concern the sequential dates of arrest of a parolee since release from prison. In nosocomial investigations multiple times of the infection following surgery may occur.

Multiple failures times also occur in the context of clustered data. In the diabetic retinopathy study, a pair of failure times was potentially observed representing the time to blindness in the right eye and left eye of the patient. One eye was treated with laser photocoagulation while the other eye served as control. In veterinary studies, a natural clustering occurs when the unit of analysis is the litter, and failure times of individual animals must be analyzed jointly.

A convenient representation of an individual's event history is a finite-state stochastic process in continuous-time $X = \{X(t) : t \geq 0\}$ where $X(t)$ is the state occupied at time t (Gardiner et al., 2006). The state space $E = \{1, 2, \ldots, m\}$ labels distinct states (health conditions or events) that may be occupied over time. Hence, the multiple durations are the sequence of sojourns spent in states. Typically, E comprises several transient states such as "remission" and "relapse," or "well" and "ill," and one or more absorbing states (e.g., "dead"). A transient state is one which if entered will be exited after a finite sojourn, while an absorbing state is never left once entered. The analog of survival time is the time to absorption from a transient state. Survival models have just two states, "alive" or "dead" and therefore only a single possible transition from "alive" to "dead." In general interest lies in the transition probabilities,

$P_{hj}(s,t) = P[X(t) = j \mid X(s) = h]$, $s \leq t$, $h, j \in \mathrm{E}$ and how these probabilities are affected by covariates $\mathbf{z}(0)$ known at $t = 0$ or by the entire past history of time-dependent covariates $\{\mathbf{z}(u) : u < s\}$.

We can also describe an individual's course of events as follows. At time $T_0 = 0$ the initial state is X_0. After a duration T_1 the patient exits the initial state making a transition to state X_1. The next transition occurs at time T_2 with passage to state X_2, and so on. This defines a sequence $\{(X_n, T_n) : n \geq 0\}$ of states and transition times. If at the n-th transition an absorbing state is entered then $X_{n+1} = X_n$ and $T_{n+1} = \infty$. The sojourn times or durations in the transient states are $T_1, T_2 - T_1, T_3 - T_2, \ldots$. The two descriptions are equivalent. From the process X, we can define $\{(X_n, T_n) : n \geq 0\}$ using the forward recurrence time $W(t) = \inf\{s \geq 0 : X(t+s) \neq X(t)\}$, that is, the first time after t that a patient leaves the state $X(t)$. If the state is absorbing, we set $W(t) = \infty$. Define inductively, $T_0 = 0, X_0 = X(0)$ and $T_{n+1} = T_n + W(T_n), n \geq 0$. If $W(T_n) < \infty$, then $X_{n+1} = X(T_{n+1})$; otherwise $X_{n+1} = X_n$. Therefore, viewed at time t, the next transition out of the present state $X(t)$ will occur at time $t + W(t)$ and the time elapsed since the last transition is $L(t) = \inf\{s \geq 0 \colon X(t-s) \neq X(t)\}$, called the backward recurrence time or the current duration. The time interval $[t - L(t), t + W(t)]$ captures the current sojourn.

Some simplifying assumptions are needed to make the mathematical development tractable. Assuming X is a Markov process restricts the dependency of (X_n, T_n) on past information. For instance, given the states the sojourn times are independent with the conditional distribution of the present sojourn $T_{n+1} - T_n$ depending only on the present state of occupation X_n and destination state X_{n+1}. It also entails that the states constitute a Markov chain. For purposes of modeling covariate effects, we may consider the natural extensions of the single event survival models such as the *multiplicative intensity model* (Andersen et al., 1993).

The purpose of this chapter is to provide an overview of the application of multistate models in analyzing multiple failure times. We consider models that accommodate covariates, both fixed and time-dependent. Section 6.2 describes the salient features of a multistate model and how covariates are incorporated into the model. The main application in Section 6.3 is the use of SAS software in estimating a 3-state wellness–illness–death model with both fixed and time-dependent covariates. If the terminal event (death) is the focus of interest, the time of the intermediate event (illness) might be viewed as a time-dependent covariate, or the occurrences of the two events illness and death could be analyzed jointly. We provide the computational details for these two approaches using the SAS PHREG procedure taking advantage of some recent enhancements.

The data set used for illustration is a sample of 137 leukemia patients who underwent bone marrow transplantation described in Klein and Moeschberger (1997) where several features of this data set are extensively discussed. Several articles have addressed estimation of probabilities of events associated with a patient's disease progression (Klein et al., 2000; Keiding et al., 2001; Klein and Shu, 2002). Similar models are used to study mortality in patients with liver cirrhosis where bleeding from the esophageal varices is an intermediate event that may affect prognosis (Andersen et al., 2000). Gardiner et al. (2006) apply a 3-state model to examine episodes of normal and impaired physical function in cancer patients from the initiation of treatments and use this framework to estimate costs of treatment. However, in this example multiple transitions between the normal and impaired states are possible in patients before the terminal event (death). In all examples, noninformative censoring of event times can occur because the observational period might not capture the full event histories of all patients.

6.2 Multistate Model

Let $X = \{X(t) : t \geq 0\}$ denote a stochastic process in continuous-time with finite state space $E = \{1, 2, \ldots, m\}$. The process is called a *nonhomogeneous Markov process* if for all $0 \leq s \leq t$ and $h, j \in E$, $P[X(t) = j \mid X(s) = h, X(u) : u < s] = P[X(t) = j \mid X(s) = h] = P_{hj}(s,t)$. The Markov assumption restricts the future development of X given the past to only the most recent past. The internal history $\{\mathcal{F}_t : t \geq 0\}$ is generated from the σ-algebra $\mathcal{F}_t = \sigma\{X(u) : u \leq t\}$ that is usually augmented by information on covariates measured at $t = 0$.

6.2.1 Transition Probabilities and Intensities

Associated with the transition probabilities $P_{hj}(s,t)$ are the transition intensities, $\alpha_{hj}(t) = \lim_{\Delta t \downarrow 0} P_{hj}(t, t+\Delta t)/\Delta t, h \neq j$, with $\alpha_{hh} = -\sum_{j \neq h} \alpha_{hj}$. The hazard rate for a sojourn in progress in state h at time t is $-\alpha_{hh}(t)$. Given that a transition out of state h occurs at time t, this transition is to state j with probability $\alpha_{hj}(t)/(-\alpha_{hh}(t))$. Let $\mathbf{P} = \{P_{hj} : h, j \in \mathrm{E}\}$, $\boldsymbol{\alpha} = \{\alpha_{hj} : h, j \in \mathrm{E}\}$ be matrices and $\mathbf{I}$ be the identity matrix. Just as the hazard rate determines the survival distribution, the transition intensities determine the transition probabilities by the product–integral relationship $\mathbf{P}(s,t) = \prod_{s<u\leq t}(\mathbf{I} + \boldsymbol{\alpha}(u)du)$. In computations from event history data, this is computed as a product of matrices. The $\mathbf{P}(s,t)$ also satisfy the differential equation, $\mathbf{P}'(s,t) = \mathbf{P}(s,t)\boldsymbol{\alpha}(t)$ subject to $\mathbf{P}(s,s) = \mathbf{I}$ where the prime denotes differentiation with respect to t. When $\boldsymbol{\alpha}$ is a continuous function the formal solution is $\mathbf{P}(s,t) = \exp(\int_s^t \boldsymbol{\alpha}(u)du)$ where $\exp(\mathbf{A})$ is the matrix exponential of the square matrix $\mathbf{A}$ (Golub and Van Loan, 1996).

Several special cases of Markov processes are obtained by restricting the dependence of the intensities $\alpha_{hj}(t)$ on t. When α_{hj} are constants the process X is a *homogeneous Markov process*. Durations are then exponentially distributed and conditionally independent given the states. When $\alpha_{hj}(t) = \alpha_{hj}(L(t))$ depends only on elapsed time $d = L(t)$ since entering the state h, the process is called a (*homogeneous*) *semi-Markov process*. In general semi-Markov processes there is explicit dependence of α_{hj} on two time scales, chronological time t and the elapsed time $d = L(t)$. If the bivariate process $\{(X(t), L(t)) : t \geq 0\}$ has the Markov property, the transition probabilities will have four time arguments, $P_{hj}(s, t, u, v) = P[X(t) = j, L(t) \leq v \mid X(s) = h, L(s) = u]$. In practice one tries to maintain the original Markov structure of X and model the duration dependence through time-varying covariates.

Associated with X is a counting process $N_{hj}(t)$ which denotes the number of direct transitions from state h to j in the time interval $[0, t]$, $N_{hj}(t) = \#\{s \leq t : X(s-) = h, X(s) = j\}, h \neq j$. The cumulative information revealed up to time t is the σ-algebra $\mathcal{F}_t$ generated by $\{N_{hj}(s), s \leq t, h \neq j, h, j \in \mathrm{E}\}$ and $X(0)$ (including covariate information known at $t = 0$). The indicator function $Y_h(t) = [X(t-) = h]$ denotes whether the process is in state h just before time t. Then with respect to the filtration $\{\mathcal{F}_t : t \geq 0\}$, the counting processes $\{N_{hj}, h \neq j, h, j \in \mathrm{E}\}$ have random intensity processes $\{\lambda_{hj}, h \neq j, h, j \in \mathrm{E}\}$ where $\lambda_{hj}(t) = \alpha_{hj}(t) Y_h(t)$. Moreover, $M_{hj}(t) = N_{hj}(t) - \int_0^t Y_h(u)\alpha_{hj}(u)du$, $h \neq j, h, j \in E$ are zero-mean local square-integrable martingales.

If observation of X is ceased after some random time U, independent of X, we denote the censored process by $N_{hj}(t \wedge U)$ and the state indicator by $Y_h(t) = [X(t-) = h,\ U \geq t]$. Then, with respect to the filtration generated by $\{N_{hj}(s \wedge U), Y_h(s) : s \leq t, h \neq j, h, j \in \mathrm{E}\}$ the aforementioned martingale property still obtains. In the sequel, we will assume that censoring has been accommodated in this way.

6.2.2 Regression Models

To incorporate heterogeneity across patients we let the transition intensities depend on a covariate vector $\mathbf{z}(t)$ through a Cox-regression model $\alpha_{hj}(t, \mathbf{z}(t)) = \alpha_{hj0}(t) \exp(\beta_{hj}' \mathbf{z}(t))$ where $\alpha_{hj0}(t)$ is an unknown baseline intensity and the regression coefficients β_{hj} are specific to the transition $h \to j$. We can recast this model as $\alpha_{hj}(t, \mathbf{z}(t)) = \alpha_{hj0}(t) \exp(\beta' \mathbf{z}_{hj}(t))$ in terms of a type-specific $p \times 1$ covariate vector $\mathbf{z}_{hj}$ computed from $\mathbf{z}$ and an associated composite regression vector β.

The covariates generate their own history $\mathcal{G}_t = \sigma\{\mathbf{z}(u) : u \leq t\}$ and therefore the observed history is $\mathcal{H}_t = \mathcal{F}_t \vee \mathcal{G}_t$—the minimum σ-algebra generated by $\mathcal{F}_t$ and $\mathcal{G}_t$. Informally, our definition of transition intensity is then $\alpha_{hj}(t, \mathbf{z}(t)) = \lim_{\Delta t \downarrow 0} P[X(t + \Delta t) = j \mid X(t) = h, \mathcal{G}_t]/\Delta t$, $h \neq j$. To maintain the martingale property on $M_{hj}(t) = N_{hj}(t) - \int_0^t Y_h(u)\alpha_{hj0}(u) \exp(\beta' \mathbf{z}_{hj}(u))du$ we assume that $\mathbf{z}(t)$ is predictable (with respect to $\{\mathcal{F}_t : t \geq 0\}$). In words, this essentially means that the values of all covariates at t are known just before t, being influenced only by information from the strict past—that is, $\mathbf{z}(t)$ is $\mathcal{F}_{t-}$-measurable.

Time-dependent covariates have implications for inference (Andersen, 1986; Andersen and Keiding, 2002). A partial likelihood function formed by conditioning on the strict past $\mathcal{H}_{t-}$ is sufficient for estimating parameter β. However, estimating transition probabilities $\mathbf{P}(s, t \mid \mathbf{z}(t))$ is no longer feasible without additional knowledge of the evolution of $\mathbf{z}(t)$. An assumption of exogeneity of $\mathbf{z}(t)$ with respect to the underlying event times allows one to interpret $\alpha_{hj}(t, \mathbf{z}(t))$. See Heckman and Singer (1985), Lancaster (1990) and van den Berg (2001) for a discussion of exogeneity. With a fixed covariate profile $\mathbf{z}_0$, we can estimate $\mathbf{P}(s, t \mid \mathbf{z}_0)$ following estimation of β from Cox regression model and the integrated baseline intensities $A_{hj0}(t) = \int_0^t \alpha_{hj0}(u)du$. For most applications in biomedical studies, modeling with multiple states is still feasible when the time-dependent covariates are discrete variables. However, there are situations in which joint modeling of the continuous covariate observations and the underlying event-history processes $\{N_{hj}(t), h \neq j, t \geq 0\}$ would be necessary (Hogan and Laird, 1997; Wulfsohn and Tsiatis, 1997; Henderson et al., 2000).

6.2.3 Estimation of Transition Probabilities

Andersen et al. (1993) pioneered an elegant asymptotic theory for estimators of β, $A_{hj}(t \mid \mathbf{z}_0)$ and $P_{hj}(s, t \mid \mathbf{z}_0)$ where $\mathbf{z}_0$ denotes a fixed covariate profile. For each of n patients in a study we observe processes of the type described. For the i-th patient the basic covariate vector is $\mathbf{z}_i(t)$, the initial state $X_i(0)$, the state indicator $Y_{hi}(t) = [X_i(t-) = h, U_i \geq t]$ and $N_{hji}(t) = \#\{s \leq t \wedge U_i : X_i(s-) = h, X_i(s) = j\}, h \neq j$. Conditional on $\{\mathbf{z}_i, X_i(0) \colon 1 \leq i \leq n\}$ assume the processes $\{X_i(t) \colon t \in \mathcal{T}\}$ are independent and that the model $\alpha_{hj}(t, \mathbf{z}(t)) = \alpha_{hj0}(t) \exp(\beta' \mathbf{z}_{hj}(t))$ holds for each individual with the same baseline intensities. Let $N_{hj}(t) = \sum_{i=1}^n N_{hji}(t)$ and $Y_h(t) = \sum_{i=1}^n Y_{hi}(t)$ be the aggregated processes over the sample.

An estimator $\hat{\beta}$ of β is derived by maximizing the generalized Cox partial likelihood given by

$$\prod_{i=1}^{n} \prod_{t} \prod_{h \neq j} \left\{ \frac{Y_{hi}(t) \exp(\beta' \mathbf{z}_{hji}(t))}{\sum_{k=1}^{n} Y_{hk}(t) \exp(\beta' \mathbf{z}_{hjk}(t))} \right\}^{\Delta N_{hji}(t)}.$$

The integrated baseline intensities $A_{hj0}(t)$ are estimated by

$$\hat{A}_{hj0}(t) = \int_0^t \frac{[Y_h(u) > 0]}{S_{hj}^{(0)}(\hat{\beta}, u)} dN_{hj}(u), \quad h \neq j$$

where $S_{hj}^{(0)}(\hat{\beta}, t) = \sum_{i=1}^n Y_{hi}(t) \exp(\hat{\beta}' \mathbf{z}_{hji}(t))$. Then we get $\hat{A}_{hj}(t \mid \mathbf{z}_0) = \hat{A}_{hj0}(t, \hat{\beta}) \exp(\hat{\beta}' \mathbf{z}_{hj0})$, $h \neq j$ and $\hat{A}_{hh}(t \mid \mathbf{z}_0) = -\sum_{j \neq h} \hat{A}_{hj}(t|\mathbf{z}_0)$. The Aalen–Johansen estimator of the transition probabilities is $\hat{\mathbf{P}}(s, t \mid \mathbf{z}_0) = \prod_{s < u \leq t} (\mathbf{I} + d\hat{\mathbf{A}}(u \mid \mathbf{z}_0))$. If a transition occurs at time u, then $\mathbf{I} + d\hat{\mathbf{A}}(u \mid \mathbf{z}_0)$ is the matrix whose (h, j)-th element is $\exp(\beta' \mathbf{z}_{hj0}) \Delta N_{hj}(u) / S_{hj}^{(0)}(\hat{\beta}, u)$ if $h \neq j$ and equal to $1 - \sum_{j \neq h} \exp(\beta' \mathbf{z}_{hj0}) \Delta N_{hj}(u) / S_{hj}^{(0)}(\hat{\beta}, u)$ if $h = j$. Under some regularity conditions the estimators $\hat{\beta}, \hat{\mathbf{A}}(t \mid \mathbf{z}_0)$ and $\hat{\mathbf{P}}(s, t \mid \mathbf{z}_0)$ are $\sqrt{n}$-consistent and asymptotically normal.

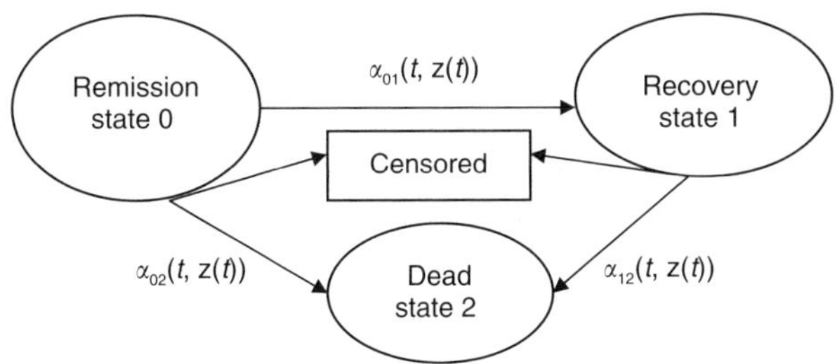

FIGURE 6.1
Three-state transition model.

6.2.4 Example

The application that we discuss in the next section uses a 3-state model with states labeled 0 = remission, 1 = recovery, and 2 = dead/relapse. Only forward transitions $0 \to 1, 0 \to 2, 1 \to 2$ are allowed as shown in Figure 6.1. Since censoring might occur from states 0 or 1, additional paths are shown from these states. However, they should not be construed as transitions.

Suppressing the dependence on a covariate profile $\mathbf{z}$, the differential equations (with differentiation with respect to t), $\mathbf{P}'(s,t) = \mathbf{P}(s,t)\boldsymbol{\alpha}(t)$ are then explicitly

$$\begin{aligned}
P'_{00}(s,t) &= -P_{00}(s,t)(\alpha_{01}(t) + \alpha_{02}(t)),\\
P'_{11}(s,t) &= -P_{11}(s,t)\alpha_{12}(t),\\
P'_{01}(s,t) &= P_{00}(s,t)\alpha_{01}(t) - P_{01}(s,t)\alpha_{12}(t).
\end{aligned}$$

The solution is

$$\begin{aligned}
P_{00}(s,t) &= \exp\left(-\int_s^t (\alpha_{01}(u) + \alpha_{02}(u))du\right),\\
P_{11}(s,t) &= \exp\left(-\int_s^t \alpha_{12}(u)du\right),\\
P_{01}(s,t) &= \int_s^t P_{00}(s,u)\alpha_{01}(u)P_{11}(u,t)du,
\end{aligned}$$

and

$$P_{02} = 1 - P_{00} - P_{01}, \quad P_{12} = 1 - P_{11}.$$

The hazard rate at time t for stay in state 0 is $\alpha_{01}(t)+\alpha_{02}(t)$. Starting in state 0 at time $t = 0$, exit out of this state occurs at time $T_1 = \inf\{t > 0 : X(t) \neq 0\}$. Hence $P[T_1 > t \mid X_0 = 0] = P_{00}(0,t)$. Similarly, measured from entry into state 1, T_{12} is the length of stay in state 1. This has hazard function $\alpha_{12}(t)$. If a transition occurs at $t = T_1$, the transition is to state 1 with probability $\alpha_{01}(t)/(\alpha_{01}(t)+\alpha_{02}(t))$, and to state 2 with probability $\alpha_{02}(t)/(\alpha_{01}(t)+\alpha_{02}(t))$.

Overall survival time is defined by $T = \inf\{t > 0 : X(t) = 2\}$. Its two conditional survival distributions are $S_{02}(t) = P[T > t \mid X_0 = 0] = P_{00}(0,t) + P_{01}(0,t) = 1 - P_{02}(0,t)$ and $S_{12}(t) = P[T > t \mid X_0 = 1] = P_{12}(0,t)$. If $\pi_h = P[X_0 = h]$ denotes initial distribution, the unconditional distribution is $S_2(t) = \pi_0 S_{02}(t) + \pi_1 S_{12}(t)$. The intensity (hazard) corresponding to $S_{02}(t)$ is

$$\alpha(t) = -\frac{d[\log S_{02}(t)]}{dt} = -\left(\frac{P'_{00}(0,t) + P'_{01}(0,t)}{P_{00}(0,t) + P_{01}(0,t)}\right).$$

Using the above differential equations leads to

$$\alpha(t) = \alpha_{02}(t)\frac{P_{00}(0,t)}{P_{00}(0,t) + P_{01}(0,t)} + \alpha_{12}(t)\frac{P_{01}(0,t)}{P_{00}(0,t) + P_{01}(0,t)}.$$

This death hazard is a combination of the two intensities $\alpha_{02}(t), \alpha_{12}(t)$. We can interpret $P_{00}(0,t)/(P_{00}(0,t) + P_{01}(0,t))$ as the conditional probability of being in state 0 at time t, given survival up to time t and that one starts in state 0 at time 0. Similarly, $P_{01}(0,t)/(P_{00}(0,t) + P_{01}(0,t))$ is the conditional probability of being in state 1 at time t, given survival up to time t and that one starts in state 0 at time 0.

If $\alpha_{02}(t) = \alpha_{12}(t) = \alpha(t)$we get $P_{00}(s,t) + P_{01}(s,t) = \exp(-\int_s^t \alpha(u)du) = P_{11}(s,t)$ and therefore $P_{02}(s,t) = P_{12}(s,t)$. In general with fixed covariates $\mathbf{z}$, even if the intensities $\alpha_{hj}(t,\mathbf{z})$ have proportional hazards, $\alpha_{hj}(t,\mathbf{z}) = \alpha_{hj,0}(t)\exp(\beta'\mathbf{z}_{hj})$ it does not necessarily yield a proportional hazards model of the intensity $\alpha(t,\mathbf{z})$.

Consider starting in state 0 and the role of the intermediate event time of recovery, $T_{01} = \inf\{t > 0 : X(t) = 1\}$ on survival. For $s < t$, $P[T \le t \mid T_{01} > s, T > s] = P_{02}(s,t)$ and $P[T \le t \mid T_{01} \le s, T > s] = P_{12}(s,t)$. The direct derivation uses the differential equations for $P_{hj}(s,t)$. We have

$$\begin{aligned}
P[T \le t \mid T_{01} > s, T > s] &= \int_s^t P_{00}(s,u)\{\alpha_{02}(u) + \alpha_{01}(u)P_{12}(u,t)\}du \\
&= \int_s^t P_{00}(s,u)[\alpha_{02}(u) + \alpha_{01}(u)]du \\
&\quad - \int_s^t P_{00}(s,u)\alpha_{01}(u)\{1 - P_{12}(u,t)\}du \\
&= \int_s^t -P'_{00}(s,u)du - P_{01}(s,t) \\
&= 1 - P_{00}(s,t) - P_{01}(s,t) = P_{02}(s,t) \\
P[T \le t \mid T_{01} \le s, T > s] &= \int_s^t P_{11}(s,u)\alpha_{12}(u)du \\
&= \int_s^t -P'_{11}(s,u)du = 1 - P_{11}(s,t) = P_{12}(s,t).
\end{aligned}$$

In the case of a homogeneous Markov process, the intensities are constants (in time). This leads to simple closed expressions for the transition probabilities. We obtain

$$P_{00}(s,t) = \exp(-(\alpha_{01} + \alpha_{02})(t-s)), \quad P_{11}(s,t) = \exp(-\alpha_{12}(t-s))$$

and

$$P_{01}(s,t) = \frac{\alpha_{01}}{\alpha_{01} + \alpha_{02} - \alpha_{12}} \{\exp(-\alpha_{12}(t-s)) - \exp(-(\alpha_{01} + \alpha_{02})(t-s))\}.$$

The survival distribution $S_{02}(t)$ can be expressed as a mixture of two distributions. The first is an exponential distribution with parameter $\lambda_0 = \alpha_{01} + \alpha_{02}$ for stay in state 0, and the second is the sum of this distribution and an independent exponential distribution with parameter $\lambda_1 = \alpha_{12}$ for stay in state 1. Let $q_{01} = \alpha_{01}/(\alpha_{01} + \alpha_{02})$—the probability of exit from state 0 to state 1. Then

$$S_{02}(t) = (1 - q_{01})e^{-\lambda_0 t} + q_{01} \left\{ \frac{\lambda_1 e^{-\lambda_0 t}}{\lambda_1 - \lambda_0} + \frac{\lambda_0 e^{-\lambda_1 t}}{\lambda_0 - \lambda_1} \right\}.$$

6.3 Application

The data set for this illustration is taken from Klein and Moeschberger (1997). The data set comprises 137 patients who underwent allogeneic bone marrow transplants for treatment of leukemia from March 1, 1984 to June 30, 1989. Patients were in one of three risk categories on the basis of their disease status before treatment. These groups were: (1) ALL—acute lymphoblastic leukemia, (2) AML—acute myelotic leukemia, low risk, and (3) AML—high risk. Covariates measured at the time of transplant include patient and donor age, gender, and cytomegalovirus immune status (CMV).

Several events may occur following transplantation. We focus on two events, death/relapse combined, and the intermediate event called platelet recovery when the platelet count returns to a normal level from a depressed level that usually occurs following surgery. Platelet recovery is a binary time-dependent covariate when one is only interested in modeling the death/relapse intensity. Alternatively, we may set up a state transition model with three states. State 0 is the initial state (remission) for all patients at time $t = 0$ and therefore $\pi_0 = P[X_0 = 0] = 1$. Platelet recovery is state 1 ($=X_1$) and we combine relapse or death into a single terminal state (X_2, state 2). This leads to the possible transitions $0 \to 1, 0 \to 2, 1 \to 2$ as shown in Figure 6.1. A patient who is still in remission at last follow-up time has not undergone any of these transitions. A patient whose platelets have recovered to normal levels by the time of last follow-up has undergone the $0 \to 1$ transition. A patient who relapsed or died during the study period would have either the $0 \to 2$ or $1 \to 2$ transition.

A general multiplicative intensity model for the three transitions is

$$\alpha_{01}(t, \mathbf{z}(t)) = \alpha_{01,0}(t) \exp(\beta_{01}\mathbf{z}(t))$$
$$\alpha_{02}(t, \mathbf{z}(t)) = \alpha_{02,0}(t) \exp(\beta_{02}\mathbf{z}(t))$$
$$\alpha_{12}(t, \mathbf{z}(t)) = \alpha_{12,0}(t) \exp(\beta_{12}\mathbf{z}(t)),$$

where our notation indicates allowance for transition-type specific baseline intensities $\alpha_{hj,0}(t)$ and regression parameters β_{hj}. Therefore, we could in effect analyze each transition separately, or as we shall demonstrate later by a single invocation of SAS PHREG. Preparation of the appropriate SAS data set depends on whether we are interested in the two events separately or only on the terminal event (dead/relapse). In the latter case, a single record per patient file can be used with time to platelet recovery as a time-dependent covariate. The regression model for the hazard function for survival of this 2-state model (ignoring state 1) is $\alpha(t, \mathbf{z}(t)) = \alpha_0(t) \exp(\beta\mathbf{z}(t))$.

6.3.1 Description of Variables in the Bone Marrow Transplant Study

The base data set BMT_SH is arrayed as a single record per patient. Variables names, a brief description and coding are in Table 6.1. There are two sets of factors. The first set of factors is measured on the patient. These are the disease risk group membership (DGROUP), an indicator (FAB) for classification grade M4 or M5 for AML patients, and an indicator (MTX) of whether the patient was given a graft-versus-host disease prophylactic (methotrexate with cyclosporine). The second set of factors is based on a combination of patient and donor characteristics, involving patient and donor sex (PSEX, DSEX), patient and donor CMV status (PSTATUS, DSTATUS), and patient and donor age (PAGE, DAGE).

Of the 137 patients who received a transplant at time $t = 0$, 120 subsequently had platelet recovery (PRI $= 1$). Of these 120 patients, 67 died or relapsed (DFI $= 1$) and 53 were alive in remission at last follow-up. Of the 17 patients who had no platelet recovery (PRI $= 0$), 16 died or relapsed and one was alive.

6.3.2 Analysis of Disease-Free Survival

We analyze disease-free survival (TFREEST) using a Cox-regression model for the overall hazard $\alpha(t, \mathbf{z}(t)) = \alpha_0(t) \exp(\beta'\mathbf{z}(t))$. For each patient the time-dependent indicator PLSTATUS(t) for platelet recovery status at time t is created as follows. For patients without platelet recovery (PRI $= 0$) define PLSTATUS(t) $= 0$; patients with platelet recovery, set PLSTATUS(t) $= 0$ if $t <$ TRETP and PLSTATUS(t) $= 1$ if $t \geq$ TRETP. All other covariates are assessed at $t = 0$.

TABLE 6.1

Variables in BMT_SH Data Set

Variable Name	Description	Coding
ID	Patient identification	
TFREEST	Disease-free survival time	Time in days to relapse, death, or end of study
DFI	Disease-free survival indicator	1 = dead or relapsed, 0 = alive disease free
PRI	Platelet recovery indicator	1 = platelet recovered, 0 = no platelet recovery during study
TRETP	Days to platelet recovery	Time is days to platelet recovery, if PRI = 1. Otherwise, TREP = .
DGROUP	Disease group	1 = ALL (acute lymphobastic leukemia) 2 = AML-low risk (acute myeloctic leukemia) 3 = AML-high risk
PAGE	Patient age in years	
DAGE	Donor age in years	
PSEX	Patient sex	1 = Male, 0 = Female
DSEX	Donor sex	1 = Male, 0 = Female
PSTATUS	Patient CMV status	1 = CMV Positive, 0 = CMV Negative
DSTATUS	Donor CMV Status	1 = CMV Positive, 0 = CMV Negative
FAB	FAB Grade	1 = Grade 4 or 5 and AML, 0 = otherwise
MTX	Methotrexate used as a Graft-Versus-Host-Disease Prophylactic	1 = Yes, 0 = No

Our primary focus is on the impact of disease risk groups (DGROUP) and platelet recovery on disease-free survival. Preliminary analyses show that patient and donor sex, and patient and donor CMV status do not have a significant effect on survival. However, with patient and donor age there seems to be a strong interaction. In what follows we will only consider the two age variables (PAGE, DAGE), FAB and MTX. Since the effect of these covariates might differ before and after platelet recovery, the model for $\alpha(t, \mathbf{z}(t))$ has regression coefficients specific to the two periods.

For example, consider the base model with DGROUP and PLSTATUS. For DGROUP we use two coefficients (β_1, β_2) for (AML-high risk, AML-low risk), with the group ALL as referent, a coefficient β_3 for PLSTATUS and two coefficients (β_4, β_5) for the crossed effect DGROUP$\times$PLSTATUS. Then $\alpha_0(t)$ is the hazard for the ALL group before platelet recovery, and $\alpha_0(t)\exp(\beta_3)$ is the hazard after platelet recovery. Therefore, $\exp(\beta_3)$ is the relative hazard for a patient in the ALL group at time t after platelet recovery compared with a patient in the ALL group at time t whose platelets have not recovered to normal levels. The other parameters are identified as shown by the other comparisons in Table 6.2.

TABLE 6.2

Regression Parameters

Platelet Recovery	Comparison Risk Groups	Relative Hazard
Before	AML-high risk vs. ALL	$\exp(\beta_1)$
Before	AML-low risk vs. ALL	$\exp(\beta_2)$
After	AML-high risk vs. ALL	$\exp(\beta_1 + \beta_4)$
After	AML-low risk vs. ALL	$\exp(\beta_2 + \beta_5)$

6.3.3 Estimation of the Base Model

To enhance the presentation the following formats may be used:

```
proc format;
value dgroup 1 = 'ALL' 2 = 'AML low risk'
  3 = 'AML high risk';
value dfi 1 = 'dead or relapsed' 0 = 'alive disease free';
value pri 1 = 'platelets returned to normal'
  0 = 'platelets never returned to normal';
value fab 1 = 'FAB grade 4 or 5 and AML' 0 =
'otherwise'; value mtx 1 = 'yes' 0 = 'no';
run;
```

The base model with the parametrization given in Table 6.2 is estimated using the SAS TPHREG (experimental) procedure in SAS version 9.1.3. This is an enhancement of the SAS PHREG procedure allowing for CLASS and CONTRAST statements. Future releases of SAS will likely fold these options into PHREG. The following syntax fits the base model. Some options are redundant but are included for clarity.

```
proc tphreg data = BMT_SH;
class dgroup(ref = 'ALL')/param = ref;
model tfreest*dfi(0) = dgroup|plstatus/ties = breslow;
if pri = 0 or (pri = 1 and tfreest<tretp) then plstatus = 0;
else plstatus = 1;
format dgroup dgroup.;
run;
```

An estimator $\hat{\beta}$ of β is derived by maximizing the generalized Cox partial likelihood given by

$$\prod_{i=1}^{n} \prod_{t} \left\{ \frac{Y_i(t) \exp(\beta' \mathbf{z}_i(t))}{\sum_{k=1}^{n} Y_k(t) \exp(\beta' \mathbf{z}_k(t))} \right\}^{\Delta N_i(t)}$$

where $Y_i(t) = 1$ if the i-th patient is at risk of death/relapse at time $t-$, and $Y_i(t) = 0$ otherwise. Also $N_i(t) = 1$ if the event death or relapse has occurred by time t in the i-th patient; if not, $N_i(t) = 0$.

TABLE 6.3

Contrast Estimates and Confidence Intervals for Base Model

Contrast	Relative Hazard	95% LCL	95% UCL	p-Value
Before recovery: AML-high risk vs. ALL	2.305	0.592	8.966	0.2284
Before recovery: AML-low risk vs. ALL	2.883	0.706	11.778	0.1402
After recovery: AML-high risk vs. ALL	1.355	0.765	2.402	0.2978
After recovery: AML-low risk vs. ALL	0.450	0.241	0.840	0.0122

The time-dependent platelet status recovery indicator must be created within the procedure. There are 83 event times in the data set. At each event time t = TFREEST, the data is scanned to determine the function (of β) $\sum_{k=1}^{n} Y_k(t)\exp(\beta'\mathbf{z}_k(t))$ where

$$\beta'\mathbf{z}(t) = \beta_1 \text{AML}_H + \beta_2 \text{AML}_L + \beta_3 \text{PLSTATUS}(t) + \beta_4 \text{AML}_H \times \text{PLSTATUS}(t) + \beta_5 \text{AML}_L \times \text{PLSTATUS}(t),$$

is evaluated for each patient. AML_L and AML_H are, respectively, indicators for the AML-low risk and AML-high risk groups.

To obtain estimates and 95% confidence intervals for the relative hazards in Table 6.2, we specify the vectors L for the parametric functions $L'\beta$. The syntax to obtain Table 6.3 uses CONTRAST and ODS statements. [The experimental TPHREG procedure is used throughout. Its options will be available in the PHREG procedure in a future release of the SAS/STAT Software.]

```
ods output contrastestimate = contrasts;
proc tphreg data = BMT_SH;
class dgroup(ref = 'ALL')/param = ref;
model tfreest*dfi(0) = dgroup|plstatus/ties = breslow;
if pri = 0 or (pri = 1 and tfreest<tretp) then
  plstatus = 0;
else plstatus = 1;
format dgroup dgroup.;
contrast 'Before Recovery: AML-high risk vs ALL'
                   DGROUP 1 0/estimate = exp;
contrast 'Before Recovery: AML-low risk vs ALL'
                   DGROUP 0 1/estimate = exp;
contrast 'After Recovery: AML-high risk vs ALL'
                  DGROUP 1 0 dgroup*plstatus 1 0/
                    estimate = exp;
contrast 'After Recovery: AML-low risk vs ALL'
                  DGROUP 0 1 dgroup*plstatus 0 1/
                    estimate = exp;
run;
```

Table 6.3 can be generated from the following syntax that creates an RTF file from the CONTRASTS data set. This article was produced as a WORD file that allowed easy insertion of the RTF file within the document with minor additional editing.

```
ods rtf file = "C:\Documents and Settings\My Documents\
  contrasts.rtf"
      style = styles.journal;
proc print data = contrasts noobs label
            style(DATA) = {font = ("TimesNewRoman", 3.5,
                           medium roman normal)}
            style(TABLE) = {frame = box}
            style(HEADER) = {font = ("TimesNewRoman", 3.5,
                            bold roman normal)};
format estimate lowerlimit upperlimit F6.3
  probchisq pvalue6.4;
var contrast estimate lowerlimit upperlimit probchisq;
label estimate = 'Relative Hazard' lowerlimit = '95% LCL'
  upperlimit = '95% UCL' probchisq = 'p-value';
run;
ods rtf close;
```

6.3.4 Variable Selection

The constellation of variables is the time invariant effects DGROUP, FAB, MTX, PAGE, DAGE and PAGE × DAGE, the time-dependent effect PLSTATUS and its interaction with all the preceding effects. We will use a forward selection process (SELECTION = FORWARD) with significance level for entry set at 10% (SLENTRY = .10). All effects are subject to the hierarchy requirement. With HIERARCHY = MULTIPLE a single main effect can enter the model or an interaction can enter the model together with all the effects that are contained in the interaction. In contrast with HIERARCHY = SINGLE, only one effect can enter at each step, subject to the model hierarchy requirement. For example, this means that the interaction DGROUP × PLSTATUS can enter the model only if both the main effects are already in the model.

SAS uses the score statistic (instead of partial likelihood ratio statistic) to assess the order of variable entry. In this example there are 13 effects—six time invariant effects, one time-dependent effect PLSTATUS, and its six interactions. The score statistic with the smallest p-value consistent with the entry significance level determines the effects that will be entered first. The multiple hierarchy option considers multiple degrees of freedom tests for interactions and all effects that contain it. At the first step, the following are entered: PAGE, DAGE, PAGE × DAGE, PLSTATUS, PLSTATUS × PAGE, PLSTATUS × DAGE and PLSTATUS × PAGE × DAGE. The second step assesses the remaining six effects for entry to augment the first model. This results in DGROUP and PLSTATUS × DGROUP being added. Step 3 then adds FAB and PLSTATUS × FAB. The only effects remaining are MTX and

MTX × PLSTATUS. The score test p-values are .46 for MTX alone, and .53 for MTX and MTX × PLSTATUS. Because they are above our significance level for entry (.10) the selection process ends.

The syntax for running the selection procedure is:

```
proc tphreg data = BMT_SH;
class dgroup(ref = 'ALL') fab(ref = 'otherwise')
  mtx(ref = 'no')/param = ref;
model tfreest*dfi(0) = dgroup|plstatus fab|plstatus
  page|dage|plstatus mtx|plstatus/selection = forward
  slentry = .10 hierarchy = multiple ties = breslow details;
if pri = 0 or (pri = 1 and tfreest<tretp) then plstatus = 0;
else plstatus = 1;
page = page-28;
dage = dage-28;
format dgroup dgroup. fab fab. mtx mtx.;
run;
```

REMARKS

The patient and donor age variables are centered at their median values (=28 years).

Had we set our entry criterion at .50, MTX would have entered at step 4. With only MTX × PLSTATUS left to consider, its score test p-value is .39, and so it too would enter the model.

With HIERARCHY = SINGLE, interactions are considered for entry only after their constituent single effects have entered the model. This option results in a model with PLSTATUS, DGROUP, FAB, PLSTATUS × DGROUP, and PLSTATUS × FAB. As noted earlier (PAGE, DAGE) are not individually significant unless their interaction is present. Therefore, the single hierarchy option would fail to select the patient/donor age effects.

The parameter estimates in the final model are shown in Table 6.4. Estimates of the relative hazard for selected contrasts are in Table 6.5. Our parametrization makes the baseline hazard $\alpha_0(t)$ the hazard for the ALL group before platelet recovery for a patient and donor at the median age (=28 years) and FAB = otherwise. With respect to disease-free survival, there is no significant difference between each AML group and the ALL group before platelet recovery. However, after platelet recovery the AML-low risk group has significant better survival than the ALL group (RH = 0.18, 95% CL: [0.08, 0.41]).

6.3.5 Using a Multiple Record, Counting Process-Style Input

An alternative computing strategy uses a file that represents the transition records of each patient according to the schema in Figure 6.1. All patients begin at $t = 0$ (TSTART) in state 0. A patient whose platelets did not return to normal levels (PRI = 0) experiences the transition $0 \rightarrow 2$ provided the

TABLE 6.4

Parameter Estimates in Final Model

Parameter	Class Value	Parameter Estimate	Standard Error	p-Value
DGROUP	AML-high risk	1.1071	1.2242	0.3658
DGROUP	AML-low risk	1.3073	0.8186	0.1103
PLSTATUS		−0.3062	0.6936	0.6589
PLSTATUS*DGROUP	AML-high risk	−1.8675	1.2908	0.1479
PLSTATUS*DGROUP	AML-low risk	−3.0374	0.9257	0.0010
FAB	FAB grade 4 or 5 and AML	−1.2348	1.1139	0.2676
PLSTATUS*FAB	FAB grade 4 or 5 and AML	2.4535	1.1609	0.0346
PAGE		−0.1538	0.0545	0.0048
DAGE		0.1166	0.0434	0.0072
PAGE*DAGE		0.0026	0.0019	0.1814
PLSTATUS*PAGE		0.1933	0.0588	0.0010
PLSTATUS*DAGE		−0.1470	0.0480	0.0022
PLSTATUS*PAGE*DAGE		0.0001	0.0023	0.9561

PLSTATUS = 0, before platelet recovery; PLSTATUS = 1, after platelet recovery.

TABLE 6.5

Contrast Estimates and Confidence Intervals from Final Model*

Contrast	Relative Hazard	95% LCL	95% UCL	p-Value
Before recovery: AML-high risk vs ALL	3.026	0.275	33.333	0.3658
Before recovery: AML-low risk vs ALL	3.696	0.743	18.391	0.1103
After recovery: AML-high risk vs ALL	0.467	0.210	1.039	0.0620
After recovery: AML-low risk vs ALL	0.177	0.077	0.409	<.0001

* Adjusted for patient and donor age, FAB.

patient is not censored. An indicator IND02 labels whether or not the transition occurred, and the ending time is TSTOP = TFREEST. A stratum label 02 is created for this record. A patient whose platelets returned to normal levels (PRI = 1) experiences the transition 0 → 1. This occurs at TSTOP = TRETP. An indicator IND01 = 1 labels that this transition occurred and a stratum label 01 is created for this record. A second record is created to represent the next transition 1 → 2 starting at TSTART = TRETP and ending time at TSTOP = TFREEST. The indicator IND12 labels whether the transition occurred: IND12 = 1, otherwise IND12 = 0. A stratum label 12 is created for this record. For each patient, IND01, IND02, IND12 are all zero unless the corresponding transition occurred.

This technique of representing the event-history data with time-dependent covariates was popularized by Therneau and Grambsch (2000). It works well when the time-dependent covariates take on a finite number of values. Each interval (TSTART, TSTOP) denotes an interval of risk for the patient with covariate measured at the beginning of the interval. At TSTOP some covariate values might change or the event status changes. If overall disease-free survival is the event of interest, a censoring variable STATUS indicates whether or not the transition resulted in this event. Therefore, STATUS = 1 for IND02 = 1 or IND12 = 1 and STATUS = 0 otherwise. Note that we do not set STATUS = 1 for IND01 = 1 because platelet recovery is not being considered an event in this analysis.

The following data step creates the extra 120 records for patients who had platelet recovery (PRI = 1). These records have PLSTATUS = 1 for the period representing the transition 1 → 2. Otherwise, PLSTATUS = 0 for all records. There is one patient in the data set whose time to platelet recovery is zero (TRETP = 0). This creates a null interval for the transition 0 → 1. In the syntax below this is prevented by adding 1 day to TSTOP.

```
data bmt_LG;
set bmt_SH(keep = id tretp tfreest pri dfi dgroup
  fab page dage mtx);
retain tstop;
IND01 = 0; IND12 = 0; IND02 = 0;
if pri = 0 then do;
tstart = 0; tstop = tfreest; plstatus = 0; stratum = '02';
       if dfi = 1 then IND02 = 1; else IND02 = 0;
status = IND02;
if tstop = tstart then tstop = tstart+1;
      output; end;
if pri = 1 then do;
tstart = 0; tstop = tretp; plstatus = 0; stratum = '01';
               IND01 = 1;
status = 0; /*Platelet recovery not regarded as event*/
if tstop = tstart then tstop = tstart+1;
output;
tstart = tstop; tstop = tfreest;
  plstatus = 1; stratum = '12';
       if dfi = 1 then IND12 = 1; else IND12 = 0;
status = IND12;
       output; end;
run;
```

The count of events among the 137 patients is easily tracked using the created indicators (Table 6.6). Of the 120 patients who had platelet recovery (transition 0 → 1) 67 died or relapsed (IND12 = 1), and 53 were censored. Of the other 17 patients (IND01 = 0), 16 died or relapsed and 1 was censored. Note that STATUS = 1 for the event of interest—died or relapsed, a total of 83 events.

Table 6.6 may be generated by the following syntax and printing of the output LIST data set.

```
proc format;
value plstatus 1 = 'after' 0 = 'before';
run;
```

TABLE 6.6

Count of Events among 137 Transplant Patients

Plstatus	Stratum	Status	IND01	IND12	IND02	Frequency
Before	01	0	1	0	0	120
Before	02	0	0	0	0	1
Before	02	1	0	0	1	16
After	12	0	1	0	0	53
After	12	1	1	1	0	67

TABLE 6.7

Parameter Estimates from Final Model (Multiple Record File)

Parameter	Class Value	Class Value	Parameter Estimate	Standard Error	p-Value
DGROUP	AML-low risk		1.3160	0.8188	0.1080
DGROUP	AML-high risk		1.1251	1.2252	0.3585
PLSTATUS	After		−0.2866	0.6956	0.6803
DGROUP*PLSTATUS	AML-low risk	After	−3.0469	0.9259	0.0010
DGROUP*PLSTATUS	AML-high risk	After	−1.8859	1.2916	0.1443
FAB	FAB grade 4 or 5 and AML		−1.2444	1.1126	0.2634
PLSTATUS*FAB	After	FAB grade 4 or 5 and AML	2.4644	1.1594	0.0335
PAGE			−0.1535	0.0546	0.0049
DAGE			0.1163	0.0435	0.0075
PAGE*DAGE			0.0026	0.0019	0.1801
PAGE*PLSTATUS	After		0.1931	0.0588	0.0010
DAGE*PLSTATUS	After		−0.1467	0.0481	0.0023
PAGE*DAGE*PLSTATUS	After		0.0001	0.0023	0.9585

```
ods output list = list;
proc freq data = bmt_LG;
tables plstatus*stratum*status*ind01*ind02*ind12/list;
format plstatus plstatus.;
run;
```

Because PLSTATUS is a class variable in the data set bmt_LG, we can readily obtain the results shown previously in Tables 6.4 and 6.5 using appropriate options in the CLASS statement. Table 6.7 is derived from the PARMS data set created by

```
ods output parameterestimates = parms;
proc tphreg data = bmt_LG namelen = 25 multipass;
class dgroup(descending) plstatus fab/param = glm;
```

```
   model (tstart, tstop)*status(0) = dgroup|plstatus
                                     fab|status page|dage|
                                     plstatus/ties = breslow;
   page = page-28;dage = dage-28;
   format dgroup dgroup. fab fab. plstatus plstatus.;
   run;
```

A full parametrization is requested through the global PARAM = GLM option. The local DESCENDING option on DGROUP makes the ALL group the referent. The default ordering makes the highest formatted value the referent that would be "AML-low risk." In the data set PARMS, all referent categories have degrees of freedom 0 (DF = 0) and therefore can be eliminated in the subsequent print procedure creating Table 6.7.

There appears to be minor differences in the parameter estimates shown here and those in Table 6.4. This is because PHREG terminated at the optimized partial −2 log likelihood value of 683.254 whereas the previous approach ended at a value of 683.042. The differences remain approximately the same under other ties-handling options (TIES = EFRON, or EXACT).

6.3.6 Plotting Survival Curves

From the death or relapse intensity model $\alpha(t, \mathbf{z}(t)) = \alpha_0(t)\exp(\beta\mathbf{z}(t))$the cumulative baseline hazard

$$A_0(t) = \int_0^t \alpha_0(u)du$$

is estimated by

$$\hat{A}_0(t) = \int_0^t \frac{[Y(u) > 0]}{S^{(0)}(\hat{\beta}, u)} dN(u)$$

where

$$S^{(0)}(\hat{\beta}, t) = \sum_{i=1}^{n} Y_i(t)\exp(\hat{\beta}'\mathbf{z}_i(t)), \quad Y(t) = \sum_{i=1}^{n} Y_i(t) \quad \text{and} \quad N(t) = \sum_{i=1}^{n} N_i(t).$$

The next step is the estimation of survival function $S(t \mid \mathbf{z}_0) = \exp(-A_0(t)\exp(\beta'\mathbf{z}_0))$ at a specified (fixed) covariate $\mathbf{z}_0$. In the final model (Table 6.7) we will estimate survival in the three disease groups for patients with PAGE = 28, DAGE = 28 and FAB = 0. The profile must also specify the platelet status indicator PLSTATUS. In the data set bmt_LG, this is explicitly created. Platelet recovery time range is 1 to 100 days, with median 18 days for the 120 patients who had platelet recovery. In the 83 patients who died or relapsed, the event time range is 1–2204 days, with median 183 days. Only three events occurred after 750 days.

The COVAR data set contains the six profiles (three disease groups with and without platelet recovery) that are fixed. Values for all covariates in the final model must be specified together with formats. This could be created by

```
proc sort data = BMT_LG out = covar(keep = dgroup plstatus)
  nodupkey;
by DGROUP PLSTATUS;
run;

data covar;
set covar;
page = 0; dage = 0; fab = 0;
format dgroup dgroup. fab fab. plstatus plstatus.;
run;
```

The PAGE and DAGE variables are set to zero because the following syntax estimates the baseline cumulative hazard at the desired profile by centering the two covariates.

```
proc tphreg data = bmt_LG namelen = 25 multipass noprint;
class dgroup(descending) plstatus fab/param = glm;
model (tstart, tstop)*status(0) = dgroup|plstatus
  fab|plstatus
page|dage|plstatus/rl ties = breslow;
baseline out = survival_est covariates = covar lower = LCL
  survival = survival upper = UCL/cltype = loglog
  method = ch nomean;
page = page-28;
dage = dage-28;
format dgroup dgroup. fab fab. plstatus plstatus.;
run;
```

The data set SURVIVAL_EST contains the survival estimates and 95% confidence limits for all six profiles. Because $\hat{\Lambda}_0(t)$ is a step function with values changing at the event times (there are 76 distinct times), each survival curve is estimated at the same grid of event times according to $\hat{S}(t \mid \mathbf{z}_0) = \exp(-\hat{\Lambda}_0(t)\exp(\hat{\beta}'\mathbf{z}_0))$. Figure 6.2 depicts the survival estimates (through 750 days) in the ALL and AML low risk groups without platelet recovery. The following syntax was used:

```
goptions reset = global colors = (black blue red
  green purple)
      gunit = pct cback = white ctext = black
        ftext = "Garamond"
      htitle = 4 htext = 3 hsize = 7 in vsize = 7 in
        offshadow = (-.5 -.5);
```

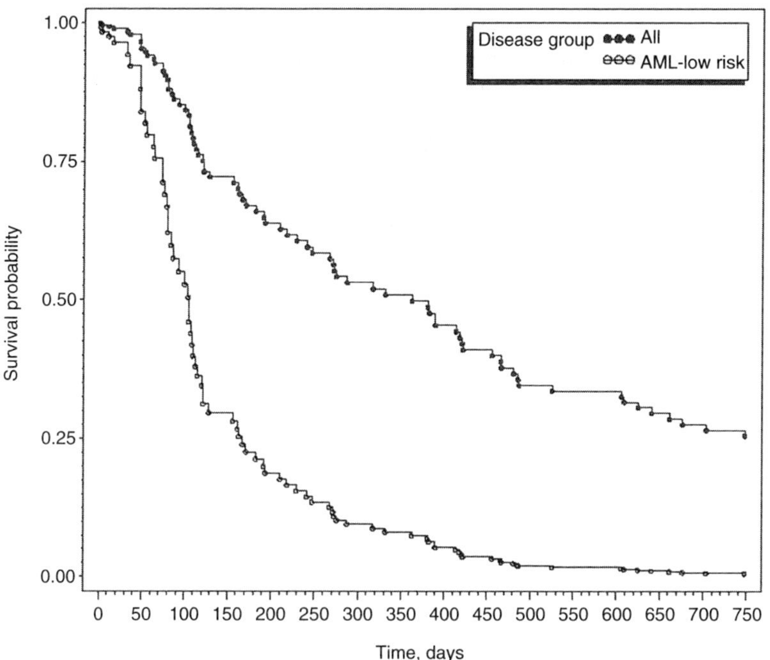

FIGURE 6.2
ALL and AML-low risk, without recovery.

```
symbol1 interpol = stepjl value = dot color = red h = 1;
symbol2 interpol = stepjl value = circle color = blue h = 1;
axis1 label = (angle = 90 'SURVIVAL PROBABILITY')
  order = (0 to 1 by .25) offset = (2 2);
axis2 label = ('TIME, Days') order = (0 to 750 by 50)
  offset = (2 2);
legend1 across = 1 label = ('Disease Group') cshadow = gray
  frame mode = protect position = (inside right top)
  offset = (-2,-2);

proc gplot data = survival_est;
where dgroup in (1, 2) and plstatus = 0;
plot survival*tstop = dgroup/vaxis = axis1
  haxis = axis2 legend = legend1;
format dgroup dgroup. plstatus plstatus.;
title 'Figure 6.2: ALL and AML-low risk, without recovery';
run; quit;
```

The CLTYPE = loglog option computes the 95% pointwise confidence intervals (LCL, UCL) for the survival curves based on the $\log(-\log S(t \mid \mathbf{z}_0))$ transformation. This is slightly more accurate than the default option based on the log transformation. Figure 6.3 plots the survival curves and confidence

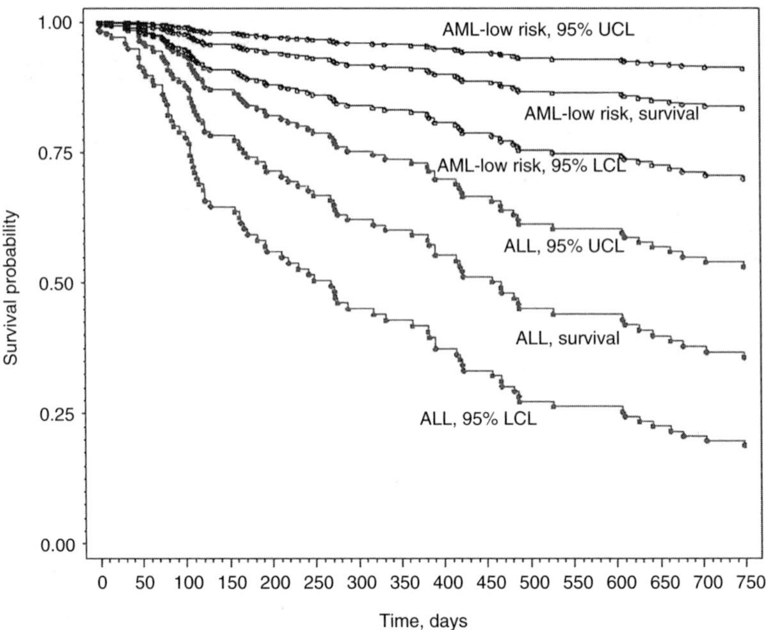

FIGURE 6.3
ALL and AML-low risk with platelet recovery: survival estimates and 95% pointwise confidence intervals.

intervals in the ALL and AML-low risk groups with platelet recovery. For plotting, a data set AFTER was created.

```
data after;
merge survival_est(keep = dgroup plstatus tstop plstatus
  survival lcl ucl
where = (dgroup = 1 and plstatus = 1)
rename = (survival = surv_ALL lcl = lcl_ALL ucl = ucl_ALL))
survival_est(keep = dgroup plstatus survival lcl ucl
where = (dgroup = 2 and plstatus = 1)
rename = (survival = surv_AML_low lcl = lcl_AML_low
  ucl = ucl_AML_low))
survival_est(keep = dgroup plstatus survival lcl ucl
where = (dgroup = 3 and plstatus = 1)
rename = (survival = surv_AML_high lcl = lcl_AML_high
  ucl = ucl_AML_high));
drop dgroup plstatus;
run;
```

The six plots of survival curves and confidence limits are created using GPLOT with additional symbol statements similar to those used in Figure 6.2.

The annotation uses a data set ANNO with labels assigned to positions on the plot by visual inspection of the (x, y) coordinates.

```
data anno;
length text $ 22 color function style $ 8;
retain function 'LABEL' position '4' hsys '3' ysys '2'
  xsys '2' size 2 style "zapfb";
text = 'ALL, 95% LCL'; x = 500; y = .25;
  color = 'red'; output;
text = 'ALL, survival'; x = 600; y = .40;
  color = 'red'; output;
text = 'ALL, 95% UCL'; x = 600; y = .58;
  color = 'red'; output;
text = 'AML low risk, 95% LCL'; x = 500; y = .70;
  color = 'blue'; position = '2'; output;
text = 'AML low risk, survival'; x = 600; y = .80;
  color = 'blue'; position = '2'; output;
text = 'AML low risk, 95% UCL'; x = 500; y = .96;
  color = 'blue'; position = '2'; output;
run;

proc gplot data = after;
plot (lcl_all surv_all ucl_all lcl_aml_low
  surv_aml_low ucl_aml_low)*tstop/annotate = anno
  overlay vaxis = axis1 haxis = axis2;
title 'Figure 6.3: ALL and AML-low risk with platelet
  recovery';
title2 j = c '95% pointwise confidence intervals';
run;
quit;
```

6.3.7 Multiple Events and Stratum-Specific Analysis

The regression model that we have considered has a single overall hazard function for the intensity of death or relapse, $\alpha(t, \mathbf{z}(t)) = \alpha_0(t)\exp(\beta\mathbf{z}(t))$ where $\mathbf{z}(t)$ is a large covariate vector that captures the variables DGROUP, FAB, PAGE, DAGE and PAGE $\times$ DAGE for the periods before and after platelet recovery through the time-dependent indicator PLSTATUS. However, platelet recovery was not regarded as an "event," but as a covariate. We now turn to a full event-specific analysis of three transitions shown in Figure 6.1.

Now regard platelet recovery also as an event and consider the three-state model with transition-specific covariates:

$$\alpha_{01}(t, \mathbf{z}(t)) = \alpha_{01,0}(t)\exp(\beta_{01}\mathbf{z}(t))$$

$$\alpha_{02}(t, \mathbf{z}(t)) = \alpha_{02,0}(t)\exp(\beta_{02}\mathbf{z}(t))$$

$$\alpha_{12}(t, \mathbf{z}(t)) = \alpha_{12,0}(t)\exp(\beta_{12}\mathbf{z}(t)).$$

Because the baseline intensities $\alpha_{01,0}, \alpha_{12,0}, \alpha_{02,0}$ and regression parameters $\beta_{01}, \beta_{12}, \beta_{02}$ are specific to each transition type, we may analyze each transition separately. However, with a small expansion of the data file all three analyses can be obtained from a single invocation of PHREG.

The syntax below creates the file bmt_EXP. The key additions to bmt_LG are records for patients who had the $0 \rightarrow 2$ transition (16 events, 1 censored). They are also considered at risk for the $0 \rightarrow 1$ transition, but all were censored and thus STATUS $= 0$. Similarly, patients who had the $0 \rightarrow 1$ transition (120 events) are considered at risk for the $0 \rightarrow 2$ transition, but all were censored. The variable STATUS indicates whether or not the events "platelet recovery," "death/relapsed" were observed.

```
data bmt_EXP;
set bmt_SH(keep = id tretp tfreest pri dfi dgroup fab page
  dage mtx);
retain tstop;
IND01 = 0; IND12 = 0; IND02 = 0;
if pri = 0 then do;
tstart = 0; tstop = tfreest; plstatus = 0; stratum = '02';
       if dfi = 1 then IND02 = 1; else IND02 = 0;
status = IND02;
if tstop = tstart then tstop = tstart+1;
      output;
stratum = '01'; IND01 = 0; IND02 = 0; status = 0; output;
end;

if pri = 1 then do;
tstart = 0; tstop = tretp; plstatus = 0; stratum = '01';
             IND01 = 1;
status = 1; /**** platelet recovery regarded as event ****/
if tstop = tstart then tstop = tstart+1;
output;

stratum = '02'; IND01 = 0; IND02 = 0; status = 0; output;
tstart=tstop; tstop=tfreest; plstatus=1; stratum='12';
       if dfi = 1 then IND12 = 1; else IND12 = 0;
status = IND12;
       output;
end;
run;
```

To estimate the transition-specific regression model with PHREG we need a STRATA statement for the different baseline intensities. All covariate effects appear as crossed effects with the stratum variable in order to get type-specific estimates. The variable PLSTATUS created above for clarity is no

TABLE 6.8
Number of Patients, Events, and Censored Events

Stratum	Total	Event	Censored
01	137	120	17
02	137	16	121
12	120	67	53
Records	394	203	191

longer necessary. The syntax for invocation of PHREG is:

```
proc tphreg data = bmt_EXP namelen = 25 multipass;
class stratum dgroup(descending)fab/param = glm;
strata stratum;
format dgroup dgroup. fab fab.;
model (tstart, tstop)*status(0) = stratum*dgroup stratum*fab
                                  stratum*page stratum*dage
                                  stratum*page*dage/
                                  ties = breslow;
page = page-28; dage = dage-28;
run;
```

Table 6.8 is derived from the default output "Summary of the Number of Event and Censored Values" using appropriate ODS statements. Stratum 01 labels the $0 \rightarrow 1$ transition. A total of 137 patients are at risk, 120 events (i.e., platelet recovery) are observed and 17 are censored for this event. For the $0 \rightarrow 2$ transition, of 137 patients at risk, 16 events (i.e., died or relapsed) are observed and 1 patient is censored for this event. The $1 \rightarrow 2$ transition concerns only the 120 patients with platelet recovery. Of these, 67 events are observed with the rest being censored.

Exactly the same results as in Table 6.9 would be obtained by analyzing each transition separately. For example, for the transition $0 \rightarrow 1$ all patients are at risk. The censoring indicator for the event "platelets recovered" is IND01. Accordingly, our syntax is then

```
proc tphreg data = bmt_exp namelen = 25;
where stratum in ('01');
class dgroup(descending)fab/param = glm;
format dgroup dgroup. fab fab.;
model (tstart, tstop)*IND01(0) = dgroup fab page dage
                                 page*dage/ties = breslow ;
page = page-28; dage = dage-28;
run;
```

TABLE 6.9

Parameter Estimates from Type-Specific Transition Model

Parameter	Class Value	Class Value	Parameter Estimate	Standard Error	p-Value
STRATUM*DGROUP	01	AML-low risk	0.3337	0.2496	0.1812
STRATUM*DGROUP	01	AML-high risk	0.1428	0.2978	0.6316
STRATUM*FAB	01	FAB grade 4 or 5 and AML	−0.1079	0.2360	0.6475
PAGE*STRATUM	01		0.0156	0.0161	0.3310
DAGE*STRATUM	01		−0.0143	0.0138	0.3016
PAGE*DAGE*STRATUM	01		−0.0016	0.0009	0.0694
STRATUM*DGROUP	02	AML-low risk	1.4664	0.9174	0.1100
STRATUM*DGROUP	02	AML-high risk	1.4477	1.3334	0.2776
STRATUM*FAB	02	FAB grade 4 or 5 and AML	−1.7567	1.3214	0.1837
PAGE*STRATUM	02		−0.1616	0.0620	0.0091
DAGE*STRATUM	02		0.1258	0.0475	0.0081
PAGE*DAGE*STRATUM	02		0.0032	0.0021	0.1303
STRATUM*DGROUP	12	AML-low risk	−1.7160	0.4255	<.0001
STRATUM*DGROUP	12	AML-high risk	−0.7565	0.4075	0.0634
STRATUM*FAB	12	FAB grade 4 or 5 and AML	1.2115	0.3223	0.0002
PAGE*STRATUM	12		0.0387	0.0218	0.0753
DAGE*STRATUM	12		−0.0292	0.0205	0.1540
PAGE*DAGE*STRATUM	12		0.0027	0.0012	0.0305

The output would match the top part of Table 6.9 with stratum class value = 01. For the transition $0 \rightarrow 2$, the risk stratum is 02 and censoring indicator IND02. For the transition $1 \rightarrow 2$, the risk stratum is 12 and censoring indicator IND12. Because there are a number of tied platelet recovery times, a better ties-handling likelihood for the $0 \rightarrow 1$ transition than the default Breslow likelihood is the Efron likelihood (TIES = EFRON) or exact likelihood (TIES = EXACT).

In Section 6.3.4, we analyzed factors associated with the overall death of relapse intensity, treating platelet recovery status as a time-dependent covariate. We found no significant effect of MTX use. If the impact of MTX on each of the transitions is now assessed we find that it has a very significant influence on platelet recovery intensity, but not on the two death or relapse intensities. In the next section we will add MTX to the covariates for the $0 \rightarrow 1$ transition.

6.3.8 Common Baseline Death or Relapse Intensities

The death/relapse intensities $\alpha_{12}(t, \mathbf{z}(t))$ and $\alpha_{02}(t, \mathbf{z}(t))$ are now modeled with common baseline intensity $\alpha_{0,0}(t)$. We will still maintain separate regression coefficients $\beta_{01}, \beta_{12}, \beta_{02}$ for the transition types. The model is

$$\alpha_{01}(t, \mathbf{z}(t)) = \alpha_{01,0}(t) \exp(\beta_{01}\mathbf{z}(t))$$
$$\alpha_{02}(t, \mathbf{z}(t)) = \alpha_{0,0}(t) \exp(\beta_{02}\mathbf{z}(t))$$
$$\alpha_{12}(t, \mathbf{z}(t)) = \alpha_{0,0}(t) \exp(\beta_{12}\mathbf{z}(t)).$$

Because they share a common baseline intensity we must analyze the transitions $0 \rightarrow 2$ and $1 \rightarrow 2$ jointly. Suppose we make $\alpha_{0,0}(t)$ represent the ALL group without regard to platelet recovery status. First, modify the data set bmt_exp by

```
data common;
set bmt_exp(where = (stratum in ('02' '12')));
if dgroup = 1 then plc = 0;
else plc = plstatus;
run;
```

The PLC indicator will be used as a class variable. It cannot be created by programming statements within the TPHREG procedure. To fit the model, both DGROUP and FAB must be sub-grouped by PLC, the latter because we want FAB = 0 to refer to a single group within ALL. The two age variables and their interactions are sub-grouped by the levels of PLSTATUS. The following syntax will fit the model.

```
proc tphreg data = common multipass;
class plc plstatus(descending) dgroup(descending)
  fab/param = glm;
format dgroup dgroup. fab fab. plc plstatus plstatus.;
model (tstart, tstop)*status(0) = dgroup*plc fab*plc
  page*plstatus dage*plstatus page*dage*plstatus/
  ties = efron;
page = (page-28);
dage = (dage-28);
run;
```

Table 6.10 replaces the portion of Table 6.9 that corresponds to the transitions $0 \rightarrow 2$ and $1 \rightarrow 2$. With MTX included in the model for $\alpha_{01}(t, \mathbf{z}(t)) = \alpha_{01,0}(t) \exp(\beta_{01}\mathbf{z}(t))$ a separate invocation of PHREG is used (Table 6.11). All three transitions could be analyzed in one single call of PHREG by creating a data set in which MTX is identically zero for the $0 \rightarrow 2$ and $1 \rightarrow 2$ transitions but retains its values for the $0 \rightarrow 1$ transition. Also, a two-level stratum variable is easily created through formatting of $0 \rightarrow 2$ and $1 \rightarrow 2$ into one stratum and $0 \rightarrow 1$ as the other stratum. The syntax is similar to that used in Section 6.3.7 leading to Table 6.9.

TABLE 6.10

Parameter Estimates from Transition Model for 0 → 2 and 1 → 2 with a Common Baseline Intensity

Parameter	Class Value	Class Value	Parameter Estimate	Standard Error	p-Value
PLC*DGROUP	Before	AML-low risk	1.5442	0.6358	0.0152
PLC*DGROUP	Before	AML-high risk	1.3148	1.1598	0.2569
PLC*FAB	Before	FAB grade 4 or 5 and AML	−1.2465	1.1152	0.2637
PAGE*PLSTATUS	Before		−0.1599	0.0538	0.0030
DAGE*PLSTATUS	Before		0.1195	0.0436	0.0062
PAGE*DAGE*PLSTATUS	Before		0.0028	0.0019	0.1422
PLC*DGROUP	After	AML-low risk	−1.7625	0.4184	<0.0001
PLC*DGROUP	After	AML-high risk	−0.7906	0.3991	0.0476
PLC*FAB	After	FAB grade 4 or 5 and AML	1.2234	0.3224	0.0001
PAGE*PLSTATUS	After		0.0404	0.0216	0.0612
DAGE*PLSTATUS	After		−0.0308	0.0203	0.1299
PAGE*DAGE*PLSTATUS	After		0.0027	0.0012	0.0292

TABLE 6.11

Parameter Estimates from Model for Platelet Recovery (0 → 1 Transition)

Parameter	Class Value	Parameter Estimate	Standard Error	p-Value
MTX	Yes	−1.1970	0.2386	<0.0001
DGROUP	AML-high risk	−0.0817	0.2960	0.7824
DGROUP	AML-low risk	0.0742	0.2553	0.7713
FAB	FAB grade 4 or 5 and AML	−0.3481	0.2431	0.1522
PAGE		0.0332	0.0169	0.0495
DAGE		−0.0247	0.0150	0.0985
PAGE*DAGE		−0.0027	0.0010	0.0074

6.3.9 Calculating Transition Probabilities

Using the model specified in the previous section, we now consider the computation of the transition probabilities $P_{hj}(s,t \mid \mathbf{z})$ for a specified covariate profile $\mathbf{z}$. If instead our model had baseline intensities and covariates specific to each transition, our task would be relatively easy for some of these probabilities. For example, $P_{00}(s,t \mid \mathbf{z}) = \exp(-\int_s^t (\alpha_{01}(u,\mathbf{z}) + \alpha_{02}(u,\mathbf{z}))du)$ could be computed from the cumulative intensities $A_{01}(t,\mathbf{z}) = \int_0^t \alpha_{01}(u,\mathbf{z})du$ and $A_{02}(t,\mathbf{z}) = \int_0^t \alpha_{02}(u,\mathbf{z})du$ by estimating the transitions $0 \to 1$ and $0 \to 2$ separately. Similarly, from $A_{12}(t,\mathbf{z}) = \int_0^t \alpha_{12}(u,\mathbf{z})du$ we can get $P_{11}(s,t \mid \mathbf{z}) = \exp(-(A_{12}(t,\mathbf{z}) - A_{12}(s,\mathbf{z})))$. For $P_{01}(s,t \mid \mathbf{z})$ we could use $P_{01}(s,t \mid \mathbf{z}) = \int_s^t P_{00}(s,u \mid \mathbf{z})dA_{01}(u,\mathbf{z})P_{11}(u,t \mid \mathbf{z})$. These are the Nelson–Aalen estimators.

A general method is to use the product integral $\mathbf{P}(s,t \mid \mathbf{z}) = \prod_{u\in(s,t]}(\mathbf{I}+d\mathbf{A}(u))$. The two computational methods will give numerically comparable but not necessarily the same results. This is similar to the difference between the Kaplan–Meier (product-limit) estimator of survival and the Nelson–Aalen estimator of the survival function for a single event.

As an example of the calculations that can be carried out in SAS, consider the model of the previous section leading to the parameter estimates in Tables 6.10 and 6.11, and the probabilities $P_{02}(s,t \mid \mathbf{z}), P_{12}(s,t \mid \mathbf{z})$ as a function of s with t held fixed at 24 months. Then $P_{02}(s,24 \mid \mathbf{z})$ is the forecast of death or relapse by 24 months, given that at current time s platelet recovery has not occurred; $P_{12}(s,24 \mid \mathbf{z})$is the forecast of death or relapse by 24 months, given that at current time s platelet recovery has already occurred. We compute these probabilities for the three DGROUP specifying FAB = 0, PAGE = 28 and DAGE = 28. We also consider MTX use because it appears in the model for platelet recovery.

1. ALL Group

According to our model and covariate specification $\alpha_{02}(t,\mathbf{z}) = \alpha_{0,0}(t) = \alpha_{12}(t,\mathbf{z})$ and $\alpha_{01}(t,\mathbf{z}) = \alpha_{01,0}(t)\exp(\beta_{01}\text{MTX})$. As shown in Section 6.2.4, irrespective of MTX use we get $P_{02}(s,24 \mid \mathbf{z}) = P_{12}(s,24 \mid \mathbf{z})$.

2. AML-low risk, AML-high risk Groups

Here

$$\alpha_{02}(t,\mathbf{z}) = \alpha_{0,0}(t)\exp(\beta_{02,1}\text{AML}_L + \beta_{02,2}\text{AML}_H),$$
$$\alpha_{12}(t,\mathbf{z}) = \alpha_{0,0}(t)\exp(\beta_{12,1}\text{AML}_L + \beta_{12,2}\text{AML}_H) \quad \text{and}$$
$$\alpha_{01}(t,\mathbf{z}) = \alpha_{01,0}(t)\exp(\beta_{01}\text{MTX} + \beta_{01,1}\text{AML}_L + \beta_{01,2}\text{AML}_H),$$

where AML_L and AML_H are indicators for low- and high-risk AML groups.

We will focus on the calculations in the AML-low risk group. Create a data set COVAR00 with the covariate profiles of interest. This is used in the BASELINE statement in PHREG when estimating the death or relapse transitions jointly. COVAR00 could be extracted using

```
proc sort data = common out = covar00(keep = plc plstatus
   dgroup) nodupkey; by dgroup plc;
run;

data covar00;
set covar00;
fab = 0; page = 0; dage = 0;
format dgroup dgroup. fab fab. plc plc. plstatus plstatus.;
run;
```

COVAR00 has five profiles with the required covariates and formats to estimate the cumulative intensities from the model of Table 6.10. Only one additional statement is required in the PHREG call:

```
baseline covariates = covar00 out = cumint_00
  logsurv = logsurv;
```

The output data set cumint_00 has the estimates of $A_{02}(t, \mathbf{z})$ and $A_{12}(t, \mathbf{z})$ in the variable LOGSURV (but with negative sign). We reverse the sign in a short data step and change the time scale to months instead of days.

```
data cumint_00;
set cumint_00;
cum_haz = -logsurv;
time_m = (tstop/365.25)*12; /*time in months*/
run;
```

The next step is to get the estimates of $A_{01}(t, \mathbf{z})$ from the model of Table 6.11. This is done in an analogous manner with a prespecified COVAR01 data set with 6 DGROUP by MTX profiles. A data set cumint_01 as above is created.

In STATS1 below, we extract the profile information for AML-low risk, and no MTX use. This will be used to compute all the estimates $P_{hj}(s, 24 \mid \mathbf{z})$ through the product-integral formula using an IML routine. Note that MTX is set to missing for STRATUM 02 and 12 because this covariate does not appear in our model for the death or relapse transitions.

```
data stats1;
set cumint_00(keep = time_m cum_haz dgroup plc where =
  (dgroup = 2 and plc = 0) in = one)
  /*0→2 without recovery*/
    cumint_00(keep = time_m cum_haz dgroup plc where =
      (dgroup = 2 and plc = 1) in = two)
      /*1→2 with recovery*/
    cumint_01(keep = mtx time_m cum_haz dgroup where =
      (dgroup = 2 and MTX = 0) in = three); /*no MTX*/
if one then stratum = '02'; if two then stratum = '12';
  if three then stratum = '01';
run;
```

Two additional data steps will create STATS3 with the jumps $\Delta A_{hj}(t, \mathbf{z})$. These steps, the IML routine and the syntax for plotting the estimated probabilities, are summarized in the Appendix. For AML low risk, with MTX use the STATS1 data set needs to be recreated (changing MTX $= 0$ to MTX $= 1$) and all the subsequent steps rerun.

Figure 6.4 shows the plot of the estimated probabilities over the range of 4 months. For patients whose platelets have recovered to normal levels, the forecast for death or relapse at 24 months is fairly low. For patients without

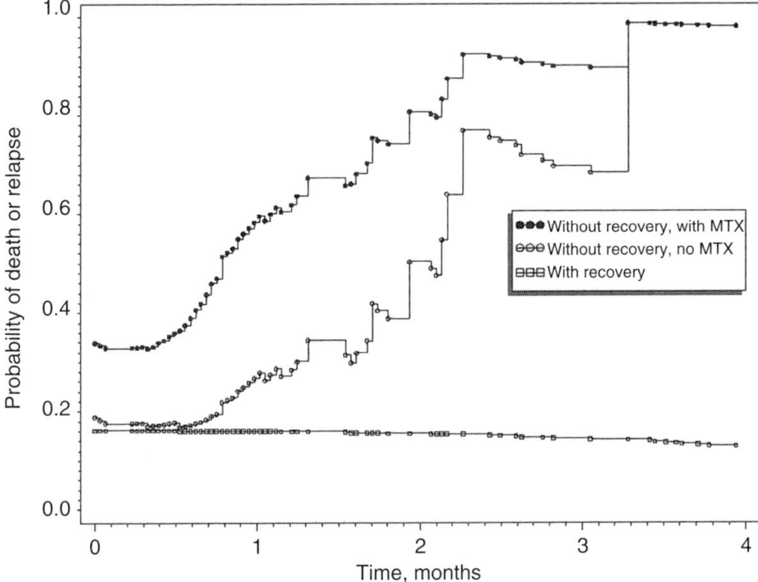

FIGURE 6.4
$P_{02}(\mathrm{u},24)$ and $P_{12}(u,24)$ for AML low risk.

recovery the probability of death or relapse increases sharply after about $1\frac{1}{2}$ months in patients with MTX use having worse outcome. From Table 6.11 we see that MTX use negatively impacts platelet recovery.

6.4 Concluding Remarks

In this chapter, we have demonstrated the use of SAS software in analyzing multiple failure times. The context is a multistate Markov model that describes the events that a patient might experience over time. Event types become synonymous with states of the process. In our example of leukemia patients who underwent bone marrow transplantation time is measured from the start of follow-up when all patients are assumed to be in remission (state 0). The terminal event is death or relapse (state 2) with platelet recovery (state 1) being an intermediate event. Using the counting process-style input we assessed the impact of DGROUP and other covariates on the transition intensities of death or relapse with platelet recovery ($1 \to 2$) and without platelet recovery ($0 \to 2$). In Section 6.3.7, our transition model has baseline intensities and regression parameters specific to each transition type. The results in Table 6.9 could also be obtained by analyzing each transition individually with separate invocations of PHREG. By expanding our data set to contain individual records for each patient at risk of each event type, a single invocation of PHREG is sufficient. In this format, one can easily test hypotheses of the effect of covariates across event types.

In Section 6.3.8, we assume a common baseline intensity for the $0 \rightarrow 2$ and $1 \rightarrow 2$ transitions and again we suggest a joint analysis with suitable modification of the input data set. Next, we show how transition probabilities could be calculated from the PHREG output with some basic matrix manipulations using SAS/IML.

The use of the multistate representation comes with some caveats, particularly in using it for time-dependent covariates. Here we have a single time-dependent covariate—the platelet recovery indicator PLSTATUS(t) that can change value from 0 to 1 only. If we had modified our model of Section 6.3.8 so that the transitions $1 \rightarrow 2$ and $0 \rightarrow 2$ have proportional intensities with respect to PLSTATUS(t) this covariate is now endogenous as noted in Section 6.2.2. Our methods still apply to time-dependent covariates that can have only a finite set of values so that the multistate representation could be made. With covariates that could change continuously this is clearly infeasible. Joint modeling of the covariate and the event-history processes are needed. Several articles addressing this problem (Hogan and Laird, 1997; Wulfsohn and Tsiatis, 1997; Henderson et al., 2000) generally have one covariate process and a single-event counting process such as mortality.

There are other analyses of multiple events times that we have not discussed here. For example, in studies where a natural clustering of units exists the events times should be analyzed at the cluster level acknowledging the likely correlation between event occurrences within the cluster. With recurrent event data where each unit could experience a number of repeated events of the same type such as episodes of a disease, or recurrences of tumors a state-specific analysis is possible by labeling the sequence of events $0 \rightarrow 1 \rightarrow 2 \rightarrow \cdots$. The PHREG procedure can also be applied in these analyses. It might be surmised from our example that the main effort lies in preparing the requisite data set that the procedure would use in creating the correct risk sets for fitting the intended model.

The PHREG procedure is a powerful tool in alleviating the computational burden of analyzing event-history data. Future releases of the software will likely address modeling of multivariate survival times through random effects and frailty models.

Appendix

SAS Steps for Computing $P(s,24)$ in AML-Low Risk

```
proc sort data = stats1; by stratum mtx; run;

data stats2; /** Subsetting to AML low risk
  with/without MTX **/
set stats1(where = (dgroup = 2 and (mtx = 0 or mtx = .))
  keep = stratum dgroup mtx time_m cum_haz);
by stratum;
cumhaz_ = lag(cum_haz);
if first.stratum then cumhaz_ = 0;
del_haz = cum_haz-cumhaz_;
run;

proc sort data = stats2; by time_m;
run;

proc transpose data = stats2 out = stats3(drop = _name_)
   prefix = DELA;
by time_m;
var del_haz;
id stratum;
run;

%/*********************************************************/
/* Jumps in cumulative intensities A00, A01, A02   */
/* A11, A12                                         */
/*********************************************************/

data stats3;
set stats3;
array s{*} dela01 dela02 dela12;
do i = 1 to dim(s);
if s{i} = . then s{i} = 0;
end;
drop i;
dela00 = -(dela01+dela02);
dela11 = -dela12;
dela10 = 0;
run;

%/*********************************************************/
/*** IML STEPS TO COMPUTE P(u,24)                    ***/
/*********************************************************/
```

```
proc iml;

use stats3(where = (time_m< = 24));
read all var{time_m} into TIME;
read all var{DELA00 DELA01 DELA02 DELA10 DELA11 DELA12}
  into A;
A[,1] = A[,1]+1;
A[,5] = A[,5]+1;
P_ = I(3);
P = shape(0,3,3);


varnames = {T P00 P01 P02 P10 P11 P12 P20 P21 P22};
nc = ncol(varnames);
ET = J(nrow(TIME),nc,0);
        do i = nrow(A) to 2 by -1;
P[1,] = A[i,{1 2 3}];
P[2,] = A[i,{4 5 6}];
P[3,] = {0 0 1};
P_ = P*P_; /** This is important **/


E = rowvec(P_);
ET[i-1,] = TIME[i-1,1]||E;
      end;


nr = nrow(TIME);
ET[nr,1] = TIME[nr,1];
ET[nr, 2:10] = {1 0 0 0 1 0 0 0 1};
mattrib ET colname = varnames;


create ESTIMATE from ET[colname = varnames];
           append from ET;
close estimate;
quit;


data estimate1(label = 'AML low risk, no MTX');
set estimate;
  label P12 = 'AML high risk with recovery'
        P02 = 'AML high risk wihout recovery';
run;
```

** Repeat routine to create estimate2 for AML-low risk, with MTX **

SAS Steps for Plotting

** Create AML_low to contain all estimates of $P_{02}(s, 24 \mid \mathbf{z})$ and $P_{12}(s, 24 \mid \mathbf{z})$**

```
data AML_low;
set estimate1(keep = T P02 in = one rename = (P02 = P))
    estimate2(keep = T P02 in = two rename = (P02 = P))
    estimate2(keep = T P12 in = three rename = (P12 = P));

if one then do; MTX = 0; PLC = 0; GRP = 1; end;
if two then do; MTX = 1; PLC = 0; GRP = 2; end;
if three then do; PLC = 1; GRP =  3; end;
run;

proc format;
value grp 1 = 'without recovery, no MTX'
          2 = 'without recovery, with MTX'
          3 = 'with recovery';
run;

goptions reset = global colors = (black blue red
  green purple)
     gunit = pct cback = white ctext = black
       ftext = "Garamond"
     htitle = 3 htext = 2.5 hsize = 7 in vsize = 7 in
       offshadow = (-.5 -.5);
symbol1 interpol = stepjl value = dot color = black h = 1;
symbol2 interpol = stepjl value = dot color = purple h = 1;
symbol3 interpol = stepjl value = dot color = blue h = 1;

axis1 label = (angle = 90 'Probability of Death or Relapse')
  order = (0 to 1 by .2) offset = (2 2);
axis2 label = ('TIME, months') order = (0 to 4 by 1)
  offset = (2 2);

legend1 across = 1 label = none cshadow = gray
  mode = protect frame position = (inside middle right)
  offset = (-2,2);

proc gplot data = AML_low;
plot P*T = grp/vaxis = axis1 haxis = axis2 legend = legend1;
title 'Figure: P02(u,24) and P12(u,24)';
title2 j = c 'AML low risk';
format grp grp.;
run;
quit;
```

Acknowledgment

This study was supported by the Agency for Healthcare Research & Quality under grant 1R01 HS14206.

References

Andersen PK. Time-dependent covariates and Markov processes. In: Moolgavkar SH, Prentice RL, Eds. *Modern Statistical Methods in Chronic Disease Epidemiology*. New York: Wiley, 1986:82–103.

Andersen PK, Borgan O, Gill RD, Keiding N. *Statistical Models Based on Counting Processes*. New York: Springer-Verlag, 1993.

Andersen PK, Esbjerg S, Sorensen TIA. Multi-state models for bleeding episodes and mortality in liver cirrhosis. *Statistics in Medicine* 2000;19(4):587–599.

Andersen PK, Keiding N. Multi-state models for event history analysis. *Statistical Methods in Medical Research* 2002;11(2):91–115.

Gardiner JC, Luo Z, Bradley CJ, Sirbu CM, Given CW. A dynamic model for estimating changes in health status and costs. *Statistics in Medicine* 2006;25(21): 3648–3667.

Gardiner JC, Luo Z, Liu L, Bradley CJ. A stochastic framework for estimation of summary measures in cost-effectiveness analyses. *Expert Review of Pharmacoeconomics & Outcomes Research* 2006;6(3):347–358.

Golub GH, Van Loan CF. *Matrix Computations. Third Edition*. Baltimore, MD: The Johns Hopkins University Press, 1996.

Heckman JJ, Singer B, Eds. *Longitudinal Analysis of Labor Market Data*. Cambridge, UK: Cambridge University Press, 1985.

Henderson R, Diggle P, Dobson A. Joint modelling of longitudinal measurements and event time data. *Biostatistics* 2000;1(4):465–480.

Hogan JW, Laird NM. Mixture models for the joint distribution of repeated measures and event times. *Statistics in Medicine* 1997;16(1–3):239–257.

Keiding N, Klein JP, Horowitz MM. Multi-state models and outcome prediction in bone marrow transplantation. *Statistics in Medicine* 2001;20(12):1871–1885.

Klein JP, Moeschberger ML. *Survival Analysis: Techniques for Censored and Truncated Data*. New York: Springer-Verlag, 1997.

Klein JP, Szydlo RM, Craddock C, Goldman JM. Estimation of current leukaemia-free survival following donor lymphocyte infusion therapy for patients with leukaemia who relapse after allografting: application of a multistate model. *Statistics in Medicine* 2000;19(21):3005–3016.

Klein JP, Shu YY. Multi-state models for bone marrow transplantation studies. *Statistical Methods in Medical Research* 2002;11(2):117–139.

Lancaster T. *The Econometric Analysis of Transition Data*. Cambridge, UK: Cambridge University Press, 1990.

SAS/STAT Software: The PHREG Procedure. Version 9.1.3. SAS Institute Inc, Cary, NC.

Therneau TM, Grambsch PM. *Modeling Survival Data: Extending the Cox Model*. New York: Springer-Verlag, 2000.

Van den Berg GJ. Duration Models: Specification, Identification and Multiple Durations. In: Heckman JJ, Leamer E, Eds. *Handbook of Econometrics, Vol. 5*. New York: North-Holland, 2001:3381–3460.

Wulfsohn MS, Tsiatis AA. A joint model for survival and longitudinal data measured with error. *Biometrics* 1997;53(1):330–339.

7

Mixed Effects Models for Longitudinal Virologic and Immunologic HIV Data

Florin Vaida, Pulak Ghosh, and Lin Liu

CONTENTS

7.1 Statistical Issues in Modeling HIV Data

More than 20 years after its initial outburst, HIV/AIDS pandemic continues to be one of the most important threats of global public health, with an estimated 39 million people infected world wide and 4 million new infections each year (UNAIDS, 2006). Modeling HIV data proved to be a challenging area of research and has led to numerous developments in the design and analysis of clinical trials and observational studies. The introduction of highly active antiretroviral treatment in the mid-1990s led to a dramatic reduction in the rates of death or development of AIDS. The severity of infection has subsequently been measured by immunological and virological markers, especially concentration of CD4 cells and HIV-1 viral RNA (viral load, VL) in the plasma. Challenges to modeling these markers are (1) the measurements are longitudinal; (2) the models are in general complex and rarely can be reduced to linear models; and (3) these markers are affected by several factors, including development of viral resistance, start or stopping of treatment, nonadherence to treatment, choice of treatment, and treatment toxicity. These factors, alone or in combination, affect the validity of simple statistical models, may induce informative dropout, errors in variables, censored observations, and so forth. Furthermore, in contrast to randomized clinical trials, observational studies data are subject to confounding, such as the decision when to start or change treatment, and dropout.

Recent work dealing with some of these issues includes Brown et al. (2005)—joint modeling of CD4, HIV-1 RNA, and time to AIDS or death; Foulkes and DeGruttola (2003)—modeling of viral resistance; Wu and Ding (1999)—modeling of HIV-1 response to antiretroviral treatment; see also Hughes (2000).

In this chapter, we will present some of our recent work in modeling longitudinal HIV-1 RNA and CD4 data using mixed effects models, with a focus on the statistical and computational challenges encountered. For the use of mixed models in the context of survival analysis models, see Xu and Donohue (2007) in this volume.

7.2 Mixed-Effects Models for Censored HIV-1 RNA Data

Linear and nonlinear mixed effects (LME and NLME respectively, referred jointly as N/LME) models have a long history in modeling HIV data, dating back to the early work of DeGruttola et al. (1991). Carlin (1996) modeled CD4 counts in a Bayesian framework. Wu and Ding (1999) used NLME for HIV-1 RNA following start of treatment; Fitzgerald et al. (2002) used NLME to model viral rebound owing to treatment failure. Guo and Carlin (2004) added explicit modeling of the dropout process. See also Ghosh and Vaida (2007) and the citations within. Mixed effects models allow for subject-level estimation and have a simple and attractive interpretation. In addition, when the modeling is done carefully they yield valid, unbiased, and efficient inference, in the presence of dropout occurring at random.

A special challenge presents the censored HIV-1 RNA values. In untreated HIV-1 individuals the large HIV-1 concentrations may be above the limit of detection of the assay. We will analyze such a dataset in Section 7.2.2. Conversely, antiretroviral treatment leads typically to "viral suppression" in plasma. Unfortunately, the HIV is not eradicated, but its concentration decreases below the limit of detection of the assay. It is important that this type of censoring be accounted for in the statistical analysis. Earlier work used ad hoc methods, such as replacing the censored values with half the detection limit (Wu and Ding, 1999). Fitzgerald et al. (2002) proposed multiple imputation to deal with censoring, and Hughes (1999) proposed a Monte Carlo Expectation-Maximization (MCEM) algorithm for estimating the maximum likelihood estimator (MLE) of the LME with censored response (LMEC). Vaida et al. (2007) extend this work to N/LME with censored response (N/LMEC) and improve the computation of this algorithm, including efficient block-sampling at the Monte Carlo E-step, improved numeric implementation and automatic monitoring and stopping of the algorithm. They also compare the three methods (ad hoc, multiple imputation, and MCEM) and show that MCEM estimation is the most statistically efficient.

Vaida et al. (2007) used closed-form formulas at the E-step for clusters with one or two censored observations. Building on this work, in this section we will discuss an implementation of the EM algorithm for N/LMEC with improved speed and precision. The main advantage of this algorithm is that it does not require Monte Carlo simulation; the E-step is available in closed form, requiring only the multinormal cummulative distribution function (CDF). This in turn is computed in R using the `mvtnorm` package (Genz, 1992).

7.2.1 EM Algorithm for N/LME with Censored Response

Consider the Laird–Ware model

$$y_i = X_i\beta + Z_ib_i + e_i, \quad b_i \sim N(0, \sigma^2 D), \quad e_{ij} \sim N(0, \sigma^2), \tag{7.1}$$

$i = 1, \ldots, m$, where $e_i = (e_{i1}, \ldots, e_{in_i})'$ and b_i, e_i are independent for all i and independent of each other. D is a positive definite matrix depending on a vector of parameters γ. Put $n = \sum_{i=1}^{m} n_i$, $\sigma^2 D = \Psi$ and $V_i = \mathrm{var}(y_i) = Z_i\Psi Z_i' + \sigma^2 I$. The response y_{ij} is not fully observed for all i, j. Assuming left censoring, let the observed data for the ith subject be (Q_i, C_i), where Q_i represents the vector of uncensored readings or censoring levels, and C_i the vector of censoring indicators

$$y_{ij} \le Q_{ij} \qquad \text{if } C_{ij} = 1; \quad y_{ij} = Q_{ij} \quad \text{if } C_{ij} = 0. \tag{7.2}$$

In the EM we update β, σ^2 with $\{y_{ij} : C_{ij} = 1\}$ as missing data, and Ψ using $\{y_{ij} : C_{ij} = 1\}$ and b_i as missing data. Decompose $D^{-1} = \Delta'\Delta$ and write: $\delta = (\beta', b_1', \ldots, b_m')'$, $\tilde{y} = (\tilde{y}_1', \ldots, \tilde{y}_m')'$, where

$$\begin{pmatrix} \tilde{y}_i & \tilde{X}_i & \tilde{Z}_i \end{pmatrix} = \begin{pmatrix} y_i & X_i & Z_i \\ 0 & 0 & \Delta \end{pmatrix}, \quad \text{and} \quad M = \begin{pmatrix} \tilde{X}_1 & \tilde{Z}_1 & & \\ \vdots & & \ddots & \\ \tilde{X}_m & & & \tilde{Z}_m \end{pmatrix}. \tag{7.3}$$

The M-step updates for the MLE (Vaida et al., 2007) are as follows:

$$\hat{\delta} = (M'M)^{-1}M'E(\tilde{y}), \tag{7.4}$$

$$\hat{\sigma}^2 = \frac{1}{n}\|E(\tilde{y}) - M\hat{\delta}\|^2 + \frac{1}{n}\sum_{i=1}^{m}\mathrm{tr}\{\mathrm{var}(y_i)\} - \frac{1}{n}\sum_{i=1}^{m}\mathrm{tr}\{W_iZ_i'\mathrm{var}(y_i)Z_i\}, \tag{7.5}$$

$$\hat{\Psi} = \frac{1}{m}\sum_{i=1}^{m}E(b_i)E(b_i)' + \frac{1}{m}\sum_{i=1}^{m}\mathrm{var}(b_i), \tag{7.6}$$

where $W_i = (Z_i'Z_i + D^{-1})^{-1}$, $E(b_i) = W_iZ_i'\{E(y_i) - X_i\beta\}$, $\text{var}(b_i) = \sigma^2 W_i + W_iZ_i'\text{var}(y_i)Z_iW_i$, and $E(y_i)$, $\text{var}(y_i)$ are the mean and variance conditional on $\{C_i, Q_i; i = 1, \ldots, m\}$, taken at the current parameter value $\theta = (\beta, \sigma^2, D)$. The update for unstructured Ψ is given by Equation 7.6. If Ψ is diagonal the right-hand side of Equation 7.6 is replaced by the diagonal matrix with same diagonal elements as Equation 7.6.

The computations use dimension reduction based on QR decomposition, which takes advantage of the sparse nature of the matrix M (Pinheiro and Bates, 2000). The key feature is that the number of columns of the matrices to be decomposed does not increase with the number of clusters m or the number of data points N.

From Equations 7.4 through 7.6 it is clear that the E-step reduces to the computation of $E(y_i|Q_i, C_i, \theta)$ and $\text{var}(y_i|C_i, Q_i, \theta)$. These are determined as follows. Partition y_i into the observed and censored parts: $y_i' = (y_i^{o\prime}, y_i^{c\prime})$, that is, $C_{ij} = 0$ for all elements in y_i^o, and 1 for all elements in y_i^c; write accordingly $Q_i' = (Q_i^{o\prime}, Q_i^{c\prime})$. Ignoring censoring for the moment, we have that marginally $y_i \sim N(X_i\beta, \Sigma = \sigma^2(I + Z_iDZ_i'))$. Then $y_i^o \sim N(X_i^o\beta, \Sigma_{oo})$, $y_i^c|y_i^o \sim N(\mu_i, S_i)$, where

$$\mu_i = X_i^c\beta + \Sigma_{co}\Sigma_{oo}^{-1}(y_i^o - X_i^o\beta),$$
$$S_i = \Sigma_{cc} - \Sigma_{co}\Sigma_{oo}^{-1}\Sigma_{oc},$$

and

$$\Sigma = \begin{pmatrix} \Sigma_{oo} & \Sigma_{oc} \\ \Sigma_{co} & \Sigma_{cc} \end{pmatrix}.$$

It follows that

$$E(y_i|Q_i, C_i, \theta) = ((y_i^o)', (\mu_i^c)')', \quad \text{var}(y_i|Q_i, C_i, \theta) = \begin{pmatrix} 0 & 0 \\ 0 & S_i^c \end{pmatrix},$$

where $\mu_i^c = E(U)$, $S_i^c = \text{var}(U)$, and $U = (y_i^c|Q_i^c, y_i^o)$ follows a multinormal distribution $N(\mu_i, S_i)$ left-truncated at Q_i^c. Let B_i be a diagonal matrix with diagonal elements equal to the square roots of the corresponding diagonal elements in S_i. Put $X = B_i^{-1}(U - \mu_i)$. Then X has a multinormal distribution $N(0, R_i)$ left-truncated at $a_i = B_i^{-1}(Q_i^c - \mu_i)$ and $R_i = B_i^{-1}S_iB_i^{-1}$ is the correlation matrix corresponding to S_i. Then $\mu_i^c = B_iE(X) + \mu_i$, $S_i^c = B_i\text{var}(X)B_i$ and calculation of μ_i, S_i^c reduces to computing the mean and variance of X. Closed-form formulas for $E(X)$, $\text{var}(X)$ were developed by Tallis (1961) and Finney (1962). They depend on the multinormal CDF, of dimension smaller than or equal to the dimension of X, or the number of censored observations in the cluster. The multinormal CDF is available in `R` through the `pmvnorm()` function from the `mvtnorm` package (Genz, 1992; R Development core Team, 2006), which is called in our computation routine. See Vaida and Liu (2007) for further details.

The variance of the MLE $\hat{\theta}$, estimated at convergence, is adjusted for the censored information using Louis' formula (Orchard and Woodbury, 1972; Louis, 1982). The variance of the fixed effects in the approximate MLE is given (Hughes, 1999) by

$$\text{var}(\hat{\beta}) = \left(\sum_{i=1}^{m} \{X_i' V_i^{-1} X_i - X_i' V_i^{-1} \text{var}(y_i | Q_i, C_i) V_i^{-1} X_i\} \right)^{-1}. \qquad (7.7)$$

7.2.1.1 The Likelihood Function

Put $\Phi_n(u; A)$ and $\phi(u; A)$ be respectively the left-tail probability (componentwise) and the probability density function of the $N(0, A)$ distribution, computed at u. Let $\alpha_i = P(y_i^c < Q_i^c | y_i^o) = \Phi_{n_i^c}(a_i; R_i)$. The likelihood for cluster i is given by

$$\begin{aligned} L_i &= P(Q_i | C_i, \theta) \\ &= P(y_i^c \leq Q_i^c | y_i^o = Q_i^o, \theta) P(y_i^o = Q_i^o | \theta) \\ &= \alpha_i \, \phi(Q_i^o - X_i^o \beta; \Sigma_{oo}). \end{aligned}$$

Therefore, the log-likelihood function for the observed data is given by

$$l(\theta) = \sum_{i=1}^{m} \{\log \alpha_i \; + \log \phi_{n_i^o}(Q_i^o - X_i^o \beta; \Sigma_{oo})\}.$$

This can be computed at each step of the EM algorithm without additional computational burden, since α_i's are computed at the E-step. The log-likelihood can be used to monitor the convergence of the algorithm. Alternatively, Vaida et al. (2007) monitor convergence using the objective function

$$f_o(\theta) = -N/2\{1 + \log(2\pi\sigma^2)\} + m/2 \log |D^{-1}| - 1/2 \sum_{i=1}^{m} \log |Z_i' Z_i + D^{-1}|,$$

which is the log-likelihood of the linear mixed model without censoring, with β profiled out (Pinheiro and Bates, 2000, Chapter 2).

7.2.1.2 Nonlinear Case

The N/LME (Lindstrom and Bates, 1990; Pinheiro and Bates, 2000) is given by

$$y_{ij} = f(\beta, b_i) + e_{ij}, \qquad (7.8)$$

where $f(\beta, b_i) = f(\beta, b_i, x_{ij})$ is a nonlinear function of the fixed β and random effect b_i; x_{ij} is a vector of covariates, and b_i and e_{ij} are given by Equation 7.1. The approximate MLE $(\hat{\beta}, \hat{\sigma}^2, \hat{\gamma})$ and predictors for the random effects $\hat{b}_i$ are computed by iterative linearization (L) of the conditional mean function. The L-step yields the LME $w_i = X_i^*\beta + Z_i^*b_i + e_i, i = 1, \ldots, m$, where

$$w_i = y_i - \left\{ f_i^* - \left(\frac{\partial f_i^*}{\partial \beta} \right) \beta^* - \left(\frac{\partial f_i^*}{\partial b_i} \right) b_i^* \right\}, \tag{7.9}$$

$X_i^* = \partial f_i^* / \partial \beta$, $Z_i^* = \partial f_i^* / \partial b_i$, y_i is the n_i-vector dependent variable for the ith subject, f_i, e_i are respectively the corresponding mean function and error n_i-vectors, and the terms marked with an asterisk are computed with the current parameters (β^*, b_i^*). For censored response the linearized model is an LME with censored data, which is solved as in the previous section. More precisely, the algorithm iterates to convergence between L, E, and M steps. See Vaida et al. (2007) for more details.

7.2.2 HIV-1 Viral Load Setpoint for Acutely Infected Subjects

Our study concerns untreated individuals with acute HIV infection. Longitudinal HIV RNA measurements were taken on 320 subjects from the Acute Infection and Early Disease Research Program (AIEDRP), a large, ongoing, multicenter observational study, who were identified as HIV positive close to the time of infection. In contrast with HIV-1 RNA data for subjects on antiretroviral treatment (e.g., Vaida et al., 2007), some observations here are right-censored, since during the acute stage of infection the large HIV RNA observations may lay *above* the limit of quantification of the assay. This limit is between 75,000 and 500,000 cp/mL, depending on the assay. The time of infection was estimated at 24 days before first positive HIV RNA sample or detectable serum p24 antigen test. We included HIV RNA in the first 180 days of follow-up and only up to the start of antiretroviral treatment. In the absence of treatment following acute infection, the HIV RNA decreases and then varies around a setpoint value. This setpoint value may differ between individuals, and is of central interest here. The viral setpoint characterizes the severity of infection, it may relate to the strength of the subject's immune system and it may predict clinical progression of the disease. The subjects had between 1 and 14 observations: 129 had one, 82 had two, and 109 had three or more observations. Of the 830 recorded observations, 185 (22%) were above the limit of quantification of the assay (right-censored). The individual profiles and a smooth mean of the observed data are included in Figure 7.1. The smooth curve agrees qualitatively with the postulated shape of the HIV RNA trajectory for acutely infected patients. There is possible indication of a continuing viral decay rather than stabilization to a setpoint, with the caveat that the observed mean curve may be biased owing to the exclusion of the censored values and to differential follow-up (see, e.g., Diggle et al., 2002, Chapter 11). It is clear that the viral setpoint values differ from subject to subject.

Our analysis considers three models for these data. We started by fitting a four-parameter logistic model taking into account the censoring information.

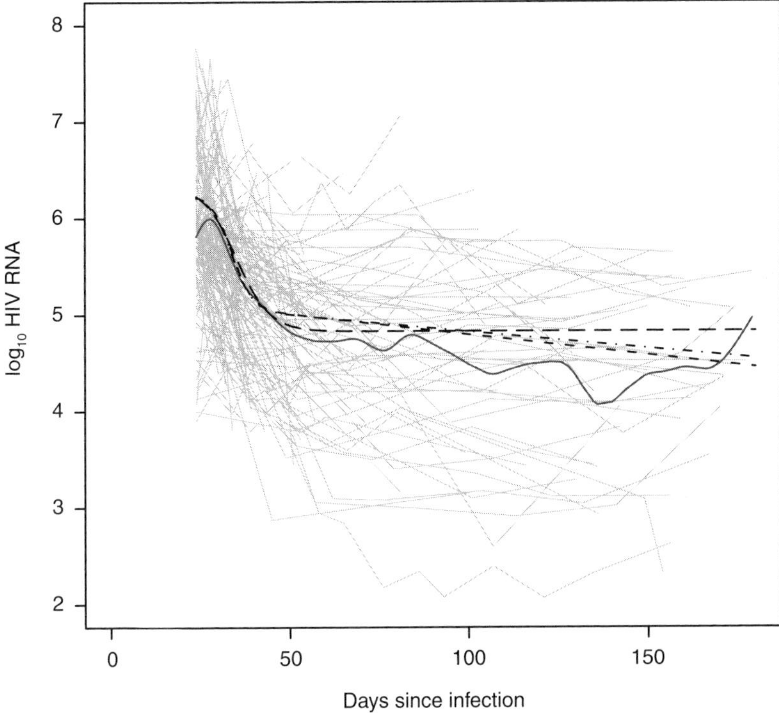

FIGURE 7.1
AIEDRP data and model fits from (1) random intercept logistic model (– –); (2) random intercept logistic model with linear decrease after 50 days (- -); (3) logistic model with random intercept and random linear decrease after 50 days (– · –). Solid line: a smooth fit of the observed data with censored observations excluded.

The model is

$$y_{ij} = \alpha_{1i} + \alpha_2[1 + \exp\{(t_{ij} - \alpha_3)/\alpha_4\}]^{-1} + e_{ij}, \tag{7.10}$$

where y_{ij} is the $\log_{10}$ HIV RNA for subject i at time t_{ij}. This is an inverted S-shaped curve, with the constant value for the later times representing the subject-specific setpoint. The parameters α_{1i} and α_2 are the setpoint value and the decrease from the maximum HIV RNA; α_4 is a scale parameter modeling the rate of decline, and α_3 is a location parameter indicating the time of achieving the HIV RNA midpoint value. In order to force the parameters to be positive, we reparametrized the model to $\beta_{1i} = \log(\alpha_{1i}), \beta_k = \log(\alpha_k), k = 2, 3, 4$. The setpoint α_{1i} was taken to be random: $\beta_{1i} = \beta_1 + b_i, b_i \sim N(0, \sigma_{b1}^2)$. It is tempting to consider models including random effects for β_3 and β_4, but there are not enough available data in the acute (earliest) phase of infection to allow for inclusion of these random parameters.

The plot of model residuals against time shows a relatively good fit (Figure 7.2), but it suggests that the model does not capture a time trend

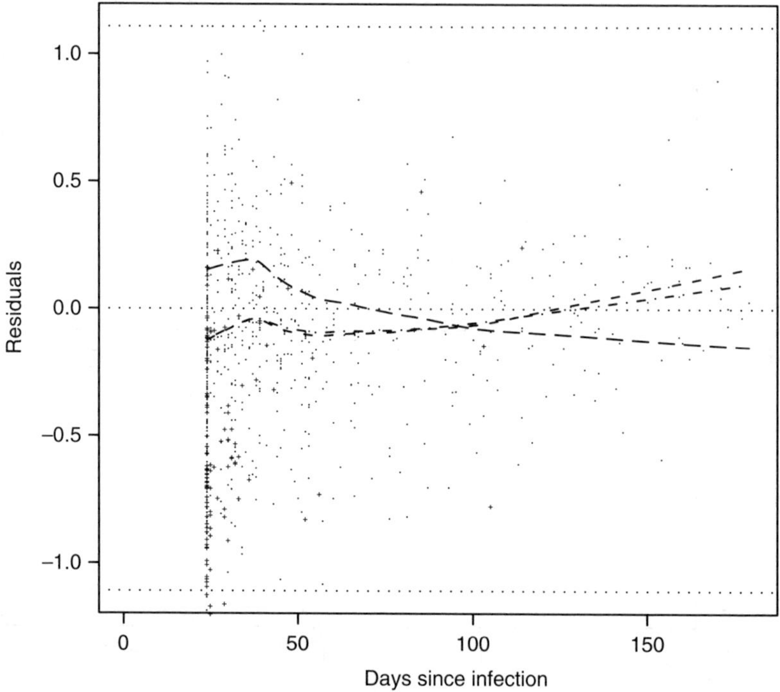

FIGURE 7.2
AIEDRP data: smooth means of residuals from (1) random intercept logistic model (– –); (2) random intercept logistic model with linear decrease after 50 days (- -); (3) logistic model with random intercept and random linear decrease after 50 days (– · –). The residuals from model (3) appear as points; the right-censored residuals appear as "+."

in the data after day 50 since infection and an initial increase in viral load (see also Figure 7.1). In addition, a variogram of the residuals (Figure 7.3) indicates long-term autocorrelation, which may be either due to bias in modeling the mean term or to genuine serial autocorrelation beyond the random intercept, unaccounted for in the model.

To address the bias concern we added a linear term after day 50 in the second model

$$y_{ij} = \alpha_{1i} + \alpha_2[1 + \exp\{(t_{ij} - \alpha_3)/\alpha_4\}]^{-1} + \alpha_5(t_{ij} - 50) + e_{ij} \tag{7.11}$$

The residuals' plot (Figure 7.2) indicates a better overall fit, but the variogram (Figure 7.3) shows that the serial autocorrelation is not properly accounted for. This suggests a third model, by adding a *random slope* after day 50

$$y_{ij} = \alpha_{1i} + \alpha_2[1 + \exp\{(t_{ij} - \alpha_3)/\alpha_4\}]^{-1} + \alpha_{5i}(t_{ij} - 50) + e_{ij}. \tag{7.12}$$

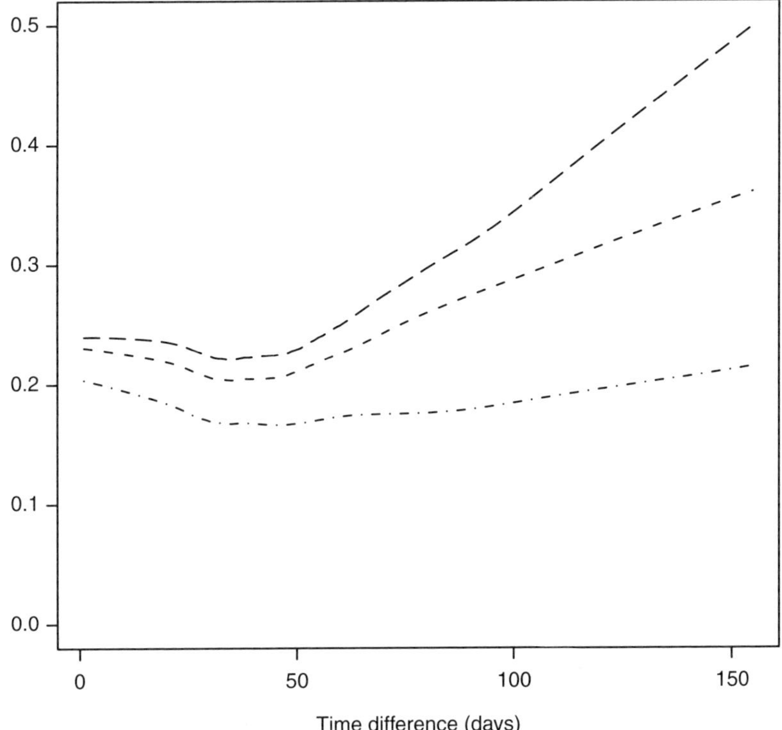

FIGURE 7.3
AIEDRP data: variogram from model residuals from (1) random intercept logistic model (– –); (2) random intercept logistic model with linear decrease after 50 days (- -); (3) logistic model with random intercept and random linear decrease after 50 days (— · —).

As in Equation 7.10, we have $\log(\alpha_{1i}) = \beta_{1i} = \beta_1 + b_{1i}$, $\beta_k = \log(\alpha_k)$ for $k = 2, 3, 4$, but $\alpha_{5i} = \beta_5 + b_{5i}$, in order to allow for increasing HIV RNA trajectories after day 50. Also, (b_{1i}, b_{5i}) are assumed to be iid, multivariate normal with unrestricted variance matrix. The model fit is slightly better than that in the second model, with the smooth mean residual curve in Figure 7.2 closer to zero, and fitted values between the fitted values of the first two models. More importantly, the residuals show no serial correlation in the variogram (Figure 7.3).

The results of the analysis are in Table 7.1. We can use the last model with reasonable confidence for predictions of viral load. For example, at 6 months since infection the average viral load is 4.55 $\log_{10}$ units (in contrast, the set-point model (7.10) estimates this at 4.83. The individual 6-month viral load estimates vary between 1.63 and 6.65, with 5th and 95th quantiles at 3.37 and 5.50. The average slope after day 50 was negative, $\beta_5 = -0.0035 \log_{10}$ HIV/day, with 95% CI $(-0.0063, -0.0006)$. However, the individual slopes α_{5i} included positive values, with 5th and 95th quantiles of -0.0070 and 0.0004.

TABLE 7.1

Analysis of Primary HIV Infection

	Setpoint Model		Five-Parameter Model	
	Estimate	**SE**	**Estimate**	**SE**
β_1	1.575	0.014	1.609	0.014
β_2	0.4240	0.0933	0.1441	0.0950
β_3	3.561	0.034	3.526	0.024
β_4	1.547	0.228	1.060	0.267
β_5			-3.48×10^{-3}	1.43×10^{-3}
σ	0.554		0.512	
σ_{b1}	0.139		0.133	
σ_{b5}			7.10×10^{-3}	
ρ_{b12}			0.17	

The parameters are from the random-intercept logistic model and logistic model with random intercept and random linear decrease after 50 days, respectively.

7.3 Random Changepoint Modeling of CD4 Cells Rebound

In most HIV patients who initiate and sustain highly active antiretroviral treatment (HAART), the viral load decreases sharply to undetectable levels. At the same time, the immune system recovery is marked by an increase in the CD4 T-cell count. Several authors noted two stages in the CD4 count increase: a sharper increase following start of treatment, until 6–18 weeks after treatment, followed by a slower, gradual increase (Bennett et al., 2002; Deeks et al., 2004; Bosch et al., 2006). There is no clear agreement in the literature on the time of change of the CD4 slope. One possible explanation is that the CD4 profiles may be different for different individuals. To account for these differences following start of HAART we propose a random changepoint model for the CD4 counts, see also Ghosh and Vaida (2007). A second important aspect concerns patient dropout. The longitudinal CD4 profiles are censored at the time of the subject going off study treatment. This is a potentially informative dropout mechanism. We model the informative dropout jointly with the CD4 count outcome, similar to Guo and Carlin (2004). The parameters of the dropout time distribution include the time of changepoint and the random effects of the CD4 model. We take a Bayesian approach to estimation. This has the major computational advantage of parameter estimation using Markov chain Monte Carlo (MCMC). Inference is based on the posterior distribution, which is assessed from the MCMC sample. As discussed by Gelman et al. (2004), the various sources of parameter uncertainty are accounted for, in contrast to the standard asymptotic methods of classical inference. Also, see Lopes et al. (2007) in this volume. We use the deviance information criterion (DIC) (Spiegelhalter et al., 2002) for model selection.

Our work is motivated by the analysis of the AIDS clinical trial ACTG 398. We introduce the general model first, followed by the data analysis.

7.3.1 A Hierarchical Bayesian Changepoint Model

Let $y_i = (y_{i1}, \ldots, y_{in_i})'$ be the response vector (CD4 cell counts) for individual i, at time $t_i = (t_{i1}, \ldots, t_{in_i})'; i = 1, \ldots, n; j = 1, 2, \ldots, n_i$. A square root transformation of the CD4 counts improves symmetry and normality of the distribution around the mean.

The first stage of the hierarchical model assumes the following changepoint ("broken stick") regression:

$$y_{ij} = \beta_1 + \beta_2(t_{ij} - K_i)_- + \beta_3(t_{ij} - K_i)_+ + b_i + e_{ij}, \tag{7.13}$$

where K_i is the changepoint for subject i, $(t_{ij} - K_i)_-$ equals $(t_{ij} - K_i)$ if $t_{ij} < K_i$ (i.e., before the changepoint) and 0 otherwise; similarly, $(t_{ij} - K_i)_+ = t_{ij} - K_i$ for $t_{ij} \geq K_i$ and 0 otherwise; $\beta = (\beta_1, \beta_2, \beta_3)'$ is the vector of fixed effects, b_i is the random effects for subject i, and e_{ij} is the error.

The parameters β_1, β_2 are the estimated fixed effect intercept and slope before the changepoint K_i, and β_3 is the slope after changepoint. The changepoint for the ith subject K_i is assumed to be unknown. At the *second stage* the random subject effects and errors are defined as

$$b_i \sim N(0, \sigma_b^2), \quad e_{ij} \sim N(0, \sigma^2), \tag{7.14}$$

independently of each other.

At the *third stage*, we model the dropout due to going off study treatment. As discussed earlier, this kind of dropout is informative (Wu and Carroll, 1988; Little and Rubin, 2002). Thus, for accurate estimation of the changepoints we jointly model the longitudinal marker and the survival censoring process (Faucett and Thomas, 1996; Touloumi et al., 1999; Lyles et al., 2000). Here the time to dropout is assumed to be exponentially distributed, with intensity parameters related to the longitudinal process through sharing the individual parameters of Equation 7.13. Specifically, Let T_i^s denote the dropout time for subject i; let C_i denote the "censoring time" for T_i^s, here 48 weeks, the length of follow-up. The observed survival data consists of $T_i = \min(T_i^s, C_i)$ and the event indicator variable δ_i taking the value 1 if $T_i = T_i^s$ and 0 otherwise. We assume that T_i^s has an exponential distribution with mean λ_i^{-1}, where

$$\lambda_i = \exp\{\alpha_0 + \alpha_1 \log(K_i) + \alpha_2 b_i\}. \tag{7.15}$$

Note that the roles of T_i^s and C_i seem at first counterintuitive, since when $T_i = T_i^s$ the longitudinal vector y_i is censored owing to subject dropout, whereas when $T_i = C_i$, the vector y_i is complete.

The association between the longitudinal and the survival processes arises in Equation 7.15 in two ways. One is through the changepoint and the other is through the sharing of the random effect b_i. The strength and significance of the association is measured by α_1 and α_2, with values of 0 indicating no association with $\log K_i$ and b_i, respectively. Other forms for λ_i are possible.

We arrived at Equation 7.15 through a model selection process, see Ghosh and Vaida (2007) for more details.

Finally, at *stage four* we define the priors for the parameters. We assume inverse-gamma priors to the measurement error variance $\sigma^2 \sim \text{IG}(a,b)$, the random effect variance $\sigma_b^2 \sim \text{IG}(c,d)$. We assign exponential before the individual changepoints $K_i \sim \exp(\gamma)$, where $\gamma \sim \text{Gamma}(a_k, b_k)$. The fixed effects are assigned a normal prior, $\beta_j \sim N(\mu_\beta, \sigma_\beta)$, $j = 1,2,3$. The association parameters are also assumed as a normal prior, $\alpha_j \sim N(\mu_\alpha, \sigma_\alpha)$.

7.3.2 Model Selection Using Deviance Information Criterion

In Bayesian data analysis, model comparison and selection are needed for at least two reasons: (1) finding the "best" model, or subset of models, which describe the data, and (2) studying the sensitivity of the results to prior specification. In the absence of prior information this means that the specified prior distribution needs to be reasonably uninformative. On a more philosophical level, how model selection is done depends on what is the purpose of inference. Usually this entails model prediction, rather than simple fit of the existing data, or even testing whether certain covariates are significantly associated with the outcome.

We use here the DIC of Spiegelhalter et al. (2002). This criterion is a Bayesian equivalent of the Akaike information criterion (AIC) (Akaike, 1973; Burnham and Anderson, 2002). For mixed effects models, Vaida and Blanchard (2005) show that DIC is related to their conditional AIC. DIC consists of two components, a term that measures goodness-of-fit and a penalty term for increasing model complexity: $\text{DIC} = \bar{D} + p_D$. The first term, $\bar{D}$, is defined as the posterior expectation of the deviance:

$$\bar{D} = E_{\theta|y}[D(\theta)] = E_{\theta|y}[-2\ln f(y|\theta)].$$

The better the model fits the data, the smaller is the value of $\bar{D}$. The second component, p_D, measures the complexity of the model by the *effective number of parameters* and is defined as the difference between the posterior mean of the deviance and the deviance evaluated at the posterior mean $\bar{\theta}$ of the parameters:

$$p_D = \bar{D} - D(\bar{\theta}) = E_{\theta|y}[D(\theta)] - D(E_{\theta|y}[\theta]) = E_{\theta|y}[-2\ln f(y|\theta)] + 2\ln f(y|\bar{\theta}). \tag{7.16}$$

Equation 7.16 shows that p_D can be regarded as the expected excess of the true over the estimated residual information in data y conditional on θ. Hence, we can interpret p_D as the expected reduction in uncertainty due to estimation. Rearranging Equation 7.16 gives $\bar{D} = D(\bar{\theta}) + p_D$. As a result, DIC can be represented as $\text{DIC} = D(\bar{\theta}) + 2p_D$. A smaller DIC indicates better model fit. As a rule of thumb, analogously to AIC, a difference larger than 10 between the DIC values of two competing models is overwhelming evidence in favor of the better model (Burnham and Anderson, 2002).

7.3.3 Analysis of ACTG 398 Data

The data set we analyze here is from AIDS Clinical Trials Group (ACTG) 398 (Hammer et al., 2002). The goal of the trial was to assess whether adding a second protease inhibitor (PI) to a PI-containing regimen would improve outcomes in patients who had already failed a PI-containing regimen. Patients were randomized to one of four treatment arms and followed for 48 weeks. The CD4 counts were measured at weeks 0 (baseline), 2, 4, 8, 16, 24, 32, 40, and 48 for all subjects. About half the patients had prior experience to the nonnucleoside reverse transcriptase inhibitors class of drugs (NNRTI), other than the study drug Efavirenz. All subjects received Efavirenz, which, as it turned out, was importantly related to virologic failure. In the first 48 weeks of the study, 317 subjects had virologic failure, whereas 164 maintained a good virologic response. We work with these 164 subjects; they have 985 CD4 observations ranging from 3 to 8 observations per subject. The mean value and median of the CD4 counts are 17.21 and 17.03, respectively. The covariates of interest included in this analysis are treatment (the three dual PI arms combined and placebo), NNRTI experience, and baseline $\log_{10}$ viral load. Figure 7.4 shows the longitudinal CD4 profiles for the 164 subjects.

ACTG 398 had high toxicity rates owing to high drug burden and to the advanced stage of infection in the study population; 46% of the subjects went off study-treatment (stopping at least one drug) owing to toxicity by week 48. When a subject goes off-study-treatment this affects immediately their viral load and CD4 cell count. In our analysis the interest is in modeling the effect

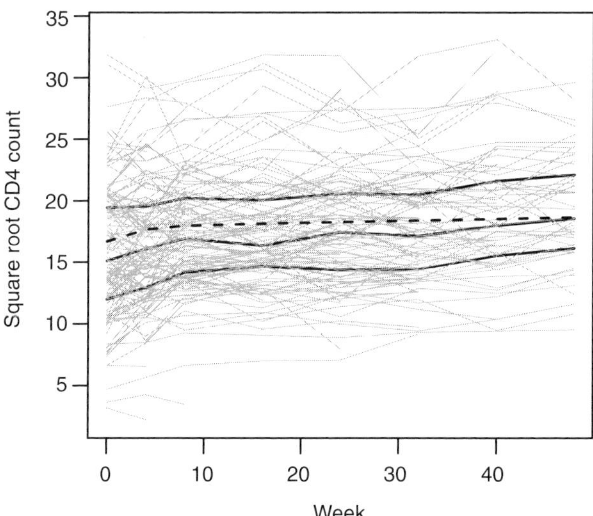

FIGURE 7.4
CD4 profiles for the 164 subjects in the study (thin lines). The solid lines mark the median and quartiles of the observed data at each time point. The dashed line is the median predicted CD4 for all subjects based on Model 6.

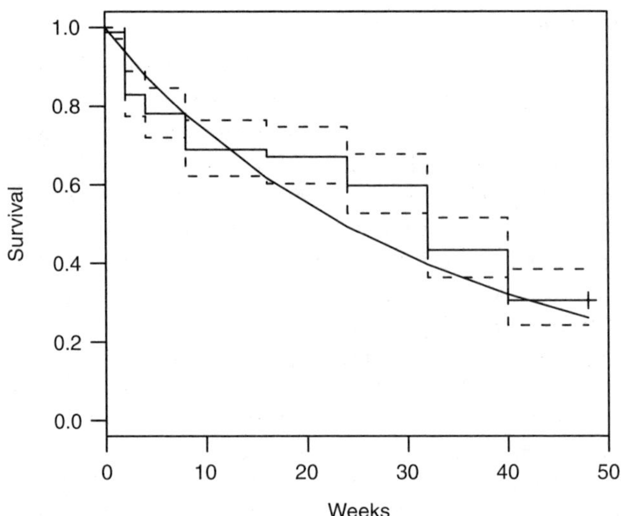

FIGURE 7.5
Dropout times for the 164 subjects in the study: Kaplan–Meier curves and model fits from the marginal dropout distribution of Model 6.

of the actual antiviral treatment on CD4 for subjects that sustain control of viral infection, so the CD4 values were censored at the time of off-study-treatment. On the basis of the main analysis results we only consider two treatment groups, the combination of three dual-PI arms and the PI-placebo arm. Of the 164 drop-off times, 50 (30%) were censored at week 48. Figure 7.5 shows the dropout curve using the Kaplan–Meier estimate and using the fits from the final joint model.

The hyperparameters in the prior distributions for the parameters in the model were chosen so that the priors are uninformative. In particular, we took

$$\beta_j \sim N(0, 1000), \quad \text{independent} \tag{7.17}$$

$$\alpha_j \sim N(0, 1000), \quad \text{independent} \tag{7.18}$$

$$\sigma^2 \sim IG(0.001, 0.001), \tag{7.19}$$

$$\sigma_b^2 \sim IG(0.001, 0.001), \tag{7.20}$$

$$\gamma \sim G(0.1, 0.1). \tag{7.21}$$

One advantage of the Bayesian hierarchical modeling is its ability to estimate multidimensional parameters using MCMC methods. We used the noncommercial statistical software WinBUGS (Spiegelhalter et al., 2003) to obtain the posterior samples (Figure 7.6). Using convergence diagnostic tool of Gelman and Rubin (1992) and the quantile plots, we concluded that 10,000 iterations were sufficient for the burn-in-period.

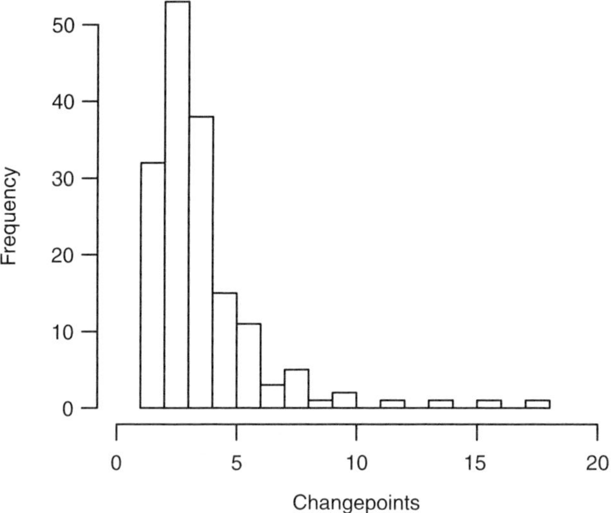

FIGURE 7.6
Histogram of the posterior mean values of the changepoints K_i ACTG 398.

We considered several joint models for y_{ij} and for the exponential parameter of T_i^s, λ_i, as follows:

Model 1: Independent models for survival and longitudinal measures, that is, Equation 7.13 and $\lambda_i = \lambda$.

Model 2: Longitudinal model given by Equation 7.13 and $\lambda_i(t) = \exp(\alpha_0 + \alpha_1 K_i)$.

Model 3: Longitudinal model given by Equation 7.13 and $\lambda_i(t) = \exp(\alpha_0 + b_i)$.

Model 4: Longitudinal model given by Equation 7.13 and

$$\lambda_i(t) = \exp\{\alpha_0 + \alpha_1 \log(K_i)\}. \tag{7.22}$$

Model 5: Fixed changepoint model, that is, longitudinal model given by Equation 7.13 with $K_i = K$, and survival model Equation 7.15.

Model 6: Longitudinal model given by Equation 7.13 (random changepoint), and survival model (7.15).

Model 7: Same as Model 6, but with the three covariates included, that is, Equation 7.23 below and Equation 7.15,

$$y_{ij} = \beta_1 + \beta_2(t_{ij} - K_i)_- + \beta_3(t_{ij} - K_i)_+ + \beta_4 x_{4i} + \beta_5 x_{5i} + \beta_6 x_{6i} + b_i + e_{ij}, \tag{7.23}$$

where x_{4i}, x_{5i}, x_{6i} are baseline $\log_{10}$ HIV-1 RNA, treatment (dual PI versus PI-placebo), and NNRTI experience, respectively.

Model 8: Same as Model 6 but with $\log_{10}$ HIV-1 RNA as a covariate.

Model 9: A random intercept and slope model for y_{ij},

$$y_{ij} = \beta_1 + \beta_2(t_{ij} - K_i)_- + \beta_3(t_{ij} - K_i)_+ + b_{0i} + b_{1i}t_{ij} + e_{ij},$$

and survival model (7.15).

Model 10: Longitudinal model given by Equation 7.1, and survival model

$$\lambda_i(t) = \exp\{\alpha_0 + \alpha_1 I(t > K_i)\}. \tag{7.24}$$

Model 5 has a fixed changepoint for the (square root) CD4, whereas all other models have random changepoint. Models 1–6 share the same mean function for the CD4; Model 7 and Model 8 include also baseline covariates. Model 8 includes only $\log_{10}$ HIV-1 RNA, which is the only covariate statistically significant in Model 7. Model 9 includes a random slope for the square root CD4. Models 1–6 and 8 have an increasingly complex model for the drop-off time. Model 1 assumes uninformative censoring mechanism for the CD4. In Model 3 the correlation is induced by b_i, whereas in Models 2 and 4 the correlation is given by K_i and $\log(K_i)$ respectively, linearly on the $\log(\lambda_i)$ scale. Finally, in Models 5–9 the correlation is given by both b_i and $\log(K_i)$, as in Equation 7.15. Model 10 has piecewise-constant hazard for each subject, before and after the CD4 changepoint K_i.

Table 7.1 describes the model comparison for these models, and includes the DIC and the effective number of parameters p_D. Model 5 (fixed changepoint) has least support, DIC = 4593.2, at least 100 larger than all other models. Model 1 (uninformative censoring) is next least supported, DIC = 4482.7. Including b_i, K_i, or $\log(K_i)$ in $\log(\lambda_i)$ (Models 2–4) slightly reduces the DIC. Including random slope (Model 9) does not improve the DIC. A large improvement is achieved by Model 6, DIC = 4457.4 where both b_i and $\log(K_i)$ are included, as in Equation 7.15. Adding covariates to the CD4 in Models 7 and 8 does not improve the DIC over Model 6 (the difference in DIC is moderately large, 7.4 and 9.2, respectively).

Model 10 is a special case, because its implementation (with piecewise constant hazard of dropout) requires additional "nodes," or parameters in the model. We used the "zero trick," or the point process representation of the survival process, as suggested in Spiegelhalter et al. (2003). As mentioned in the previous section, the DIC in this case includes these additional nodes "in focus," and therefore it is not directly comparable with the DIC of the other models. The computation of the relevant DIC is nontrivial, it requires integration of the additional nodes, and is an open area of research. For comparison with Model 6 we also ran Model 6 using the equivalent "zero trick" implementation as for Model 10, and we compared the DIC values from this implementation (Table 7.1). In this comparison the more parsimonious Model 6 has the lower DIC value.

Thus, Model 6 is the favored model. The number of effective degrees of freedom p_D is the smallest for the fixed changepoint model, $p_D = 165.3$. Interestingly, the 164 random effects b_i in this model are counted almost as full parameters. The "full status" of the b_i indicates that the random effects model fit will be similar to one in which the random parameters are treated as fixed. Vaida and Blanchard (2005) show an example of a LME model where such a random effects model has a better fit than the corresponding model with the random parameters treated as fixed effects, and much better than a model where the random effects are completely ignored and discuss the issue of counting the random effects. The random changepoint accounts for an additional 27–35 degrees of freedom, $p_D = 190.8$–202.6 for the other models. The fixed changepoint model places K at 6.8 weeks (standard deviation, SD = 1.1 weeks). The other seven models have very similar posterior means of the changepoint values, which is placed, on average, at 3.6 weeks. However, the credible interval for K in the fixed changepoint model overlaps with that for the average K in the random changepoint models. Figure 7.3 shows the histogram of the posterior means for the changepoints K_i.

The parameter estimates for Model 6 are in Table 7.2. Although Model 7 has a higher DIC, since the influence of the covariates may be of interest nonetheless, we report in Table 7.2 the parameter estimates from this model as well. Except for the intercept, the two models give similar parameter values. For Model 6, the initial slope is of about 0.40 (in square root CD4 concentrations per week), which translates in a first-week increase of about 13.7 cells/μL, and a first-month increase of about 43 cells/μL. The subsequent increase, following the changepoint, is much smaller, of 0.017, which translates into a

TABLE 7.2

Model Comparison: DIC Values for Models 1–10

	DIC	p_D	$\bar{K}$ (SD)
Model 1	4482.7	201.4	3.6 (1)
Model 2	4481.3	199.5	3.8 (1.3)
Model 3	4481.1	202.6	3.3 (0.9)
Model 4	4470	191.7	3.406 (0.9)
Model 5	4593.2	165.3	6.8 (1.1)
Model 6	4457.4	200.6	3.6 (1.2)
Model 7	4463	190.8	3.5 (1.2)
Model 8	4464.2	191	3.4 (1.2)
Model 9	4481.2	198.6	3.5 (1)
Model 6*	7137.4	238.7	
Model 10*	7755.9	201.3	

p_D is the effective degrees of freedom, and $\bar{K}$ is the average (and standard deviation) of the posterior mean changepoint for the 164 subjects, in weeks. DICs for models marked with * include additional parameters and are only comparable with each other.

TABLE 7.3

Parameter Estimates: Posterior Means (and 95% Posterior Intervals) for Models 6 and 7

	Model 6		Model 7	
β_1	16.89	(16.1, 17.8)	24.6	(19.0, 29.0)
β_2	0.40	(0.399, 0.481)	0.436	(0.404, 0.503)
β_3	0.017	(0.002, 0.027)	0.018	(0.008, 0.028)
β_4	—	—	−1.90	(−2.84, −0.77)
β_5	—	—	1.23	(−0.28, 2.85)
β_6	—	—	0.90	(−0.64, 2.54)
σ_b	5.06	(4.40, 5.58)	4.73	(4.20, 5.33)
σ	1.59	(1.50, 1.68)	1.59	(1.50, 1.68)
α_0	−3.475	(−4.01, −3.19)	−3.71	(−4.07, −3.27)
α_1	−0.035	(−0.456, 0.352)	−0.030	(−0.127, 0.055)
α_2	−0.066	(−0.107, −0.019)	−0.060	(−0.109, −0.014)

first-month increase following the changepoint of less than 1 cell/μL. Both slopes, β_2 and β_3, are significantly positive, that is the 95% posterior credible interval of each of these parameters does not include zero (Table 7.3). The residual and between-subject standard deviations are 1.54 and 4.96, respectively, which indicates that 8.3% of the variability is due to random error, and 86.4% is due to between-subjects variation. The remaining 5.2% of the variability is accounted by the random K_i.

For Model 7, the covariates parameters show that the square root CD4 counts are significantly associated with baseline $\log_{10}$ HIV-1 viral load, with a −1.90 reduction in response, or about −68 cells/μL, per one $\log_{10}$ increase in viral load, but it is not significantly associated with study treatment or with NNRTI experience.

Turning to the parameters of the drop-off times, we note that neither α_1 nor α_2 is zero, indicating a correlation between longitudinal and dropout model. However, whereas α_1 is not statistically significant, α_2 is significantly different from 0. The point estimates for both are negative, suggesting that lower baseline CD4 values and earlier changepoint values are associated with earlier drop-off times.

A visual inspection of various model parameters showed that the posterior densities are smooth and unimodal. The trace plots indicated good mixing and convergence (not included). DIC gave credible values with positive and meaningful p_D values.

7.4 Discussion

Using the the two statistical analyses presented in this chapter, we have illustrated some of the complexities occurring in HIV research and provided potential approaches which can be used. Much work is still under way in HIV

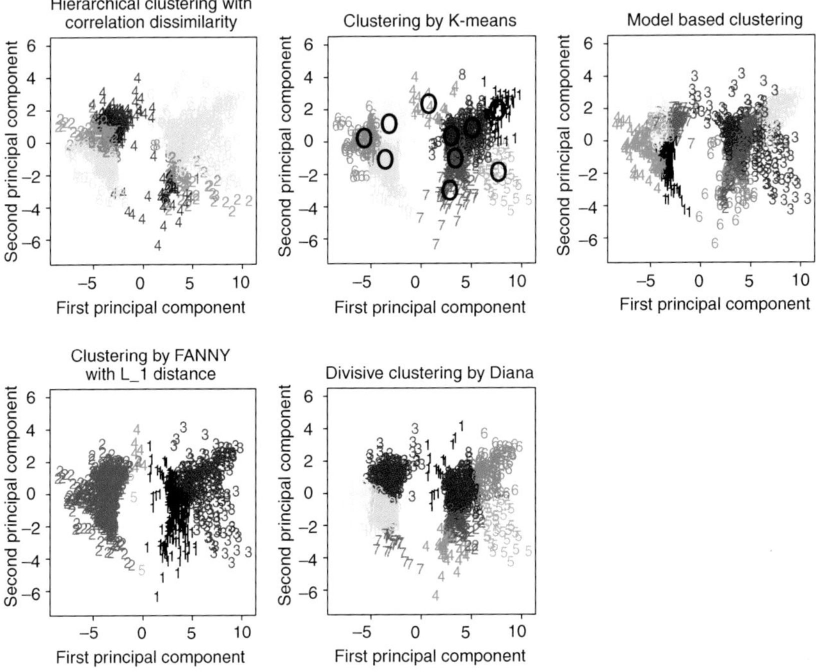

COLOR FIGURE 1.3
The genes were clustered into seven groups using their expression profiles during sporulation of yeast; five different clustering algorithms were attempted. (Adapted from Datta, S. and Arnold, J. (2002). In *Advances in Statistics, Combinatorics and Related Areas*, C. Gulati, Y.-X. Lin, S. Mishra, and J. Rayner, (Eds.), World Scientific, 63–74.)

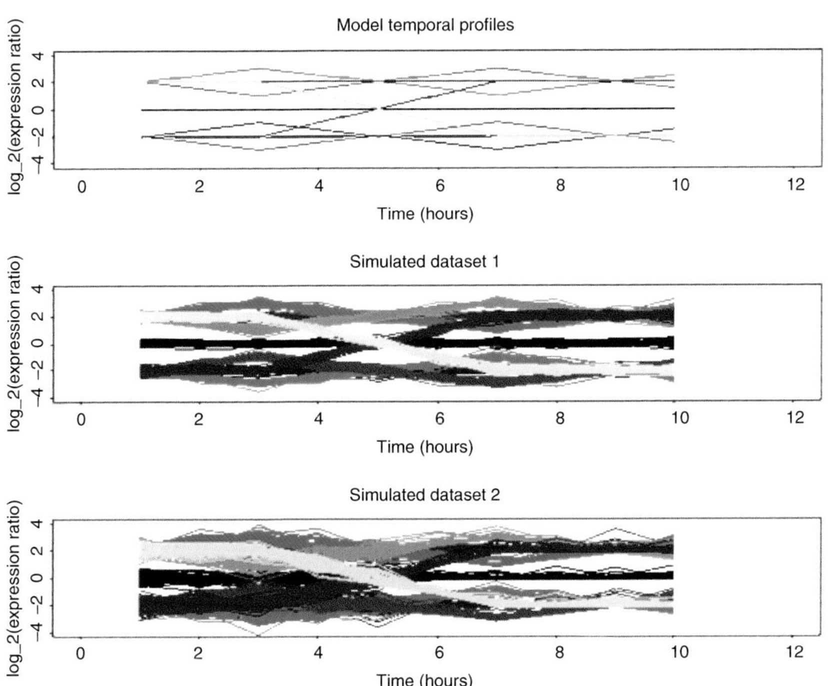

COLOR FIGURE 1.4
The average proportion of nonoverlap measure for various clustering algorithms applied to simulated data sets.

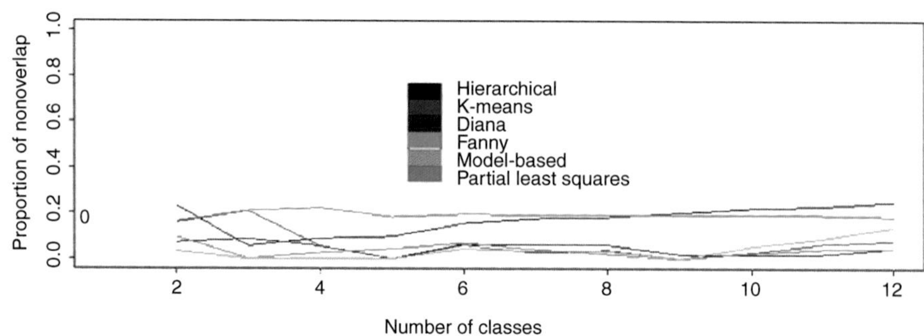

Simulated dataset 2

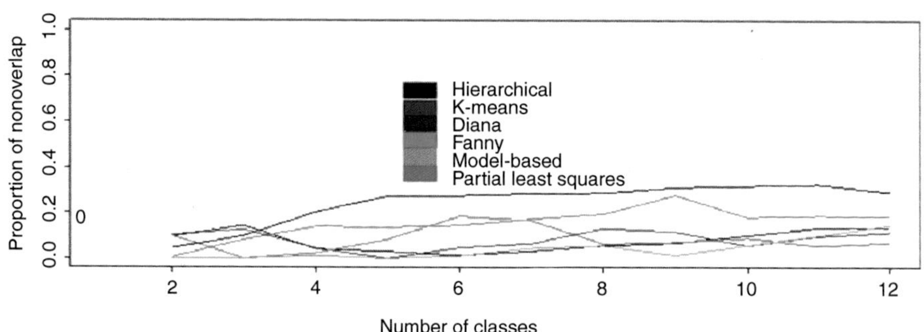

COLOR FIGURE 1.5
Two simulated datasets of gene expressions were created by adding random noise to a model profile.

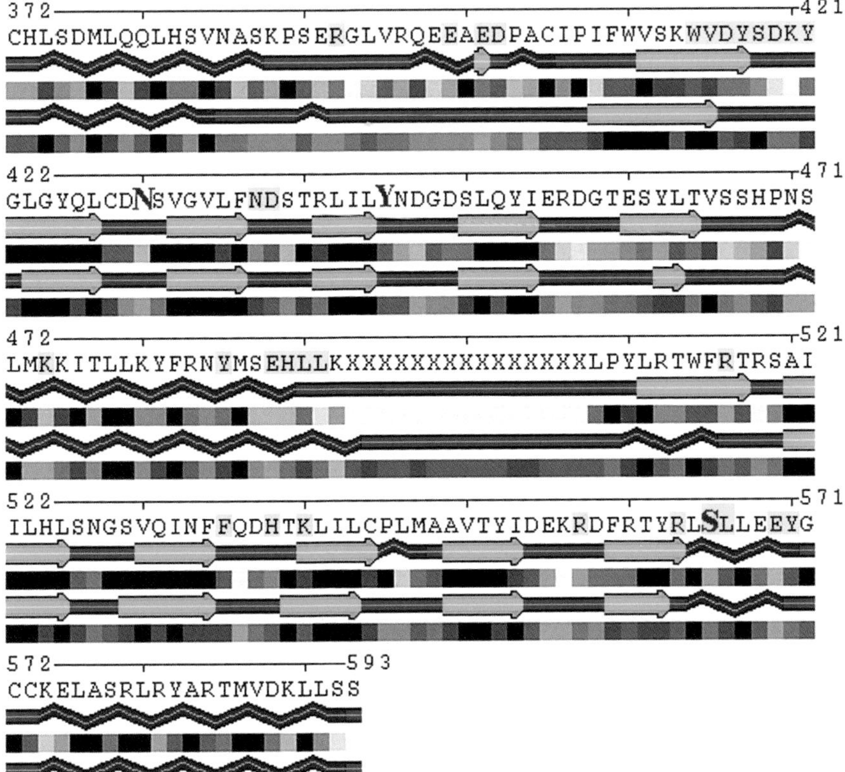

COLOR FIGURE 2.1

Comparison of experimentally observed (PDB structure lq4k, chain A, two upper rows) and predicted (using the SABLE (From R. Adamczak, A. Porollo and J. Meller, *Proteins*, 56, 753–767, 2004.) server, lower rows) structures of polo kinase PIki. Helices are indicated using red braids, beta-strands are indicated using green arrows and loops are shown in blue. The relative solvent accessibility is represented by shaded boxes, with black boxes corresponding to fully buried residues. Sites located in known protein–protein interaction interfaces are highlighted using yellow, whereas residues corresponding to polymorphic sites are highlighted in red and Xs represent fragments unresolved in the crystal structure. Figure generated using the POLYVIEW server (http://polyview.cchmc.org).

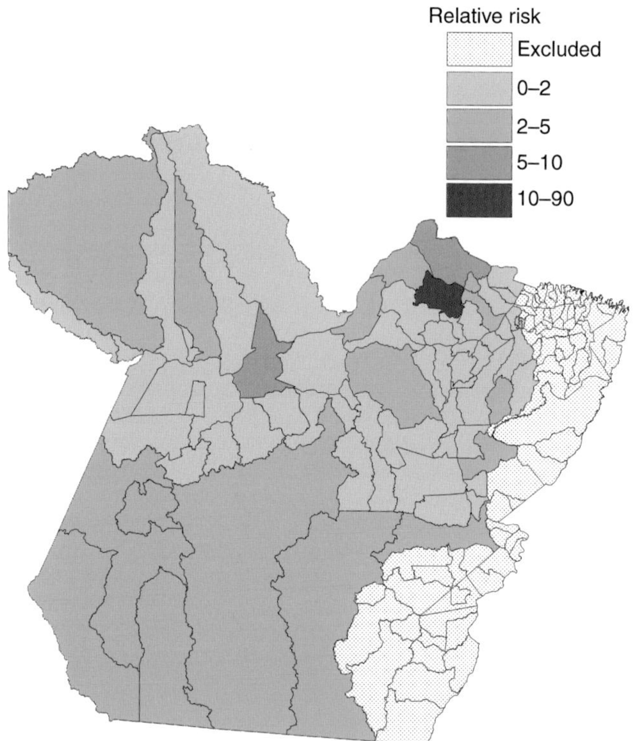

COLOR FIGURE 8.1
Malaria in Pará: log-relative risk's March 1997 posterior median when $\beta_t \sim N(0, \sigma_t^2 I_n)$ and $\sigma_t^2 \sim$ Log-normal($\log(\sigma_{t-1}^2), \tau^2$). For instance, Anajás county highest risk may be owing to its proximity to several rivers and to the island of Marajó.

research as well as in longitudinal data analysis. We will mention here only two directions of active research: (1) modeling the dropout process (see Diggle et al., 2002, Chapter 11 and Hogan et al., 2004 for an overview) and (2) smooth modeling using p-splines and mixed effects, see for example, Durban et al. (2005). Our own current work investigates using a semiparametric survival model for dropout and joint modeling of CD4 and HIV-1 RNA longitudinal profiles.

References

Akaike, H. (1973). Information theory and an extension of the maximum likelihood principle. In *International Symposium on information Theory*, B. N. Petrov and F. Csaki, eds., pp. 267–81, Akademia Kiado, Budapest.

Bennett, K. K., DeGruttola, V. G., Havlir, D. V., and Richman, D. D. (2002). Baseline predictors of CD4 T-lymphocyte recovery with combination antiretroviral therapy. *JAIDS* **31**, 20–26.

Bosch, R. J., Wang, R., Vaida, F., Lederman, M. M., and Albrecht, M. A. (2006). Changes in the slope of the CD4 count increase after initiation of potent antiretroviral treatment. *JAIDS* **43**, 433–435.

Brown, E., Ibrahim, J. G., and DeGruttola, V. (2005). A flexible B-spline model for multiple longitudinal biomarkers and survival. *Biometrics* **61**, 64–73.

Burnham, K. P. and Anderson, D. R. (2002). *Model Selection and Multimodel Inference: A Practical Information—Theoretic Approach*, 2nd ed. Springer, New York.

Carlin, B. (1996). Hierarchical longitudinal modeling. In *Markov Chain Monte Carlo in Practice*, W. R. Gilks, S. Richardson, and D. J. Spiegelhalter, eds., pp. 303–320, Chapman & Hall, London.

Deeks, S., Kitchen, C., Liu, L., Guo, H., Gascon, R., Narváez, A. B., Hunt, P., et al. (2004). Immune activation set point during early HIV infection predicts subsequent CD4+ T-cell changes independent of viral load. *Blood* **104**, 4, 942–947.

DeGruttola, V., Lange, N., and Dafni, U. (1991). Modeling the progression of HIV infection. *Journal of the American Statistical Association* **74**, 829–836.

Diggle, P. J., Heagerty, P., Liang, K.-Y., and Zeger, S. L. (2002). *Analysis of Longitudinal Data*. 2nd ed. Oxford University Press, New York.

Durban, M., Harezlak, J., Wand, M. P., and Carroll, R. J. (2005). Simple fitting of subject-specific curves for longitudinal data. *Statistics in Medicine* **24**, 1153–1167.

Faucett, C. L. and Thomas, D. C. (1996). Simultaneouly modeling censored survival data and repeatedly measured covariates: A Gibbs sampling approach. *Statistics in Medicine* **15**, 1663–1685.

Finney, D. J. (1962). Cummulants of truncated multinormal distribution. *Journal of the Royal Statistical Society, Series B* **24**, 535–536.

Fitzgerald, A. P., DeGruttola, V. G., and Vaida, F. (2002). Modeling viral rebound using non-linear mixed effects models. *Statistics in Medicine* **21**, 2093–2108.

Foulkes, A. S. and DeGruttola, V. (2003). Characterizing classes of antiretroviral drugs by genotype. *Statistics in Medicine* **22**, 2637–2655.

Gelman, A., Carlin, J. B., Stern, H. S., and Rubin, D. B. (2004). *Bayesian Data Analysis*. 2nd ed. Chapman & Hall, London.

Gelman, A. and Rubin, D. B. (1992). Inference from iterative simulation using multiple sequences. *Statistical Science* **7**, 457–472.

Genz, A. (1992). Numerical computation of multivariate normal probabilities. *Journal of Computational and Graphical Statistics* **1**, 141–150.

Ghosh, P. and Vaida, F. (2007). Random changepoing modeling of HIV immunologic responses (in press). *Statistics in Medicine* **26**, 2074–2087.

Guo, X. and Carlin, B. P. (2004). Separate and joint modeling of longitudinal and event time data using standard computer packages. *The American Statistician* **58**, 16–24.

Hammer, S. M., Vaida, F., Bennett, K. K., Holohan, M. K., Sheiner, L., Eron, J. J., et al. (2002). Dual vs single protease inhibitor therapy following artiretroviral treatment failure. A randomized Trial. *Journal of the American Medical Association* **288**, 169–180.

Hogan, J. W., Roy, J., and Korkontzelou, C. (2004). Handling drop-out in longitudinal studies. *Statistics in Medicine* **23**, 1455–1497.

Hughes, J. P. (1999). Mixed effects models with censored data with applications to HIV RNA levels. *Biometrics* **55**, 625–629.

Hughes, M. (2000). Analysis and design issues for studies using censored biomarker measurements, with an example of viral load measurements in HIV clinical trials. *Statistics in Medicine* **19**, 3171–3191.

Lindstrom, M. J. and Bates, D. M. (1990). Nonlinear mixed effects models for repeated measures data. *Biometrics* **46**, 673–687.

Little, R. J. A. and Rubin, D. B. (2002). *Statistical Analysis with Missing Data*. 2nd ed., Wiley, Hoboken, New Jersey.

Lopes, H. F., Muller, P., and Ravishanker, N. (2007). Bayesian computational methods in biomedical research. In *Computational Methods in Biomedical Research,* R. Khattree and D. N. Naik, eds., pp. 211–260.

Louis, T. A. (1982). Finding the observed information matrix when using the EM algorithm. *Journal of the Royal Statistical Society, Series B* **44**, 226–233.

Lyles, R. H., Lyles, C. M., and Taylor, D. J. (2000). Random regression models for human immunodeficiency virus ribonucleic acid data subject to left censoring and informative drop-outs. *Journal of the Royal statistical Soceity, Series C* **49**, 485–497.

Orchard, T. and Woodbury, M. A. (1972). A missing information principle, theory and application. In *Proceedings of the 6th Berkeley Symposium on Mathematical Statistics and Probability*, vol. 1, 679–715.

Pinheiro, J. C. and Bates, D. M. (2000). *Mixed-Effects Models in S and S-PLUS*. Springer, New York.

R Development Core Team (2006). *R: A Language and Environment for Statistical Computing*. R Foundation for Statistical Computing, Vienna, Austria. ISBN 3-900051-07-0.

Spiegelhalter, D., Thomas, A., Best, N., and Lunn, D. (2003). WinBUGS user manual, version 1.4. Available at `http://www.mrc-bsu.cam.ac.uk/bugs`.

Spiegelhalter, D. J., Best, N. G., Carlin, B. P., and van der Linde, A. (2002). Bayesian measures of model complexity and fit. *Journal of the Royal Statistical Society, Series B* **64**, 1–34.

Tallis, G. M. (1961). The moment generating function of the truncated multi-normal distribution. *Journal of the Royal Statistical Society, Series B* **23**, 223–229.

Touloumi, G., Pocock, S. J., Babiker, A. G., and Darbyshire, J. H. (1999). Estimation and comparison of rates of change in longitudinal studies with informative drop-outs. *Statistics in Medicine* **18**, 1215–1233.

UNAIDS (2006). *Report On the Global HIV/AIDS Epidemic*. Geneva, Switzerland.

Vaida, F. and Blanchard, S. (2005). Conditional Akaike information for mixed effects models. *Biometrika* **92**, 2, 351–370.

Vaida, F., Fitzgerald, A., and DeGruttola, V. (2007). *Computational Statistics and Data Analysis*, **51**, 5718–5730.

Vaida, F. and Liu, L. (2007). Fast implementation for normal mixed effects models with censored response. In *Proceedings of the 22nd International Workshop on Statistical Modelling*, J. del Castillo, A. Espinal, and P. Puig, eds., Barcelona, Spain, 578–583.

Wu, H. and Ding, A. (1999). Population HIV-1 dynamics *in vivo*: Applicable models and inferential tools for virological data from AIDS clinical trials. *Biometrics* **55**, 410–418.

Wu, M. C. and Carroll, R. J. (1988). Estimation and comparison of changes in the presence of informative right censoring by modeling the censoring process. *Biometrics* **44**, 175–188.

Xu, R. and Donohue, M. (2007). Proportional hazards mixed models and applications. In *Computational Methods in Biomedical Research*, R. Khattree and D. N. Naik, eds., 297–322.

8

Bayesian Computational Methods in Biomedical Research

Hedibert F. Lopes, Peter Müller, and Nalini Ravishanker

CONTENTS

8.1 Introduction

This article gives a survey of Bayesian techniques useful for biomedical applications. Given the extensive use of Bayesian methods, especially with the recent advent of Markov Chain Monte Carlo (MCMC), we can never do exhaustive justice. Nevertheless, we have made an attempt to present different Bayesian applications from different viewpoints and differing levels of complexity. We start with a brief introduction to the Bayesian paradigm in Section 8.2, and give some basic formulas. In Section 8.3, we review conjugate Bayesian analysis in both the static and dynamic inferential frameworks, and give references to some biomedical applications. The conjugate Bayesian approach is often insufficient for handling complex problems that arise in several applications. The advent of sampling-based Bayesian methods has opened the door to carrying out inference in a variety of settings. They must of course be used with care, and with sufficient understanding of the underlying stochastics. In Section 8.4, we present details on the algorithms most commonly used in Bayesian computing and provide exhaustive references. Sections 8.5–8.8 show illustrations of Bayesian computing in biomedical applications that are of current interest.

8.2 A General Framework for Bayesian Modeling

Assume an investigator is interested in understanding the relationship between cholesterol levels and coronary heart disease. Both classical and Bayesian statistics start by describing the relative likelihood of possible observed outcomes as a probability model. This probability model is known as the sampling model or likelihood, and is usually indexed by some unknown parameters. For example, the parameters could be the odds of developing coronary heart disease at different levels of cholesterol. Bayesian inference describes uncertainty about the unknown parameters by a second probability model. This probability model on the parameters is known as the prior distribution. Together, the sampling model and the prior probability model describe a joint probability model on the data and parameters. In contrast, classical statistics proceed without assuming a probability model for the parameters. The prior probability model describes uncertainty on the parameters before observing any data. After observing data, the prior distribution is updated using rules of probability calculus (Bayes' rule). The updated probability distribution on the parameters is known as the posterior distribution and contains all relevant information on the unknown parameters. From a Bayesian perspective, all statistical inference can be deduced from the posterior distribution by reporting appropriate summaries. In the rest of this section, we review the formal rules of probability calculus that are used to carry out inference, as well as model adequacy, model selection, and prediction.

8.2.1 Discrete Case

Suppose $A_1, \ldots, A_K$ are K disjoint sets and suppose $\pi_i = P(A_i)$ is the prior probability assigned to this event, $0 \leq \pi_i \leq 1$, $\sum_{i=1}^{K} \pi_i = 1$. Consider n observable events $B_1, \ldots, B_n$. Let $p(B_j \mid A_i)$ denote the relative likelihood of the events B_j under the events A_i (sampling model). The conditional probability of A_i given the observed events is from Bayes' theorem

$$P(A_i|B_j) = \frac{P(B_j|A_i)P(A_i)}{P(B_j)}, \tag{8.1}$$

where $P(B_j)$ is the marginal probability of observing B_j and is

$$P(B_j) = \sum_{i=1}^{K} P(B_j|A_i)P(A_i). \tag{8.2}$$

We often write this posterior probability as $P(A_i|B_j) \propto P(B_j|A_i)P(A_i)$. In this discrete case, the notion of Bayesian learning (updating) is described by

$$P(A_i|B_j, B_j^*) \propto P(B_j^*, B_j|A_i)P(A_i) = P(B_j^*|B_j, A_i)P(A_i|B_j). \tag{8.3}$$

EXAMPLE 1

A simple biomedical application discussed in Gelman et al. (2004) and Sorensen and Gianola (2002) deals with obtaining the probability that a woman *XYZ* is a carrier of the gene causing the genetic disease hemophilia. Double recessive women (*aa*) and men who carry the *a* allele in the

X-chromosome manifest the disease. If a woman is a carrier, she will transmit the a allele with probability 0.5, and ignoring mutation, will not transmit the disease if she is not a carrier. Suppose a woman is not hemophilic, her father and mother are unaffected, but her brother is hemophilic (their mother must be a carrier of a). We wish to determine the probability that XYZ is a carrier. Suppose A_1 and A_2 respectively denote the events that XYZ is a carrier, and that she is not a carrier, with (prior) probabilities $P(A_1) = P(A_2) = 0.5$. In terms of a discrete random variable θ, with $\theta = 1$ denoting she is a carrier and $\theta = 0$ indicating she is not, the prior distribution on θ is $P(\theta = 1) = P(\theta = 0) = 0.5$. The prior odds in favor of the mother not being a carrier is $P(\theta = 0)/P(\theta = 1) = 1$. Suppose information is also provided that neither of the two sons of XYZ has the disease. For $i = 1, 2$, let Y_i be a random variable assuming value 1 if the ith son has the disease and value 0 if he does not. Assuming Y_1 and Y_2 are independent conditional on θ, we have that given $\theta = 1$,

$$P(Y_1 = 0, Y_2 = 0|\theta = 1) = P(Y_1 = 0|\theta = 1)P(Y_2 = 0|\theta = 1) = 0.5 \times 0.5 = 0.25.$$

Also, given $\theta = 0$,

$$P(Y_1 = 0, Y_2 = 0|\theta = 0) = P(Y_1 = 0|\theta = 1)P(Y_2 = 0|\theta = 1) = 1 \times 1 = 1.$$

The posterior distribution of θ can be written for $j = 0, 1$ as

$$P(\theta = j|Y_1 = 0, Y_2 = 0) = \frac{P(\theta = j)P(Y_1 = 0, Y_2 = 0|\theta = j)}{\sum_{i=0}^{1} P(\theta = i)P(Y_1 = 0, Y_2 = 0|\theta = i)}$$

so that given that neither son is affected, the probability that XYZ is a carrier is 0.2 and that she is not a carrier is 0.8. The posterior odds in favor of XYZ not being a carrier of hemophilia is

$$P(\theta = 0|Y_1 = 0, Y_2 = 0)/P(\theta = 1|Y_1 = 0, Y_2 = 0) = 4.$$

8.2.2 Continuous Case

Let $\boldsymbol{\theta} = (\theta_1, \ldots, \theta_k)$ be a k-dimensional vector of unknown parameters ($k \geq 1$), and suppose that a priori beliefs about $\boldsymbol{\theta}$ are given in terms of the probability density function (pdf) $\pi(\boldsymbol{\theta})$ (prior). Let $\mathbf{y} = (y_1, \ldots, y_n)$ denote an n-dimensional observation vector whose probability distribution depends on $\boldsymbol{\theta}$ and is written as $p(\mathbf{y}|\boldsymbol{\theta})$ (sampling model). Both $\boldsymbol{\theta}$ and $\mathbf{y}$ are assumed to be continuous-valued. To make probability statements about $\boldsymbol{\theta}$ given $\mathbf{y}$, the posterior density using Bayes' theorem is defined as (Berger, 1985)

$$\pi(\boldsymbol{\theta}|\mathbf{y}) = \frac{\pi(\boldsymbol{\theta}, \mathbf{y})}{m(\mathbf{y})} = \frac{\pi(\boldsymbol{\theta})p(\mathbf{y}|\boldsymbol{\theta})}{m(\mathbf{y})}, \tag{8.4}$$

where $m(\mathbf{y}) = \int \pi(\boldsymbol{\theta})p(\mathbf{y}|\boldsymbol{\theta})d\boldsymbol{\theta}$ is the marginal density of $\mathbf{y}$ and does not depend on $\boldsymbol{\theta}$. Also called the predictive distribution of $\mathbf{y}$, $m(\mathbf{y})$ admits the marginal likelihood identity $m(\mathbf{y}) = \pi(\boldsymbol{\theta})p(\mathbf{y}|\boldsymbol{\theta})/\pi(\boldsymbol{\theta}|\mathbf{y})$, and plays a useful role in Bayesian decision theory, empirical Bayes methods, and model selection. Recall that the likelihood function $l(\boldsymbol{\theta}|\mathbf{y}) = p(\mathbf{y}|\boldsymbol{\theta})$, regarded as a function of $\boldsymbol{\theta}$. We think of the prior–posterior relationship as *Posterior* $\propto$ *Prior* $\times$ *Likelihood*, and write

$$\pi(\boldsymbol{\theta}|\mathbf{y}) \propto \pi(\boldsymbol{\theta})l(\boldsymbol{\theta}|\mathbf{y}). \tag{8.5}$$

It is often convenient to work with the logarithm of the likelihood, $L(\boldsymbol{\theta}|\mathbf{y})$. Bayesian inference involves moving from a prior distribution on $\boldsymbol{\theta}$ before observing $\mathbf{y}$ to a posterior distribution $\pi(\boldsymbol{\theta}|\mathbf{y})$ for $\boldsymbol{\theta}$, and in general, consists of obtaining and interpreting $\pi(\boldsymbol{\theta}|\mathbf{y})$ through plots (contour and scatter), numerical summaries of posterior location and dispersion (mean, mode, quantiles, standard deviation, interquartile range), credible intervals (also called highest posterior density (HPD) regions), and hypotheses tests (see Lee, 1997; Congdon, 2003; Gelman et al., 2004). The sequential use of Bayes' theorem is instructive. Given an initial set of observations $\mathbf{y}$, and a posterior density (Equation 8.5), suppose we have a second set of observations $\mathbf{z}$ distributed independently of $\mathbf{y}$, it can be shown that the posterior $\pi(\boldsymbol{\theta}|\mathbf{y},\mathbf{z})$ is obtained from using $\pi(\boldsymbol{\theta}|\mathbf{y})$ as the prior for $\mathbf{z}$, that is,

$$\pi(\boldsymbol{\theta}|\mathbf{y},\mathbf{z}) \propto \pi(\boldsymbol{\theta}|\mathbf{y})l(\boldsymbol{\theta}|\mathbf{z}). \tag{8.6}$$

EXAMPLE 2

Lopes et al. (2003) consider hematologic, that is, blood count data from a cancer chemotherapy trial. For each patient in the trial we record white blood cell count over time as the patient undergoes the first cycle of a chemotherapy treatment. Patients are treated at different doses of the chemotherapy agent(s). The main concern is inference about the number of days that the patient is exposed to a dangerously low white blood cell count. We proceed with a parametric model for the white blood cell profile over time. In other words, we assume initially a constant baseline count, followed by a sudden drop when chemotherapy is initiated, and finally a slow S-shaped recovery back to baseline after the chemotherapy. The profile is indexed by a 7-dimensional vector of random effects (see Section 8.3.3) that parameterize a nonlinear regression curve that reflects these features. Let $\boldsymbol{\theta}_i$ denote this 7-dimensional vector for patient i. Let $f(t;\ \boldsymbol{\theta}_i)$ denote the value at time t for the profile indexed by $\boldsymbol{\theta}_i$. Let $y_{ij}, j = 1, \ldots, n_i$ denote the observed blood counts for patient i on (known) days t_{ij}. We assume a nonlinear regression with normal residuals

$$y_{ij} = f(t_{ij}; \boldsymbol{\theta}_i) + \epsilon_{ij}, \tag{8.7}$$

with a normal distributed residual error, $\epsilon_{ij} \sim N(0, \sigma^2)$. For simplicity we assume that the residual variance σ^2 is known. Model (8.7) defines the sampling model. The model is completed with a prior probability model $\pi(\boldsymbol{\theta}_i)$. In other words, the prior reflects the judgment of likely initial white blood counts, the extent of the drop during chemotherapy, and the typical speed of recovery. Let $\mathbf{y}_i = (y_{ij}, j = 1, \ldots, n_i)$ denote the observed blood counts for patient i. Using Bayes' theorem we update the prior $\pi(\boldsymbol{\theta}_i)$ to the posterior $\pi(\boldsymbol{\theta}_i|\mathbf{y}_i)$. Instead of reporting the 7-dimensional posterior distribution, inference is usually reported by posterior summaries for relevant functions of the parameters. For example, in this application an important summary is the number of days that the patient has white blood cell count below a critical threshold. Let $f(\boldsymbol{\theta}_i)$ denote this summary. We plot $\pi(f(\boldsymbol{\theta}_i)|\mathbf{y}_i)$. In this short description we only discussed inference for one patient, i. The full model includes submodels for each patient, $i = 1, \ldots, n$, linked by a common prior $\pi(\boldsymbol{\theta}_i)$. The larger model is referred to as a hierarchical model (see Section 8.3.1). The prior $\pi(\boldsymbol{\theta}_i)$ is also known as the random effects distribution. It is usually indexed with additional unknown (hyper-) parameters $\boldsymbol{\phi}$ and might include a regression on patient specific covariates $\mathbf{x}_i$, in summary $\pi(\boldsymbol{\theta}_i|\boldsymbol{\phi}, \mathbf{x}_i)$. The covariate vector $\mathbf{x}_i$ includes the treatment dose for patient i. The model is completed with a prior probability model $\pi(\boldsymbol{\phi})$ for the hyperparameter ϕ. In summary, the full hierarchical model is

$$\begin{aligned} \text{likelihood:} \quad & y_{ij} = f(t_{ij}; \boldsymbol{\theta}_i) + \epsilon_{ij}, \quad i = 1, \ldots, n, \quad j = 1, \ldots, n_i, \\ \text{prior:} \quad & \boldsymbol{\theta}_i \sim p(\boldsymbol{\theta}_i \mid \boldsymbol{\phi}, \mathbf{x}_i), \quad \boldsymbol{\phi} \sim \pi(\boldsymbol{\phi}). \end{aligned}$$

In the context of this hierarchical model, inference of particular interest is the posterior predictive distribution $\pi(\boldsymbol{\theta}_{n+1}|\mathbf{y}_1, \ldots, \mathbf{y}_n, \mathbf{x}_{n+1})$. This distribution is used to answer questions of the type: "What is the maximum dose that can be given and still bound the probability of more than 4 days below the critical lower threshold at less than 5%?."

8.2.3 Prior and Posterior Distributions

A prior distribution represents an assumption about the nature of the parameter $\boldsymbol{\theta}$, and clearly has an impact on posterior inference. Early Bayesian analyses dealt with conjugate priors, and this is explored further in Section 8.3. A prior density $\pi(\boldsymbol{\theta})$ is said to be proper if it does not depend on the data and integrates to 1. Bayesian inference is often subject to a criticism that posterior inference might be affected by choice of a subjective, injudicious prior, especially if the sample size is small or moderate. Considerable effort at defining an objective prior, whose contribution relative to that of the data is small, is often made. An extensive literature exists on approaches for specifying objective or noninformative priors (Berger and Bernardo 1992; Bernardo and Smith 1994). The uniform prior, $\pi(\theta) = 1/(b-a)$ for $\theta \in (a, b)$, is the most commonly used noninformative (vague) prior. Note that a uniform prior for a continuous parameter θ on $(-\infty, \infty)$ is improper, that is, the integral of the pdf is not finite. Although it is generally acceptable to use an improper prior, care must be exercised in applications to verify that the resulting posterior is proper. A class of improper priors proposed by Jeffreys (1961) is based on using Fisher's information measure through $\pi(\boldsymbol{\theta}) \propto |I(\boldsymbol{\theta})|^{1/2}$, where $I(\boldsymbol{\theta}) = E\{-\partial^2 L(\boldsymbol{\theta}|\mathbf{y})/\partial\boldsymbol{\theta}\partial\boldsymbol{\theta}'\}$.

EXAMPLE 3
When $y_1, \ldots, y_n$ are independent and identically distributed (iid) $N(\mu, \sigma^2)$ with known σ^2 and $\pi(\mu) = c$, for some constant c, then it is easily verified that the posterior of μ is $\pi(\mu|\mathbf{y}) \propto \exp\{-n(2\sigma^2)^{-1}(\mu - \overline{y})^2\}$, which integrates to $\sqrt{2\pi\sigma^2/n}$ and is proper. The *Jeffreys' prior* is $\pi(\mu) \propto \sqrt{n/\sigma^2} = c$.

See Bernardo (1979) for a discussion of reference priors, Sivia (1996) for examples of maximum entropy priors, and Robert (1996) and Congdon (2003) for discrete mixtures of parametric densities and Dirichlet process priors (DPP) with applications for smoothing health outcomes (Clayton and Kaldor, 1987) and modeling sudden infant death syndrome (SIDS) death counts (Symons et al., 1983). Prior elicitation (Kass and Wasserman, 1996) continues to be an active area of research.

Posterior inference will be robust if it is not seriously affected by the choice of the model (likelihood, prior, loss function) assumptions and is insensitive to inputs into the analysis (Kadane, 1984; Berger et al., 2000); see Sivaganesan (2000) for a detailed review of global robustness based on measures such as the Kullback–Leibler distance. Robustness measures are closely related to the class of priors used for analysis. Typically, a class of priors is chosen by first specifying a single prior p_0 and then choosing a suitable neighborhood Λ to reflect our uncertainty about p_0, such as ϵ-contamination classes, density bounded classes, density ratio classes, and so forth. For instance, an ϵ-contamination class is $\Lambda = \{p\colon p = (1-\epsilon)p_0 + \epsilon q;\ q \in Q\}$, where Q is a set of probability distributions that are possibly deviations from p_0. Kass et al. (1989) have described approximate methods to assess sensitivity whereas Gustafson (1996) has discussed an 'informal' sensitivity analysis to compare inference on a finite set of alternative priors.

8.2.4 Predictive Distribution

The predictive distribution of $\mathbf{y}$ accounts both for the uncertainty about $\boldsymbol{\theta}$ and the residual uncertainty about $\mathbf{y}$ given $\boldsymbol{\theta}$, and as such, enables us to check model (prior, likelihood, loss function) assumptions. Predictive inference about an unknown observable $\tilde{y}$ is described through the posterior predictive distribution

$$p(\widetilde{y}|\mathbf{y}) = \int p(\widetilde{y}, \boldsymbol{\theta}|\mathbf{y})\,\mathrm{d}\boldsymbol{\theta} = \int p(\widetilde{y}|\boldsymbol{\theta})\pi(\boldsymbol{\theta}|\mathbf{y})\,\mathrm{d}\boldsymbol{\theta}, \tag{8.8}$$

where the second identity assumes that $\widetilde{y}$ and $\mathbf{y}$ are independent conditional on $\boldsymbol{\theta}$.

EXAMPLE 4
Suppose $y_1, \ldots, y_n$ iid $N(\theta, \sigma^2)$, $\theta|\sigma^2 \sim N(\theta_0, \lambda_0^{-1}\sigma^2)$ and $\sigma^2 \sim \text{Inv-}\chi^2(\alpha_0, \sigma_0^2)$, that is, $\sigma^{-2} \sim \chi^2(\alpha_0, \sigma_0^2)$,

$$\pi(\theta|\sigma^2) = (2\pi\lambda_0^{-1}\sigma^2)^{-1/2}\exp\{-0.5\lambda_0(\theta - \theta_0)^2/\sigma^2\}$$
$$\pi(\sigma^2) = (0.5\alpha_0)^{0.5\alpha_0}\Gamma^{-1}(0.5\alpha_0)\sigma_0^{\alpha_0}\sigma^{-(\alpha_0+2)}\exp\{-0.5\alpha_0\sigma_0^2/\sigma^2\},$$

so that the joint prior $\pi(\theta, \sigma^2)$ is given by their product as N-Inv-$\chi^2(\theta_0, \lambda_0^{-1}\sigma^2; \alpha_0, \sigma_0^2)$ and the joint posterior is N-Inv-$\chi^2(\theta_1, \lambda_1^{-1}\sigma_1^2; \alpha_1, \sigma_1^2)$ where $\theta_1 = (\lambda_0 + n)^{-1}[\lambda_0\theta_0 + n\overline{y}]$, $\lambda_1 = \lambda_0 + n$, $\alpha_1 = \alpha_0 + n$ and

$$\sigma_1^2 = \alpha_1^{-1}[\alpha_0\sigma_0^2 + (n-1)s^2 + (\lambda_0 + n)^{-1}\lambda_0 n(\overline{y} - \theta_0)^2].$$

Suppose a noninformative prior specification $\pi(\theta, \sigma^2) \propto (\sigma^2)^{-1}$ is used, the joint posterior specification is given by $\pi(\theta|\sigma^2, \mathbf{y})$ is $N(\overline{y}, n^{-1}\sigma^2)$ and $\pi(\sigma^2|\mathbf{y})$ is Inv-$\chi^2(n-1, s^2)$.

Gelman et al. (2004) describe an application of a normal hierarchical model to a meta-analysis, whose goal is to make combined inference from data on mortality after myocardial infarction in 22 clinical trials, each consisting of two groups of heart attack subjects randomly allocated to receive or not receive beta-blockers; see Rubin (1989) for more details on Bayesian meta-analysis.

8.2.5 Model Determination

Model determination consists of model checking (for adequate models) and model selection (for best model); see Gamerman and Lopes (2006, Chapter 7) and Gelman et al. (2004). Classical and Bayesian model choice methods would involve comparison of measures of fit with the current fitted data or cross-validatory fit to out-of-sample data. Formal Bayesian model assessment is based on the marginal likelihoods from J models M_j ($j = 1, \ldots, J$) with (1) parameter vector $\boldsymbol{\theta}_j$ whose prior density is p_j, and (2) with prior model probability $P(M_j)$, with $\sum_j P(M_j) = 1$. Given data, the posterior model probability and the probability of the data conditional on the model (Gelfand and Ghosh, 1994) are respectively

$$P(M_j|\mathbf{y}) = P(M_j)\frac{\int l(\boldsymbol{\theta}_j|\mathbf{y})\pi(\boldsymbol{\theta}_j)d\boldsymbol{\theta}_j}{\sum_{l=1}^{J}[P(M_l)\int l(\boldsymbol{\theta}_l|\mathbf{y})\pi(\boldsymbol{\theta}_l)d\boldsymbol{\theta}_l}$$

$$P(\mathbf{y}|M_j) = m_j(\mathbf{y}) = \int l(\boldsymbol{\theta}_j|\mathbf{y})\pi(\boldsymbol{\theta}_j)d\boldsymbol{\theta}_j. \tag{8.9}$$

The Bayes factor for two distinct models M_1 and M_2 is the ratio of the marginal likelihoods $m_1(\mathbf{y})$ and $m_2(\mathbf{y})$, that is,

$$\frac{P(\mathbf{y}|M_1)}{P(\mathbf{y}|M_2)} = \frac{P(M_1|\mathbf{y})}{P(M_2|\mathbf{y})} \times \frac{P(M_2)}{P(M_1)}. \tag{8.10}$$

See Kass and Raftery (1995) and Pauler et al. (1999) for an application to variance component models. For applications with improper priors, Bayes factors cannot be defined and several other model selection criteria have been proposed, such as the pseudo Bayes factor (Geisser, 1975), intrinsic Bayes factor (Berger and Pericchi, 1996), and so forth. Model averaging is another option in finding the best inference; see Hoeting et al. (1999) for a review. These methods are most effective with the sampling-based approach to Bayesian inference described in Section 8.4.

8.2.6 Hypothesis Testing

The Bayesian approach to hypothesis testing of a simple $H_0 : \boldsymbol{\theta} \in \Theta_0 = \{\theta_0\}$ versus simple $H_1 : \boldsymbol{\theta} \in \Theta_1 = \{\theta_1\}$, where $\Theta = \Theta_0 \cup \Theta_1$, is more straightforward than the classical approach; it consists of making a decision based on the magnitudes of the posterior probabilities $P(\theta \in \Theta_0|\mathbf{y})$ and $P(\theta \in \Theta_1|\mathbf{y})$, or using the Bayes factor in favor of H_0 versus H_1 as

$$BF = \frac{\pi_1 P(\theta \in \Theta_0|\mathbf{y})}{\pi_0 P(\theta \in \Theta_1|\mathbf{y})},$$

where $\pi_0 = P(\theta \in \Theta_0)$ and $\pi_1 = P(\theta \in \Theta_1)$ are the prior probabilities. The Bayesian p-value is defined as the probability that the replicated data could be more extreme than the observed data, as measured by the test statistic T:

$$\begin{aligned} p_B &= P(T(\mathbf{y}^{\text{rep}}, \boldsymbol{\theta}) \geq T(\mathbf{y}, \boldsymbol{\theta})|\mathbf{y}) \\ &= \int\int I_{T(\mathbf{y}^{\text{rep}}, \boldsymbol{\theta}) \geq T(\mathbf{y}, \boldsymbol{\theta})} p(\mathbf{y}^{\text{rep}}|\boldsymbol{\theta}) p(\boldsymbol{\theta}|\mathbf{y}) \mathrm{d}\mathbf{y}^{\text{rep}}\, \mathrm{d}\boldsymbol{\theta}. \end{aligned} \tag{8.11}$$

Bayesian decision analysis involves optimization over decisions in addition to averaging over uncertainties (Berger, 1985). An example on medical screening is given in Gelman et al. (2004), Chapter 22.

8.3 Conjugate or Classical Bayesian Modeling

Conjugate Bayesian analysis was widely prevalent until the advent of an efficient and feasible computing framework to handle more complicated applications. A class $\mathcal{P}$ of prior distributions for $\boldsymbol{\theta}$ is naturally conjugate for a class of sampling distributions $\mathcal{F}$ if $\mathcal{P}$ is the set of all densities with the same functional form as the likelihood, and if for all densities $p(\cdot|\boldsymbol{\theta}) \in \mathcal{F}$ and all priors $p(\cdot) \in \mathcal{P}$, the posterior $p(\boldsymbol{\theta}|\mathbf{y})$ belongs to $\mathcal{P}$. It is well known that sampling distributions belonging to an exponential family have natural conjugate prior distributions. Specifically, suppose the sampling distribution for $\mathbf{y}$ and the prior distribution for the parameter $\boldsymbol{\theta}$ have the forms

$$\begin{aligned} p(\mathbf{y}|\boldsymbol{\theta}) &\propto g(\boldsymbol{\theta})^n \exp[\boldsymbol{\phi}(\boldsymbol{\theta})'\mathbf{t}(\mathbf{y})], \\ \pi(\boldsymbol{\theta}) &\propto g(\boldsymbol{\theta})^\eta \exp[\boldsymbol{\phi}(\boldsymbol{\theta})'\boldsymbol{\nu}], \end{aligned} \tag{8.12}$$

where $\mathbf{t}(\mathbf{y})$ is a sufficient statistic for $\boldsymbol{\theta}$; the posterior density for $\boldsymbol{\theta}$ has the form

$$\pi(\boldsymbol{\theta}|\mathbf{y}) \propto g(\boldsymbol{\theta})^{(\eta+n)} \exp[\boldsymbol{\phi}(\boldsymbol{\theta})'(\boldsymbol{\nu} + \mathbf{t}(\mathbf{y})]. \tag{8.13}$$

Although resulting computations are simple and often available analytically in closed forms, it has been shown that exponential families are in general the only classes of sampling distributions that have natural conjugate priors.

There are several applications in biomedical areas. A useful model in epidemiology for the study of incidence of diseases is the Poisson model. Suppose $y_1, \ldots, y_n$ is a random sample from a Poisson(μ) distribution, so that $l(\mu|\mathbf{y}) \propto \mu^{\sum y_i} \exp(-n\mu)$, which is in the exponential family. Suppose $\pi(\mu) \sim \text{Gamma}(\alpha, \beta)$ (with shape α and scale β), then the posterior also belongs to the same family. Nonconjuacy is preferable or necessary to handle most complicated problems that arise in practice. Further, analytical results may not be available and we must use simulation methods, as described in Section 8.4. Static and dynamic linear modeling offer a versatile class of models that may be handled using simple computational approaches under standard distributional assumptions.

8.3.1 Linear Modeling

Since the discussion of Bayesian inference for the linear model with a single-stage hierarchical prior structure (Lindley and Smith, 1972), great strides have been made in using Bayesian techniques for hierarchical linear, generalized linear, and nonlinear mixed modeling. Their two-stage hierarchical normal linear model supposes

$$\begin{aligned} &\mathbf{y}|\boldsymbol{\theta}_1 \sim N(\mathbf{A}_1\boldsymbol{\theta}_1, \mathbf{C}_1), \boldsymbol{\theta}_1|\boldsymbol{\theta}_2 \sim N(\mathbf{A}_2\boldsymbol{\theta}_2, \mathbf{C}_2) \text{ and} \\ &\boldsymbol{\theta}_2|\boldsymbol{\theta}_3 \sim N(\mathbf{A}_3\boldsymbol{\theta}_3, \mathbf{C}_3), \end{aligned}$$

where additionally, $\boldsymbol{\theta}_3$ is a known k_3-dimensional vector, and $\mathbf{A}_1$, $\mathbf{A}_2$, $\mathbf{A}_3$, $\mathbf{C}_1$, $\mathbf{C}_2$, and $\mathbf{C}_3$ are known positive definite matrices of appropriate dimensions. The posterior distribution of $\boldsymbol{\theta}_1$ given $\mathbf{y}$ is then $N(\mathbf{Dd}, \mathbf{D})$ where

$$\begin{aligned} \mathbf{D}^{-1} &= \mathbf{A}_1'\mathbf{C}_1^{-1}\mathbf{A}_1 + [\mathbf{C}_2 + \mathbf{A}_2\mathbf{C}_3\mathbf{A}_2']^{-1} \\ \mathbf{d} &= \mathbf{A}_1'\mathbf{C}_1^{-1}\mathbf{y} + [\mathbf{C}_2 + \mathbf{A}_2\mathbf{C}_3\mathbf{A}_2']^{-1}\mathbf{A}_2\mathbf{A}_3\boldsymbol{\theta}_3. \end{aligned} \tag{8.14}$$

The mean of the posterior distribution is seen to be a weighted average of the least squares estimate $(\mathbf{A}_1'\mathbf{C}_1^{-1}\mathbf{A}_1)^{-1}\mathbf{A}_1'\mathbf{C}_1^{-1}\mathbf{y}$ of $\boldsymbol{\theta}_1$ and its prior mean $\mathbf{A}_2\mathbf{A}_3\boldsymbol{\theta}_3$, and is a point estimate of $\boldsymbol{\theta}_1$. The three-stage hierarchy can be extended to several stages. Smith (1973) examined the Bayesian linear model in more detail and studied inferential properties. There is an extensive literature on the application of these methods to linear regression and analysis of designed experiments in biomedical research. Classical Bayesian inference for univariate linear regression to model responses $\mathbf{y} = (y_1, \ldots, y_n)$ as a function of an observed predictor matrix $\mathbf{X}$ stems from

$$\mathbf{y}|\boldsymbol{\beta}, \sigma^2, \mathbf{X} \sim N_n(\mathbf{X}\boldsymbol{\beta}, \sigma^2\mathbf{I}); \quad \pi(\boldsymbol{\beta}, \sigma^2|\mathbf{X}) \propto \sigma^{-2} \tag{8.15}$$

where the noninformative prior specification is adequate in situations when the number of cases n is large relative to the number of predictors p. Posterior inference follows from

$$\boldsymbol{\beta}|\sigma^2, \mathbf{y} \sim N(\hat{\boldsymbol{\beta}}, \sigma^2\mathbf{V}_{\beta}); \quad \sigma^2|\mathbf{y} \sim \text{Inv} - \chi^2(n-p, s^2), \text{ where}$$

$$\hat{\boldsymbol{\beta}} = (\mathbf{X}'\mathbf{X})^{-1}\mathbf{X}'\mathbf{y}; \quad \mathbf{V}_{\beta} = (\mathbf{X}'\mathbf{X})^{-1}; \quad s^2 = (n-p)^{-1}(\mathbf{y} - \mathbf{X}\hat{\boldsymbol{\beta}})'(\mathbf{y} - \mathbf{X}\hat{\boldsymbol{\beta}}).$$

The conjugate family of prior distributions is the normal-Inv-χ^2 shown in Section 8.2. Jeffreys' prior is $\pi(\boldsymbol{\beta}, \sigma^2) \propto \sqrt{|\mathbf{I}(\boldsymbol{\beta}, \sigma^2)|} \propto (\sigma^2)^{-(p+2)/2}$. An extension to normal multivariate regression with q-variate independently distributed responses $\mathbf{y}_1, \ldots, \mathbf{y}_n$ is straightforward

$$\mathbf{y}_i|\boldsymbol{B}, \Sigma, \mathbf{x}_i \sim N_q(\mathbf{x}_i'\boldsymbol{B}, \Sigma), i = 1, \ldots, n; \quad \pi(\boldsymbol{\beta}, \sigma^2|\mathbf{X}) \propto \sigma^{-2}. \tag{8.16}$$

A noninformative prior specification is the multivariate Jeffreys prior $\pi(\boldsymbol{\beta}, \boldsymbol{\Sigma}) \propto |\boldsymbol{\Sigma}|^{-(q+1)/2}$, and the corresponding posterior distribution is $\boldsymbol{\Sigma}|\mathbf{y} \sim$ Inv-Wishart$_{n-1}(\mathbf{S})$ and $\boldsymbol{\beta}|\boldsymbol{\Sigma}, \mathbf{y} \sim N(\overline{\mathbf{y}}, n^{-1}\boldsymbol{\Sigma})$. The conjugate prior family for $(\mathbf{B}, \boldsymbol{\Sigma})$ is the normal-Inv-Wishart$(\boldsymbol{B}_0, \boldsymbol{\Sigma}/\lambda_0, \nu_0, \Lambda_0^{-1})$ distribution. Several applications exist in the literature. Buonaccorsi and Gatsonis (1988) discussed inference for ratios of coefficients in the linear model with applications to slope-ratio bioassay, comparison of the mean effects of two soporific drugs and a drug bioequivalence problem in a two-period changeover design with no carryover and with fixed subject effects. Other typical examples of hierarchical normal linear models in biomedical applications are Hein et al. (2005), Lewin et al. (2006) for gene expression data, and Müller et al. (1999) for case-control studies.

8.3.2 Dynamic Linear Modeling

In contrast to cross-sectional data, we frequently encounter situations where the responses and covariates are observed sequentially over time. It is of interest to develop inference for such problems in the context of a dynamic linear model. Let $\mathbf{y}_1, \ldots, \mathbf{y}_T$ denote p-dimensional random variables that are available at times $1, \ldots, T$. Suppose $\mathbf{y}_t$ depends on an unknown q-dimensional *state* vector $\boldsymbol{\theta}_t$ (that may again be scalar or vector-valued) through the *observation equation*

$$\mathbf{y}_t = \mathbf{F}_t\boldsymbol{\theta}_t + \mathbf{v}_t, \tag{8.17}$$

where $\mathbf{F}_t$ is a known $p \times q$ matrix, and we assume that the observation error $\mathbf{v}_t \sim N(\mathbf{0}, \mathbf{V}_t)$, with known $\mathbf{V}_t$. The dynamic change in $\boldsymbol{\theta}_t$ is represented by the *state equation*

$$\boldsymbol{\theta}_t = \mathbf{G}_t\boldsymbol{\theta}_{t-1} + \mathbf{w}_t, \tag{8.18}$$

where $\mathbf{G}_t$ is a known $q \times q$ state transition matrix, and the state error $\mathbf{w}_t \sim N(\mathbf{0}, \mathbf{W}_t)$, with known $\mathbf{W}_t$. In addition, we suppose that $\mathbf{v}_t$ and $\mathbf{w}_t$ are independently distributed. Note that θ_t is a random vector; let

$$\boldsymbol{\theta}_t|\mathbf{y}_t \sim N(\hat{\boldsymbol{\theta}}_t, \boldsymbol{\Sigma}_t) \tag{8.19}$$

represent the posterior distribution of $\boldsymbol{\theta}_t$. The *Kalman filter* is a recursive procedure for determining the posterior distribution of $\boldsymbol{\theta}_t$ in this conjugate setup, and thereby predicting $\mathbf{y}_t$ (West and Harrison, 1989). Estimation of parameters via the expectation-maximization model (EM) algorithm (Dempster et al., 1977) has been discussed in Shumway and Stoffer (2004), who provide R code for handling such models; see also http://cran.r-projrct.org/doc/packages/dlm.pdf (R package from G. Petris). Gamerman and Migon (1993) extended this to dynamic hierarchical modeling in the Gaussian framework.

Gordon and Smith (1990) used a dynamic linear model framework to model and monitor medical time series such as body-weight adjusted reciprocal serum creatinine concentrations for online monitoring under renal transplants, and daily white blood count cells levels in patients with chronic kidney disorders. Kristiansen et al. (2005) used the Kalman filter based on a double integrator for tracking urinary bladder filling from intermittent bladder volume measurements taken by an ultrasonic bladder volume monitor. Wu et al. (2003) have used a switching Kalman filter model as a real-time decoding algorithm for a neural prosthesis application, specifically for the real-time inference of hand kinematics from a population of motor cortical neurons.

8.3.3 Beyond Linear Modeling

The linear mixed model (LMM) generalizes the fixed effects models discussed above to include random effects and has wide use in biomedical applications. Let $\boldsymbol{\beta}$ and $\mathbf{u}$ denote p- and q-dimensional location vectors related to the n-dimensional observation vector $\mathbf{y}$ through the regressor matrix $\mathbf{X}$ and design matrix $\mathbf{Z}$

$$\mathbf{y} = \mathbf{X}\boldsymbol{\beta} + \mathbf{Z}\mathbf{u} + \boldsymbol{\varepsilon}; \quad \boldsymbol{\varepsilon}|\sigma_\varepsilon^2 \sim N_n(\mathbf{0}, \sigma_\varepsilon^2\mathbf{I}), \tag{8.20}$$

so that $\mathbf{y}|\boldsymbol{\beta}, \mathbf{u}, \sigma_\varepsilon^2 \sim N_n(\mathbf{X}\boldsymbol{\beta} + \mathbf{Z}\mathbf{u}, \sigma_\varepsilon^2\mathbf{I})$. Assume priors $\pi(\boldsymbol{\beta}|\sigma_\beta^2) \sim N(\mathbf{0}, \sigma_\beta^2\mathbf{B})$, $\mathbf{u}|\mathbf{V}\sigma_u^2 \sim N(\mathbf{0}, \sigma_u^2\mathbf{V}\sigma_u^2)$, where $\mathbf{B}$ and $\mathbf{V}$ are known, nonsingular matrices, and σ_u^2 and σ_β^2 are unknown hyperparameters. It is easily seen that no closed form expressions for the posterior distributions are possible, and inference for these models is feasible through the sampling-based Bayesian approach described in Section 8.4.

Bayesian analysis for generalized linear models (GLIMs) useful for analyzing non-normal data has seen rapid growth in the last two decades; see Gelfand and Ghosh (2000) for an overview and summary. West (1985) and Albert (1988) were among early discussants of a general hierarchical framework for GLIMs. Most familiar examples are logit or probit models for binary/binomial responses, loglinear models for responses of counts, cumulative logit models for ordinal categorical responses, and so forth. Suppose the responses $y_1, \ldots, y_n$ are independent with pdf belonging to an exponential family, so that the likelihood is

$$l(\boldsymbol{\theta}|\mathbf{y}) = \prod_{i=1}^{n} \exp[a^{-1}(\phi_i)\{y_i\theta_i - \psi(\theta_i)\} + c(y_i;\phi_i)], \tag{8.21}$$

where $\boldsymbol{\theta} = (\theta_1, \ldots, \theta_n)$, θ_i are unknown parameters related to the predictors $\mathbf{x}_1, \ldots, \mathbf{x}_n$ and regression parameters $\boldsymbol{\beta}$ via $\theta_i = h(\mathbf{x}_i'\boldsymbol{\beta})$ for a strictly increasing sufficiently smooth *link* function $h(\cdot)$, and $a(\phi_i)$ is known. A simple conjugate prior for $\boldsymbol{\beta}$ is $N(\boldsymbol{\beta}_0, \boldsymbol{\Sigma})$ where $\boldsymbol{\beta}_0$ and $\boldsymbol{\Sigma}$ are known, so that the posterior has the form

$$\pi(\boldsymbol{\beta}|\mathbf{y}) \propto \exp\left\{\sum_i a^{-1}(\phi_i)[y_i h(\mathbf{x}_i'\boldsymbol{\beta}) - \psi(h(\mathbf{x}_i'\boldsymbol{\beta}))] - \frac{1}{2}(\boldsymbol{\beta} - \boldsymbol{\beta}_0)'\boldsymbol{\Sigma}^{-1}(\boldsymbol{\beta} - \boldsymbol{\beta}_0)\right\}. \tag{8.22}$$

This posterior is not analytically tractable. Computational Bayesian methods discussed in Section 8.4 enable inference in complex situations such as generalized linear mixed models (GLMMs) (Clayton, 1996), nonlinear random effects models (Dey et al., 1997), models with correlated random effects and hierarchical GLMMs (Sun et al., 2000), correlated categorical responses (Chen and Dey, 2000), overdispersed GLIMs (Dey and Ravishanker, 2000), and survival data models (Kuo and Peng, 2000). Applications of such models to disease maps is discussed further in Section 8.6 (see also Waller et al., 1997).

8.4 Computational Bayesian Framework

As indicated in the previous discussion, the main ingredients of the Bayesian feast are probabilistic models (parametric or semiparametric) and prior distributions, which when combined, produce posterior distributions, predictive distributions, and summaries thereof. More specifically, recall the posterior, the predictive, and the posterior predictive distributions (Equations 8.4, 8.9, and 8.8), that is, $\pi(\boldsymbol{\theta}|\mathbf{y}) = \pi(\boldsymbol{\theta})p(\mathbf{y}|\boldsymbol{\theta})/m(\mathbf{y})$, $m(\mathbf{y}) = \int \pi(\boldsymbol{\theta})p(\mathbf{y}|\boldsymbol{\theta})\mathrm{d}\boldsymbol{\theta}$ and $p(\widetilde{y}|\mathbf{y}) = \int p(\widetilde{y}|\boldsymbol{\theta},\mathbf{y})\pi(\boldsymbol{\theta}|\mathbf{y})\mathrm{d}\boldsymbol{\theta}$. The Bayesian agenda includes, among other things, posterior modes, $\max_{\boldsymbol{\theta}} \pi(\boldsymbol{\theta}|\mathbf{y})$, posterior moments, $E_\pi[g(\boldsymbol{\theta})]$, density estimation, $\hat{\pi}(g(\boldsymbol{\theta})|\mathbf{y})$, Bayes factors, $m_0(\mathbf{y})/m_1(\mathbf{y})$, and decision making, $\max_d \int U(d,\boldsymbol{\theta})\pi(\boldsymbol{\theta}|\mathbf{y})d\boldsymbol{\theta}$.

Historically, those tasks were (partially) performed by analytic approximations, which include asymptotic approximations (Carlin and Louis, 2000),

Gaussian quadrature (Naylor and Smith, 1982) and Laplace approximations (Tierney and Kadane, 1986; Kass et al., 1988; Tierney et al., 1989). Modern, fast, and cheap computational resources have facilitated the widespread use of Monte Carlo methods, which in turn has made Bayesian reasoning commonplace in almost every area of scientific research. This trend is overwhelmingly apparent in the biomedical sciences (Sorensen and Gianola, 2002; Larget, 2005; Do et al., 2006). Initially, simple Monte Carlo schemes were extensively used for solving practical Bayesian problems, including the Monte Carlo method (Geweke, 1989), the rejection algorithm (Gilks and Wild, 1992), the weighted resampling algorithm (Smith and Gelfand, 1992), among others. Currently, one could argue that the most widely used Monte Carlo schemes are the Gibbs sampler/data augmentation algorithm (Tanner and Wong, 1987; Gelfand and Smith, 1990) and the Metropolis–Hastings (MH) algorithm (Metropolis et al., 1953; Hastings, 1970), which fall in the category of MCMC algorithms (Gilks et al., 1996). This section only briefly introduces the main algorithms. The more curious and energetic reader will find in Gamerman and Lopes (2006) and its associated webpage `www.ufrj.br/mcmc`, among other things, extensions of these basic algorithms, recent developments, didactic examples and their R codes, and an extensive and updated list of freely downloadable statistical routines and packages.

8.4.1 Normal Approximation

Let $\mathbf{m}$ be the posterior mode, that is, $\mathbf{m} = \arg\max_{\boldsymbol{\theta}} \pi(\boldsymbol{\theta}|\mathbf{y})$; a standard Taylor series expansion leads to approximating $\pi(\boldsymbol{\theta}|\mathbf{y})$ by a (multivariate) normal distribution with mean vector $\mathbf{m}$ and precision matrix $\mathbf{V}^{-1} = -(\partial^2 \log \pi(\mathbf{m}|\mathbf{y})/\partial\boldsymbol{\theta}\partial\boldsymbol{\theta}')$. This approximation can be thought of as a Bayesian version of the central limit theorem (Heyde and Johnstone, 1979; Schervish, 1995; Carlin and Louis, 2000). Finding $\mathbf{m}$ is not a trivial task and generally involves solving a set of nonlinear equations, usually by means of iterative Newton–Raphson-type and Fisher's scoring algorithms (Thisted, 1988). For most problems, specially in high dimension, normal approximations tend to produce rather crude and rough estimates.

8.4.2 Integral Approximation

One of the main tasks in Bayesian analysis is the computation of posterior summaries, such as means, variances, and other moments, of functions of $\boldsymbol{\theta}$, say $t(\boldsymbol{\theta})$, that is,

$$E[t(\boldsymbol{\theta})] = \frac{\int t(\boldsymbol{\theta}) p(\mathbf{y}|\boldsymbol{\theta}) \pi(\boldsymbol{\theta}) \, d\boldsymbol{\theta}}{\int p(\mathbf{y}|\boldsymbol{\theta}) \pi(\boldsymbol{\theta}) \, d\boldsymbol{\theta}}. \tag{8.23}$$

In this section, a brief review of the main analytic and stochastic approximations to the above integral is provided.

8.4.2.1 Quadrature Approximation

Quadrature rules approximate unidimensional integrals $\int_a^b g(\theta)d\theta$, for instance $g(\theta) = t(\theta)p(y|\theta)\pi(\theta)$ in Equation 8.23, by $\sum_{i=1}^{n} w_i g(\theta_i)$ for suitably chosen weights w_i and grid points θ_i, $i = 1, \ldots, n$. Simple quadrature rules are Simpson's and the trapezium rules. *Gaussian quadrature* are special rules for situations where $g(\theta)$ can be well approximated by the product of a polynomial and a density function. Gauss–Jacobi rule arises under the uniformity $[-1, 1]$ density, whereas Gauss–Laguerre and Gauss–Hermite rules arise under gamma and normal densities. See, for instance, Abramowitz and Stegun (1965) for tabulations and Naylor and Smith (1982) and Pole and West (1990) for Bayesian inference under quadrature. Quadrature rules are not practical even in problems of moderate dimensions.

8.4.2.2 Laplace Approximation

For $t(\boldsymbol{\theta}) > 0$, Equation 8.23 can be written in the exponential form as

$$E[t(\boldsymbol{\theta})] = \frac{\int \exp\{L^*(\boldsymbol{\theta})\}\, d\boldsymbol{\theta}}{\int \exp\{L(\boldsymbol{\theta})\}\, d\boldsymbol{\theta}}, \tag{8.24}$$

where $L^*(\boldsymbol{\theta}) = \log t(\boldsymbol{\theta}) + \log p(\mathbf{y}|\boldsymbol{\theta}) + \log \pi(\boldsymbol{\theta})$ and $L(\boldsymbol{\theta}) = \log p(\mathbf{y}|\boldsymbol{\theta}) + \log \pi(\boldsymbol{\theta})$. Better (than normal) approximations can be obtained by Taylor series expansions both in the numerator and the denominator of the previous equation. It can be shown under fairly general conditions, and when $t(\boldsymbol{\theta}) > 0$ that

$$\hat{E}_{\text{lap}}[t(\boldsymbol{\theta})] = \left(\frac{|\mathbf{V}^*|}{|\mathbf{V}|}\right)^{1/2} \exp\{L^*(\mathbf{m}^*) - L(\mathbf{m})\}, \tag{8.25}$$

where $\mathbf{m}^*$ is the value of $\boldsymbol{\theta}$ that maximizes L^* and $\mathbf{V}^*$ as minus the inverse Hessian of L^* at the point $\mathbf{m}^*$. This approximation is known as the *Laplace approximation* (Kass et al., 1988). Laplace approximations for the case where $t(\boldsymbol{\theta}) < 0$ appear in Tierney et al. (1989). The Laplace approximation tends to be poor either when the posterior is multimodal or when approximate normality fails.

8.4.2.3 Monte Carlo Integration

If a sample $\boldsymbol{\theta}_1, \ldots, \boldsymbol{\theta}_n$ from the prior $\pi(\boldsymbol{\theta})$ is available, then

$$\hat{E}_{mc}[t(\boldsymbol{\theta})] = \frac{\sum_{j=1}^{n} t(\boldsymbol{\theta}_j)p(\mathbf{y}|\boldsymbol{\theta}_j)}{\sum_{j=1}^{n} p(\mathbf{y}|\boldsymbol{\theta}_j)} \tag{8.26}$$

is a *Monte Carlo* (MC) estimator of $E[t(\boldsymbol{\theta})]$ in Equation 8.23. Limiting theory assures us that, under mild conditions on $t(\boldsymbol{\theta})$, the above MC estimator converges to its mean $E[t(\boldsymbol{\theta})]$ (Geweke, 1989).

It is well known that sampling from the prior distribution may produce poor estimates and a practically infeasible number of draws will be necessary to achieve reasonably accurate levels of approximation. The MC estimator (Equation 8.26) can be generalized for situations where, roughly speaking, a proposal density $q(\boldsymbol{\theta})$ is available that mimics $\pi(\boldsymbol{\theta}|\mathbf{y})$ in the center, dominates it in the tails, and is easy to sample from. More specifically, if a sample $\boldsymbol{\theta}_1, \ldots, \boldsymbol{\theta}_n$ from $q(\boldsymbol{\theta})$ is available (see sampling from distributions later in this section), then another MC estimator for Equation 8.23 is

$$\hat{E}_{mcis}[t(\boldsymbol{\theta})] = \frac{\sum_{j=1}^{n} t(\boldsymbol{\theta}_j)p(\mathbf{y}|\boldsymbol{\theta}_j)\pi(\boldsymbol{\theta}_j)/q(\boldsymbol{\theta}_j)}{\sum_{j=1}^{n} p(\mathbf{y}|\boldsymbol{\theta})\pi(\boldsymbol{\theta}_j)/q(\boldsymbol{\theta}_j)}. \tag{8.27}$$

This estimator is commonly known as a *Monte Carlo via Importance Sampling* (MCIS) estimator of $E[t(\boldsymbol{\theta})]$. It is easy to see that the previous MC estimator is a special case of the MCIS when $q(\boldsymbol{\theta}) = \pi(\boldsymbol{\theta})$. The proposal density q is also referred to as *importance density* and sampling from q is known as *importance sampling*. Limiting theory assures us that MC estimators, under mild conditions on $t(\boldsymbol{\theta})$ and $q(\boldsymbol{\theta})$, approximate the actual posterior expectations (see Geweke, 1989).

8.4.3 Monte Carlo-Based Inference

On the basis of the previous argument, if $\boldsymbol{\theta}_1, \ldots, \boldsymbol{\theta}_n$ is a readily available sample from the posterior $\pi(\boldsymbol{\theta}|\mathbf{y})$ then the MC approximation to $E[t(\boldsymbol{\theta})]$ would be $n^{-1}\sum_{j=1}^{n} t(\boldsymbol{\theta}_j)$. The problem is that $\boldsymbol{\theta}_1, \ldots, \boldsymbol{\theta}_n$ is rarely readily available. We now discuss two algorithms for sampling independent draws from $\pi(\boldsymbol{\theta}|\mathbf{y})$, namely, the *rejection algorithm* and the *weighted resampling algorithm*. We also discuss two iterative algorithms, namely, the *Gibbs sampler* and the *Metropolis–Hastings algorithm*, that sample from Markov chains whose limiting, equilibrium distributions are the posterior $\pi(\boldsymbol{\theta}|\mathbf{y})$. One common aspect of all these algorithms is that they all potentially use draws from auxiliary, proposal, importance densities, $q(\boldsymbol{\theta})$, whose importance are weighted against the target, posterior distribution $\pi(\boldsymbol{\theta}|\mathbf{y})$, that is, by considering weights $\pi(\boldsymbol{\theta}|\mathbf{y})/q(\boldsymbol{\theta})$ (see the explanation between Equations 8.26 and 8.27). For notational reasons, let $\tilde{\pi}(\boldsymbol{\theta}) = p(\mathbf{y}|\boldsymbol{\theta})\pi(\boldsymbol{\theta})$ be the unnormalized posterior, while $\omega(\boldsymbol{\theta}) = \tilde{\pi}(\boldsymbol{\theta})/q(\boldsymbol{\theta})$ is the unnormalized weight.

8.4.3.1 Rejection Method

When samples are easily drawn from a proposal q such that $\tilde{\pi}(\boldsymbol{\theta}) \leq Aq(\boldsymbol{\theta})$, for some finite A and for all possible values of $\boldsymbol{\theta}$, then it can be shown that the following algorithm produces independent draws from $\tilde{\pi}(\boldsymbol{\theta})$. The proposal density q is commonly known as a *blanketing density* or an *envelope density*, while A is the *envelope constant*. If both $\tilde{\pi}$ and q are normalized densities, then $A \geq 1$ and the theoretical acceptance rate is given by $1/A$. In other words, more draws are likely to be accepted the closer q is to π, that is, the closer A is to one.

Algorithm 8.1 Rejection method

1. Set $j = 1$.
2. Draw $\boldsymbol{\theta}^*$ from q and u from $U[0, 1]$.
3. Compute the unnormalized weight, $\omega(\boldsymbol{\theta}^*)$.
4. If $Au \leq \omega(\boldsymbol{\theta}^*)$, then set $\boldsymbol{\theta}_j = \boldsymbol{\theta}^*$ and $j = j + 1$.
5. Repeat steps 2, 3, and 4 while $j \leq n$.

The resulting sample $\{\boldsymbol{\theta}_1, \ldots, \boldsymbol{\theta}_n\}$ is distributed according to $\pi(\boldsymbol{\theta}|\mathbf{y})$.

8.4.3.2 Weighted Resampling Method

When A is not available or is hard to derive, *weighted resampling* is a direct alternative since it can use draws from a density q without having to find the constant A. This algorithm is more commonly known as the *sampling importance resampling* (SIR) algorithm (Smith and Gelfand, 1992). We should also mention other adaptive rejection algorithms and the renewed interest in SIR-type algorithms in sequential Monte Carlo (Doucet et al., 2001) and population Monte Carlo (Cappé et al., 2004).

Algorithm 8.2 Weighted resampling method

1. Sample $\boldsymbol{\theta}_1^*, \ldots, \boldsymbol{\theta}_m^*$ from q.
2. For $i = 1, \ldots, m$
   (a) Compute unnormalized weights: $\omega_i = \omega(\boldsymbol{\theta}_i^*)$;
   (b) Normalize weights: $w_i = \omega_i / \sum_{l=1}^{m} \omega_i$.
3. Sample $\boldsymbol{\theta}_j$ from $\{\boldsymbol{\theta}_1^*, \ldots, \boldsymbol{\theta}_m^*\}$, such that $Pr(\boldsymbol{\theta}_j = \boldsymbol{\theta}_i^*) = w_i$, for $j = 1, \ldots, n$.

For large m and n, the resulting sample $\{\boldsymbol{\theta}_1, \ldots, \boldsymbol{\theta}_n\}$ is approximately distributed according to $\pi(\boldsymbol{\theta}|\mathbf{y})$.

8.4.3.3 Metropolis–Hastings Algorithm

Instead of discarding the current rejected draw, as in the rejection algorithm, MH algorithms (Metropolis et al., 1953; Hastings, 1970) translate the rejection information into higher importance, or weight, to the previous draw. This generates an iterative chain of dependent draws (a Markov scheme). Markov chain arguments guarantee that, under fairly general regularity conditions and in the limit, such a Markov scheme generates draws from π, also known as the *target, limiting, equilibrium* distribution of the chain (Tierney, 1994).

Algorithm 8.3 Metropolis–Hastings algorithm

1. Set the initial value at $\boldsymbol{\theta}_0$ and $j = 1$.
2. Draw $\boldsymbol{\theta}^*$ from $q(\boldsymbol{\theta}_{j-1}, \cdot)$ and u from $U[0, 1]$.
3. Compute unnormalized weights and the acceptance probability

$$\omega(\boldsymbol{\theta}_{j-1}, \boldsymbol{\theta}^*) = \tilde{\pi}(\boldsymbol{\theta}^*)/q(\boldsymbol{\theta}_{j-1}, \boldsymbol{\theta}^*)$$

$$\omega(\boldsymbol{\theta}^*, \boldsymbol{\theta}_{j-1}) = \tilde{\pi}(\boldsymbol{\theta}_{j-1})/q(\boldsymbol{\theta}^*, \boldsymbol{\theta}_{j-1})$$

$$\alpha(\boldsymbol{\theta}_{j-1}, \boldsymbol{\theta}^*) = \min\left\{1, \frac{\omega(\boldsymbol{\theta}_{j-1}, \boldsymbol{\theta}^*)}{\omega(\boldsymbol{\theta}^*, \boldsymbol{\theta}_{j-1})}\right\}.$$

4. If $u \leq \alpha(\boldsymbol{\theta}_{j-1}, \boldsymbol{\theta}^*)$ set $\boldsymbol{\theta}_j = \boldsymbol{\theta}^*$, otherwise set $\boldsymbol{\theta}_j = \boldsymbol{\theta}_{j-1}$.
5. Set $j = j + 1$ and go back to step 2 until convergence is reached.

In the above algorithmic representation, the proposal density $q(\cdot, \boldsymbol{\theta})$ plays a similar role as $q(\boldsymbol{\theta})$ in both rejection and SIR algorithms. Two commonly used versions of the MH algorithm are the *random walk* MH and the *independent* MH, where $q(\boldsymbol{\phi}, \boldsymbol{\theta}) = q(|\boldsymbol{\theta} - \boldsymbol{\phi}|)$ and $q(\boldsymbol{\phi}, \boldsymbol{\theta}) = q(\boldsymbol{\theta})$, respectively. Proper choice of q and convergence diagnostics are of key importance to validate the algorithm and are, in the majority of the situations, problem-specific. See Chen et al. (2000) or Gamerman and Lopes (2006), Chapter 6, for additional technical details, references, and didactic examples.

8.4.3.4 Gibbs Sampler

Like MH algorithms, the *Gibbs sampler* is an iterative MC algorithm that takes advantage of easy to sample from, and easy to evaluate full conditional distributions that appear in several statistical modeling structures. More specifically, it is an algorithm that breaks the vector of parameters $\boldsymbol{\theta}$ into d blocks (scalar, vector or matrix) of parameters $\boldsymbol{\theta}_1, \ldots, \boldsymbol{\theta}_d$ and recursively samples $\boldsymbol{\theta}_i$ from its full conditional $\pi(\boldsymbol{\theta}_i|\boldsymbol{\theta}_{-i}, \mathbf{y})$, where $\boldsymbol{\theta}_{-i} = (\boldsymbol{\theta}_1, \ldots, \boldsymbol{\theta}_{i-1}, \boldsymbol{\theta}_{i+1}, \ldots, \boldsymbol{\theta}_d)$, for $i = 1, \ldots, d$. Its name is derived from its initial use in the context of image processing, where the posterior distribution was a Gibbs distribution (Geman and Geman, 1984), while Gelfand and Smith (1990) made it popular within the statistical community. As the number of $\boldsymbol{\theta}$ draws increases, the chain approaches its equilibrium, viz., $\pi(\boldsymbol{\theta}|\mathbf{y})$. Convergence is then assumed to hold approximately. It can be described algorithmically as follows.

Algorithm 8.4 The Gibbs sampler

1. Set the initial values at $\boldsymbol{\theta}^{(0)} = (\boldsymbol{\theta}_1^{(0)}, \ldots, \boldsymbol{\theta}_d^{(0)})'$ and $j = 1$.
2. Obtain a new value $\boldsymbol{\theta}^{(j)} = (\boldsymbol{\theta}_1^{(j)}, \ldots, \boldsymbol{\theta}_d^{(j)})'$ from $\boldsymbol{\theta}^{(j-1)}$ as follows:

$$\begin{aligned}
\boldsymbol{\theta}_1^{(j)} &\sim \pi(\boldsymbol{\theta}_1|\boldsymbol{\theta}_2^{(j-1)}, \ldots, \boldsymbol{\theta}_d^{(j-1)}),\\
\boldsymbol{\theta}_2^{(j)} &\sim \pi(\boldsymbol{\theta}_2|\boldsymbol{\theta}_1^{(j)}, \boldsymbol{\theta}_3^{(j-1)}, \ldots, \boldsymbol{\theta}_d^{(j-1)}),\\
&\vdots\\
\boldsymbol{\theta}_d^{(j)} &\sim \pi(\boldsymbol{\theta}_d|\boldsymbol{\theta}_1^{(j)}, \ldots, \boldsymbol{\theta}_{d-1}^{(j)}).
\end{aligned}$$

3. Set $j = j + 1$ and go back to step 2 until convergence is reached.

8.4.3.5 *MCMC Over Model Spaces*

Suppose models M_j, for $j \in \mathcal{J}$ (for instance, $\mathcal{J} = \{1, \ldots, J\}$), are entertained. Also, under model M_j, assume that $p(\mathbf{y}|\theta_j, M_j)$ is a probability model for $\mathbf{y}$ parameterized by a d_j-dimensional vector $\boldsymbol{\theta}_j$ (usually in $\Re^{d_j}$). Because the dimension (and interpretation) of the model parameters might vary with the models, the above MCMC algorithms cannot be directly used in order to derive, for example, approximations for $\pi(\boldsymbol{\theta}_j|\mathbf{y}, M_j)$ and, perhaps more importantly, $\Pr(M_j|\mathbf{y})$, the posterior model probability for model M_j. Nonetheless, Carlin and Chib (1995) and Green (1995) respectively proposed generalizations of the Gibbs sampler and the MH algorithm to situations where the parameter vector becomes $(j, \boldsymbol{\theta})$, where $\boldsymbol{\theta} = (\boldsymbol{\theta}_j : j \in \mathcal{J})$. The former is commonly known as the *Carlin–Chib algorithm* and the latter as the *reversible jump* MCMC (RJMCMC) algorithm. Dellaportas et al. (2002) and Godsill (2001) propose hybrid versions of these two algorithms, whereas Clyde (1999) argues that MCMC model composition (Raftery et al., 1997) and stochastic search variable selection (George and McCulloch, 1992) are particular cases of the RJMCMC algorithm. Among many others, Kuo and Song (2005) used RJMCMC for carrying out inference in dynamic frailty models for multivariate survival times (see Section 8.5.3), Waagepetersen and Sorensen (2001) used it in genetic mapping and Lopes et al. (2003) in multivariate mixture modeling of hematologic data (see Example 2). Detailed and comprehensive review of MCMC algorithms over model spaces appeared in Sisson (2005) along with an extensive list of freely available packages for RJMCMC and other transdimensional algorithms. Further details appear in Chapter 7 of Gamerman and Lopes (2006).

8.4.3.6 Public Domain Software

Until recently, a practical impediment to the routine use of Bayesian approaches was the lack of reliable software. This has rapidly changed over the last few years. The BUGS project (Spiegelhalter et al., 1999) provides public domain code that has greatly facilitated the use of Bayesian inference. The BUGS code is available from `http://www.mrc-bsu.cam.ac.uk/bugs/winbugs/contents.shtml`. The software is widely used and well tested and validated.

Another source of public domain software for Bayesian inference consists of libraries for the statistical analysis system **R** (R Development Core Team, 2006). **R** provides many libraries (packages) for Bayesian inference, including *MCMCpack* for MCMC within R, *bayesSurv* for Bayesian survival regression, *boa* for MCMC convergence diagnostics, *DP-package* for nonparametric Bayesian inference, and *BayesTree* for Bayesian additive regression trees. All packages are accessible and can be downloaded from the main **R** website `http://cran.r-project.org`.

8.5 Bayesian Survival Analysis

Bayesian methods for survival analysis to model times-to-events have been widely used in recent years in the biomedical and public health areas. The book by Ibrahim et al. (2001) is an excellent text describing various aspects of Bayesian inference. Here, we give a brief summary, with references, of a few useful areas, and then describe in detail frailty models for multivariate survival times.

8.5.1 Models for Univariate Survival Times

Univariate survival analysis assumes a suitable model for a continuous nonnegative survival time T which may be described in terms of the survival function $S(t) = P(T > t)$ or the hazard function $h(t) = -\mathrm{d}\log S(t)/\mathrm{d}t$. A parametric framework assumes that i.i.d. survival times $\mathbf{t} = (t_1, \ldots, t_n)$ follow a parametric model such as exponential, Weibull, gamma, lognormal, or poly-Weibull. The data might be complete or censored. Recall that a survival time is right (left) censored at c if its actual value is unobserved and it is only known that the time is greater than or equal to (less than or equal to) c, and is interval censored if it is only known that it lies in the interval (c_1, c_2). Given a set of p covariates (risk factors) $Z_1, \ldots, Z_p$, it is straightforward to use a standard software like BUGS to fit a suitable regression model and derive posterior and predictive distributions. A proportional hazards model specifies $h(t|\mathbf{z}) = h_0(t)\exp(\mathbf{z}_i'\boldsymbol{\beta})$ for a vector of parameters $\boldsymbol{\beta} = (\beta_1, \ldots, \beta_p)'$, and baseline hazard $h_0(.)$. Biomedical applications of parametric models have been discussed for instance in Achcar et al. (1985) and Kim and Ibrahim (2000). For situations in which the proportional hazards may be inappropriate, flexible hierarchical models with time dependent covariates have been fit so that $h(t|\mathbf{z}) = h_0(t)\exp(\mathbf{x}_i'\boldsymbol{\beta}(t)$, or even by replacing the linear term by a neural network; see Gustafson (1998), Carlin and Hodges (1999), and Faraggi and Simon (1995). Gustafson (1998) discusses data on duration of nursing home stays through the hybrid MCMC algorithm. As another alternative, a generalization of the Cox model has been discussed in Sinha et al. (1999) through a discretize hazard and time dependent regression coefficients, with application to breast cancer data (Finkelstein and Wolfe, 1985).

With recent advances in computational Bayesian methods, semiparametric and nonparametric methods have become more prevalent. The piecewise constant semiparametric hazard model defines a finite partition of the time axis into K intervals, and assumes a constant baseline hazard in each subinterval; the simplest baseline, called the piecewise exponential model, will be described under frailty modeling later in this section. Nonparametric models include use of gamma process priors (Kalbfleisch, 1978), beta process priors (Sinha, 1997), correlated gamma process priors (Mezzetti and Ibrahim, 2000), Dirichlet process priors (DPP) (Gelfand and Mallick, 1995), mixtures of DPP's or MDP models (MacEachern and Müller, 1998), and Polya tree process priors (Lavine, 1992).

8.5.2 Shared Frailty Models for Multivariate Survival Times

Dependent multivariate times to events frequently occur in several biomedical applications, and frailty models (Vaupel et al., 1979) have been extensively used for modeling dependence in such multivariate survival data. The dependence frequently arises because subjects in the same group are related to each other, or due to multiple recurrence times of a disease for the same patient, and computational Bayesian methods facilitate a variety of frailty models. The widely used shared frailty models for multivariate times to events assume that there is an unobserved random effect, known as frailty, which explains dependence that may arise owing to association among subjects in the same group, or among multiple recurrence times of an event for the same subject. Suppose the survival time of the kth subject ($k = 1, \ldots, m$) in the jth group ($j = 1, \ldots, n$) is denoted by T_{jk}, and $\mathbf{z}_{jk}$ is a fixed, possibly time dependent covariate vector of dimension p.

EXAMPLE 5

An often cited example (McGilchrist and Aisbett, 1991) consists of the kidney infection data on times to first and second occurrence of infection in 38 patients on portable dialysis machines ($n = 38$, $m = 2$). Binary variables representing, respectively, the censoring indicators for the first and second recurrences are available; occurrence of infection is indicated by 1, and censoring by 0.

The gender of the patients (0 indicating male, and 1 indicating female), is a covariate (see Ibrahim et al., 2001, Table 1.5). Other covariates, such as age and disease type of each patient, are also available with this data. However, initial analysis by McGilchrist and Aisbett, using what they referred to as a penalized partial likelihood approach, showed that the effect of these covariates on infection times was not statistically significant; hence they are omitted from the analysis.

Given the unobserved frailty parameter w_j for the jth group, the modified Cox proportional hazards shared frailty regression model (Clayton and Cuzick, 1985; Oakes, 1989) through the hazard function is

$$h(t_{jk}|w_j, \mathbf{z}_{jk}) = h_0(t_{jk}) \exp(\boldsymbol{\beta}'\mathbf{z}_{jk})w_j, \tag{8.28}$$

where $\boldsymbol{\beta}$ is the vector of regression parameters of the same dimension, $h_0(.)$ is the baseline hazard function and w_j is an individual random effect (frailty) representing common unobserved covariates and generating dependence. The event times are assumed to be conditionally independent given the shared frailty. For identifiability purposes, it is usual to require that the linear model component $\boldsymbol{\beta}'\mathbf{z}_{jk}$ has no intercept term. Let $\theta_{jk} = \exp(\boldsymbol{\beta}'\mathbf{z}_{jk})$.

8.5.2.1 Baseline Hazard Function

The baseline hazard function could be a simple constant hazard function, a Lévy process, a Gamma process, a Beta process or a correlated prior process (see Sinha and Dey, 1997 for an extensive review). A common parametric baseline hazard is the Weibull form $h_0(t_{ij}) = \lambda\gamma t_{ij}^{\gamma-1}$, for $\lambda > 0$, and $\gamma > 0$. The more flexible piecewise exponential correlated prior process baseline hazard requires that the time period is divided into g intervals, $I_i = (t_{i-1}, t_i)$ for $i = 1, \ldots, g$, where $0 = t_0 < t_1 < \cdots < t_g < \infty$, t_g denoting the last survival or censored time. The baseline hazard is assumed to be constant within each interval, that is, $\lambda_0(t_{jk}) = \lambda_i$ for $t_{jk} \in I_i$. A discrete-time martingale process is used to correlate the λ_i's in adjacent intervals, thus introducing some smoothness (Arjas and Gasbarra, 1994). Given $(\lambda_1, \ldots, \lambda_{i-1})$, specify that

$$\lambda_i|\lambda_1, \ldots, \lambda_{i-1} \sim \text{Gamma}\left(c_i, \frac{c_i}{\lambda_{i-1}}\right), \quad i = 1, \ldots, g,$$

where $\lambda_0 = 1$, so that $E(\lambda_i|\lambda_1, \ldots, \lambda_{i-1}) = \lambda_{i-1}$; let $\tilde{\lambda} = (\lambda_1, \ldots, \lambda_g)$. A small value of c_i indicates less information for smoothing the λ_i's; if $c_i = 0$, then λ_i is independent of λ_{i-1} while if $c_i \to \infty$, $\lambda_i = \lambda_{i-1}$. Regarding the choice of g, a very large value would result in a nonparametric model and produce unstable estimates of λs whereas a very small value of g would lead to inadequate model fitting. In practical situations, g is determined on the basis of the design. A random choice of g will lead to a posterior distribution with variable dimensions, which may be handled through a RJMCMC (Green, 1995) described in Section 8.4.

8.5.2.2 Frailty Specifications

Alternate parametric shared frailty specifications that have appeared in the recent literature include the gamma model (Clayton and Cuzick, 1985), the log-normal model (Gustafson, 1997), and the positive stable model (Hougaard, 2000). The gamma distribution is the most common finite mean frailty distribution, and we assume that w_j are i.i.d. Gamma$(\kappa^{-1}, \kappa^{-1})$ variables, so that the mean is 1 and the variance is the unknown κ. For identifiability, the mean of w_j's must be 1. For Bayesian inference on a shared Gamma frailty-Weibull baseline model with application to the kidney infection data, see Section 8.4.1 in Ibrahim et al. (2001). In a proportional hazards frailty model, the unconditional effect of a covariate, which is measured by the hazard ratio between unrelated subjects (i.e., with different frailties) is always less than its conditional effect, measured by the hazard ratio among subjects with the same frailty. In particular, suppose we consider two subjects from different groups and with respective covariates 0 and $\tilde{z}$; let $S_0(t)$ and $S_1(t)$ denote the corresponding unconditional survivor functions derived under this frailty specification. It has been shown that the covariate effects, as measured by the hazard ratio are always attenuated and further, $S_0(t)$ and $S_1(t)$ are usually not related through a proportional hazards model. If the frailty distribution is an infinite variance positive stable distribution, then $S_1(t)$ and $S_0(t)$ will have proportional hazards; we no longer need to choose between conditional and unconditional model specifications, since a single specification can be interpreted either way. A positive stable frailty distribution thus not only permits a proportional hazards model to apply unconditionally, but also allows for a much higher degree of heterogeneity among the common covariates than would be possible by using a frailty distribution with finite variance, such as the Gamma distribution. The frailty parameters $w_j, j = 1, \ldots, n$ are assumed to be i.i.d. for every group, according to a positive α-stable distribution. The density function of a positive stable random variable w_j is not available in closed form. However, its characteristic function is available and has the form

$$E(e^{i\vartheta w_j}) = \exp\{-|\vartheta|^{\alpha}(1 - i\text{sign}(\vartheta)\tan(\pi\alpha/2))\}, \tag{8.29}$$

where $i = \sqrt{-1}$, ϑ is a real number, $\text{sign}(\vartheta) = 1$ if $\vartheta > 0$, $\text{sign}(\vartheta) = 0$ if $\vartheta = 0$ and $\text{sign}(\vartheta) = -1$ if $\vartheta < 0$.

8.5.2.3 Positive Stable Shared Frailty Model

Although the positive stable frailty model is conceptually simple, estimation of the resulting model parameters is complicated owing to the lack of a closed form expression for the density function of a stable random variable. Qiou et al. (1999) described a Bayesian framework using MCMC for this problem with application to the kidney infection data. This was later extended by Ravishanker and Dey (2000) to include a mixture of positive stables frailty, and by Mallick and Ravishanker (2004) to a power variance family (PVF) frailty (indexed by parameters η and α). The Bayesian approach is based on an expression provided by Buckle (1995) for the joint density of n i.i.d. observations from a stable distribution by utilizing a bivariate density function $f(w_j, y|\alpha)$ whose marginal density with respect to one of the two variables gives exactly a stable density. Let $f(w_j, y|\alpha)$ be a bivariate function such that it projects $[(-\infty, 0) \times (-1/2, l_\alpha)] \cup [(0, \infty) \times (l_\alpha, 1/2)]$ to $(0, \infty)$:

$$f(w_j, y|\alpha) = \frac{\alpha}{|\alpha - 1|} \exp\left\{ - \left| \frac{w_j}{\tau_\alpha(y)} \right|^{\alpha/(\alpha-1)} \right\} \left| \frac{w_j}{\tau_\alpha(y)} \right|^{\alpha/(\alpha-1)} \frac{1}{|w_j|}, \tag{8.30}$$

where

$$\tau_\alpha(y) = \frac{\sin(\pi\alpha y + \psi_\alpha)}{\cos \pi y} \left[\frac{\cos \pi y}{\cos\{\pi(\alpha - 1)y + \psi_\alpha\}} \right]^{(\alpha-1)/\alpha},$$

$w_j \in (-\infty, \infty), y \in (-1/2, 1/2), \psi_\alpha = \min(\alpha, 2-\alpha)\pi/2$ and $l_\alpha = -\psi_\alpha/\pi\alpha$. Then

$$f(w_j|\alpha) = \frac{\alpha |w_j|^{1/(\alpha-1)}}{|\alpha - 1|} \int_{-1/2}^{1/2} \exp\left\{ - \left| \frac{w_j}{\tau_\alpha(y)} \right|^{\alpha/(\alpha-1)} \right\} \left| \frac{1}{\tau_\alpha(y)} \right|^{\alpha/(\alpha-1)} dy. \tag{8.31}$$

Denoting by $\mathcal{D}$ the triplets, $(t_{jk}, \delta_{jk}, \mathbf{z}_{jk})$, the vector of unobserved w_j's by $\mathbf{w}$ and the vector of unobserved auxiliary variables $(y_1, \ldots, y_n)$ by $\mathbf{y}$, the complete data likelihood is

$$l(\boldsymbol{\beta}, \boldsymbol{\lambda}, \alpha|\mathbf{w}, \mathbf{y}, \mathcal{D}) = \prod_{j=1}^{n} \prod_{k=1}^{m} \left[\prod_{i=1}^{g-1} \exp\{-\lambda_i \Delta_i \theta_{jk} w_j\} \right] \times \exp\{-\lambda_g (t_{jk} - t_{g-1}) \theta_{jk} w_j\} (\lambda_g \theta_{jk} w_j)^{\delta_{jk}}. \tag{8.32}$$

The observed data likelihood based on the observed data $\mathcal{D}$ is obtained by integrating out the w_j's from Equation 8.32 using the stable density expression in Equation 8.31, and corresponds to the marginal model whereas Equation 8.32 corresponds to the conditional model given the frailty. Assuming suitable priors, the joint posterior density based on the observed data likelihood is derived and appropriate MCMC algorithms (see Section 8.4) are used to generate samples from the posterior distribution through complete conditional distributions. For instance, the ratio-of-uniforms algorithm is used to generate λ_k and β_j samples, the MH algorithm is used for α, and the rejection algorithm for y_i. See (Qiou et al., 1999) for details, as well as for results corresponding to bivariate times with the kidney infection data. In

TABLE 8.1
Posterior Summary for Kidney Infection Data under Shared Frailty Models

	Frailty Model			
	Gamma	Positive Stable	PVF	Additive Stable
Parameter	**Mean (s.d.)**	**Mean (s.d.)**	**Mean (s.d.)**	**Mean (s.d.)**
β	−1.62 (0.42)	−1.06 (0.36)	−1.15 (0.88)	−1.40 (0.61)
$1/\kappa$	0.33 (0.17)	—	—	—
α	—	0.86 (0.07)	0.38 (0.35)	—
η	—	—	1.09 (1.05)	—
$\alpha 1$	—	—	—	0.35 (0.05)
$\alpha 2$	—	—	—	0.39 (0.04)
3	—	—	—	0.42 (0.04)
λ_1	0.002 (0.003)	0.001 (0.002)	0.001 (0.004)	0.012 (0.015)
λ_2	0.001 (0.003)	0.003 (0.007)	0.004 (0.016)	0.023 (0.036)
λ_3	0.001 (0.002)	0.0038	0.005 (0.019)	0.031 (0.049)
λ_4	0.001 (0.002)	0.004 (0.009)	0.005 (0.019)	0.037 (0.057)
λ_5	0.001 (0.002)	0.003 (0.008)	0.005 (0.019)	0.040 (0.060)
λ_6	0.001 (0.002)	0.004 (0.011)	0.005 (0.019)	0.045 (0.066)
λ_7	0.001 (0.002)	0.004 (0.010)	0.006 (0.020)	0.058 (0.065)
λ_8	0.001 (0.003)	0.006 (0.013)	0.008 (0.025)	0.097 (0.098)
λ_9	0.006 (0.005)	0.027 (0.022)	0.038 (0.063)	0.298 (0.196)
λ_{10}	0.360 (0.150)	1.84 (0.604)	2.60 (2.13)	2.86 (1.22)

Table 8.1, we show results from fitting the different shared frailty models to the kidney infection data.

8.5.3 Extensions of Frailty Models

There are several extensions of frailty modeling. Unlike the shared frailty model that corresponds to the assumption of a common risk dependence among multivariate times, the additive frailty model (Hougaard, 2000) allows us to handle such survival times with varying degrees of dependence, by combination of subgroups. For instance, data on time to tumorigenesis of female rats in litters was discussed by Mantel et al. (1977). Each litter had three rats; one rat was drug-treated and the other two served as control. Time of tumor appearance was recorded (death owing to other causes was considered as censoring), and the study was ended after 104 weeks. Under an additive frailty model, it is assumed that the three female rats in each of 50 litters correspond to three different frailty components instead of sharing a single random component under the shared frailty model, thereby yielding a richer dependence structure. Each subcomponent of the resulting multivariate frailty random variable is further decomposed into independent additive frailty variables, and the frailty component of each rat in every litter is the sum of the litter effect and the individual rat effect. The dependence among rats in each litter then arises owing to the litter effect, whereas the individual rat effect generates additional variability. Specifically, the additive frailty model specifies that the components of a frailty random vector are combined additively for the jth subject within the ith group, and they then act multiplicatively in the Cox proportional hazards model, that is,

$$h(t_{ij}|w_{ij}, \mathbf{z}_{ij}) = \lambda_0(t_{ij}) \exp(\beta' \mathbf{z}_{ij}) w_{ij}; \tag{8.33}$$

the dependence is generated by setting

$$w_{ij} = \mathbf{A}'_{ij} \mathbf{X}_i, \tag{8.34}$$

where $\mathbf{X}'_i = (X_{i1}, X_{i2}, \ldots, X_{is})$, and $\mathbf{A}'_{ij} = (a_{ij1}, a_{ij2}, \ldots, a_{ijs})$ is the vector of design components for the jth subject in the ith group. The other quantities in Equation 8.33 have been defined earlier. Some components in w_{ij} are shared by other subjects in the same group and thereby generate dependence. Non-shared components produce individual variability in the model. For bivariate times to events, suppressing the group index, the frailty may be expressed in the form (W_1, W_2) where W_k corresponds to the frailty variable for the kth subject, $k = 1, 2$, and is given by

$$W_1 = X_0 + X_1, \quad \text{and} \quad W_2 = X_0 + X_2,$$

where X_0, X_1, and X_2 are independently distributed positive-valued random variables. The dependence between bivariate times to events arises owing to the common term X_0, while the other two terms X_1 and X_2 generate additional unshared variability corresponding to individual random effects. Bayesian inference under this framework has been recently discussed in Mallick and Ravishanker (2006) with application to the tumorigenesis data and to the kidney infection data. Extension to vector frailty is interesting (Xue and Brookmeyer, 1996).

Hierarchical frailty models for multilevel multivariate survival data has been discussed by Gustafson (1995) in the context of data from a clinical trial of chemotherapy for advanced colorectal cancer. The data were collected from 419 patients who participated in the trial conducted at 16 clinical sites; Ibrahim et al. (2001) have described the use of the hybrid Monte Carlo method for this example. Kuo and Song (2005) have described a dynamic frailty model that assumes a subject's risk changes over time; they used RJMCMC for carrying out inference. Another interesting class of models are multivariate cure rate models (Chen et al., 2002). The cure rate model has been useful for modeling data from cancer clinical trials, where it is assumed that a certain proportion q of the population is cured, whereas the remaining $1 - q$ is not cured. For bivariate times, see Ibrahim et al. (2001, Section 5.5) for details and examples.

8.6 Disease Mapping

Disease mapping is, broadly speaking, the modeling of the spatial behavior of disease rates, as well as the identification, classification, or clustering of areas of highest (lowest) risk rates and their association to explanatory variables. Resource allocation policies and testing of epidemiologic and environmental hypotheses are standard scientific enquiries facilitated by disease mapping models. Statistical models for disease mapping are meaningfully stated as hierarchical models, such as the one characterized by Equations 8.35 through 8.37. The Bayesian approach to disease mapping problems has become commonplace over the last decade, with region-specific random effects modeled by spatially structured priors at one or several hierarchy levels in general hierarchical models (see Section 8.3.1). It is often useful to present a disease mapping model as a graphical model. This helps us discuss the model structure without distracting details [see Mollié (1996) for an example].

One standard approach in disease mapping is to locally model counts by Poisson distributions, that is,

$$y_i|\rho_i \sim \text{Poisson}(\rho_i), \tag{8.35}$$

where ρ_i is the relative risk in region i, whose structural dependence appears in a second hierarchical level, for instance, as

$$\rho_i = g(x_i'\theta + \beta_i + \epsilon_i), \tag{8.36}$$

where g is a link function (e.g., exponential), x_i' is a vector of explanatory variables, β_i is a region-specific random effect and ϵ_i is a noise term, usually $N(0, \sigma_\epsilon^2)$. The random effects coefficients β_is follow a standard conditionally autoregressive (CAR) spatial structure (Besag et al., 1991) with the conditional distribution of β_i depending on β_j for all neighboring regions j, that is,

$$\beta_i|\beta_{-i} \sim N\left(\sum_{j\in\delta_i} w_{ij}\beta_j, \sigma_\beta^2/n_i\right), \tag{8.37}$$

for $\beta_{-i} = (\beta_1, \ldots, \beta_{i-1}, \beta_{i+1}, \ldots, \beta_n)$, δ_i a set of regions adjacent to i, weighting function ω_{ij}, $n_i = \sum_{j\in\delta_i} \omega_{ij}$ and $w_{ij} = \omega_{ij}/n_i$. The most often used neighboring structure is the one that assumes that $\omega_{ij} = 1$ if i and j are neighbor counties and zero otherwise. Bernardinelli et al. (1995), Best et al. (1999), Kelsal and Wakefield (2002) and Wall (2004) are a few additional studies which use CAR prior distributions for disease mapping.

EXAMPLE 6

Nobre et al. (2005) examined the spatial and temporal behavior of malaria incidence and its relationship to rainfall over time and across counties, for several counties of Pará, one of Brazil's largest states located in the Amazon region. In 2001, for instance, Pará had around 17,000 cases of malaria. Malaria affects about 600 million persons a year worldwide and is the most common infectious disease found in Brazil's rainforest. Malaria is transmitted by mosquitoes from the *Anopheles sp* genus. Temperature and rainfall are important natural risk factors affecting life cycle and breeding of the mosquitoes. Extreme rainfall and extreme drought are both equally likely to lead to proliferation of the mosquitoes, and therefore the disease. Limited public health policies and population migration are major social risk factors. Assuming the state of Pará is divided into n contiguous counties (subregions), and that y_i are the number of malaria cases in county i, they extend the standard CAR prior by proposing the following space-time model for malaria counts:

$$\begin{aligned} \rho_{it} &= \exp\{\theta_t x_{it} + \beta_{it}\}, \\ \theta_t &\sim N(\theta_{t-1}, \tau_\theta^2), \\ \beta_t &\sim \mathcal{D}(\sigma_t^2), \end{aligned} \tag{8.38}$$

where x_{it} is a measure of rainfall and $\mathcal{D}$ is the distribution of β_t as a function of σ_t^2. They entertain four space-time models by crossing two specifications for $\mathcal{D}$, that is, $\beta_t \sim N(0, \sigma_t^2 I_n)$ and $\beta_t \sim CAR(\sigma_t^2)$, and two specifications for σ_t^2, that is, $\sigma_t^2 \sim IG(a, c)$ and $\sigma_t^2 \sim \text{Log-normal}(\log(\sigma_{t-1}^2), \tau^2)$. Figure 8.1 exhibits log-relative risks (posterior medians) for the month of March 1997 based on the third model specification, that is, $\beta_t \sim N(0, \sigma_t^2 I_n)$ and $\sigma_t^2 \sim \text{Log-normal}(\log(\sigma_{t-1}^2), \tau^2)$. The temporal structures resemble dynamic linear models and dynamic generalized linear models (West et al., 1985).

Nobre et al. (2005) generalize the models that appear in Waller et al. (1997) and Knorr-Held and Besag (1998), who analyze lung cancer mortality in the state of Ohio over the years. In addition, recent space-time studies in disease mapping are Assunção et al. (2001) who model the diffusion and prediction of Leishmaniasis and Knorr-Held and Richardson (2003) who examine surveillance data on meningococcal disease incidence. MacNab et al. (2004), Congdon (2005, Chapter 8), Best et al. (2005) and references therein provide additional discussion about the Bayesian and empirical Bayesian estimation in disease mapping.

8.7 Bayesian Clinical Trials

Berry (2006) argues that a Bayesian approach is natural for clinical trial design and drug development. An important advantage is that the Bayesian approach allows for gradual updates of knowledge, rather than restricting the process to updating in large discrete steps measured in trials or phases.

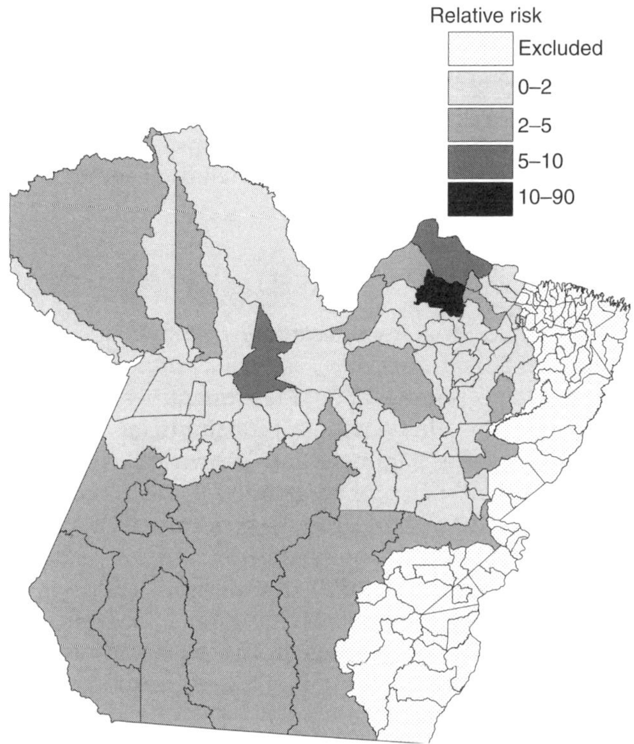

FIGURE 8.1
(See color insert following page 207.) Malaria in Pará: log-relative risk's March 1997 posterior median when $\beta_t \sim N(0, \sigma_t^2 I_n)$ and $\sigma_t^2 \sim$ Log-normal$(\log(\sigma_{t-1}^2), \tau^2)$. For instance, Anajás county highest risk may be owing to its proximity to several rivers and to the island of Marajó.

The process of updating information under a Bayesian approach is "specifically tied to decision making, within a particular trial, within a drug development program, and within establishing a company's portfolio of drugs under development" (Berry, 2006). He argues that the therapeutic areas in which the clinical end points are observed early should benefit most from a Bayesian approach. Bayesian methods are particularly useful for statistical inference related to diseases such as cancer in which there is a burgeoning number of biomarkers available for assessing the disease's progress (Berry, 2006). An example is presented of a Phase II neoadjuvant HER2/neu-positive breast cancer trial conducted at M. D. Anderson Cancer Center, in which 164 patients were randomized to two treatment arms, chemotherapy with and without trastuzumab (Herceptin; Genentech), and where the primary endpoint was pathological complete response (pCR) of the tumor. In the middle of the trial designed from a frequentist perspective and with the protocol specifying no interim analyses, data available to assess pCR on 34 patients showed that the trastuzumab arm had dramatic improvement, that is, 4 of 16 control patients (25%) and 12 of 18 trastuzumab patients (67%) experienced a pCR. The Bayesian predictive probability of the standard frequentist statistical significance when 164 patients had been treated was computed to be 95%, demonstrating the use of Bayesian analysis in conjunction with a frequentist design to override the protocol and stop the trial.

In this section we discuss principles of Bayesian clinical trial design, how it relates to frequentist design, and we explain details of some typical Bayesian clinical trial designs.

8.7.1 Principles of Bayesian Clinical Trial Design

The planning of a clinical trial involves many unknown quantities, including patient responses, that is, future data, as well as unknown parameters that are never observed. Uncertainties about these quantities are best described by defining appropriate probability models. Probability models that are defined on observable data as well as parameters are known as Bayesian models. Clinical trial designs based on such probability models are referred to as Bayesian clinical trial designs. Augmenting the Bayesian model for data and parameters with a formal description of the desired decision leads to a Bayesian decision problem. We refer to clinical trial designs based on this setup as Bayesian decision theoretic designs. Many popular designs stop short of a decision theoretic formulation. Designs that are based on a Bayesian probability model, without a formal definition of a loss function are referred to as proper Bayesian designs (Spiegelhalter et al., 2004).

Typical examples of proper Bayesian approaches are Thall et al. (1995) or Thall and Russell (1998).

EXAMPLE 7

Consider the following stylized example for an early phase trial. We assume that the observed outcome is an indicator $y_i \in \{0, 1\}$ for tumor response for the i-th patient. Let $\theta = Pr(y_i = 1)$ denote the unknown probability of tumor response. Assume that the current standard of care for the specific disease has a known success probability of $\theta_0 = 15\%$. Let n denote the number of currently enrolled patients, and let $x = \sum_{i=1}^{n} y_i$ denote the recorded number of tumor responses. Assuming a Beta prior, $\theta \sim \text{Be}(a, b)$ and a binomial sampling model, $x \sim \text{Bin}(n, \theta)$, we can at any time evaluate the posterior distribution $p(\theta \mid y_1, \ldots, y_n) = \text{Be}(a + x, b + n - x)$.

A typical proper Bayesian design could proceed with the following protocol:

- After each patient cohort, update the posterior distribution.
- Stop and recommend the experimental therapy if $Pr(\theta > 0.2 \mid y_1, \ldots, y_n) > \pi_1$.
- Stop and abandon the experimental therapy if $Pr(\theta < 0.1 \mid y_1, \ldots, y_n) > \pi_2$ or $n > n_{\max}$.

The design requires the elicitation of the prior parameters (a, b) and the choice of policy parameters (tuning parameters) $(\pi_1, \pi_2, n_{\max})$. Policy parameters are determined by matching desired operating characteristics, as shown below.

8.7.2 Operating Characteristics

The distinction between Bayesian and classical (or frequentist) design is not clear cut. Any Bayesian design can be considered and evaluated from a frequentist perspective. As before, let θ and y generically denote the parameters in the underlying probability model and the observed outcomes, respectively. Let δ denote the design. In particular, δ might include a rule to allocate patients to alternative treatment arms, a stopping rule, and a final decision. In Example 1, the design $\delta = (d, a)$ included a stopping rule $d(y) \in \{0, 1\}$ with $d = 1$ indicating stopping, and a final decision $a(y) \in \{0, 1\}$ with $a = 1$ indicating a recommendation of the experimental therapy. The recommendation might imply a decision to launch a following confirmatory trial.

For a given set of policy parameters and an assumed truth, we can evaluate frequentist error rates and other properties by evaluating repeated sampling expectations of the relevant summaries. Formally, we consider $E(g(d, \theta, y) \mid \theta)$ for an assumed truth θ. The choice of g and θ depends on the desired summary. For example, Type-I error is evaluated by setting $\theta = \theta_0$ and using $g(d, \theta, y) = I(a = 1)$. To evaluate power we would consider $g(d, \theta, y) = I(a = 1)$ for a grid of θ values with $\theta > 0.2$. Other important summaries are the expected sample size, $g(d, \theta, y) = \min_{n=1,2,\ldots,n_{\max}}\{n\colon d(y_1, \ldots, y_n) = 1\}$ and the expected number of successfully treated patients $g(d, \theta, y) = \sum y_i$. Such summaries are routinely reported as operating characteristics of a design. Formally, these summaries are evaluated by essentially ignoring the Bayesian nature of the design, and considering it as a possible frequentist design. The use of Bayes rules to construct promising candidates for a good frequentist procedure is a commonly used approach, even beyond the context of clinical trial design. It is considered good practice to report operating characteristics when proposing a Bayesian design. In most regulatory or review settings such reports are mandatory.

Besides the reporting purpose, an important use of operating characteristics is to select policy parameters. Similar to most clinical trial designs, many Bayesian designs require the selection of various policy parameters, such as $(\pi_1, \pi_2, n_{\max})$ in the earlier example. A commonly used procedure is to evaluate operating characteristics for a variety of choices of the policy parameters and fixing the final design by matching desired values for the operating characteristics. The resulting design is valid as both, a bona fide frequentist procedure as well as a coherent Bayesian design.

8.7.3 A Two-Agent Dose-Finding Design

Extensive recent reviews of Bayesian designs for early phase trials appear in Estey and Thall (2003), Berry (2005a), or Berry (2005b). A more comprehensive review, including issues beyond clinical trial design, is presented by Spiegelhalter et al. (2004).

As a typical example for a nontrivial Bayesian design we review in this section a design proposed in Thall et al. (2003). They consider a protocol for dose-finding with two agents. The response is an indicator for toxicity, $y_i \in \{0, 1\}$, for the i-th patient. For given doses $x = (x_1, x_2)$ of the two agents, let $\pi(x_1, x_2, \theta)$ denote the probability of toxicity. We assume

$$\pi(x, \theta) = \frac{\alpha_1 x_1^{\beta_1} + \alpha_2 x_2^{\beta_2} + \alpha_3 (x_1^{\beta_1} x_2^{\beta_2})^{\beta_3}}{1 + \alpha_1 x_1^{\beta_1} + \alpha_2 x_2^{\beta_2} + \alpha_3 (x_1^{\beta_1} x_2^{\beta_2})^{\beta_3}}.$$

Here $\theta = (\alpha_1, \beta_1, \alpha_2, \beta_2, \alpha_3, \beta_3)$. The model is chosen to imply parsimonious submodels for the corresponding single agent therapies with $x_1 = 0$ and $x_2 = 0$, respectively. This allows us to include available substantial prior information for (α_1, β_1) and (α_2, β_2). In the application the two agents are gemcitibine and cyclophosphamide, two chemotherapy agents that are extensively used in the treatment of various cancers. Without loss of generality, assume that both agents are available at doses $x_j \in \{0, 0.1, \ldots, 0.9, 1.0\}$.

The proposed design proceeds in two stages. In the first stage we escalate the dose of both agents along a predefined grid $D_1 = \{x^1, \ldots, x^k\}$ of dose pairs (x_1^j, x_2^j). For example, the predefined grid could be $(x_1^j, x_2^j) = (j \cdot 0.1, j \cdot 0.1)$, $j = 1, \ldots, 8$. Patients are assigned in cohorts of K patients, using for example, $K = 2$. Let $Y_i = (y_1, \ldots, y_i)$ denote the recorded responses of the first i patients. After each patient cohort we evaluate the posterior distribution $p(\theta \mid Y)$ for all i currently enrolled patients. The posterior on θ implies a posterior estimated toxicity surface $E(\pi(\theta, x) \mid Y_i)$. Subject to an overdose control, patients in the next cohort are assigned to the dose combination x^j that is closest to a desired target level of toxicity π^*. Overdose control means that no dose combination x^j on the grid can be skipped, that is, patients can only be assigned to x^j when earlier patients were earlier assigned to x^{j-1}.

After a predetermined number of patients, say n_1, the design switches to the second stage. In the second stage, we drop the restriction to the grid D_1. In other words, the assumption is that stage one has approximately identified a dose combination (x_1^*, x_2^*) on the grid D_1 with the desired toxicity, and we can now vary the doses x_1 and x_2 of the two agents to optimize cancer killing and learning about the toxicity surface. This optimization is carried out among all dose pairs with the same a posteriori estimated toxicity level, $E(\pi(x_1, x_2, \theta) \mid Y) \approx \pi^*$. Here Y are all responses that were observed up to now. Cancer killing is approximated as $\lambda(x_1 - x_1^*) + (x_2 - x_2^*)$. This is based on the assumption that the cancer-killing effect of agent 1 is stronger than that of agent 2 by a factor λ, and that the cancer-killing effects of both agents are additive and proportional to the dose. Learning about θ is formalized as the log of the determinant of the Fisher information matrix.

In summary, the design has several policy parameters, the choice of the stage one grid D_1, the sample size n_1, the cohort size K, and other parameters that control details that we did not describe above. All policy parameters are chosen by matching desired frequentist operating characteristics.

8.8 Microarray Data Analysis

8.8.1 Introduction

High-throughput assays like microarrays, serial analysis of gene expression (SAGE), and protein mass spectrometry are becoming increasingly more commonly used in medical and biological research. Microarrays and SAGE experiments measure mRNA expression, whereas mass/charge spectra record protein abundance. An excellent review of the experimental setup for all three formats, and related statistical methods appears in Baggerly et al. (2006). See also Datta et al. (2007) and Naik and Wagner (2007).

Microarray experiments are by far the most commonly used of the three mentioned high-throughput assays. Many published methods for statistical inference (after completed preprocessing) focus on the stylized setup of two-group comparisons. Two-group comparisons are experiments that record gene expression for samples under two biologic conditions of interest and seek to identify those genes that are differentially expressed. Most microarray experiments involve more complicated designs. But the two-group comparison serves as a good canonical example. Popular methods include the use of two-sample t-tests and nonparametric permutation tests, applied to one gene at a time. Several methods have been proposed to adjust the resulting p values for multiplicities and compile a list of differentially expressed genes. This includes popular methods based on controlling (frequentist) expectation of the false discovery rate (FDR) (Benjamini and Hochberg, 1995; Storey, 2002; and Storey et al., 2004), the beta-uniform method (Pounds and Morris, 2003), or significance analysis of microarrays, popularly known as SAM (Tusher et al., 2001). Here, FDR is defined as the number of false discoveries, that is, the number of genes that are falsely reported as differentially expressed, relative to the total number of genes that are reported as differentially expressed. The beta-uniform method is based on modeling the distribution of p values across all genes, and SAM is an algorithm that uses repeated simulation to determine significance cutoffs.

In this section we review some Bayesian inference for group comparison microarray experiments and a Bayesian perspective to error control in massive multiple comparisons. One of the more popular methods is the empirical Bayes approach proposed in Efron et al. (2001). The authors describe their approach as an empirical Bayes method. They assume that data are summarized as a set of difference scores, with one score for each gene. The method assumes that these scores arise from a mixture model with submodels corresponding to differentially and nondifferentially expressed genes. The desired inference of identifying differentially expressed genes is formally described as the problem of deconvolution of this mixture. This is achieved by clever, but ad hoc methods. Parmigiani et al. (2002) assume a mixture model with three terms corresponding to over-, under-, and normal-expressed genes, using uniform distributions for over- and under-expression, and a central Gaussian for normal expression. The authors argue that further inference should be based on the imputed latent trinary indicators in this mixture. Ibrahim et al. (2002) propose a model with an explicit threshold to accommodate the many genes that are not strongly expressed, and proceed then with a mixture model including a point mass for the majority of not-expressed genes and a log normal distribution for expressed genes. Another class of methods is based on a Gamma/Gamma hierarchical model developed in Newton et al. (2001). The model includes parameters that are interpreted as latent indicators for differential expression. Other recent proposals based on mixture models and model selection include Ghosh (2004), Ishwaran and Rao (2003), Tadesse et al. (2003), Broet et al. (2002), Dahl (2003), Tadesse et al. (2005), Hein et al. (2005), and Lewin et al. (2006) develop approaches based on hierarchical models. Frigessi et al. (2005) develop a hierarchical model based on a detailed description of the physical process, including the details of hybridization, and so forth.

In this section we review three of the mentioned approaches as typical examples for this literature.

8.8.2 The Gamma/Gamma Hierarchical Model

In general, a model for microarray group comparison experiments should be sufficiently structured and detailed to reflect the prior expected levels of noise, a reasonable subjective judgment about the likely numbers of differentially expressed genes, and some assumption about dependencies, if relevant. It should also be easy to include prior data when available. One model that achieves these desiderata is that introduced in Newton et al. (2001) and Newton and Kendziorski (2003).

The model does not include details of the data cleaning process, including spatial dependence of measurement errors across the microarray, correction for misalignments, and so forth. Although such details are important, it is difficult to automate inference in the form of a generally applicable probability model. We feel normalization and standardization are best dealt with on a case by case basis, exploiting available information about the experimental setup. The remaining variability resulting from preprocessing and normalization can be subsumed as an aggregate in the prior description of the expected noise. So in the following discussion we assume that the data are appropriately standardized and normalized and that the noise distribution implicitly includes these considerations. See, for example, Tseng et al. (2001), Baggerly et al. (2001), or Yang et al. (2002) for a discussion of the process of normalization.

We focus on the comparison of two conditions and assume that data will be available as arrays of appropriately normalized intensity measurements X_{ij} and Y_{ij} for gene i, $i = 1, \ldots, n$, and experiment j, $j = 1, \ldots, J$, with X and Y denoting the intensities in the two conditions. Newton et al. (2001) propose probabilistic modeling for the observed gene frequencies in a single-slide experiment.

For each gene i we record a pair (X_i, Y_i) of intensities corresponding to transcript abundance of a gene in the two samples. The true unknown mean

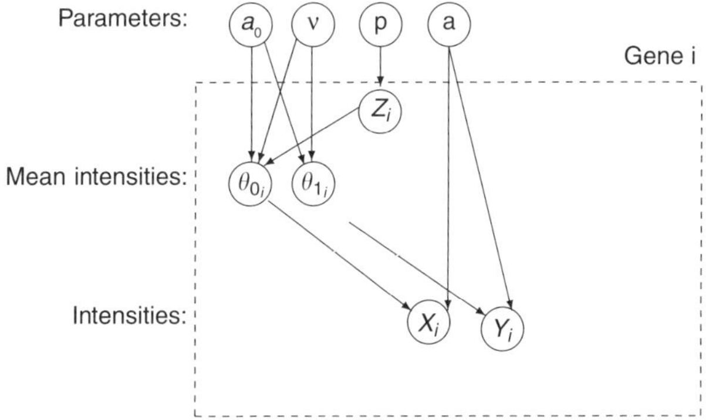

FIGURE 8.2
A model for differential gene expression from Newton et al. (2001) for fluorescence intensity measurements in a single-slide experiment. For each gene, that is, each spot on the chip, we record a pair (X, Y) of intensities corresponding to transcript abundance of a gene in both samples. The true unknown mean expression values are characterized by θ_{0i} and θ_{1i}. The aim of the experiment is to derive inference about equality of θ_{1i} and θ_{0k}. Uncertainty about θ_{0i} and θ_{1i} is described by parameters (a_0, ν). The variable z_i is a binomial indicator for equal mean values, that is, equal expression, associated with a probability p. All information about differential expression of gene i is contained in the posterior distribution for z_i.

expression levels are denoted by θ_{0i} and θ_{1i}. Other parameters, like scale or shape parameters, are denoted a. The aim of the experiment is to derive inference about the ratio θ_{1i}/θ_{0i}. Uncertainty about θ_{0i} and θ_{1i} is described by hyperparameters (a_0, ν, p). A latent variable $z_i \in \{0, 1\}$ is an indicator for unequal mean values for gene i, that is, equal expression. We use $z_i = 0$ to indicate equal expression, and $z_i = 1$ to indicate differential expression. Figure 8.2 summarizes the structure of the probability model. Conditional on the observed fluorescence intensities, the posterior distribution on z_i contains all information about differential expression of gene i. Let $r_i = X_i/Y_i$ denote the observed relative expression for gene i, and let $\eta = (a_0, \nu, p, a)$. Newton et al. (2001) gives the marginal likelihood $p(r_i \mid z_i, \eta)$, marginalized with respect to θ_{1i} and θ_{0i}. They proceed by maximizing the marginal likelihood for η by an implementation of the EM algorithm.

A simple hierarchical extension allows inference for multiple arrays. We assume repeated measurements X_{ij} and Y_{ij}, $j = 1, \ldots, J$, to be conditionally independent given the model parameters. We assume a Gamma-sampling distribution for the observed intensities X_{ij}, Y_{ij} for gene i in sample j,

$$X_{ij} \sim \text{Ga}(a, \theta_{0i}) \quad \text{and} \quad Y_{ij} \sim \text{Ga}(a, \theta_{1i}).$$

The scale parameters are gene-specific random effects $(\theta_{0i}, \theta_{1i})$. The model includes an a priori positive probability for lack of differential expression

$$Pr(\theta_{0i} = \theta_{1i}) = Pr(z_i = 0) = p.$$

Conditional on latent indicators z_i for differential gene expression, $z_i = I(\theta_{0i} \neq \theta_{1i})$, we assume conjugate gamma random effects distributions

$$\theta_{0i} \sim \text{Ga}(a_0, \nu)$$
$$(\theta_{1i}|z_i = 1) \sim \text{Ga}(a_0, \nu) \quad \text{and} \quad (\theta_{1i}|z_i = 0) \sim \mathbb{I}_{\theta_{0i}}(\theta_{1i}). \tag{8.39}$$

The model is completed with a prior for the parameters $(a, a_0, \nu, p) \sim \pi(a, a_0, \nu, p)$ We fix ν, assume a priori independence and use marginal gamma priors for a_0 and a, and a conjugate beta prior for p.

Let $X_i = (X_{ij},\ j = 1, \ldots, J)$ and $Y_i = (Y_{ij},\ j = 1, \ldots, J)$. The above model leads to a closed form marginal likelihood $p(X_i, Y_i|\eta)$ after integrating out θ_{1i}, θ_{0i}, but still conditional on $\eta = (p, a, a_0)$.

Availability of the closed form expression for the marginal likelihood greatly simplifies posterior simulation. Marginalizing with respect to the random effects reduces the model to the 3-dimensional marginal posterior $p(\eta \mid y) \propto p(\eta) \prod_i p(X_i, Y_i|\eta)$. Conditional on currently imputed values for η we can at any time augment the parameter vector by generating $z_i \sim p(z_i \mid \eta, X_i, Y_i)$ as simple independent Bernoulli draws, if desired. This greatly simplifies posterior simulation.

8.8.3 A Nonparametric Bayesian Model for Differential Gene Expression

Do et al. (2005) proposed a semiparametric Bayesian approach to inference for microarray group comparison experiments. Following the setup in Efron et al. (2001), they assume that the data are available as a difference score z_g for each gene, $g = 1, \ldots, G$. For example, the difference score z_g could be a two-sample t-statistic computed with the measurements recorded for gene g on all arrays, arranged in two groups by biologic condition. The summary is a t-statistic only in name, that is, we do not assume that the sampling model for the statistic is t-distribution under the null hypothesis. Instead, inference proceeds by assuming that the difference scores z_g arise by independent sampling from some unknown distribution f_1 for differentially expressed genes; and from an unknown distribution f_0 for nondifferentially expressed genes. For a reasonable choice of difference scores, the distribution f_0 should be a unimodal distribution centered at zero. The distribution f_1 should be bimodal with symmetric modes to the left and right of zero corresponding to over- and under-expressed genes. Of course, the partition into differentially and nondifferentially expressed genes is unknown. Thus, instead of samples from f_0 and f_1, we can only work with the samples generated from a mixture $f(z) = p_0 f_0(z) + (1 - p_0) f_1(z)$. Here p_0 is an unknown mixture weight. The desired inference about differential expression for each gene is formalized as a deconvolution of this mixture. We proceed by defining prior probability models on the unknown f_0 and f_1. Probability models on random distributions are traditionally known as nonparametric Bayesian models. See, for example, Müller and Quintana (2004) for a review of nonparametric Bayesian methods. We argue that the marginal posterior probability of gene g being differentially expressed can be evaluated as posterior expectation of $P_g \equiv (1 - p_0)f_1(z_g)/f(z_g)$.

8.8.4 The Probability of Expression Model

The Probability of Expression (POE) model is described in Parmigiani et al. (2002). A key feature of the model is the use of trinary indicators for over- and under-expression. In particular, let y_{gt} denote the observed gene expression for gene i in sample j, with $i = 1, \ldots, n$ and $j = 1, \ldots, J$. We introduce a latent variable $e_{ij} \in \{-1, 0, 1\}$ and assume the following mixture of normal and uniform model, parameterized with $\theta_S = (a_j, \mu_i, s_i, \kappa_i^-, \kappa_i^+)$:

$$p(y_{gt} \mid e_{ij}) \sim f_{eg}(y_{gt}) \quad \text{with} \begin{cases} f_{-1i} = U(-\kappa_i^- + a_j + \mu_i, a_j + \mu_i) \\ f_{0i} = N(a_j + \mu_i, s_i) \\ f_{1i} = U(a_j + \mu_i, a_j + \mu_i + \kappa_i^+). \end{cases} \tag{8.40}$$

In other words, we assume that the observed gene expressions arise from a mixture of a normal distribution and two uniform distributions defined to model overdispersion relative to the normal. Conditional on the parameters and the latent indicators e_{ij}, we assume that the observed gene expressions y_{ij} are independent across genes and samples. The interpretation of the normal component is as a baseline distribution for gene i, and the two uniform terms as the distribution in samples where gene i is over- and under-expressed, respectively. In Equation 8.40, a_j, $j = 1, \ldots, J$ are sample-specific effects, allowing inference to adjust for systematic variation across samples, μ_i are gene-specific effects that model the overall prevalence of each gene, and κ_i^- and κ_i^+ parameterize the overdispersion in the tails. Finally, s_i is the variance of the baseline distribution for gene i. Parmigiani et al. (2002) define (conditionally) conjugate priors for μ_i, s_i and κ_i^+ and κ_i^-. For the slide-specific effect we impose an identifiability constraint $a_j \sim N(0, \tau^2)$, i.i.d., subject to $\sum a_j = 0$.

8.8.5 Multiplicity Correction: Controlling False Discovery Rate

High-throughput gene expression experiments often give rise to massive multiple comparison problems. We discuss related issues in the context of microarray group comparison experiments. Assume that for genes, $i = 1, \ldots, n$, for large n, we wish to identify those that are differentially expressed across two biologic conditions of interest. From a classical perspective, multiple comparisons require an adjustment of the error rate, or, equivalently, an adjustment of the nominal significance level for each comparison. This is achieved, for example, in the Bonferroni correction or by the Benjamini and Hochberg (1995) correction mentioned in Section 8.8.1.

It can be argued that Bayesian posterior inference already accounts for multiplicities, and no further adjustment is required (Scott and Berger, 2006).

The argument is valid for the evaluation of posterior probabilities of differential expression. In a hierarchical model, the reported posterior probabilities are correctly adjusted for the multiplicities. But reporting posterior probabilities only solves half the problem. We still need to address the second step of the inference problem, namely, the identification of differentially expressed genes. Berry and Hochberg (1999) discuss this perspective.

This identification is most naturally discussed as a decision problem. The formal statement of a decision problem requires a probability model $p(\theta, y)$ for all unknown quantities, including parameters θ and data y, a set of possible actions, $\delta \in D$, and a loss function $L(\delta, \theta, y)$ that formalizes the relative preferences for decision δ under hypothetical outcomes y and assumed parameter values θ. The probability model could be, for example, the hierarchical gamma/gamma model described in Section 8.8.2. The action is a vector of indicators, $\delta = (\delta_1, \ldots, \delta_n)$ with $\delta_i = 1$ indicating that the gene is reported as differentially expressed. We write $\delta(y)$ when we want to highlight the nature of δ as a function of the data. Let $r_i \in \{0, 1\}$ denote an indicator for the (unknown) truth. The r_i are part of the parameter vector θ. Usually, the probability model includes additional parameters besides r. It can be argued (Robert, 2001) that a rational decision maker should select the rule $\delta(y)$ that maximizes L in expectation. The relevant expectation is the probability model on θ conditional on the observed data, leading to the optimal rule

$$\delta^*(y) = \arg\min_{\delta} \int L(\delta, \theta, y)\, p(\theta \mid y)\, \mathrm{d}\theta.$$

Let $v_i = E(r_i \mid y)$ denote the marginal posterior probability of gene i being differentially expressed. The assumption of nonzero prior probabilities, $0 < p(r_i = 1) < 1$, ensures nontrivial posterior probabilities. In Müller et al. (2004) we show that for several reasonable choices of $L(\delta, \theta, y)$ the optimal rule is of the form $\delta_i^*(y) = I(v_i > t)$. In other words, the optimal decision is to report all genes with marginal probability of differential expression beyond a certain threshold t as differentially expressed. The value of the threshold depends on the specific loss function. The optimal rule δ^* is valid for several loss functions defined in Müller et al. (2004). Essentially, all are variations of basic 0–1 loss functions. Let

$$\mathrm{FD} = \sum \delta_i\,(1 - r_i) \quad \text{and} \quad \mathrm{FN} = \sum (1 - \delta_i) r_i$$

denote false discovery and negative counts, and let

$$\mathrm{FDR} = \mathrm{FD} / \sum \delta_i \quad \text{and} \quad \mathrm{FNR} = \mathrm{FN} / \sum (1 - \delta_i)$$

denote false discovery and false negative rates. The definitions FD(R) and FN(R) are summaries of parameters, r, and data, $\delta(y)$. Taking an expectation with respect to y and conditioning on r, one would arrive at the usual definition of FDRs, as used, among many others, in Benjamini and Hochberg (1995),

Efron and Tibshirani (2002), Storey (2002), and Storey et al. (2004). Instead we use posterior expectations, defining $\overline{\text{FD}} = E(\text{FD} \mid y)$, and so forth. See, Genovese and Wasserman (2002, 2004) for a discussion of posterior expected FDR. Using these posterior summaries we define the following losses: $L_N(\delta, z) = c\,\overline{\text{FD}} + \overline{\text{FN}}$, and $L_R(\delta, z) = c\,\overline{\text{FDR}} + \overline{\text{FNR}}$. The loss function L_N is a natural extension of $(0, 1, c)$ loss functions for traditional hypothesis testing problems (Lindley, 1971). Alternatively, we consider bivariate loss functions that explicitly acknowledge the two competing goals: $L_{2R}(\delta, z) = \overline{\text{FNR}}$, subject to $\overline{\text{FDR}} < \alpha_R$, and $L_{2N}(\delta, z) = \overline{\text{FN}}$, subject to $\overline{\text{FD}} < \alpha_N$. Under all four loss functions, L_N, L_R, L_{2R}, and L_{2N}, the nature of the optimal rule is δ^*. See Müller et al. (2004) for the definition of the thresholds.

One can argue that not all false negatives and all discoveries are equally important. False negatives of genes that are massively differentially expressed are more serious than only marginally differentially expressed genes. To formalize this notion we need to assume that the probability model includes parameters that can be interpreted as extent of differential expression, or strength of the signal. Assume that the model includes parameters m_i, $i = 1, \ldots, n$, with $m_i > 0$ if $r_i = 1$ and $m_i = 0$ if $r_i = 0$. For example, in the gamma/gamma hierarchical model of Section 8.8.2 a reasonable definition would use $m_i = \log(\theta_{i1}/\theta_{i0})$. Assuming parameters m_i that can be interpreted as level of differential expression for gene i, we define

$$L_\rho(\rho, \delta, z) = -\sum \delta_i m_i + k \sum (1 - \delta_i) m_i + c \sum \delta_i.$$

For $c > 0$, the optimal solution is easily found as $\delta_i^* = I\{\bar{m}_i \geq c/(1+k)\}$. For more discussion and alternative loss functions see, for example, Müller et al. (2007).

8.9 Summary

In this chapter, we have reviewed the basic framework of Bayesian statistics and typical inference problems that arise in biomedical applications. In summary, Bayesian inference can be carried out for any problem that is based on a well-defined probability framework. In particular, Bayesian inference requires a likelihood, that is, a sampling model of the observable data conditional on assumed values of the parameters, and a prior, that is, a prior judgment about the parameters that is formalized as a probability model. As long as the likelihood and the prior are available for any set of assumed parameter values, one can in principle implement Bayesian inference. Evaluation up to a constant is sufficient for MCMC posterior simulation with MH chains.

The main advantage of the Bayesian framework that inference is based on a principled and coherent approach. In particular, even for complicated setups with hierarchical models, multiple studies, mixed data types, delayed responses, complicated dependence and so forth, as long as the investigator is willing to write down a probability model, one can carry out Bayesian inference.

There are some important limitations of the Bayesian approach. We always need a well-defined probability model. There are (almost) no genuinely nonparametric Bayesian methods (a class of models known as "nonparametric Bayesian models" are really random functions, that is, probability models on infinite dimensional spaces). For example, the proportional hazards rate model for event time data has no easy Bayesian equivalence. Even simple approaches like kernel density estimation, or loess smoothing have no simple Bayesian analogs. Another limitation of Bayesian inference that arises from the need for well-defined probability models is the difficulty to define good model validation schemes. Principled Bayesian model comparison is always relative to an assumed alternative model. There is no easy Bayesian equivalence of a simple chi-square test of fit. Although there are several very reasonable Bayesian model validation approaches, none is based on first principles. Another great limitation of Bayesian inference is the sensitivity to the prior probability model. This is usually not a problem when the goal of the data analysis is prediction or parameter estimation. Results about posterior asymptotics assure us that the impact of the prior choice will eventually wash out. However, the same is not true for model comparison. Bayes factors are notoriously sensitive to prior assumptions, and there is no easy way to avoid this.

In summary, the increasing notion that the advantages outweigh the limitations has led to an increasingly greater use of Bayesian methods in several areas of application including biomedical applications. Increasingly more complex inference problems and increasingly more expensive data collection in experiments or clinical studies require that we make the most of the available data. For complex designs, Bayesian inference may often be the most feasible way to proceed.

References

Abramowitz, M. and Stegun, I. A. (eds). 1965. *Handbook of Mathematical Functions*, National Bureau of Standards, Washington.

Achcar, J. A., Brookmeyer, R. and Hunter, W. G. 1985. An application of Bayesian analysis to medical follow-up data. *Statistics in Medicine* 4, 509–520.

Albert, J. 1988. Computational methods using a Bayesian hierarchical generalized linear model. *Journal of the American Statistical Association* 83, 1037–1044.

Arjas, E. and Gasbarra, D. 1994. Nonparametric Bayesian inference for right-censored survival data, using the Gibbs sampler. *Statistica Sinica* 4, 505–524.

Assunção, R. M., Reis, I. A. and Oliveira, C. D. L. 2001. Diffusion and prediction of Leishmaniasis in a large metropolitan area in Brazil with a Bayesian space-time model. *Statistics in Medicine* 20, 2319–2335.

Baggerly, K. A., Coombes, K. R., Hess, K. R., Stivers, D. N., Abruzzo, L. V. and Zhang, W. 2001. Identifying differentially expressed genes in cDNA microarray experiments. *Journal of Computational Biology* 8, 639–659.

Baggerly, K. A., Coombes, K. R. and Morris, J. S. 2006. An introduction to high-throughput bioinformatics data. In *Bayesian Inference for Gene Expression and Proteomics*, K.-A. Do, P. Müller and M. Vannucci, eds., 1–34. Cambridge University Press, Cambridge.

Benjamini, Y. and Hochberg, Y. 1995. Controlling the false discovery rate: A practical and powerful approach to multiple testing. *Journal of the Royal Statistical Society, Series B* 57, 289–300.

Berger, J. O. 1985. *Statistical Decision Theory and Bayesian Analysis*. Berlin: Springer-Verlag.

Berger, J. O. and Bernardo, J. M. 1992. On the development of reference priors. In *Bayesian Statistics 4*, J. M. Bernardo, J. O. Berger, A. P. Dawid and A. F. M. Smith, eds., 35–60. Oxford University Press, Oxford.

Berger, J. O., Insua, D. R. and Ruggeri, F. 2000. Bayesian robustness. In *Robust Bayesian Analysis*, D. R. Insua and F. Ruggeri, eds., 1–32, Springer-Verlag, New York.

Berger, J. O. and Pericchi, L. R. 1996. The intrinsic Bayes factor for model selection and prediction. *Journal of the American Statistical Association* 91, 109–122.

Bernardinelli, L., Clayton, D. and Montomoli, C. 1995. Bayesian estimates of disease maps: How important are priors? *Statistics in Medicine* 14, 2411–2431.

Bernardo, J. M. 1979. Reference posterior distributions for Bayesian inference. *Journal of the Royal Statistical Society, Series B* 41, 113–147.

Bernardo, J. M. and Smith, A. F. M. 1994. *Bayesian Theory*. New York: Wiley.

Berry, D. 2005(a). The Bayesian principle: Can we adapt? yes! *Stroke* 36, 1623–1624.

——— 2005(b). Clinical trials: Is the Bayesian approach ready for prime time? yes! *Stroke* 36, 1621–1622 [commentary].

——— 2006. A guide to drug discovery: Bayesian clinical trials. *Nature Reviews Drug Discovery* 5, 27–36.

Berry, D. A. and Hochberg, Y. 1999. Bayesian perspectives on multiple comparisons. *Journal of Statistical Planning and Inference* 82, 215–227.

Besag, J., York, J. and Mollié, A. 1991. Bayesian image restoration, with applications in spatial statistics (with discussion). *Annals of the Institute of Statistical Mathematics* 43, 1–59.

Best, N. G., Arnold, R. A., Thomas, A., Waller, L. A. and Conlon, E. M. 1999. Bayesian models for spatially correlated disease and exposure Data. In *Bayesian Statistics 6*, J. M. Bernardo, J. O. Berger, A. P. Dawid and A. F. M. Smith (eds.), 131–156. Oxford University Press, Oxford.

Best, N. G., Richardson, S. and Thomson, A. 2005. A comparison of Bayesian spatial for disease mapping. *Statistical Methods in Medical Research*, 14, 35–59.

Broet, P., Richardson, S. and Radvanyi, F. 2002. Bayesian hierarchical model for identifying changes in gene expression from microarray experiments. *Journal of Computational Biology* 9, 671–683.

Buckle, D. J. 1995. Bayesian inference for stable distributions. *Journal of the American Statistical Association* 90, 605–613.

Buonaccorsi, J. P. and Gatsonis, C. A. 1988. Bayesian inference for ratios of coefficients in a linear model. *Biometrics* 44, 87–101.

Cappé O., Guillin, A., Marin, J. M. and Robert, C. P. 2004. Population Monte Carlo. *Journal of Computational and Graphical Statistics* 13, 907–929.

Carlin, B. P. and Chib, S. 1995. Bayesian model choice via Markov chain Monte Carlo methods. *Journal of the Royal Statistical Society, Series B* 57, 473–484.

Carlin, B. P. and Hodges, J. S. 1999. Hierarchical proportional hazards regression models for highly stratified data. *Biometrics* 55, 1162–1170.

Carlin, B. P. and Louis, T. 2000. *Bayes and Empirical Bayes Methods for Data Analysis*. Chapman & Hall/CRC, London.

Chen, H.-H., Shao, Q. M. and Ibrahim, J. G. 2000. *Monte Carlo Methods in Bayesian Computation*. Springer-Verlag, New York.

Chen, M.-H. and Dey, D. K. 2000. Bayesian analysis for correlated ordinal data models. In *Generalized Linear Models: A Bayesian Perspective*, D. K. Dey, S. K. Ghosh and B. K. Mallick, eds., 133–155. Marcel Dekker, New York.

Chen, M.-H., Ibrahim, J. G. and Sinha, D. 2002. Bayesian inference for multivariate survival data with a surviving fraction. *Journal of Multivariate Analysis* 80, 1011–1026.

Clayton, D. 1996. Generalized linear mixed models. In *Markov Chain Monte Carlo in Practice*, W. R. Gilks, S. Richardson and D. J. Spiegelhalter, eds., 275–301, Chapman & Hall, New York.

Clayton, D. and Cuzick, J. 1985. Multivariate generalizations of the proportional hazards model. *Journal of the Royal Statistical Society A* 148, 82–117.

Clayton, D. and Kaldor, J. 1987. Empirical Bayes estimates of age-standardised relative risks for use in disease mapping. *Biometrics* 43, 671–681.

Clyde, M. A. 1999. Bayesian model averaging and model search strategies. In *Bayesian Statistics 6*, J. M. Bernardo, J. O. Berger, A. P. Dawid and A. F. M. Smith, eds., 157–185. Oxford University Press, Oxford.

Congdon, P. 2003. *Applied Bayesian Modeling*. Wiley, New York.

Congdon, P. 2005. *Bayesian Models for Categorical Data*. Wiley, New York.

Dahl, D. 2003. Modeling differential gene expression using a Dirichlet process mixture model. In *2003 Proceedings of the American Statistical Association, Bayesian Statistical Sciences Section*. American Statistical Association, Alexandria, VA.

Datta, S., Datta, S., Parresh, R. S. and Thompson, C. M. 2007. Microarray data analysis. In *Computational Methods in Biomedical Research*, R. Khattree and D. N. Naik, eds., this volume, pp. 1–44.

Dellaportas, P., Forster, J. J. and Ntzoufras, I. 2002. On Bayesian model and variable selection using MCMC. *Statistics and Computing* 12, 27–36.

Dempster, A. P., Laird, N. M. and Rubin, D. B. 1977. Maximum likelihood from incomplete data via the EM algorithm (with discussion). *Journal of the Royal Statistical Society Series B* 39, 1–38.

Dey, D. K., Chen, M.-H. and Chang, H. 1997. Bayesian approach for the nonlinear random effects model. *Biometrics* 53, 1239–1252.

Dey, D. K. and Ravishanker, N. 2000. Bayesian approaches for overdispersion in generalized linear models. In *Generalized Linear Models: A Bayesian Perspective*, D. K. Dey, S. K. Ghosh and B. K. Mallick (eds.), 73–84. Marcel Dekker, New York.

Do, K., Müller, P. and Tang, F. 2005. A Bayesian mixture model for differential gene expression. *Journal of the Royal Statistical Society C* 54, 627–644.

Do, K., Müller, P. and Vannucci, M. 2006. *Bayesian Inference for Gene Expression and Proteomics*. Cambridge University Press, Cambridge.

Doucet, A., de Freitas, N. and Gordon, N. 2001. *Sequential Monte Carlo Methods in Practice*. Springer-Verlag, New York.

Efron, B. and Tibshirani, R. 2002. Empirical bayes methods and false discovery rates for microarrays. *Genet Epidemiology*, 23, 70–86.

Efron, B., Tibshirani, R., Storey, J. D. and Tusher, V. 2001. Empirical Bayes analysis of a microarray experiment. *Journal of the American Statistical Association*, 96, 1151–1160.

Estey, E. and Thall, P. 2003. New designs for phase 2 clinical trials. *Blood* 102, 442–448.

Faraggi, D. and Simon, R. 1995. A neural network model for survival data. *Statistics in Medicine* 14, 73–82.

Finkelstein, D. M. and Wolfe, R. A. 1985. A semiparametric model for regression analysis of interval-censored survival data. *Biometrics* 41, 933–945.

Frigessi, A., van de Wiel, M., Holden, M., Glad, I., Svendsrud, D. and Lyng, H. 2005. Genome-wide estimation of transcript concentrations from spotted cDNA microarray data. *Nucleic Acids Research—Methods Online* 33, e143.

Gamerman, D. and Lopes, H. F. 2006. *Markov Chain Monte Carlo: Stochastic Simulation for Bayesian Inference* (2nd ed.). Chapman & Hall/CRC, Boca Raton.

Gamerman, D. and Migon, H. S. 1993. Dynamic hierarchical models. *Journal of the Royal Statistical Society, Series B* 55, 629–642.

Geisser, S. 1975. The predictive sample reuse method with application. *Journal of the American Statistical Association* 70, 320–328, 350.

Gelfand, A. E. and Ghosh, M. 2000. Generalized linear models: A Bayesian view. In *Generalized Linear Models: A Bayesian Perspective*, D. K. Dey, S. K. Ghosh and B. K. Mallick (eds.), 3–21. Marcel Dekker, New York.

Gelfand, A. E. and Ghosh, S. 1994. Model choice: A minimum posterior predictive loss approach. *Biometrika* 85, 1–11.

Gelfand, A. E. and Mallick, B. K. 1995. Bayesian analysis of proportional hazards models built from monotone functions. *Biometrics* 51, 843–852.

Gelfand, A. E. and Smith, A. F. M. 1990. Sampling-based approaches to calculating marginal densities. *Journal of the American Statistical Association* 85, 398–409.

Gelman, A., Carlin, J. B., Stern, H. S. and Rubin, D. B. 2004. *Bayesian Data Analysis*, Chapman & Hall/CRC, New York.

Geman, S. and Geman, D. 1984. Stochastic relaxation, Gibbs distributions and the Bayesian restoration of images. *IEEE Transactions on Pattern Analysis and Machine Intelligence* 6, 721–741.

Genovese, C. R. and Wasserman, L. 2002. Operating characteristics and extensions of the FDR procedure. *Journal of the Royal Statistical Society, Series B* 64, 499–517.

Genovese, C. R. and Wasserman, L. 2004. A stochastic process approach to false discovery control. *Annals of Statistics* 32, 1035–1061.

George, E. I. and McCulloch, R. E. 1992. Variable selection via Gibbs sampling. *Journal of the American Statistical Association*, 79, 677–683.

Geweke, J. 1989. Bayesian inference in econometric models using Monte Carlo integration. *Econometrica* 57, 1317–1339.

Ghosh, D. 2004. Mixture models for assessing differential expression in complex tissues using microarray data. *Bioinformatics* 20, 1663–1669.

Gilks, W. R., Richardson, S. and Spiegelhalter, D. J. 1996. *Markov Chain Monte Carlo in Practice*, Chapman & Hall, London.

Gilks, W. R. and Wild, P. 1992. Adaptive rejection sampling for Gibbs sampling. *Applied Statistics* 41, 337–348.

Godsill, S. J. 2001. On the relationship between Markov chain Monte Carlo methods for model uncertainty. *Journal of Computational and Graphical Statistics* 10, 1–19.

Gordon, K. and Smith, A. F. M. 1990. Modeling and monitoring biomedical time series. *Journal of the American Statistical Association* 85, 328–337.

Green, P. J. 1995. Reversible jump Markov chain Monte Carlo computation and Bayesian model determination. *Biometrika* 82, 711–732.

Gustafson, P. 1995. A Bayesian analysis of bivariate survival data from a multicentre cancer clinical trial. *Statistics in Medicine* 14, 2523–2535.

Gustafson, P. 1996. Robustness considerations in Bayesian analysis. *Statistical Methods in Medical Research* 5, 357–373.

Gustafson, P. 1997. Large hierarchical Bayesian analysis of multivariate survival data. *Biometrics* 53, 230–243.

Gustafson, P. 1998. Flexible Bayesian modeling for survival data. *Lifetime Data Analysis* 4, 281–299.

Hastings, W. K. 1970. Monte Carlo sampling methods using Markov chains and their applications. *Biometrika* 57, 97–109.

Hein, A., Richardson, S., Causton, H., Ambler, G. and Green, P. 2005. Bgx: A fully Bayesian integrated approach to the analysis of affymetrix genechip data. *Biostatistics* 6, 349–373.

Heyde, C. C. and Johnstone, I. M. 1979. On asymptotic posterior normality of stochastic processes. *Journal of the Royal Statistical Society, Series B* 41, 184–189.

Hoeting, J., Madigan, D., Raftery, A. E. and Volinsky, C. 1999. Bayesian model averaging (with discussion). *Statistical Science* 14, 382–417.

Hougaard, P. 2000. *Analysis of Multivariate Survival Data*. Springer-Verlag, New York.

Ibrahim, J., Chen, M.-H. and Gray, R. 2002. Bayesian models for gene expression with DNA microarray data. *Journal of the American Statistical Association*, 97, 88–100.

Ibrahim, J. G., Chen, M.-H. and Sinha, D. 2001. *Bayesian Survival Analysis*. Springer-Verlag, New York.

Ishwaran, H. and Rao, J. S. 2003. Detecting differentially expressed genes in microarrays using Bayesian model selection. *Journal of the American Statistical Association* 98, 438–455.

Jeffreys, H. 1961. *Theory of Probability*. Oxford University Press, Oxford.

Kadane, J. B. (ed.) 1984. *Robustness of Bayesian Analysis*. North-Holland, Amsterdam.

Kalbfleisch, J. D. 1978. Nonparametric Bayesian analysis of survival time data. *Journal of the Royal Statistical Society, Series B* 40, 214–221.

Kass, R. E. and Raftery, A. 1995. Bayes factors and model uncertainty. *Journal of the American Statistical Association* 90, 773–795.

Kass, R. E., Tierney, L. and Kadane, J. B. 1988. Asymptotics in Bayesian computation (with discussion). In *Bayesian Statistics 3*, J. M. Bernardo, M. H. DeGroot, D. V. Lindley, A. F. M. Smith, eds., 261–278. Oxford University Press, Oxford.

Kass, R. E., Tierney, L. and Kadane, J. B. 1989. Approximate methods for assessing influence and sensitivity in Bayesian analysis. *Biometrika* 76, 663–674.

Kass, R. E. and Wasserman, L. 1996. The selection of prior distributions by formal rules. *Journal of the American Statistical Association* 91, 1343–1370.

Kelsal, J. and Wakefield, J. 2002. Modeling Spatial Variation in Disease Risk: A Geostatistical Approach, *Journal of the American Statistical Association* 97, 692–701.

Kim, S. W. and Ibrahim, J. G. 2000. On Bayesian inference for proportional hazards models using noninformative priors. *Lifetime Data Analysis* 6, 331–341.

Knorr-Held, L. and Besag, J. 1998. Modeling Risk from a disease in time and space. *Statistics in Medicine* 17, 2045–2060.

Knorr-Held, L. and Richardson, S. 2003. A hierarchical model for space-time surveillance data on meningococcal disease incidence. *Applied Statistics* 52, 169–183.

Kristiansen, N. K., Sjöström, S. O. and Nygaard, H. 2005. Urinary bladder volume tracking using a Kalman filter. *Medical and Biological Engineering and Computing* 43, 331–334.

Kuo, L. and Peng, F. 2000. A mixture-model approach to the analysis of survival data. In *Generalized Linear Models: A Bayesian Perspective*, D. K. Dey, S. K. Ghosh and B. K. Mallick (eds.), 195–209. Marcel Dekker, New York.

Kuo, L. and Song, C. 2005. A new time varying frailty model for recurrent events. *Proceedings of the ASA section on Bayesian Statistical Science 2005, American Statistical Association*.

Larget, B. 2005. Introduction to Markov chain Monte Carlo methods in molecular evolution. In *Statistical Methods in Molecular Evolution*, R. Nielsen (ed.), 45–62. Springer-Verlag, New York.

Lavine, M. 1992. Some aspects of Polya-tree distributions for statistical modeling. *Annals of Statistics* 20, 1222–1235.

Lee, P. M. 1997. *Bayesian Statistics: An Introduction*. Wiley, New York.

Lewin, A., Richardson, S., Marshall, C., Glazier, A. and Aitman, T. 2006. Bayesian modeling of differential gene expression. *Biometrics*, 62, 1–9.

Lindley, D. V. 1971. *Making decisions*. John Wiley & Sons, New York.

Lindley, D. V. and Smith, A. F. M. 1972. Bayes estimtes for the linear model. *Journal of the Royal Statistical Society, Series B* 34, 1–41.

Lopes, H. F., Müller, P. and Rosner, G. L. 2003. Bayesian meta-analysis for longitudinal data models using multivariate mixture priors. *Biometrics* 59, 66–75.

MacEachern, S. N. and Müller, P. 1998. Estimating mixture of Dirichlet process models. *Journal of Computational and Graphical Statistics* 7, 223–238.

MacNab, Y. C., Farrell, P. J., Gustafson, P. and Wen, S. 2004. Estimation in Bayesian Disease mapping. *Biometrics* 60, 865–873.

McGilchrist, C. A. and Aisbett, C. W. 1991. Regression with frailty in survival analysis. *Biometrics* 47, 461–466.

Mallick, M. and Ravishanker, N. 2004. Multivariate survival analysis with PVF frailty models. In *Advances in Ranking and Selection, Multiple Comparisons, and Reliability, with Applications*, N. Balakrishnan, N. Kannan and H. N. Nagaraja (eds.), 369–384. Birkhauser, Boston.

Mallick, M. and Ravishanker, N. 2006. Additive positive stable frailty models. *Methodology and Computing in Applied Probability* 8, 541–558.

Mantel, N., Bohidar, N. R. and Ciminera, J. L. 1977. Mantel–Haenszel analyses of litter-matched time-to-response data, with modifications for recovery of interlitter information. *Cancer Research* 37, 3863–3868.

Metropolis, N., Rosenbluth, A. W., Rosenbluth, M. N., Teller, A. H. and Teller, E. 1953. Equation of state calculations by fast computing machine. *Journal of Chemical Physics* 21, 1087–1091.

Mezzetti, M. and Ibrahim, J. G. 2000. Bayesian inference for the Cox model using correlated gamma process priors. *Technical Report*. Department of Biostatistics, Harvard School of Public Health.

Mollié, A. 1996. Bayesian mapping of disease. In *Markov Chain Monte Carlo in Practice*, W. R. Gilks, S. Richardson and D. J. Spiegelhalter (eds.), 359–379. Chapman & Hall, New York.

Müller, P., Parmigiani, G. and Rice, K. 2007. FDR and Bayesian multiple comparisons rules. In *Bayesian Statistics 8*, J. Bernardo, S. Bayarri, J. Berger, A. Dawid, D. Heckerman, A. Smith and M. West (eds.), to appear. Oxford University Press, Oxford, 349–370.

Müller, P., Parmigiani, G., Robert, C. and Rouseau, J. 2004. Optimal sample size for multiple testing: The case of gene expression microarrays. *Journal of the American Statistical Association* 99, 990–1001.

Müller, P., Parmigiani, G., Schildkraut, J. and Tardella, L. 1999. A Bayesian hierarchical approach for combining case-control and prospective studies. *Biometrics* 55, 258–266.

Müller, P. and Quintana, F. A. 2004. Nonparametric Bayesian data analysis. *Statistical Science* 19, 95–110.

Naik, D. N. and Wagner, M. 2007. Protein profiling for disease proteomics with mass spectrometry: Computational challenges. In *Computational Methods in Biomedical Research*, R. Khattree and D. N. Naik (eds.), this volume, 103–130.

Naylor, J. C. and Smith, A. F. M. 1982. Application of a method for the efficient computation of posterior distributions. *Applied Statistics* 31, 214–225.

Newton, M. A. and Kendziorski, C. M. 2003. Parametric empirical Bayes methods for micorarrays. In *The Analysis of Gene Expression Data: Methods and Software*, G. Parmigiani, E. S. Garrett, R. Irizarry and S. L. Zeger (eds). Springer Verlag, New York, 254–271.

Newton, M. A., Kendziorski, C. M., Richmond, C. S., Blattner, F. R. and Tsui, K. W. 2001. On differential variability of expression ratios: Improving statistical inference about gene expression changes from microarray data. *Journal of Computational Biology* 8, 37–52.

Nobre, A. A., Schmidt, A. M. and Lopes, H. F. 2005. Spatio-temporal models for mapping the incidence of malaria in Pará. *Environmetrics* 16, 291–304.

Oakes, D. 1989. Bivariate survival models induced by frailties. *Journal of the American Statistical Association* 84, 487–493.

Parmigiani, G., Garrett, E. S., Anbazhagan, R. and Gabrielson, E. 2002. A statistical framework for expression-based molecular classification in cancer. *Journal of the Royal Statistical Society, Series B* 64, 717–736.

Pauler, D. K., Wakefield, J. C. and Kass, R. E. 1999. Bayes factors for variance component models. *Journal of the American Statistical Association* 94, 1241–1253.

Pole, A. and West, M. 1990. Efficient Bayesian Learning in Non-linear dynamic models. *Journal of Forecasting*, 9, 119–136.

Pounds, S. and Morris, S. 2003. Estimating the occurence of false positives and false negatives in microarray studies by approximating and partitioning the empirical distribution of p-values. *Bioinformatics* 19, 1236–1242.

Qiou, Z., Ravishanker, N. and Dey, D. K. 1999. Multivariate survival analysis with positive stable frailties. *Biometrics* 55, 637–644.

R Development Core Team 2006. *R: A Language and Environment for Statistical Computing*, R Foundation for Statistical Computing: Vienna, Austria.

Raftery, A. E., Madigan, D. and Hoeting, J. A. 1997. Bayesian model averaging for linear regression models. *Journal of the American Statistical Association* 92, 179–191.

Ravishanker, N. and Dey, D. K. 2000. Multivariate survival models with a mixture of positive stable frailties. *Methodology and Computing in Applied Probability* 2, 293–308.

Robert, C. 2001. *The Bayesian Choice*. Springer-Verlag, New York.

Robert, C. P. 1996. Mixtures of distributions: inference and estimation. In Markov Chain Monte Carlo in Practice, W. R. Gilks, S. Richardson and D. J. Spiegelhalter, eds., Chapman & Hall/CRC, New York, 441–461.

Rubin, D. B. 1989. A new perspective on meta-analysis. In *The Future of Meta-Analysis*, K. W. Wachter and M. L. Straf (eds.). Russell Sage Foundation, New York, 155–166.

Schervish, M. J. 1995. *Theory of Statistics*. Springer-Verlag, New York.

Scott, J. and Berger, J. 2006. An exploration of aspects of Bayesian multiple testing. *Technical Report*. Duke University, Institute of Statistics and Decision Sciences.

Shumway, R. H. and Stoffer, D. S. 2004. *Time Series Analysis and its Applications*. Springer-Verlag, New York.

Sinha, D. 1997. Time discrete Beta-process model for interval censored survival data. *Canadian Journal of Statistics* 25, 445–456.

Sinha, D., Chen, M.-H. and Ghosh, S. K. 1999. Bayesian analysis and model selection for interval-censored survival data. *Biometrics* 55, 585–590.

Sinha, D. and Dey, D. K. 1997. Semiparametric Bayesian analysis of survival data. *Journal of the American Statistical Association* 92, 1195–1313.

Sisson, S. A. 2005. Transdimensional Markov chains: A decade of progress and future perspectives. *Journal of the American Statistical Association* 100, 1077–1089.

Sivaganesan, S. 2000. Global and local robustness: Uses and limitations. In *Robust Bayesian Analysis*, D. R. Insua and F. Ruggeri (eds.), 89–108, Springer-Verlag, New York.

Sivia, D. S. 1996. *Data Analysis. A Bayesian Tutorial*. Oxford University Press, Oxford.

Smith, A. F. M. 1973. A general Bayesian linear model. *Journal of the Royal Statistical Society, Series B* 35, 67–75.

Smith, A. F. M. and Gelfand, A. E. 1992. Bayesian statistics without tears: A sampling–resampling perspective. *American Statistician* 46, 84–88.

Sorensen, D. and Gianola, D. 2002. *Likelihood, Bayesian and MCMC Methods in Quantitative Genetics*. Springer-Verlag, New York.

Spiegelhalter, D. J., Abrams, K. R. and Myles, J. P. 2004. *Bayesian approaches to clinical trails and health care evaluation*. John Wiley & Sons, Chichester, UK.

Spiegelhalter, D. J., Thomas, A. and Best, N. G. 1999. *WinBUGS Version 1.2 User Manual*, MRC Biostatistics Unit, Cambridge, UK.

Storey, J. 2002. A direct approach to false discovery rates. *Journal of the Royal Statistical Society, Series B* 64, 479–498.

Storey, J. D., Taylor, J. E. and Siegmund, D. 2004. Strong control, conservative point estimation and simultaneous conservative consistency of false discovery rates: A unified approach. *Journal of the Royal Statistical Society B* 66, 187–205.

Sun, D., Speckman, P. L. and Tsutakawa, R. K. 2000. Random effects in generalized linear mixed models (GLMM's). In *Generalized Linear Models: A Bayesian Perspective*, D. K. Dey, S. K. Ghosh and B. K. Mallick (eds.), Marcel Dekker, New York, 3–36.

Symons, M. J., Grimson, R. C. and Yuan, Y. C. 1983. Clustering of rare events. *Biometrics* 39, 193–205.

Tadesse, M., Sha, N. and Vannucci, M. 2005. Bayesian variable selection in clustering high-dimensional data. *Journal of the American Statistical Association* 100, 602–617.

Tadesse, M. G., Ibrahim, J. G. and Mutter, G. L. 2003. Identification of differentially expressed genes in high-density oligonucleotide arrays accounting for the quantification limits of the technology. *Biometrics* 59, 542–554.

Tanner, M. A. and Wong, W. 1987. The calculation of posterior distributions by data augmentation (with discussion). *Journal of the American Statistical Association* 82, 528–550.

Thall, P. F., Millikan, R. E., Müller, P. and Lee, S.-J. 2003. Dose-finding with two agents in phase I oncology trials. *Biometrics* 59, 487–496.

Thall, P. F. and Russell, K. E. 1998. A strategy for dose-finding and safety monitoring based on efficacy and adverse outcomes in phase I/II clinical trials. *Biometrics* 54, 251–264.

Thall, P. F., Simon, R. M. and Estey, E. H. 1995. Bayesian sequential monitoring designs for single-arm clinical trials with multiple outcomes. *Statistics in Medicine* 14, 357–379.

Tierney, L. 1994. Markov chains for exploring posterior distributions (with discussion). *Annals of Statistics* 22, 1701–1762.

Tierney, L. and Kadane, J. B. 1986. Accurate approximations for posterior moments and marginal densities. *Journal of the American Statistical Association* 81, 82–86.

Tierney, L., Kass, R. E. and Kadane, J. B. 1989. Fully exponential Laplace approximations for expectations and variances of nonpositive functions. *Journal of the American Statistical Association* 84, 710–716.

Thisted, R. A. 1988. *Elements of Statistical Computing*, Chapman & Hall, New York.

Tseng, G. C., Oh, M. K., Rohlin, L., Liao, J. and Wong, W. 2001. Issues in cDNA microarray analysis: Quality filtering, channel normalization, models of variations and assessment of gene effects. *Nucleic Acids Research* 29, 2549–2557.

Tusher, V. G., Tibshirani, R. and Chu, G. 2001. Significance analysis of microarrays applied to the ionizing radiation response. *Proceedings of the National Academy of Sciences* 98, 5116–5121.

Vaupel, J. W., Manton, K. G. and Stallard, E. 1979. The impact of heterogeneity in individual frailty on the dynamics mortality. *Demography* 16, 439–454.

Waagepetersen, R. and Sorensen, D. 2001. A tutorial on reversible jump MCMC with a view toward applications in QTL-mapping. *International Statistical Review* 69, 49–61.

Wall, M. M. 2004. A close look at the spatial correlation structure implied by the CAR and SAR models. *Journal of Statistical Planning and Inference* 121, 311–324.

Waller, L. A., Carlin, B. P., Xia, H. and Gelfand, A. E. 1997. Hierarchical spatio-temporal mapping of disease rates. *Journal of the American Statistical Association* 92, 607–617.

West, M. 1985. Generalized linear models: Scale parameters, outlier accommodation and prior distributions. In *Bayesian Statistics*, J. M. Bernardo, Morris H. Degroot, D. V. Lindley and A. F. M. Smith, eds., Oxford University Press, Oxford, 2, 531–557.

West, M. and Harrison, P. J. 1989. *Bayesian Forecasting and Dynamic Models*. Springer-Verlag, New York.

West, M., Harrison, P. J. and Migon, H. 1985. Dynamic generalised linear models and Bayesian forecasting (with discussion). *Journal of the American Statistical Association* 80, 73–97.

Wu, W., Black, M., Mumford, D., Gao, Y., Bienenstock, E., and Donoghue, J. 2003. A switching Kalman filter model for the motor cortical coding of hand motion. In *Proc. IEEE Engineering in Medicine and Biology Society Conference*, 2083–2086.

Xue, X. and Brookmeyer, R. 1996. Bivariate frailty model for the analysis of multivariate survival time. *Lifetime Data Analysis* 2, 277–289.

Yang, Y. H., Dudoit, S., Luu, P., Lin, D. M., Peng, V., Ngai, J. and Speed, T. 2002. Normalization for cDNA microarray data: A robust composite method addressing single and multiple slide systematic variation. *Nucleic Acids Research* 30, e15.

9

Sequential Monitoring of Randomization Tests

Yanqiong Zhang and William F. Rosenberger

CONTENTS

9.1 Introduction

Sequential monitoring has become a hallmark of the well-conducted randomized clinical trial. The statistician computes interim values of the test statistic for the primary outcome, and a decision is made to continue the trial or stop, based on these interim assessments of efficacy. There is a large literature on group sequential monitoring of population-based inference procedures; see Jennison and Turnbull (2000) for a comprehensive overview of the subject. There is no corresponding literature on sequential monitoring using randomization-based inference, which is rooted historically and arises naturally from the randomized nature of the clinical trial. Here we attempt to rectify that, by presenting what we know (so far) about sequential monitoring of randomization tests. This work represents a sketch of the doctoral thesis of Zhang (2004), and focuses not on the theory, but on how to perform sequential monitoring of randomization tests in practice. A sequential monitoring plan necessarily involves computing the joint asymptotic distribution of sequentially computed statistics. The mathematical formulation of the procedure therefore is complicated and involves multidimensional integration and test statistics that are messy to compute. Although we try to minimize the mathematical content of this paper, it is necessary to present the formulas for what needs to be computed and the conditions under which such formulas are appropriate.

It is our contention (and perhaps a controversial one) that although randomization has become an entrenched part of the biostatistician's culture, it is unfortunately often treated in a lackadaisical manner. In particular, great arguments over the appropriateness of incorporating randomization in the analysis have been subsumed by standard analyses using SAS. Here, we take the approach that randomization tests provide a nonparametric, assumption-free test of the treatment effect in a clinical trial, and indeed arise naturally from the structure of the clinical trial. Thus, randomization-based analyses should be conducted as a matter of course, either as a complement to population-based analyses or as a stand-alone primary analysis.

9.1.1 Sequential Monitoring under Population Model

Let us first consider a two-sample case with sample sizes N_A, N_B for each treatment. Let $X_1, \ldots, X_{N_A}$ be a set of independent and identically distributed random variables assumed to be normally distributed with mean μ_x and variance σ^2. Similarly, let $Y_1, \ldots, Y_{N_B}$ be a set of independent and identically distributed random variables assumed to be normally distributed with mean μ_y and variance σ^2. The usual test statistic of the hypothesis $H_0 : \mu_x = \mu_y$

against the alternative hypothesis $H_A : \mu_x > \mu_y$ in a nonsequential design is

$$Z = \frac{\bar{X}_{N_A} - \bar{Y}_{N_B}}{\sigma\sqrt{1/N_A + 1/N_B}}.$$

In practice, the sample sizes may not be equal. Let us first assume that $N_A = N_B = N/2$. In order to estimate the difference, $\delta = \mu_x - \mu_y$, using $\hat{\delta} = \bar{X}_{N_A} - \bar{Y}_{N_B}$, we have Fisher's information $I = (\text{Var}(\hat{\delta}))^{-1} = 1/\sigma^2(1/N_A + 1/N_B) = N/4\sigma^2$.

The concepts of information fraction and Brownian motion are applicable in the two sample case. We can write the test statistic at the end of the study as

$$Z_1 = \frac{\bar{X}_{N_A} - \bar{Y}_{N_B}}{\sigma\sqrt{1/N_A + 1/N_B}}.$$

If we monitor the data after N_{A1} and N_{B1} patients have responded, then the interim test can be based on the test statistic

$$Z_\tau = \frac{\bar{X}_{N_{A1}} - \bar{Y}_{N_{B1}}}{\sigma\sqrt{1/N_{A1} + 1/N_{B1}}},$$

and the Fisher's information is $1/\sigma^2(1/N_{A1} + 1/N_{B1})$. Thus, the information fraction will be defined as

$$\tau = \frac{(1/N_{A1} + 1/N_{B1})^{-1}}{(1/N_A + 1/N_B)^{-1}}.$$

Note that τ may be different from the fraction of patients available, however, in practice, the difference may be negligible. Define $B_\tau = Z_\tau\sqrt{\tau}$, for $\tau \in [0, 1]$, which are known as the B-values (Lan and Wittes, 1988). The stochastic process B_τ forms a discretized Brownian motion, which provides the theoretical basis for sequential monitoring under population model set up.

Now we consider the design of a clinical trial with a plan to monitor the data $K > 1$ times. The critical problem is to decide the boundary values such that

$$P_{H_0}(\text{reject } H_0) = P_{H_0}(Z_{\tau_1} \geq b_1, \text{ or } Z_{\tau_2} \geq b_2, \text{ or } \cdots Z_{\tau_K} \geq b_K) = \alpha,$$

where $\tau_1, \ldots, \tau_K$ are the information fractions at each inspection time and α is the prespecific type I error rate. Suppose that the interim inspections happen with equal increments in information such that $\tau_1 = 1/K$, $\tau_2 = 2/K, \ldots, \tau_K = 1$.

- Pocock (1977) employed $b_i = c$, for $1 \leq i \leq K$, such that

$$P_{H_0}(Z_{\tau_i} \geq c \text{ for some } i = 1, \ldots, K) = \alpha.$$

A constant boundary for the Z-values at each interim inspection is used. As the number of inspection times K increases, the boundary c increases also.

- O'Brien and Fleming (1979) used a constant boundary for the B-values instead of Z-values.

$$P_{H_0}(B_{\tau_i} \geq c \text{ for some } i = 1, \dots K) = \alpha.$$

In terms of Z-values,

$$P_{H_0}\left(Z_{\tau_i} \geq c\sqrt{\frac{K}{i}} \text{ for some } i = 1, \dots K\right) = \alpha.$$

O'Brien-Fleming boundaries are more strict at early inspections, while Pocock boundaries are constant at each inspection.

Lan and DeMets (1983) introduced the concept of the *spending function.* A spending function $\alpha(\tau)$, is defined as a nondecreasing function, for $\tau \in [0,1]$, with $\alpha(0) = 0$ and $\alpha(1) = \alpha$. The function $\alpha(\tau)$ specifies the type I error rate allowed to be spent at the time with information fraction τ. The spending function approach provides flexibility in sequential monitoring, and the inspections do not need to be prespecified. Removing the restriction of fixed and equally spaced inspections, Lan and DeMets gave the equivalent spending function for both Pocock and O'Brien-Fleming approaches. The O'Brien-Fleming spending function can be approximated by

$$\alpha(\tau) = 2(1 - \Phi(z_{\alpha/2}/\sqrt{\tau})),$$

where Φ is the cumulative distribution function of standard normal random variable; the Pocock spending function can be approximated by

$$\alpha(\tau) = \alpha \log(1 + (e - 1)\tau).$$

To carry out sequential monitoring using the spending function approach, one decomposes the rejection regions into disjoint regions. For K inspections,

$$\begin{aligned}
&P(Z_{\tau_1} \geq b_1) = \alpha(\tau_1),\\
&P(Z_{\tau_1} < b_1, Z_{\tau_2} \geq b_2) = \alpha(\tau_2) - \alpha(\tau_1).\\
&\vdots\\
&P(Z_{\tau_1} < b_1, \dots, Z_{\tau_{K-1}} < b_{K-1}, Z_{\tau_K} \geq b_K) = \alpha(\tau_K) - \alpha(\tau_{K-1}).
\end{aligned}$$

We can easily calculate b_1 if we specify the value $\alpha(\tau_1)$. When solving for b_2, we only need the joint distribution of (Z_{τ_1}, Z_{τ_2}). Note that $\text{cov}(Z_{\tau_1}, Z_{\tau_2}) = (\tau_1/\tau_2)^{1/2}$. Then b_2 can be calculated by numerical integration. We then solve for b_3 using the values of b_1 and b_2. The process continues until b_K has been solved.

9.1.2 Randomization Tests

Randomization tests have been thoroughly debated in the literature (e.g., Basu, 1980). As discussed in Rosenberger and Lachin (2002), randomization tests are a useful alternative to, or a complement to, traditional population model-based methods.

Under a population model, patients are assumed to be randomly sampled from a population. Suppose we have two treatments, A and B, and we are testing whether the two treatments are different. It is commonly assumed that the responses of patients are randomly sampled from two populations of responses and hence are internally and identically distributed (i.i.d.) observations from a given distribution. Standard population-based test statistics are then used, most available in SAS.

However, clinical trials do not follow a random sampling model. There are many factors that will make the selection nonrandom. Even if subjects are recruited at random, they can only be represented by some undefined patient population, an *invoked population*. Homogeneous population models are still questionable even under the invoked population. Owing to the lack of a homogeneous population assumption, the permutation test or randomization test is a useful alternative. Under the null hypothesis, we assume that treatment A and treatment B have no difference with respect to the responses of the subjects. The responses observed are assumed to be fixed under the null hypothesis and the sequence of treatment assignments is random. This differs from a population model, which assumes the equality of parameters of interest.

A broad family of randomization tests is that of the linear rank tests, which are defined by

$$S_n = \sum_{i=1}^{n} (a_{in} - \bar{a}_n) T_i,$$

where a_{in} is a score function based on response of the jth subject, $\bar{a}_n$ is the mean score, $T_i, i = 1, \ldots, n$ are 1 if treatment A is assigned and -1 if treatment B is assigned. The treatment assignments are the only random element, and their distribution is computed on the basis of the particular randomization procedure used. Under the randomization null hypothesis, the set of observed responses is assumed to be a set of deterministic values that are unaffected by treatment. That is, under the null, each subject's observed responses is what would have been observed regardless of whether treatment A or B had been assigned.

Before we discuss more about linear rank tests, we will introduce different randomization procedures most often used in randomized clinical trials. These are given in detail in Chapter 3 of Rosenberger and Lachin (2002). The simplest form of a randomization procedure is the toss of a fair coin, called *complete randomization*. This is rarely used in practice, but induces a sequence of i.i.d. Bernoulli trials that is easy to analyze. Restricted randomization is employed when we desire to maintain balance between treatment assignments. There are three types of restricted randomization procedures that are used in practice: the permuted block design, Wei's urn design (1977, 1978), and Efron's biased coin design (1971). Of these, Efron's biased coin design does not yield an asymptotically normal randomization test (Smythe and Wei, 1983), so current methodology for sequential monitoring, involving monitoring a Gaussian process, does not apply. Therefore, we will not consider Efron's biased coin design in this work.

The permuted block design forces balance between two treatments. Assume there are M blocks and $n = mM$ subjects, where M and m are positive integers. Within a block, one forces balance using one of two methods: the random allocation rule, where one draws the m treatment assignments sequentially without replacement from an urn with exactly $m/2$ treatment A balls and $m/2$ treatment B balls, or the truncated binomial design, which uses complete randomization until one treatment has $m/2$ subjects and then assigns all remaining subjects to the opposite treatment. We will assume the random allocation rule is used within blocks. Note that imbalances can occur only when the final block in the randomization is unfilled.

Wei (1977, 1978) developed an urn model, which is denoted by $UD(\alpha, \beta)$. One can view it as following: an urn contains α A balls and α B balls. A ball is drawn and replaced and the patient is assigned the corresponding treatment. A ball of type $i = A, B$ generates the addition of β balls of the opposite type. Mathematically, the urn design can be defined by

$$P(T_i = 1|T_1, \ldots, T_{i-1}) = \frac{\alpha + \beta N_B(i-1)}{2\alpha + \beta(i-1)}, \quad i \geq 2,$$

and $P(T_1 = 1) = 1/2$, where $N_B(i)$ is the number of assignments to treatment B at the ith assignment. Note that the $UD(\alpha, 0)$ is complete randomization. Wei (1978) proposed a *general urn design*, given by

$$P(T_i = 1|T_1, \ldots, T_{i-1}) = p\left(\frac{D_{i-1}}{i-1}\right),$$

where $D_i = N_A(i) - N_B(i)$, p is a nonincreasing function, $p(x) + p(-x) = 1$ and $p(x) = 1/2 + p'(0)x + B(x)x^2$ with $\sup_{|x|\leq 1} |B(x)| < \infty$. For the symmetry of this design, $P(T_1 = 1) = 1/2$. *Smith's class* of design (Smith, 1984) uses

$$p(x) = \frac{(1-x)^r}{(1+x)^r + (1-x)^r}.$$

It turns out that $r = -2p'(0)$, and when $r = 0$, we have complete randomization, when $r = 1$, we have Wei's urn design with $\alpha = 0$, and when $r = 2$, we have Atkinson's (1982) design. This dynamic design offers a compromise between exact balance and complete randomization.

When important covariates are known to be related with the primary outcome, stratification is sometimes employed. For example, in multicenter clinical trials, randomization is often stratified by clinical center. The common techniques for stratified randomization is the *stratified block design*, which uses a permuted block design within each stratum, and the *stratified urn design*, which uses Wei's urn design, or its generalization, in each stratum.

The randomization test p-value is determined by permuting all possible treatment assignment sequences and recalculating the test statistic, along with its associated probability depending on the particular randomization procedure. The p-value then is the probability of obtaining a result as or more extreme than the observed test statistic value. Since we are performing sequential monitoring based on asymptotic Gaussian processes, we compute the randomization p-value based on the normal distribution. The p-value can be evaluated on the basis of different reference sets. An *unconditional reference set* is the set of all possible permutations; *conditional reference set* includes only those sequences with the same number of allocations on each treatment as observed. On the basis of the different reference sets, we have two types of randomization tests. One is the *unconditional test* and the other is the *conditional test*.

9.1.3 Outline

In Section 9.2, we will introduce commonly used score functions and discuss their asymptotic properties. In Section 9.3, asymptotic properties of the linear rank test will be discussed. As we proceed, the theoretical basis in this paper is the asymptotic distribution of the linear rank test. Also, we will mention the asymptotic results of the joint distribution in the sequential setup. In Section 9.4, we define a randomization-based concept of information fraction. Also, the calculation for different randomization procedures will be provided. In Section 9.5, general theory for K-inspection unconditional tests will be discussed. In Section 9.6, calculation details will be presented. In Section 9.7, the commonly used three designs will be presented. In Section 9.8, conditional tests with only one interim inspection will be considered. Finally, in Section 9.9, we will have a summary.

9.2 Score Functions

The beauty of the linear rank formulation for the randomization test is that so many of our standard population-based tests fall into that framework by choosing an appropriate score function. Smythe and Wei (1983) were probably the first to investigate the asymptotic distribution of such tests. The key condition for the asymptotic normality of the randomization test for the randomization procedures we have described is the *Lindeberg condition*, which is defined as

$$\lim_{n\to\infty} \frac{\max_{1\le i\le n}(a_{in}-\bar{a}_n)^2}{\sum_{i=1}^{n}(a_{in}-\bar{a}_n)^2} = 0.$$

It essentially says that no individual absolute score can grow too large relative to the sum of all the absolute scores. We now describe various score functions and discuss when the Lindeberg condition is satisfied.

For a clinical trial with binary response data, one can assign *binary scores* $a_{in} = 1$ or 0. Usually, we center a_{in} by subtracting the proportion of 1's. As long as the number of 1's have the same magnitude as the total number responses, the Lindeberg condition holds for the binary scores.

For a clinical trial with continuous outcomes, The *simple rank scores* are given by

$$a_{\text{in}} = \frac{r_{\text{in}}}{n+1} - \frac{1}{2},$$

where r_{in} are the integer ranks. It is simple to show that the Lindeberg condition always holds for the simple rank score functions. The linear rank test using the simple rank scores is the randomization-based analog of the Wilcoxon rank-sum test. As an alternative for continuous data, the *van der Waerden scores* are defined as

$$a_{\text{in}} = \Phi^{-1}\left(\frac{r_{\text{in}}}{n+1}\right),$$

where Φ is the standard normal distribution function. The linear rank test using van der Waerden scores is asymptotically equivalent to the normal scores test (Lehmann, 1986). The Lindeberg condition also holds for the van der Waerden scores.

For survival data, in the usual notation of survival analysis, $\tau_1, \ldots, \tau_n$ are the event times of patients $1, \ldots, n$. We have n distinct ordered survival times $\tau_{(1)}, \ldots, \tau_{(n)}$ corresponding to the scores called the *logrank or Savage scores*, given by

$$a_{\text{in}} = E(X_{(i)}) - 1,$$

where $X_{(1)}, \ldots, X_{(n)}$ are order statistics from unit exponential random variables (Prentice, 1978). They yield a randomization-based equivalent of the logrank test (when there are no ties and no censoring). It is easily verified that the Lindeberg condition also holds for logrank score function (Zhang and Rosenberger, 2005).

With censored data, we observe a pair of data (Y_i, δ_i), where $\delta_i = 1$ for the patient with an event and $\delta_i = 0$ for a censored patient. Let $\tau_{(1)} < \tau_{(2)} < \cdots < \tau_{(M)}$ denote the M ordered event times and R_m be the number of patients still at risk before $\tau_{(m)}$. The *censored logrank scores* are defined as

$$a_{\text{in}} = \delta_i - \sum_{m=1}^{i} \frac{\delta_m}{R_m}.$$

The score function a_{in} is centered and Lindeberg condition does not always hold as for the logrank score function in the case of censoring. As shown in Zhang and Rosenberger (2005), when the limit for the ratio of the number of events to the total number of subject is larger than 0, the Lindeberg condition holds for the censored logrank scores. In this case, the randomization test using the censored logrank scores is a randomization-based analog of the usual logrank test with censoring. However, there is an implicit assumption that censoring is deterministic and unrelated to treatment assignment.

9.3 Asymptotic Results for the Unconditional Randomization Test

9.3.1 Asymptotic Results in the Nonsequential Case

We have introduced the Lindeberg condition in Section 9.2, and we have seen that most commonly used score functions satisfy the Lindeberg condition. As in Rosenberger and Lachin (2002), if the scores satisfy the Lindeberg condition, under most of the randomization procedures, such as complete randomization, the random allocation rule, and the general urn design, the normalized statistic under the null hypothesis will converge to a standard normal distribution. For most commonly used score functions, such as the simple rank scores, van der Waerden scores, and logrank scores, we will have the asymptotic normality of the test statistic. For binary score and censored logrank scores, under the mild conditions mentioned in Section 9.2, asymptotic normality will hold for most of the designs. The normalized statistic is defined as

$$W_n = \frac{S_n}{\sqrt{\mathbf{a}_\mathbf{n}' \Sigma \mathbf{a}_\mathbf{n}}},$$

where $\mathbf{a}_\mathbf{n}$ is the vectorized score function and Σ is the variance–covariance matrix of the randomization sequence. In general, while Σ will be unknown for some randomization procedures, for complete randomization, $\Sigma = I$ and

$$W_n = \frac{S_n}{\sqrt{\sum_{i=1}^n a_{in}^2}};$$

for the random allocation rule, as $\Sigma = n/(n-1)I - 1/(n-1)J$, where I is an identity $n \times n$ matrix and J is $n \times n$ with all elements of 1's, thus

$$W_n = \frac{S_n}{\sqrt{n/(n-1)\sum_{i=1}^n a_{in}^2}} \approx \frac{S_n}{\sqrt{\sum_{i=1}^n a_{in}^2}}.$$

For the general urn design, Σ does not have a closed form. However, an asymptotically equivalent normalization (Wei et al., 1986) is given by

$$W_n = \frac{S_n}{\sqrt{\sum_{i=1}^{n} b_{in}^2}},$$

where b_{in} is a modified score function, defined as

$$b_{in} = a_{in} - r \sum_{l=i+1}^{n} \left[\frac{a_{ln}}{l-1} \prod_{j=i}^{l-2} (1 - r/j) \right], \quad i = 1, \ldots, n,$$

where $r = -2p'(0) > 0$. Note that when $r = 0$, we have complete randomization, and we have an identical formula for W_n.

9.3.2 Asymptotic Results for *K* Inspections

We now move to sequential monitoring with K inspections. Let $N_1, N_2, \ldots, N_K$ denote the total number of subjects observed at each inspection, $S_1, S_2, \ldots, S_K$ denote the linear rank test statistic and $W_1, W_2, \ldots, W_K$ denote the standardized linear rank test statistic. Asymptotic joint normality is needed to employ the Lan-DeMets spending function approach. Suppose we have specific $\alpha_1, \alpha_2, \ldots, \alpha_K$, such that $\alpha_1 + \alpha_2 + \cdots + \alpha_K = \alpha$, where α is the prespecified type I error rate. Then

$$\begin{aligned}
&P(W_1 \geq d_1) = \alpha_1, \\
&P(W_1 < d_1, W_2 \geq d_2) = \alpha_2 - \alpha_1, \\
&\vdots \\
&P(W_1 < d_1, W_2 < d_2, \ldots, W_{K-1} < d_{K-1}, W_K \geq d_K) = \alpha_K - \alpha_{K-1},
\end{aligned}$$

where $d_1, d_2, \ldots, d_K$ are the sequential boundaries that we need to compute. If we have the asymptotic distribution of W_1, we can determine the value of d_1 using α_1; if we have the asymptotic joint distribution of (W_1, W_2), by the value of d_1 we obtained in the first step, we can determine d_2. Continuing this process, we will have the values of $d_k, 1 \leq k \leq K$ if we have the asymptotic joint distribution of $(W_1, W_2, \ldots, W_K)$.

Owing to previous unpublished work of Wei and Smythe (1983) on asymptotic results for the general urn design, we can find the joint asymptotic distribution of the K sequentially computed test statistics. Suppose that at kth inspection, there are N_k subjects with corresponding scores of $a_{jk}, j = 1, \ldots, N_k$. Let $\lambda_k = (b_{1k}, b_{2k}, \ldots, b_{N_k k}, 0, \ldots, 0)$, $1 \leq k \leq K$, where λ_k is a $N_K \times 1$ vector. Also, let $\Sigma = \lambda_{\mathbf{k}}' \lambda_{\mathbf{l}}$, a $K \times K$ matrix, the asymptotic variance–covariance matrix of $(S_1, S_2, \ldots, S_K)$; let $R = \lambda_{\mathbf{k}}' \lambda_{\mathbf{l}} / \sqrt{\lambda_{\mathbf{k}}' \lambda_{\mathbf{k}}} \sqrt{\lambda_{\mathbf{l}}' \lambda_{\mathbf{l}}}$ be the asymptotic correlation matrix. If some conditions on the score functions are satisfied, $\Sigma^{-1/2}(S_1, \ldots, S_K)$, or $R^{-1/2}(W_1, W_2, \ldots, W_K)$, converges in distribution to a multivariate normal random vector with mean $\mathbf{0}$ and identity variance–covariance matrix.

At the same time, even though the results are developed for the general urn design, they also apply to the random allocation rule; the only difference is $\lambda_k = (a_{1k}, a_{2k}, \ldots, a_{N_k k}, 0, \ldots, 0), 1 \leq k \leq K$. For the simple case of the complete randomization, since $a_{ik} = b_{ik}$ for $1 \leq i \leq N_k$, $\lambda_k = (a_{1k}, a_{2k}, \ldots, a_{N_k k}, 0, \ldots, 0)$.

For the score functions we discussed, the simple rank scores and the van der Waerden scores satisfy the conditions, which imply the joint asymptotic normality of the test statistics with these score functions under the procedures we mentioned, such as the general urn design and the random allocation rule. For binary scores, as $N_k - N_{k-1} \to \infty, k = 1, \ldots, K$, as long as the number of 1's in between each inspection goes to infinity and has the same magnitude of $N_k - N_{k-1}$, the joint asymptotic normality holds for the binary scores. For the logrank test, as long as the number of events occurring between the $(k-1)$th and kth inspection, has the same order as $N_k - N_{k-1}$, the conditions are satisfied. Multivariate normality cannot be rigorously established for the case of bounded $N_k - N_{k-1}$ with our method, but if $N_k - N_{k-1}$ is of moderate size, the multivariate normal approximation should be reasonable.

We conclude that in most cases of practical interest, there is no problem with the joint asymptotic normality of sequentially computed randomization tests.

EXAMPLE 1

We now show an example of how to use these results to conduct the monitoring plan. Suppose we use complete randomization with two inspections, one interim inspection and one final inspection. Assume we have total of N_2 subjects in the trial and we have observed N_1 subjects in the interim inspection. Both N_1 and N_2 are large numbers. Suppose we specify $\alpha_1 = 0.025$ and $\alpha = 0.05$ for the interim inspection and final inspection. For complete randomization, $W_1 = \sum_{i=1}^{N_1} a_{i1} T_i / \sqrt{\sum_{i=1}^{N_1} a_{i1}^2}$ and $W_2 = \sum_{i=1}^{N_2} a_{i2} T_i / \sqrt{\sum_{i=1}^{N_2} a_{i2}^2}$ are the test statistics. Assume we are using van der Waerden scores for the test statistic.

Step 1: At the interim inspection,

$$P(W_1 \geq d_1) = \alpha_1 = 0.025.$$

As the asymptotic distribution of W_1 is known as discussed in previous sections, we have

$$W_1 = \frac{S_1}{\sqrt{\sum_{i=1}^{N_1} a_{i1}^2}} \sim N(0, 1).$$

Also, for the van der Waerden score function, $\sum_{i=1}^{N_1} a_{i1}^2 \approx N_1$. Thus, we can calculate the boundary $d_1 = 1.96$. At the same time, based on the observed score function and randomization sequence, we can compute the observed W_1. If the observed $W_1 \geq d_1$, then stop the trial; otherwise, continue the trial until we have accrued N_2 patients.

Step 2: At the final inspection,

$$P(W_1 < d_1, W_2 \geq d_2) = \alpha - \alpha_1 = 0.025.$$

From the discussion of the joint asymptotic results of (W_1, W_2), for complete randomization, we have $\lambda_1 = (a_{11}, a_{21}, \ldots, a_{N_1 1}, 0, \ldots, 0)_{N_2 \times 1}$ and $\lambda_2 = (a_{12}, a_{22}, \ldots, a_{N_2 2})$; thus

$$\Sigma = \begin{pmatrix} \sum_{i=1}^{N_1} a_{i1}^2 & \sum_{i=1}^{N_1} a_{i1}a_{i2} \\ \sum_{i=1}^{N_1} a_{i1}a_{i2} & \sum_{i=1}^{N_2} a_{i2}^2 \end{pmatrix}.$$

Let

$$\rho = \frac{\sum_{i=1}^{N_1} a_{i1}a_{i2}}{\sqrt{\sum_{i=1}^{N_1} a_{i1}^2 \sum_{i=1}^{N_2} a_{i2}^2}},$$

thus

$$R = \begin{pmatrix} 1 & \rho \\ \rho & 1 \end{pmatrix}.$$

As $R^{-1/2}(W_1, W_2)' \sim N(\mathbf{0}, I)$, we obtain that

$$\begin{aligned} &P(W_1 < 1.96, W_2 \geq d_2) \\ &\quad = \frac{1}{2\pi\sqrt{1-\rho^2}} \int_{d_2}^{\infty} \int_{-\infty}^{1.96} \exp\left(-\frac{1}{2(1-\rho^2)}(w_1 - \rho w_2)^2\right) \\ &\qquad \times \exp\left(-\frac{w_2^2}{2}\right) \mathrm{d}w_1 \mathrm{d}w_2 \\ &\quad = \frac{1}{2\pi} \int_{d_2} \Phi\left(\frac{1.96 - \rho w_2}{\sqrt{1-\rho^2}}\right) \exp\left(-\frac{w_2^2}{2}\right) \mathrm{d}w_2. \end{aligned}$$

The above term is a decreasing function of d_2 only. So setting the above term equal to 0.025 will yield a unique solution. We can use numerical integration to find the solution. We then compare the observed W_2 and d_2 and draw conclusions. Note in practice, to calculate d_2, we do not have to do the numerical integration exactly. Both SAS IML and the R package mvtnorm, provide the multivariate normal distribution function, which makes the calculation much easier.

9.4 Information Fraction

In sequential monitoring, the amount of statistical information accumulated is a measure of how far a trial has progressed. Under a population model, the information obtained by N_k would be defined according to the Fisher's information for the estimator of the parameter of interest. This is asymptotically equivalent to the inverse of the asymptotic variance of the estimator. For tests expressed as a partial sum rather than a mean, the expression for the information is proportional to the variance of the sum, with both increasing in sample size. However, Fisher's information can be formally defined only in a population model context. We need to develop an analog of Fisher's information. Since the linear rank test involves the sum of the scores, the information fraction can be roughly understood as the proportion of the variance of the currently accumulated samples to the variance of all samples planned. The information fraction with N_k subjects observed can therefore given by

$$t_k = \frac{\mathbf{a}_\mathbf{k}' \Sigma \mathbf{a}_\mathbf{k}}{\mathbf{a}_\mathbf{K}' \Sigma \mathbf{a}_\mathbf{K}},$$

where Σ is variance–covariance matrix of the treatment assignments, and $\mathbf{a}_\mathbf{k}$ is the vector of score functions at the kth inspection (Rosenberger and Lachin, 2002).

In general, Σ will not be known for some randomization procedures. For the general urn design, the exact form of the variance–covariance form is unknown, and we will use the asymptotic variance as an estimate in the information fraction. Usually, we do not know the total score vector $\mathbf{a}_\mathbf{K}$ at the interim time points. Thus, it is necessary to estimate the total information or employ a surrogate measure of the total information to calculate the information fraction. We assume that all score functions are centered and scaled.

We will now calculate the information fraction under different randomization procedures such as complete randomization, and general urn design. For more complicated designs such as the permuted block design, stratified block design, and stratified urn design, detailed information will be given in Section 9.6.

9.4.1 Complete Randomization

For complete randomization, using simple rank scores, van der Waerden scores, and logrank scores without censoring, $t_k = N_k/N_K$. Since the treatment variance–covariance matrix is the identity matrix, the order of the rank of the responses does not affect the information accumulated. So we can obtain an exact result for the total information even though we have not observed the other $N_K - N_k$ observations.

For the censored logrank scores, suppose the number of events between the $(k-1)$th and kth inspections have the same order of $N_k - N_{k-1}$, then information fraction is defined as

$$t_k = \frac{\sum_{i=1}^{N_k} a_{ik}^2}{\sum_{i=1}^{N_K} a_{iK}^2} \approx \frac{D_k}{D_K},$$

where $D_k, 1 \le k \le K$ are the number of events by the kth inspection. Note that in Lan and Lachin (1990), they also discuss the information fraction concept of the logrank test with censoring which is computed as the ratio of the expected number of events. Since the expected number of events is not observable, in practice the information fraction is estimated by ratio of the number of events, which is the same as the result we obtained above. Their approach of estimating the information fraction can be applied here. In a maximum information trial design, if the total number of events is known, we can calculate the information fraction at each inspection. In a maximum duration trial, if the survival times of patients are exponentially distributed, the information fraction can be estimated as the fraction of total patient exposure. See details in Lan and Lachin (1990).

Since Σ is asymptotically equivalent for the random allocation rule and complete randomization, the information fraction is the same for complete randomization and the random allocation rule.

9.4.2 General Urn Design

For the general urn design, after we have observed N_k responses, we have the score function a_{ik}. To calculate the information, we still need the variance–covariance matrix of treatment allocation and the final score function a_{iK}. However, we know neither of them at the kth inspection. As we have shown that the asymptotic normality of (S_k, S_K), thus we can use the asymptotic variances to estimate the information. The information fraction at the kth inspection is given as

$$t_k = \frac{\mathbf{b}_\mathbf{k}' \mathbf{b}_\mathbf{k}}{\mathbf{b}_\mathbf{K}' \mathbf{b}_\mathbf{K}} = \frac{\sum_{i=1}^{N_k} b_{ik}^2}{\sum_{i=1}^{N_K} b_{iK}^2},$$

where $\mathbf{b}_\mathbf{k}'$ is vector version of the modified score function from a_{ik}. To calculate the information fraction, we need to estimate the total information $Var(S_K) = \sum_{i=1}^{N_K} b_{iK}^2$. It can be shown that, for the scores satisfying Lindeberg condition, it is therefore conservative to estimate the information fraction by replacing the total information $\sum_{i=1}^{N_K} b_{iK}^2$ with $\sum_{i=1}^{N_K} a_{iK}^2$. Thus, at the kth interim time point, the estimated information fraction can be written as

$$t_k = \frac{\sum_{i=1}^{N_k} b_{ik}^2}{\sum_{i=1}^{N_K} a_{iK}^2}.$$

For most of the rank-related score functions, we can calculate $\sum_{i=1}^{N_K} a_{iK}^2$. For example, for simple rank scores, $\sum_{i=1}^{N_K} a_{iK}^2 \approx N_K/12$; for van der Waerden scores, $\sum_{i=1}^{N_K} a_{iK}^2 \approx N_K$. For binary scores, we will estimate the information fraction as

$$\frac{\sum_{i=1}^{N_k} b_{ik}^2}{\sum_{i=1}^{N_K} a_{iK}^2},$$

and we again assume the same proportion of 1's and 0's will be observed at the end of the trial as at the midpoint.

9.5 *K*-Inspection Unconditional Tests

In this section, we will focus on sequential monitoring using the unconditional test. All the asymptotic distribution are considered using the reference set with all possible permutations. The spending function approach will be applied to control the type I error rate spent at each inspection. Let S_k, $1 \leq k \leq K$, be the linear rank test statistic at the kth inspection and $W_k, 1 \leq k \leq K$, will be the standardized test statistic at the kth inspection. The same asymptotic distribution properties for $(W_1, W_2, \ldots, W_K)$ will hold as in Section 9.3.

Suppose we take the spending function approach of Lan and DeMets (1983). Let $t_k, 1 \leq k \leq K$ be the information fraction at the kth inspection. To perform sequential monitoring, we have

$$\begin{aligned}
&P(W_1 \geq d_1) = \alpha(t_1),\\
&P(W_1 < d_1, W_2 \geq d_2) = \alpha(t_2) - \alpha(t_1),\\
&\vdots\\
&P(W_1 < d_1, W_2 < d_2, \ldots, W_{K-1} < d_{K-1}, W_K \geq d_K) = \alpha(t_K) - \alpha(t_{K-1}),
\end{aligned}$$

where $d_1, d_2, \ldots, d_K$ are the sequential boundaries that need to be determined. Note that this is different from the approach we described in Section 9.3. Instead of specifying $\alpha_1, \ldots, \alpha_K$ at each inspection before conducting the trial, we use the spending function approach, where the spending function is an increasing function of the information fraction and the information fraction is a measure of the proportion of total information accrued.

To conduct sequential monitoring using the spending function approach, at each inspection we need to estimate the information fraction according to the score function used. Using the chosen spending function, we can calculate the type I error rate we spend at each inspection.

TABLE 9.1

Boundary Values for Simple Rank Example

Spending Function	$\alpha(1/2)$	$\alpha(1)-\alpha(1/2)$	d_1	d_2
O'Brien-Fleming	0.0056	0.0444	2.538	1.679
Pocock	0.0310	0.0190	1.866	1.965

EXAMPLE 2

For a simple illustration, suppose we are conducting a trial with one interim inspection and one final inspection, where we have observed N_1 and N_2 patients at each inspection and $N_2 = 2N_1$. For the simple rank scores, the information fraction will be $t_1 = N_1/N_2 = 1/2$, and we obtain

$$P(W_1 \geq d_1) = \alpha\left(\tfrac{1}{2}\right),$$
$$P(W_1 < d_1, W_2 \geq d_2) = \alpha(1) - \alpha\left(\tfrac{1}{2}\right).$$

Assuming $\alpha(1) = 0.05$, for different spending functions, such as O'Brien-Fleming and Pocock, we can calculate the type I error rate spent at each inspection as listed in the following table. By the asymptotic distribution of W_1 and $\alpha(1/2)$, d_1 can be easily calculated. At the interim inspection, we obtain $a_{i1}, i = 1, \ldots, N_1$ according to the responses of N_1 subjects. If $W_1 \geq d_1$, we will reject the null hypothesis and stop the trial. Otherwise, we will continue the trial and observe all N_2 subjects. At the final inspection, we obtain the score function $a_{i2}, i = 1, \ldots N_2$ according to the responses from all the subjects. At the same time, the correlation coefficient ρ can be calculated. By the asymptotic distribution of (W_1, W_2), d_2 can be determined by the numerical integration. Assume that we have $\rho = 0.5$. In Table 9.1, we have calculated the corresponding boundary values d_1 and d_2 for different spending functions.

Here, the boundary value d_1 can be easily obtained and will change according to the information fraction and different spending functions. The boundary value d_2 will control not only the type I error rate spent at final inspection, but also the boundary value d_1 and correlation coefficient ρ between W_1 and W_2.

9.6 Discussion of Calculations

In this section, we will discuss calculation of the boundary values. As we have mentioned before, both the R package *mvtnorm* and SAS *probmvn* provide a way to calculate multivariate normal distribution probabilities. Details can be seen in Genz (1992, 1993).

For the procedures described in Section 9.5, to perform sequential monitoring, we need to calculate the boundary values $d_1, d_2, \ldots, d_K$, assuming

$\alpha(t_1), \ldots, \alpha(t_K)$ are known at each inspection. In this section, our discussion will focus on the appropriate score function and randomization sequence such that joint asymptotic normality holds for $W_1, \ldots, W_l$, where $1 \leq l \leq K$.

At each inspection, we can calculate the information fraction t_i and $\alpha(t_i)$ by using the appropriate spending function. Thus, $\alpha(t_1), \alpha(t_2)-\alpha(t_1), \ldots, \alpha(t_K)-\alpha(t_{K-1})$ are known at each inspection time. Thus, from

$$P(W_1 \geq d_1) = \alpha(t_1),$$

we can decide $d_1 = \Phi^{-1}(1 - \alpha(t_1))$, the $(1 - \alpha(t_1))$th percentile of normal distribution. From

$$P(W_1 < d_1, W_2 \geq d_2) = \alpha(t_2) - \alpha(t_1),$$

the direct formula for d_2 is not available, and we need to use iteration to obtain the boundary value d_2. Similarly, we can determine the boundary values $d_3, d_4, \ldots, d_K$. The procedure for calculation of d_2 is as follows.

As we know that

$$P(W_1 < d_1, W_2 \geq d_2) \leq P(W_2 \geq d_2),$$

the true value for d_2 will be less than $\Phi^{-1}(1 - (\alpha_2 - \alpha_1))$.

- Choose the initial upper value $u_d = \Phi^{-1}(1 - (\alpha_2 - \alpha_1))$; let $d_2 = u_d$. Calculate the diff $= \alpha_2 - \alpha_1 - P(W_1 < \Phi^{-1}(1 - \alpha_1), W_2 \geq d_2)$. Let $d_2 = d_2 - \delta_1$ and usually $\delta_1 = 0.01$. Repeat the above process until diff < 0. Let u_d be the last d_2 such that the diff > 0 and l_d be the first d_2 such that diff < 0.
- Let d_2 be the average of u_d and l_d. Calculate *diff*, and if diff > 0, then $u_d = d_2$, else if diff < 0, then $l_d = d_2$. Continue this process until the absolute value of diff < 0.0001, the prespecified accuracy.

Also, we consider the two-sided test case. From

$$P(|W_1| \geq d_1) = \alpha(t_1),$$

we can determine $d_1 = \Phi^{-1}(1-\alpha(t_1)/2)$, the $(1-\alpha(t_1)/2)$ percentile of normal distribution; from

$$P(|W_1| < d_1, |W_2| \geq d_2) = \alpha(t_2) - \alpha(t_1),$$

which is equivalent to

$$P(|W_1| < d_1, W_2 \geq d_2) = (\alpha(t_2) - \alpha(t_1))/2.$$

The direct formula for d_2 is not available, and a similar iteration method will give us the boundary value d_2. Continuing this way, we can determine the boundary values $d_3, d_4, \ldots, d_K$.

Our program in R simplifies calculation when we have the information fractions and the correlation matrix of W_is. Also, we have a SAS macro **LRT** to perform all the calculations, and the input from the user are the specifications for the trial, such as the total number of inspections, the score function, the random sequence, spending function, and the response data from the trial. We will explain their use in detail in the following subsections.

9.6.1 R Calculations

Let us use Example 2 in the last section, as we have the correlation matrix of W_1 and W_2 and the information fraction at the interim inspection $t_1 = 0.5$. If we choose the O'Brien-Fleming spending function, we will do the following:

```
>t<-c(0.5,1)
>corr<-matrix(c(1,0.5,0.5,1),2,2)
>boundary(obrien(t),corr)
$d
[1] 2.537988 1.678991
```

The *boundary* function has two inputs, one is the type I error rate that could be spent at each inspection, and the other is the correlation matrix of W_i's. In this example, the function *obrien*(t) calculates the α_1 and $\alpha_2 - \alpha_1$ from the information fraction sequence t. Also, user could choose to use the Pocock spending function as

```
>boundary(pocock(t),corr)
$d
[1] 1.866214 1.964978
```

Kim and DeMets's (1987) spending functions are also available to use. For details, user can refer to the program in Appendix A.

9.6.2 SAS Calculations

The SAS macro **LRT** provides a whole process calculation with only necessary inputs from the user. For example,

```
%LRT(          k= 3,
              nk= %str(20//40//60),
          datain1= datadir.seq,
          datain2= datadir.score,
             rand=CR,
```

```
         score=srank,
          col1=response,
          col2=,
             r=,
            sp=obrien,
         sided=one,
         alpha=0.05,
         debug=n);
```

here $k = 3$ means that the total number of inspections will be three. Also, the first inspection will be performed when 20 patient responses have been observed; the second inspection will be performed when 40 responses have been observed; the last inspection will be performed when total 60 responses have been observed. The parameter *datain*1 will provide the randomization treatment sequence with column name *TRT*, and $TRT = -1$ when treatment A is assigned and $TRT = 1$ when treatment B is assigned. The parameter *datain*2 will provide the patient's real response data and the parameter *col*1 tells the program the column name for patient response. The parameter *rand* will provide the information of the randomization sequences, and the possible choices will be **CR**, complete randomization, **RAR**, the random allocation rule, and **GUD**, the generalized urn design. The parameter *score* will provide the score function used, and thc possible choices will be **SRANK**, simple rank scores, **VWDEN**, van der Waerden scores, and **LOGRANK**, logrank scores. The parameter *col*2 needs to be specified only when logrank score is chosen. It specifies the column name of the censor variable in *datain*2. The parameter r needs to be specified when the generalized urn design is used and $r = -2p'(0)$. The parameter *sp* specifies the spending function and the possible choices are **OBRIEN**, **POCOCK**, **KIM1**(kim spending function with $\theta = 1$), **KIM2** (kim spending function with $\theta = 2$). The parameter *sided* specifies either a one-sided or two-sided test and the possible choices are **ONE**, **TWO**. The type I error rate in the trial is α and the default value is 0.05. If $debug = y$, it will keep some calculation files and the user has a way to track the result. If $debug = n$, only necessary output will be shown.

A good practice is to have a startup.sas file and generate the program directory and data directory. The startup.sas file specifies the path that stores the programs and the path that stores your data files. A sample startup.sas is as follows:

```
%let path=%str(C:\XXXXX);
%let pgmdir=&path\program;
%let datadir=&path\data;
libname datadir "&datadir"
```

In the program directory, one can store the score.sas, sp.sas, probmvn.sas, and lrt1.sas as in Appendix B. In the data directory, one can store the randomization sequence data and the analysis data that can be used to generate the score function.

In the following, we will show a sample result from randomly generated data that is attached in Appendix C. The above example SAS call will be used.

```
                    The SAS System

                COVAR

         1 0.6779251 0.5633854
0.6779251          1 0.8435999
0.5633854 0.8435999          1

                  D

3.3211931 2.1625538 1.6808799
```

The SAS output provides the iteration process. It also provides the correlation matrix for the test statistics and the boundary values d_1, d_2, d_3. In this mock-up example, $d_1 = 3.3212$, $d_2 = 2.1626$ and $d_3 = 1.6809$. At the same time, it will output a SAS dataset called interim.sas7bdat in the data directory, which is shown as follows:

s	v	w	t	diffst
0.8571428571	1.5079365079	0.6980100609	0.3118105999	0.0004481675
2.3902439024	3.1704342653	1.3424027786	0.6556428276	0.0150487912
1.3606557377	4.8352593389	0.6187829964	1	0.0345030413

where s records the linear rank statistics S_1, S_2, S_3; v records the variances for S_1, S_2, S_3; w records the standardized test statistics W_1, W_2, W_3; t is the estimated information fractions at each inspection; and *diffst* records the type I error rate that could be spent at each inspection by the given spending function.

9.7 Unconditional Test for Three Common Designs

There are three commonly used randomization procedures: the permuted block design, stratified block design, and stratified urn design. In this section, we consider sequential monitoring of unconditional randomization tests for these procedures. The full methodology and theoretical developments are given in Zhang, Rosenberger, and Smythe (2007).

Since we are now considering randomization sequences and scores within blocks within strata, the notation will become necessarily more complex. We will adopt the following notation for these special designs. In the nonsequential setting, for the complete randomization procedure, a_{in} is the score function for the ith patient relative to the total n patients. For more complicated designs or the sequential setting, the index n will be omitted for brevity; thus when we talk about the permuted block design, $a_{i(j)}$ is the score function for the ith patient in the jth block relative to patients in block j. For the stratified block design, $a_{i(jl)}$ is the score function for the ith patient in the jth block of stratum l relative to all patients in the jth block of stratum l. For the stratified urn design, $a_{i(l)}$ is the score function for the ith patient in stratum l relative to all patients in stratum l. In the sequential setting, for the simple randomization procedure, $a_i^{(k)}, 1 \leq k \leq K$, is the score function for patient i at the kth inspection. For both the permuted block design and the stratified block design, the notation will not change since the score function is the rank-related function relative to one block and will not change with different inspections. For the stratified urn design, $a_{i(l)}^{(k)}$ is the score function for the ith patient in stratum l at the kth inspection.

9.7.1 Properties of Randomization Tests in the Nonsequential Setting

Before we discuss the linear rank statistics under different randomization procedures in a sequential monitoring setting, let us see how these tests are carried out without interim inspections. We follow the development of Rosenberger and Lachin (2002, Chapter 8). For the random allocation rule, let a_{in} be a centered score function and $S_{\text{RAR}} = \sum_{i=1}^{n} a_{in} T_i$ be the linear rank test statistic. If the score function satisfies the Lindeberg condition, then $W_{\text{RAR}} = S_{\text{RAR}} / \left(\{n/(n-1)\} \sum_{i=1}^{n} a_{in}^2\right)^{1/2}$ will converge to standard normal under the null hypothesis as $n \to \infty$.

For the case of a permuted block randomization, using a random allocation rule with block size m within each of M blocks, let

$$S_B = \sum_{j=1}^{M} w_j \sum_{i=1}^{m} a_{i(j)} T_{i(j)},$$

$$W_B = \frac{S_B}{(\{m/(m-1)\} \sum_{j=1}^{M} w_j^2 \sum_{i=1}^{m} a_{i(j)}^2)^{1/2}},$$

where w_j is the weight to block j, $a_{i(j)}$ is the rank-related score function and $T_{i(j)}$ is the treatment assignment for the ith patient in the jth block. As $M \to \infty$, the statistic W_B is asymptotically distributed as standard normal by the usual i.i.d. central limit theorem. For the unfilled block case, as $M \to \infty$, the unfilled part is negligible, and the test statistic is still asymptotically normal.

In a stratified block design, let

$$S_{TB} = \sum_{l=1}^{L} \sum_{j=1}^{M_l} w_{jl} \sum_{i=1}^{m} a_{i(jl)} T_{i(jl)},$$

$$W_{TB} = \frac{S_{TB}}{(\{m/(m-1)\} \sum_{l=1}^{L} \sum_{j=1}^{M_l} w_{jl}^2 \sum_{i=1}^{m} a_{i(jl)}^2)^{1/2}},$$

where w_{jl} is the weight assigned to block j in stratum l, M_l is the total number of blocks in stratum l, $a_{i(jl)}$ is the score function for the ith patient in the jth block of stratum l and $T_{i(jl)}$ is the treatment assignment for the ith patient in the jth block of stratum l. For each l, $1 \leq l \leq L$, as $M_l \to \infty$, the statistic W_{TB} is asymptotically distributed as standard normal. As the total number of strata is a finite number, for the unfilled block case, the test statistic is still asymptotically normal.

For the generalized urn design, let a_{in} be the centered score function and $S_{\text{GUD}} = \sum_{i=1}^{n} a_{in} T_i$. Define a sequence of modified scores $\{b_{in}\}$ as follows:

$$b_{in} = a_{in} - r \sum_{l=i+1}^{n} \left[\frac{a_{ln}}{l-1} \prod_{j=i}^{l-2} (1 - r/j) \right], \quad i = 1, \ldots, n,$$

where $r = -2p'(0)$. Let $s_n^2 = \sum_{i=1}^{n} b_{in}^2$. Wei, Smythe, and Smith (1986) showed that, if the score function a_{in} satisfies the Lindeberg condition, then $W_{\text{GUD}} = S_{\text{GUD}}/s_n$ converges in distribution to standard normal. In general, we will be interested in the case $r = 1$, which is the often used Wei's urn design (Wei, 1977).

When a separate urn randomization is employed within each stratum, we have the following setup. For a total of L strata, let

$$S_l = \sum_{i=1}^{n} v_{li} a_{i(l)} T_{i(l)}, \quad 1 \leq l \leq L,$$

where v_{li} is the indicator variable for stratum l, $v_{li} = 1$ if patient i is in stratum l, $v_{li} = 0$, if patient i is not in stratum l, $a_{i(l)}$ is the score function for the ith patient in stratum l, and the number of patients in stratum l is $n_l = \sum_{i=1}^{n} v_{li}$. Let

$$S_T = \sum_{l=1}^{L} w_l S_l, \quad \text{and} \quad W_T = \frac{S_T}{(\sum_{l=1}^{L} w_l^2 \sum_{i=1}^{n_l} b_{i(l)}^2)^{1/2}},$$

where w_l is a weight assigned to stratum l and $\{b_{i(l)}\}$ is the modified score function corresponding to $\{a_{i(l)}\}$. As $n_l \to \infty$ as $n \to \infty$, for each stratum l, and if $\max a_{i(l)}^2 / \sum_{i=1}^{n_l} a_{i(l)}^2 \to 0$, then S_l is normally distributed within each stratum, for large n_l, and W_T is asymptotically distributed as a standard normal.

9.7.2 Development of a Monitoring Plan

For sequential monitoring with K inspections, let $N_1, N_2, \ldots, N_K$ denote the total number of patients observed at each inspection. For the permuted block design, let $M_1, M_2, \ldots, M_K$ be the total number of blocks observed at each inspection and each block with m patients. When we perform the interim inspections, we only consider the complete blocks we have observed. In the following, at the kth inspection, let $S_B^{(k)}$ denote the linear rank test for permuted block design; let $S_{TB}^{(k)}$ denote linear rank test for stratified block design; let $S_T^{(k)}$ denote linear rank test for stratified urn design.

9.7.2.1 Permuted Block Design

For the permuted block design with random allocation rule, let

$$S_B^{(k)} = \sum_{j=1}^{M_k} w_j \sum_{i=1}^{m} a_{i(j)} T_{i(j)}, \quad 1 \leq k \leq K,$$

with $N_k = mM_k$, where w_j is the weight for block j, $a_{i(j)}$ is the centered score function for the ith patient in block j and $T_{i(j)}$ is the treatment allocation for the ith patient in block j. As $M_1, \ldots, M_K \to \infty$, $\Sigma^{-1/2}(S_B^{(1)}, S_B^{(2)}, \ldots, S_B^{(K)})$ converges in distribution to a multivariate normal random vector with mean $\mathbf{0}$, and identity covariance matrix, where $\Sigma_{ii} = Var(S_B^{(i)})$, $\Sigma_{ij} = \text{Var}(S_B^{(k)})$, with $k = \min(i, j)$, and

$$\text{Var}(S_B^{(k)}) = \frac{m}{m-1} \sum_{j=1}^{M_k} w_j^2 \sum_{i=1}^{m} a_{i(j)}^2.$$

For the information fraction at the kth inspection,

$$t_k = \frac{\text{Var}(S_B^{(k)})}{\text{Var}(S_B^{(K)})} = \frac{\sum_{j=1}^{M_k} w_j^2 \sum_{i=1}^{m} a_{i(j)}^2}{\sum_{j=1}^{M_K} w_j^2 \sum_{i=1}^{m} a_{i(j)}^2}.$$

For most of the rank-related score functions, such as the simple rank scores, van der Waerden scores, and the logrank scores,

$$t_k = \frac{\text{Var}(S_B^{(k)})}{\text{Var}(S_B^{(K)})} = \frac{\sum_{j=1}^{M_k} w_j^2}{\sum_{j=1}^{M_K} w_j^2}.$$

If the same weights have been used for each block, the information fraction will turn out to be the proportion of the number of blocks observed or the proportion of the number of patients observed.

9.7.2.2 Stratified Block Design

In a stratified randomization using permuted blocks, let

$$S_{TB}^{(k)} = \sum_{l=1}^{L} \sum_{j=1}^{M_{lk}} w_{jl} \sum_{i=1}^{m} a_{i(jl)} T_{i(jl)}, \quad 1 \leq k \leq K, \qquad 1 \leq l \leq L,$$

with $N_k = \sum_{l=1}^{L} mM_{lk}$, where w_{jl} is the weight block j in stratum l, M_{lk} is the number of blocks at the kth inspection in stratum l, $a_{i(jl)}$ is the centered score function for the ith patient in the jth block of stratum l and $T_{i(jl)}$ is the treatment allocation for the ith patient in the jth block of stratum l. From the setup of the stratified block design, we will have very similar asymptotic result as the permuted block design. For the stratified block design, for each stratum l, $1 \leq l \leq L$, as $M_{l1}, \ldots, M_{lK} \to \infty$, $\Sigma^{-1/2}(S_{TB}^{(1)}, S_{TB}^{(2)}, \ldots, S_{TB}^{(K)})$ converges in distribution to a multivariate normal random vector with mean $\mathbf{0}$, and identity covariance matrix, where $\Sigma_{ii} = \text{Var}(S_{TB}^{(i)})$, $\Sigma_{ij} = \text{Var}(S_{TB}^{(k)})$, with $k = \min(i, j)$, and

$$\text{Var}(S_{TB}^{(k)}) = \frac{m}{m-1} \sum_{l=1}^{L} \sum_{j=1}^{M_{lk}} w_{jl}^2 \sum_{i=1}^{m} a_{i(jl)}^2.$$

For the information fraction at the kth inspection,

$$t_k = \frac{\text{Var}(S_{TB}^{(k)})}{\text{Var}(S_{TB}^{(K)})} = \frac{\sum_{l=1}^{L} \sum_{j=1}^{M_{lk}} w_{jl}^2 \sum_{i=1}^{m} a_{i(jl)}^2}{\sum_{l=1}^{L} \sum_{j=1}^{M_{lK}} w_{jl}^2 \sum_{i=1}^{m} a_{i(jl)}^2}.$$

For most of the rank-related score functions, such as the simple rank scores, van der Waerden scores, and the logrank scores,

$$t_k = \frac{\text{Var}(S_{TB}^{(k)})}{\text{Var}(S_{TB}^{(K)})} = \frac{\sum_{l=1}^{L} \sum_{j=1}^{M_{lk}} w_{jl}^2}{\sum_{l=1}^{L} \sum_{j=1}^{M_{lK}} w_{jl}^2}.$$

If the same weights have been used for each block and each stratum, the information fraction will be the proportion of the number of blocks observed or the proportion of the number of patients observed at the interim point.

9.7.2.3 Stratified General Urn Design

For the stratified urn design with total K inspections, let

$$S_l^{(k)} = \sum_{i=1}^{N_{lk}} a_{i(l)}^{(k)} T_{i(l)}, \qquad 1 \leq l \leq L, \quad 1 \leq k \leq K,$$

where $a_{i(l)}^{(k)}$ is the score function for patient i in stratum l at the kth inspection, and let $S_T^{(k)} = \sum_{l=1}^{L} w_l S_l^{(k)}$.

By a result from Smythe and Wei (1983), as long as for each stratum l, $N_{lk} - N_{l,k-1} \to \infty$, for $k = 1, \ldots, K$, for most of score functions we discussed in Section 9.3.2, such as the simple rank scores, van der Waerden scores, the logrank scores, the asymptotic joint normality of the test statistic can be derived. For binary scores and censored logrank scores, as for each stratum, the conditions specified in Section 9.3.2 hold, then we will also have the asymptotic joint normality of the test statistic. Let $\lambda_{lk} = (0, \ldots, 0, b_{1(l)}^{(k)}, \ldots, b_{N_{lk}(k)}^{(k)}, 0, \ldots, 0)'$, where λ_{lk} is a $N_K \times 1$ vector. Also, let $(\Sigma_{km}) = \sum_{l=1}^{L} w_l^2 \lambda_{lk}' \lambda_{lm}$, a $K \times K$ matrix. Then as $N_{lk} - N_{l,k-1} \rightarrow \infty, k = 1, \ldots, K, \Sigma^{-1/2}(S_T^{(1)}, \ldots, S_T^{(K)})'$ converges in distribution to a multivariate normal random vector with mean $\mathbf{0}$, and identity covariance matrix.

For the information fraction,

$$t_k = \frac{\sum_{l=1}^{L} w_l^2 Var(S_l^{(k)})}{\sum_{l=1}^{L} w_l^2 Var(S_l^{(K)})} = \frac{\sum_{l=1}^{L} w_l^2 \sum_{i=1}^{N_{lk}} (b_{i(l)}^{(k)})^2}{\sum_{l=1}^{L} w_l^2 \sum_{i=1}^{N_{lK}} (b_{i(l)}^{(K)})^2},$$

with $1 \leq k \leq K$. For each stratum l, we can estimate the total information accrued $\sum_{i=1}^{N_{lK}} (b_{i(l)}^{(K)})^2$ with $\sum_{i=1}^{N_{lK}} (a_{i(l)}^{(K)})^2$. The information fraction at the kth inspection is thus estimated as

$$t_k = \frac{\sum_{l=1}^{L} w_l^2 \sum_{i=1}^{N_{lk}} (b_{i(l)}^{(k)})^2}{\sum_{l=1}^{L} w_l^2 \sum_{i=1}^{N_{lK}} (a_{i(l)}^{(K)})^2}.$$

EXAMPLE 3

We repeat here verbatim the main example from Zhang, Rosenberger, and Smythe (2007), using data provided by Dr. Neal Oden, from a clinical trial using the stratified urn design.

The Supplemental Therapeutic Oxygen for Prethreshold Retinopathy of Prematurity (STOP-ROP) trial (2000) enrolled 649 infants in 30 clinical centers, who were randomly assigned to receive either supplemental therapeutic oxygen or conventional therapy to reduce the probability of progression to threshold ROP from February 1994 to March 1999. Randomization assignments were generated by the coordinating center using the urn design within strata, and the trial was stratified by center and baseline ROP severity. To illustrate our methods, we will analyze data on 156 infants from the five largest strata. We take October 31, 1996, as our interim time point. At the interim inspection, 83 infants responses have been observed, where 44 were on conventional treatment and 39 were on supplemental treatment. As the primary endpoint of this study is the progression of at least one study eye of an infant to the threshold ROP, we will use the binary score function here. As on the whole, about 50% of infants using conventional treatment progressed to threshold ROP, we will center the scores by subtracting 1/2. Suppose equal weights are applied for each stratum, then

$$S_T^{(1)} = \sum_{l=1}^{5}\sum_{i=1}^{N_{l1}} a_{i(l)}^{(1)} T_{i(l)}^{(1)} = 4.5,$$

$$Var(S_T^{(1)}) = \sum_{l=1}^{5}\sum_{i=1}^{N_{l1}} (b_{i(l)}^{(1)})^2 = 20.491;$$

thus $W_T^{(1)} = 0.994$. To estimate the information fraction at this interim time point, we will assume that 50% of total infants will progress to threshold ROP. We obtain

$$t_1 = \frac{\sum_{l=1}^{5}\sum_{i=1}^{N_{l1}} (b_{i(l)}^{(1)})^2}{\sum_{l=1}^{5}\sum_{i=1}^{N_{l2}} (a_{i(l)}^{(2)})^2} = \frac{20.491}{39} = 0.525.$$

If the O'Brien-Fleming (1979) spending function is used, $\alpha(0.525) = 0.007$, $d_1 = 2.457 \times \sqrt{20.491} = 11.122$. We would conclude that there is no sufficient evidence that the supplemental treatment is better than the conventional treatment at the interim time point and we would continue the trial. For the final inspection,

$$S_T^{(2)} = \sum_{l=1}^{5}\sum_{i=1}^{N_{l2}} a_{i(l)}^{(2)} T_{i(l)}^{(2)} = -3,$$

$$Var(S_T^{(2)}) = \sum_{l=1}^{5}\sum_{i=1}^{N_{l2}} (b_{i(l)}^{(2)})^2 = 39.684;$$

thus $W_T^{(2)} = -0.476$. As $\alpha(1)-\alpha(0.525) = 0.043$, from the variance–covariance matrix of $(W_T^{(1)}, W_T^{(2)})$, we obtain $d_2 = 1.702 \times \sqrt{39.684} = 10.722$. We conclude that there is no sufficient evidence that the supplemental treatment is better than the conventional treatment.

To compare our result with an analysis that ignores the randomization procedure, we calculate the Mantel–Haenszel test for five independent strata. At the interim time point, the stratified-adjusted Mantel–Haenszel test statistic $Z_1 = 1.747$. As $t_1 = (\sum_{l=1}^{5} 1/N_{l2})/(\sum_{l=1}^{5} 1/N_{l1}) = 0.45$, using the O'Brien-Fleming spending function, $\alpha(t_1) = 0.003$, thus $d_1 = 2.748$. A similar conclusion is made as for the randomization-based test at the interim time point. The final stratified-adjusted Mantel–Haenszel test statistic is $Z_2 = -0.681$. As $\alpha(1) - \alpha(0.45) = 0.047$, using the Browian motion process result, $cov(Z_1, Z_2) = \sqrt{t_1} = 0.671$, we obtain $d_2 = 1.654$. A similar conclusion is made as for the randomization-based test at the final time point.

9.8 Conditional Tests

In previous sections, we have discussed a strategy for sequential monitoring using unconditional linear rank tests. In this section, we will develop a similar strategy for sequential monitoring of conditional linear rank tests. The difference between the unconditional test and the conditional test is the reference set. The conditional reference set includes only those sequences with the same number of treatments assigned to A and B as were obtained in the particular randomization sequence employed. The significance level is computed on the basis of the observed number of patients in each treatment group. Many statisticians favor the conditional reference set because it excludes highly improbable sequences with large imbalances. Since it is conditional on the number of the assignments in each treatment we have observed, the variance–covariance structure will become more complicated. In the following discussion, we will focus on the general urn design. We will set up the sequential monitoring plan with only one interim inspection. When $K > 2$, the mathematical techniques we use break down, and this is a topic for further research. Details can be found in Zhang (2004) or Zhang and Rosenberger (2007).

9.8.1 Sequential Monitoring Procedures

For sequential monitoring with two inspections, let N_1, N_2 denote the total number of subjects observed at each inspection, with N_{A1}, N_{A2} for the total number of subjects receiving treatment A and N_{B1}, N_{B2} for the total number of subjects receiving treatment B. Let $D_{N_1} = N_{A1} - N_{B1}$ and $D_{N_2} = N_{A2} - N_{B2}$. Also, here we assume that $\lim_{N_2 \to \infty} N_1/N_2 = q > 0$.

After we have observed N_1 subjects,

$$P(S_1 \geq b_1 | N_{A1} = n_{A1}) = \alpha_1$$

and

$$P(S_1 < b_1, S_2 \geq b_2 | N_{A1} = n_{A1}, N_{A2} = n_{A2}) = \alpha - \alpha_1.$$

From the above formulas, it is necessary to find the distribution of S_1, given $N_{A1} = n_{A1}$, and the joint distribution of (S_1, S_2), given that $N_{A1} = n_{A1}, N_{A2} = n_{A2}$. Conditioning on the $N_{A1} = n_{A1}, N_{A2} = n_{A2}$ is equivalent to conditioning on $D_{N_1} = m_{N_1}, D_{N_2} = m_{N_2}$, where m_{N_1}, m_{N_2} are two sequences of integers with the property that

$$m_{N_1} - N_1 \text{ is even and } m_{N_1} = x_1 N_1^{1/2} + o(N_1^{1/2}),$$
$$m_{N_2} - N_2 \text{ is even and } m_{N_2} = x_2 N_2^{1/2} + o(N_2^{1/2}),$$

where x_1 and x_2 are any real numbers.

We can simplify the above procedure as

$$P(S_1 \geq b_1 | D_{N_1} = m_{N_1}) = \alpha_1$$

and

$$P(S_1 < b_1, S_2 \geq b_2 | D_{N_1} = m_{N_1}, D_{N_2} = m_{N_2}) = \alpha - \alpha_1.$$

9.8.2 Asymptotic Results

Smythe (1988) proved the conditional asymptotic normality of the linear rank test under the general urn design. Here, we will extend the idea to the bivariate case, with one interim inspection. In the following, we will list the main results of our work.

As $N_1^{-1/2} D_{N_1} = \sum_{i=1}^{N_1} N_1^{-1/2} T_i$, $N_2^{-1/2} D_{N_2} = \sum_{i=1}^{N_2} N_2^{-1/2} T_i$, let $\{c_{i1}\}$ be the modified score of $\{N_1^{-1/2}\}$ and $\{c_{i2}\}$ be the modified score of $\{N_2^{-1/2}\}$ with

$$c_{i1} = N_1^{-1/2} - r \sum_{l=i+1}^{N_1} \frac{N_1^{-1/2}}{l-1} \prod_{j=i}^{l-2} \left(1 - \frac{r}{j}\right),$$

$$c_{i2} = N_2^{-1/2} - r \sum_{l=i+1}^{N_2} \frac{N_2^{-1/2}}{l-1} \prod_{j=i}^{l-2} \left(1 - \frac{r}{j}\right).$$

Let $\{a_{i1}\}$ denote the score functions observed at the first inspection and $\{a_{i2}\}$ denote the score functions observed at the second inspection and let $\{b_{i1}\}$ and $\{b_{i2}\}$ be the corresponding modified score functions. Let $\tau^2 = (1 - 4p'(0))^{-1}$, as proved in Wei (1978),

$$\lim_{N_1 \to \infty} \sum_{i=1}^{N_1} c_{i1}^2 = \lim_{N_2 \to \infty} \sum_{i=1}^{N_2} c_{i2}^2 = \tau^2.$$

Define

$$\mu = \mu_1 + \begin{pmatrix} 0 \\ s\mu_2 \end{pmatrix},$$

where

$$\mu_1 = \begin{pmatrix} \sum_{i=1}^{N_1} b_{i1} c_{i1} x_1 / \sum_{i=1}^{N_1} c_{i1}^2 \\ \sum_{i=1}^{N_1} b_{i2} c_{i1} x_1 / \sum_{i=1}^{N_1} c_{i1}^2 \end{pmatrix},$$

and

$$s\mu_2 = \frac{\sum_{i=N_1+1}^{N_2} b_{i1}c_{i1}\left(x_1 \sum_{i=1}^{N_1} c_{i1}c_{i2} - x_2 \sum_{i=1}^{N_1} c_{i1}^2\right)}{c_{N_1,N_2}^2},$$

with $c_{N_1,N_2}^2 = \sum_{i=1}^{N_1} c_{i1}^2 \sum_{i=1}^{N_2} c_{i2}^2 - \left(\sum_{i=1}^{N_1} c_{i1}c_{i2}\right)^2$. Also

$$\Sigma = \Sigma_1 + \begin{pmatrix} 0 & 0 \\ 0 & s^2\sigma^2 \end{pmatrix},$$

where

$$\Sigma_1 = \begin{pmatrix} \sum_{i=1}^{N_1} b_{i1}^2 & \sum_{i=1}^{N_1} b_{i1}b_{i2} \\ \sum_{i=1}^{N_1} b_{i1}b_{i2} & \sum_{i=1}^{N_1} b_{i2}^2 \end{pmatrix}$$

$$-\frac{1}{\sum_{i=1}^{N_1} c_{i1}^2}\begin{pmatrix} \left(\sum_{i=1}^{N_1} b_{i1}c_{i1}\right)^2 & \left(\sum_{i=1}^{N_1} b_{i1}c_{i1}\right)\left(\sum_{i=1}^{N_1} b_{i2}c_{i1}\right) \\ \left(\sum_{i=1}^{N_1} b_{i1}c_{i1}\right)\left(\sum_{i=1}^{N_1} b_{i2}c_{i1}\right) & \left(\sum_{i=1}^{N_1} b_{i2}c_{i1}\right)^2 \end{pmatrix},$$

and

$$s^2\sigma^2 = \sum_{i=N_1+1}^{N_2} b_{i2}^2 - \frac{\left(\sum_{i=N_1+1}^{N_2} b_{i2}c_{i2}\right)^2 \sum_{i=1}^{N_2} c_{i2}^2}{c_{N_1,N_2}^2}.$$

Under some conditions on the score functions, which are satisfied by the simple rank scores, conditioning on $D_{N_1} = m_{N_1}, D_{N_2} = m_{N_2}$, the distribution of

$$\Sigma^{-1/2}\left(\begin{pmatrix} S_1 \\ S_2 \end{pmatrix} - \mu\right)$$

converges as $N_1 \to \infty, N_2 \to \infty$ to a bivariate normal distribution. We can now use bivariate normal distribution to conduct sequential monitoring.

EXAMPLE 4

Complete randomization. As we have discussed in Section 9.1, for complete randomization, we have $p(x) = 1/2$. Thus, $r = -2p'(0) = 0, b_{i1} = a_{i1}, b_{i2} = a_{i2}$, $c_{i1} = N_1^{-1/2}$ and $c_{i2} = N_2^{-1/2}$. Then

$$\mu_1 = \begin{pmatrix} \sum_{i=1}^{N_1} b_{i1}c_{i1} / \sum_{i=1}^{N_1} c_{i1}^2 x_1 \\ \sum_{i=1}^{N_1} b_{i2}c_{i1} / \sum_{i=1}^{N_1} c_{i1}^2 x_1 \end{pmatrix} = \begin{pmatrix} \sum_{i=1}^{N_1} a_{i1}x_1/\sqrt{N_1} \\ \sum_{i=1}^{N_1} a_{i2}x_1/\sqrt{N_1} \end{pmatrix},$$

and

$$s\mu_2 = \frac{\sum_{i=N_1+1}^{N_2} b_{i2}c_{i2}\left(x_1\sum_{i=1}^{N_1} c_{i1}c_{i2} - x_2\sum_{i=1}^{N_1} c_{i1}^2\right)}{c_{N_1,N_2}^2}$$
$$= \frac{\sum_{i=N_1+1}^{N_2} a_{i2}\left(-\sqrt{N_1}x_1 + \sqrt{N_2}x_2\right)}{N_2 - N_1}.$$

Thus, the mean vector

$$\mu = \begin{pmatrix} \sum_{i=1}^{N_1} a_{i1}x_1/\sqrt{N_1} \\ \sum_{i=1}^{N_1} a_{i2}x_1/\sqrt{N_1} + \sum_{i=N_1+1}^{N_2} a_{i2}\left(-\sqrt{N_1}x_1 + \sqrt{N_2}x_2\right)/(N_2 - N_1) \end{pmatrix}.$$

Assume the centered score function we are using here, then

$$\Sigma_1 = \begin{pmatrix} \sum_{i=1}^{N_1} a_{i1}^2 & \sum_{i=1}^{N_1} a_{i1}a_{i2} \\ \sum_{i=1}^{N_1} a_{i1}a_{i2} & \sum_{i=1}^{N_1} a_{i2}^2 - \left(\sum_{i=1}^{N_1} a_{i2}\right)^2/N_1 \end{pmatrix},$$

and

$$s^2\sigma^2 = \sum_{i=N_1+1}^{N_2} a_{i2}^2 - \frac{\left(\sum_{i=N_1+1}^{N_2} a_{i2}\right)^2}{N_2 - N_1},$$

thus the asymptotic variance–covariance matrix

$$\Sigma = \begin{pmatrix} \sum_{i=1}^{N_1} a_{i1}^2 & \sum_{i=1}^{N_1} a_{i1}a_{i2} \\ \sum_{i=1}^{N_1} a_{i1}a_{i2} & \sum_{i=1}^{N_2} a_{i2}^2 - \left(\sum_{i=1}^{N_1} a_{i2}\right)^2/N_1 - \left(\sum_{i=N_1+1}^{N_2} a_{i2}\right)^2/(N_2 - N_1) \end{pmatrix}.$$

For the practitioner, the group sequential procedure will be as follows. After N_1 patients, find b_1, such that

$$P(S_1 \geq b_1) = \alpha_1,$$

where S_1 has asymptotic distribution

$$N\left(\frac{\sum_{i=1}^{N_1} a_{i1}x_1}{\sqrt{N_1}}, \sum_{i=1}^{N_1} a_{i1}^2\right).$$

Then after observing N_2 patients, find b_2, such that

$$P(S_1 < b_1, S_2 \geq b_2) = \alpha_2,$$

where (S_1, S_2) has asymptotic distribution $N(\mu, \Sigma)$ with μ and Σ defined above. Note that this differs quite a bit from the unconditional approach.

EXAMPLE 5

Wei's UD(0, 1) design. For Wei's Urn UD(0, 1), $p(x) = (1 - x)/2$, thus $r = -2p'(0) = 1$. Then

$$c_{i1} = N_1^{-1/2} - \sum_{k=i+1}^{N_1} \left(\frac{N_1^{-1/2}}{k-1} \prod_{j=i}^{k-2} \left(1 - \frac{1}{j}\right) \right),$$

we obtain

$$c_{i1} = N_1^{-1/2} \frac{i-1}{N_1 - 1}, \quad \text{and} \quad \sum_{i=1}^{N_1} c_{i1}^2 \to \frac{1}{3}.$$

Similarly,

$$c_{i2} = N_2^{-1/2} \frac{i-1}{N_2 - 1}, \quad \text{and} \quad \sum_{i=1}^{N_2} c_{i2}^2 \to \frac{1}{3}.$$

Note that this matches Wei's (1978) results with $\lim_{N_1 \to \infty} c_{i1}^2 = \lim_{N_2 \to \infty} c_{i2}^2 = (1 - 4p'(0))^{-1} = 1/3$. By the asymptotic results,

$$\mu_1 = \begin{pmatrix} 3x_1 \sum_{i=1}^{N_1} (i-1) b_{i1} / ((N_1 - 1)\sqrt{N_1}) \\ 3x_1 \sum_{i=1}^{N_1} (i-1) b_{i2} / ((N_1 - 1)\sqrt{N_1}) \end{pmatrix},$$

$$\begin{aligned} s\mu_2 &= \frac{\sum_{i=N_1+1}^{N_2} b_{i2} c_{i2} \left(-x_1 \sum_{i=1}^{N_1} c_{i1} c_{i2} + x_2 \sum_{i=1}^{N_1} c_{i1}^2 \right)}{c_{N_1,N_2}^2} \\ &\approx \frac{3(-x_1 q^3 + x_2)}{1 - q^3} \frac{\sum_{i=N_1+1}^{N_2} (i-1) b_{i2}}{(N_2 - 1)\sqrt{N_2}}. \end{aligned}$$

Then the conditional mean vector

$$\mu = \mu_1 + \begin{pmatrix} 0 \\ s\mu_2 \end{pmatrix}.$$

Also

$$\Sigma_1 = \begin{pmatrix} \sum_{i=1}^{N_1} b_{i1}^2 & \sum_{i=1}^{N_1} b_{i1}b_{i2} \\ \sum_{i=1}^{N_1} b_{i1}b_{i2} & \sum_{i=1}^{N_1} b_{i2}^2 \end{pmatrix} - \frac{3}{(N_1-1)^2 N_1}$$

$$\times \begin{pmatrix} \left(\sum_{i=1}^{N_1}(i-1)b_{i1}\right)^2 & \left(\sum_{i=1}^{N_1}(i-1)b_{i1}\right)\left(\sum_{i=1}^{N_1}(i-1)b_{i2}\right) \\ \left(\sum_{i=1}^{N_1}(i-1)b_{i1}\right)\left(\sum_{i=1}^{N_1}(i-1)b_{i2}\right) & \left(\sum_{i=1}^{N_1}(i-1)b_{i2}\right)^2 \end{pmatrix},$$

$$s^2\sigma^2 \approx \sum_{i=N_1+1}^{N_2} b_{i2}^2 - \frac{3\left(\sum_{i=N_1+1}^{N_2}(i-1)b_{i2}\right)^2}{(1-q^3)(N_2-1)^2 N_2}.$$

Then the conditional variance–covariance matrix is given by

$$\Sigma = \Sigma_1 + \begin{pmatrix} 0 & 0 \\ 0 & s^2\sigma^2 \end{pmatrix}.$$

For the practitioner, the group sequential procedure will be as follows. After observing N_1 subjects responses, find b_1, such that

$$P(S_1 \geq b_1) = \alpha_1,$$

where S_1 has asymptotic distribution

$$N\left(\frac{3x_1\sum_{i=1}^{N_1}(i-1)b_{i1}}{(N_1-1)\sqrt{N_1}}, \sum_{i=1}^{N_1} b_{i1}^2 - \frac{3(\sum_{i=1}^{N_1}(i-1)b_{i1})^2}{(N_1-1)^2 N_1}\right),$$

which can also be obtained from Smythe (1988). Then after observed N_2 subjects responses, find b_2, such that

$$P(S_1 < b_1, S_2 \geq b_2) = \alpha_2,$$

where (S_1, S_2) has asymptotic joint bivariate normal distribution with mean μ and variance Σ as described above. Note that this differs from the unconditional approach. Also, only the simple rank score function will satisfy the conditions here, which is the same condition as in Smythe (1988).

9.8.3 Information Fraction for Conditional Tests

In the previous discussion, we choose α_1, α_2, such that $\alpha_1 + \alpha_2 = \alpha$, where α is prespecified. For the sequential monitoring with conditional tests, we want to introduce the information fraction concept and apply the spending function approach, as we did in Section 9.4.

Define the information fraction as t_1 since we have only one interim inspection. Conditioning on $D_{N_1} = m_{N_1}, D_{N_2} = m_{N_2}$,

$$t_1 = \frac{\text{Var}(S_1)}{\text{Var}(S_2)}.$$

For complete randomization,

$$\text{Var}(S_1) = \sum_{i=1}^{N_1} a_{i1}^2,$$

$$\text{Var}(S_2) = \sum_{i=1}^{N_2} a_{i2}^2 - \frac{\left(\sum_{i=1}^{N_1} a_{i2}\right)^2}{N_1} - \frac{\left(\sum_{i=N_1+1}^{N_2} a_{i2}\right)^2}{N_2 - N_1} \le \sum_{i=1}^{N_2} a_{i2}^2.$$

A conservative estimator of t_1 will be $\sum_{i=1}^{N_1} a_{i1}^2 / \sum_{i=1}^{N_2} a_{i2}^2$. For the simple rank scores, it will be N_1/N_2, the proportion of the subjects observed.

For the general urn design,

$$\text{Var}(S_1) = \sum_{i=1}^{N_1} b_{i1}^2 - \frac{3\left(\sum_{i=1}^{N_1} (i-1) b_{i1}\right)^2}{(N_1 - 1)^2 N_1},$$

$$\text{Var}(S_2) = \sum_{i=1}^{N_2} b_{i2}^2 - \frac{3\left(\sum_{i=1}^{N_1} (i-1) b_{i2}\right)^2}{(N_1 - 1)^2 N_1} - \frac{3\left(\sum_{i=N_1+1}^{N_2} (i-1) b_{i2}\right)^2}{(1-q^3)(N_2 - 1)^2 N_2} \le \sum_{i=1}^{N_2} b_{i2}^2.$$

Following the same approach as described in Section 9.4, a conservative estimator of the information fraction is

$$t_1 = \frac{\text{Var}(S_1)}{\sum_{i=1}^{N_2} a_{i2}^2}.$$

9.9 Conclusion

We have laid a completely new framework for the sequential monitoring of randomization tests. This has been a concise summary of the first author's doctoral dissertation, and several papers are being planned for current and future submission that will contain the mathematical details. It is hoped that this work will give the tools necessary to practicing biostatisticians to conduct a group sequential clinical trial with a randomization test as the primary outcome.

Appendix A

1. R boundary program

Appendix B

1. score.sas
2. sp.sas
3. probmvn.sas
4. lrt1.sas
5. sample startup.sas, which needs to be changed accordingly.

Appendix C

1. seq.sas7bdat
2. score.sas7bdat

References

Atkinson A. C. 1982. Optimum biased coin design for sequential clinical trials with prognostic factors. *Biometrika*, 69:61–67.

Basu D. 1980. Randomization analysis of experimental data: The fisher randomization test. *Journal of American Statistical Association*, 75:575–595.

Genz A. 1992. Numerical computation of multivariate normal probabilities. *Journal of Computational and Graphical Statistics*, 1:141–150.

Genz A. 1993. Comparison of methods for the computation of multivariate normal probabilities. *Computing Science and Statistics*, 25:400–405.

Jennison C. and Turnbull B. W. 2000. *Group Sequential Methods with Applications to Clinical Trials*. Chapman & Hall/CRC, Boca Raton.

Kim K. and DeMets D. L. 1987. Design and analysis of group sequential tests based on the type I error spending rate function. *Biometrika*, 74:149–154.

Lan K. K. G. and DeMets D. L. 1983. Discrete sequential boundaries for clinical trial. *Biometrika*, 70:659–663.

Lan K. K. G. and Lachin J. M. 1990. Implementation of group sequential logrank tests in a maximum duration trial. *Biometrics*, 46:759–770.

Lan K. K. G. and Wittes, J. 1988. The *B*-values: A tool for monitoring data. *Biometrics*, 44:579–585.

Lehmann E. L. 1986. *Nonparametrics: Statistical Methods Based on Ranks*. Holden-Day, San Francisco.

O'Brien P. C. and Fleming T. R. 1979. A multiple testing procedure for clinical trials. *Biometrics*, 35:549–556.

Pocock S. J. 1977. Group sequential methods in the design and analysis of clinical trials. *Biometrika*, 64:191–199.

Prentice R. L. 1978. Linear rank tests with right-censored data. *Biometrika*, 65:167–179.

Rosenberger W. F. and Lachin J. M. 2002. *Randomization in Clinical Trials: Theory and Practice*. Wiley, New York.

Smith R. L. 1984. Sequential treatment allocation using biased coin designs. *Journal of the Royal Statistical Society B*, 46:519–543.

Smythe R. T. 1988. Conditional inference for restricted randomization designs. *Annals of Statistics*, 16:1155–1161.

Smythe R. T. and Wei L. J. 1983. Significance tests with restricted randomization design. *Biometrika*, 70:496–500.

The STOP-ROP multicenter study group. 2000. Supplemental therapeutic oxygen for prethreshold retinopathy of prematurity (STOP-ROP), a randomized, controlled trial I: primary outcomes. *Pediatrics*, 100:295–310.

Wei L. J. 1977. A class of designs for sequential clinical trials. *Journal of the American Statistical Association*, 72:382–386.

Wei L. J. 1978. The adaptive biased coin design for sequential experiments. *The Annals of Statistics*, 6:92–100.

Wei L. J. and Smythe R. T. 1983. Two-sample repeated permutational tests with restricted randomization rules, unpublished manuscript.

Wei L. J., Smythe, R. T. and Smith R. L. 1986. *k*-treatment comparisons with restricted randomization rules in clinical trials. *Annals of Statistics*, 14:265–274.

Zhang Y. 2004. *Sequential Monitoring of Randomization Tests*. Doctoral thesis, University of Maryland Graduate School, Baltimore.

Zhang Y. and Rosenberger W. F. 2005. On asymptotic normality of the randomization-based logrank test. *Journal of Nonparametric Statistics*, 17:833–839.

Zhang Y. and Rosenberger W. F. 2007. Sequential monitoring of conditional randomization tests: generalized urn design, submitted.

Zhang Y., Rosenberger W. F. and Smythe R.T. 2007. Sequential monitoring of randomization tests: stratified randomization, *Biometrics*, in press.

10

Proportional Hazards Mixed-Effects Models and Applications

Ronghui Xu and Michael Donohue

CONTENTS

10.1 Univariate Frailty Models

For independent identically distributed (*i.i.d.*) failure time data, the proportional hazards model has been widely used since its introduction in Cox (1972). In the 1980s it began to be noted in the statistical literature that, unlike linear models, the Cox regression models are generally not nested, so that leaving out important covariates can lead to potential bias in the estimation of the regression parameters of interest (Lancaster and Nickell, 1980; Gail et al., 1984; Struthers and Kalbfleisch, 1986; Bretagnolle and Huber-Carol, 1988; Anderson and Fleming, 1995; Ford et al., 1995). In particular, the unconditional hazard rate would be underestimated if *unobserved heterogeneity* in the data was ignored (Omori and Johnson, 1993). See also Xu (1996) for a review and discussion on the topic. Frailty models are then used to model such unobserved heterogeneity in general. The term "frailty" was first used in Vaupel et al. (1979).

A univariate frailty model has the form

$$\lambda_{ij}(t) = \lambda_0(t)\exp(\beta' Z_{ij} + b_i), \tag{10.1}$$

where $\lambda_{ij}(t)$ is the hazard function for individual j in cluster i ($i = 1, \ldots, n$, $j = 1, \ldots, n_i$), $\lambda_0(t)$ is the baseline hazard, β is the vector of fixed regression effects as in the Cox (1972) model, and b_i is the random effect or, log frailty, for cluster i. We write the model for the general clustered data that will be discussed in this chapter. For *i.i.d.* data the cluster sizes are equal to one, so that each individual is a cluster. The term $\exp(b_i)$ is called the frailty, which acts multiplicatively on the hazard function. When at least some of the cluster sizes are greater than one, model (10.1) is often referred to as the shared frailty model, since individuals from the same cluster share the same frailty. In the literature the distribution of $\exp(b_i)$ is often assumed to be gamma or log-normal. The gamma distribution has mainly been used for mathematical convenience, since it provides a conjugate prior in the likelihood function where the unobservable random effects are integrated out; see below for more details. The normal distribution, being symmetric, is sometimes considered more natural for the unobserved b_i's, and will be shown later to be also appropriate for random covariate effects. Currently, programs to fit model (10.1) with gamma or log-normally distributed frailties are available in *Splus*, *R*, and *SAS*; see Therneau and Grambsch (2000) for details.

There has been a substantial literature on the frailty models (see Oakes, 1992; Hougaard, 2000). A number of these are on the gamma frailty model (Clayton and Cuzick, 1985; Klein, 1992; Nielsen et al., 1992; Murphy, 1994, 1995; Andersen et al., 1997; Parner, 1998; etc.). In particular, Klein (1992) and Andersen et al. (1997) provided the Expectation-Maximization (EM) algorithm for the nonparametric maximum likelihood estimation under the gamma frailty model. Consistency and asymptotic normality under that model was established in Murphy (1994, 1995) and Parner (1998).

For interpretation under model (10.1), note that the random effect b_i acts additively on the log hazard function, with no interaction between any covariate and the clustering indicator. Figure 10.1 is an illustration for this case, where the curves are the hypothetical log hazard functions over time. The upper five curves are from one treatment arm, with the thick curve in the middle indicating the average log hazard rate within that arm. The lower five curves represent another treatment arm with the middle curve indicating the average log hazard rate with this other arm. The first curve in the upper group and the first curve in the lower group are assumed to be from the same cluster. And the second curves in the upper and lower group, respectively, are also assumed to be from another common cluster, and so on. The figure shows that the treatment difference, in terms of log hazard, is the same across all the clusters. Model (10.1) cannot be used to model the type of varying

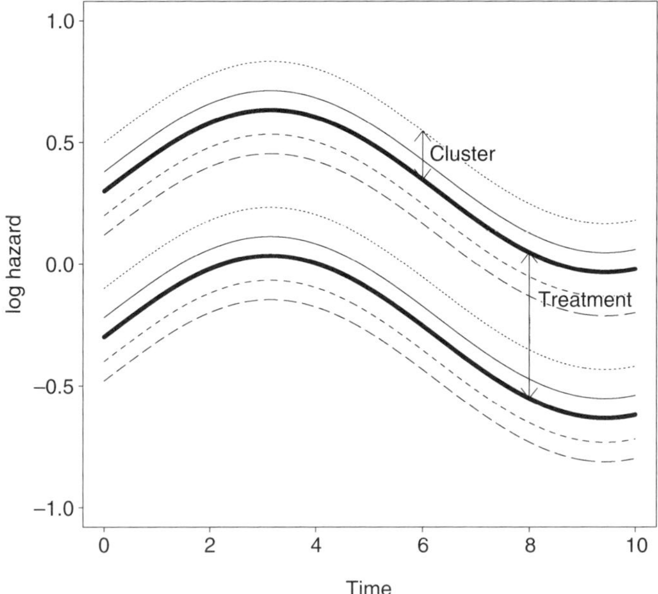

FIGURE 10.1
Additive effect of frailty and treatment on log hazard function.

treatment effect that exists in the lung cancer data below, which involves a cluster by treatment interaction.

For completeness as a review on the topic of univariate frailty models, in the rest of this section we briefly describe the EM algorithm of Klein (1992) under model (10.1) with gamma frailty distribution. Assume that $V_i = \exp(b_i) \sim \text{Gamma}(1/\alpha, \alpha)$, so that $E(V) = 1$ and $\text{Var}(V) = \alpha$. The likelihood is given in Equation 10.11 with $p(\cdot; \Sigma)$ replaced by the gamma density, and the joint log-likelihood of the observed data and the unobserved random effects is $L_1(\beta, \lambda) + L_2(\alpha)$, where

$$L_1(\beta, \lambda) = \sum_{i=1}^{n} \sum_{j=1}^{n_i} \{\delta_{ij}[\beta' Z_{ij} + \log \lambda_0(X_{ij})] - V_i \Lambda_0(X_{ij}) \exp(\beta' Z_{ij})\}, \quad (10.2)$$

$$L_2(\alpha) = -n \left\{ \frac{1}{\alpha} \log \alpha + \log \Gamma \left(\frac{1}{\alpha} \right) \right\} + \sum_{i=1}^{n} \left\{ \left(\frac{1}{\alpha} + D_i - 1 \right) \log V_i - \frac{V_i}{\alpha} \right\}, \quad (10.3)$$

and $D_i = \sum_j \delta_{ij}$ is the observed number of failures in cluster i.

To perform the E-step of the EM algorithm, we observe that owing to conjugacy, given data the V_i's are independent gamma random variables with shape parameters $A_i = 1/\alpha + D_i$ and scale parameters $C_i = 1/\alpha + \sum_j \Lambda_0(X_{ij}) \exp(\beta' Z_{ij})$. The conditional expectation of $L_1(\beta, \lambda) + L_2(\alpha)$ given the observed data is then $Q_1(\beta, \lambda) + Q_2(\alpha)$, where

$$Q_1(\beta, \lambda) = \sum_{i,j} \left\{ \delta_{ij}[\beta' Z_{ij} + \log \lambda_0(X_{ij})] - \frac{A_i}{C_i} \Lambda_0(X_{ij}) \exp(\beta' Z_{ij}) \right\}, \qquad (10.4)$$

$$\begin{aligned} Q_2(\alpha) = & - n \left\{ \frac{1}{\alpha} \log \alpha + \log \Gamma \left(\frac{1}{\alpha} \right) \right\} \\ & + \sum_{i=1}^{n} \left\{ \left(\frac{1}{\alpha} + D_i - 1 \right) [\psi(A_i) - \log(C_i)] - \frac{A_i/C_i}{\alpha} \right\}, \end{aligned} \qquad (10.5)$$

and $\psi(\cdot)$ denotes the digamma function.

The M-step is similar to that under model (10.7) described in Section 10.3: Q_1 is maximized using the Cox partial likelihood and Breslow's estimate with an offset, and Q_2 is maximized numerically. The properly estimated variance of the parameter estimate is given in Andersen et al. (1997).

10.2 Clustered Survival Data

Clustered failure time data arise from various areas of biomedical research, including genetic or familial studies, multicenter clinical trials, group-randomized trials, matched pairs designs, recurrent events, environmental studies, retrospective studies from collaborative registry protocols or meta-analysis, and so forth. As an example, the Vietnam Era Twin (VET) Registry data was used to study the genetic and environmental contributions to age at onset of psychiatric disorders such as alcohol dependence and nicotine dependence (Liu et al., 2004a,b). The VET Registry consists of a total of 7375 male–male twin pairs born between 1939 and 1957 and both twins had the experience of serving in the military during the Vietnam era. In 1992, trained interviewers from the Institute for Survey Research at Temple University invited approximately 5000 twin pairs from the Registry to participate in telephone-administrated interviews. Of these, 3372 twin pairs completed the interviews and had known zygosity. The difference between monozygotic and dizygotic twins can be used to estimate the genetic contribution to alcohol dependence, as will be shown later. In addition to clustering by the twin pairs, there are also other levels of clustering in the data including different symptom categories of alcohol dependence for the same person, as well as correlation between age at onset of alcohol dependence and age at onset of nicotine dependence for the same person. Apart from decomposition of the total variation in age at onset into attributes owing to genetic and environmental factors, it is also of interest to assess the interaction between genetic factors and other variables such as early versus late alcohol users. Notice that there is censoring in the age at onset data; that is, some twins had not developed alcohol dependence at the time of phone interview. Therefore, the conventional linear or generalized linear mixed model for this type of genetic epidemiology studies cannot be applied directly.

Another example of clustered failure time data comes from a multicenter clinical trial in lung cancer conducted by the Eastern Cooperative Oncology Group (ECOG). The study compared two different chemotherapy regimens: a standard therapy cyclophosphamide, doxorubicin and vincristine (CAV) and an alternating regimen (CAV-HEM) where cycles of CAV were alternated with altretamine (hexamethylmelamine), etoposide and methotrexate (HEM). It was found that the treatment difference between the CAV-HEM and the CAV arms varied substantially across the participating institutions (Gray, 1994, 1995). Figure 10.3 shows the estimated treatment effects from the 31 institutions in terms of the log hazard ratio under a proportional hazards model. As seen in the figure, the log hazard ratios in some institutions such as number 16 and 18 are around −0.5, whereas in some other institutions such as number 19, 28, and 29 the log hazard ratios are almost zero. In multicenter clinical trials, center-to-center variations can be caused by, despite the tightly structured protocols, different standards of practice, types of supportive care, interpretations of dose modifications, and patient populations, and so forth.

For statistical applications involving clustered data in general, random effects models provide a powerful tool. This is not only so for genetic studies such as the VET Registry study where the primary interest lies in the estimation of the variance components. When compared to marginal models where the correlation structures are not specified but where robust variance estimators of the marginal regression effect estimates are usually used, the random effects model has the capability of making cluster-specific inference in addition to the estimation of the variance components. For example, in multicenter clinical trials, estimation of center-specific treatment effects allows the investigators to further study the possible causes of different treatment effects observed at different trial centers, and therefore help understand the conditions associated with potential therapy benefits. Some of these multicenter trials tend to involve many institutions, each of which typically contributes only a few patients to a particular trial. This is the type of setting where random effects models are suitable if center effects are suspected. The use of random effects survival models in clinical trials was also advocated in Glidden and Vittinghoff (2004), Murray et al. (2004), and Sylvester et al. (2002). Random effects models, on the other hand, are usually more complex and require more computation as compared to the marginal models.

Petersen et al. (1996) and Petersen (1998) extended the univariate gamma frailty model to additive frailty models for the analysis of genetic data, where the sum of two or more gamma frailty terms acts multiplicatively on the hazard function. Petersen (1998) wrote the model as

$$\lambda_{ij}(t) = \lambda_0(t)(a_i' U_{ij}) \exp(\beta' Z_{ij}), \tag{10.6}$$

where U is a design vector of 0's and 1's, and a is a vector of gamma frailties. Petersen (1998) summarizes the use of model (6) for genetic and familial data.

Model (6), however, is unable to incorporate random effects for arbitrary covariates, or equivalently, covariate by cluster interactions as exemplified in the lung cancer data. Furthermore, as will be shown in Section 10.4, for interpretation of the random effects in genetic data, the extension of Equation 10.1 to proportional hazards mixed effects model (PHMM) of Section 10.3, has the advantage of decomposing variation in the outcome variable itself as opposed to variation in the hazard of the outcome variable under model (10.6), leading to results that are comparable to those already established in the genetic epidemiology literature.

10.3 Proportional Hazards Mixed-Effects Model

In the following we continue to use $i = 1, \ldots, n$ to indicate the clusters, and $j = 1, \ldots, n_i$ to indicate the individuals within the ith cluster. T_{ij} is the failure time for individual ij, C_{ij} is the censoring time. We observe $X_{ij} = \min(T_{ij}, C_{ij})$ and $\delta_{ij} = I(T_{ij} < C_{ij})$, and Z_{ij} is a vector of covariates possibly associated with the event time outcome.

Vaida and Xu (2000) proposed the proportional hazards model with general random effects

$$\lambda_{ij}(t) = \lambda_0(t) \exp(\beta' Z_{ij} + b_i' W_{ij}), \tag{10.7}$$

where W_{ij} is a vector of covariates that have random effects. W_{ij} is usually a subvector of Z_{ij}, plus a possible "1" for the random effect on the baseline hazard function, as in the frailty model (10.1) before. In contrast with the additive frailty model (10.6), model (10.7) generalizes the univariate frailty model by allowing a multivariate random effect with arbitrary design matrix in the log relative risk, similar to the linear, generalized linear and nonlinear mixed models. This allows incorporation of random effects on general covariates, and is useful for assessing gene and covariate interactions in genetic data, and in multicenter clinical trials for modelling center by treatment interactions as described earlier. As an example, for the lung cancer data, if a single covariate $Z = 0$ or 1 indicates one of the two treatment assignments, the following model may be used to capture the cluster-specific treatment effect

$$\lambda_{ij}(t) = \lambda_0(t) \exp\{b_{0i} + (\beta + b_{1i}) Z_{ij}\}.$$

The treatment effect from institution i is then $\beta + b_{1i}$. Figure 10.2 is a illustration of this case. In contrast to Figure 10.1, the treatment effect within one cluster is different from that within another cluster. We call model (10.7) the PHMM.

For the distribution of the random effects in Equation 10.7, it is assumed that

$$b_i \overset{iid}{\sim} N(0, \Sigma), \tag{10.8}$$

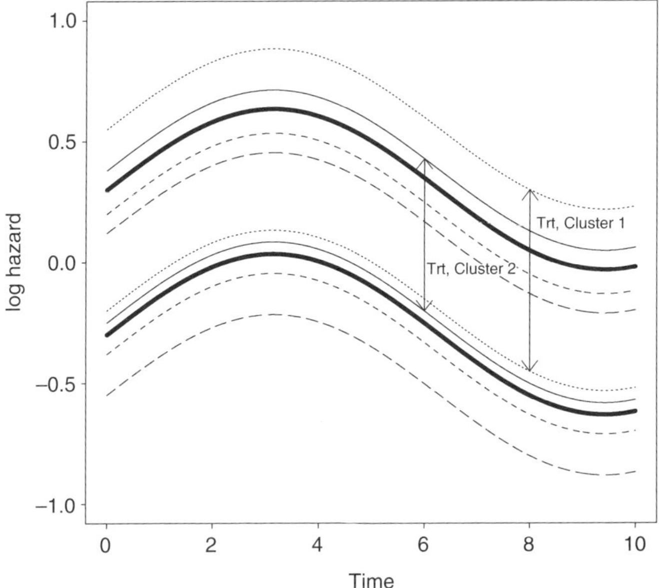

FIGURE 10.2
The effect of cluster by treatment interaction on log hazard function.

with unknown $d \times d$ covariance matrix Σ. The advantages of the normal assumption for random effects, as explained in Vaida and Xu (2000), are symmetry and scale-invariance. The gamma distribution, in particular, is not suitable here because it is not scale-invariant, so that the inference is not invariant under a change of measuring unit for the random effects; if $\exp(b)$ belongs to the gamma family and W is multiplied by a constant c, then $\exp(b/c)$ no longer belongs to the gamma family in general. The normality assumption for the random effects, however, can be replaced by other scale-invariant parametric distribution such as the multivariate t-distribution.

We note that Xue and Brookmeyer (1996) used a special case of model (10.7) with bivariate normal random effects, Li and Lin (2000) studied a model similar to Equation 10.7 with covariate measurement errors, Yau (2001) and Ma et al. (2003) considered multilevel clustering that is also a special case under Equation 10.7. Related work has also been done in the Bayesian context by Gray (1994), Gustafson (1997), Sargent (1998), and Carlin and Hodges (1999).

10.3.1 Interpretation under PHMM

Model (10.7) is in parallel with the linear, nonlinear, and generalized linear mixed-effects models. On the log hazard scale it can be written as

$$\log \lambda_{ij}(t) = \log \lambda_0(t) + \beta' Z_{ij} + b_i' W_{ij}.$$

This describes the conditional hazard, or equivalently, the conditional distribution, of the event time T of an individual as a function of the covariates and the random effects.

Model (10.7) can be written in the mathematically equivalent form of a linear transformation model

$$g(T_{ij}) = -(\beta' Z_{ij} + b_i' W_{ij}) + e_{ij}, \tag{10.9}$$

where $g(t) = \log \int_0^t \lambda_0(s)ds$, and e has a fixed extreme value distribution with variance equal to 1.645. The fixed error distribution is due to, basically, the proportional hazards assumption. When fitting model (10.9) to data, one tries to find the "best" parameter values under the constraint of the extreme value error distribution, which is equivalent to the proportional hazards constraint under model (10.7).

In view of Equation 10.9, the PHMM then decomposes the variation in the transformed event time T itself, instead of its hazard function, into contributions from the fixed covariate effects, the random effects, and a fixed error variance. Note in particular that the difference in the total variances between Equations 10.7 and 10.9 is due to the extra term e_{ij} in Equation 10.9. This is similar to the linear regression when predicting a future individual outcome, there is an extra variance term as compared to predicting the mean of a future outcome. Equation 10.9 can be useful for interpretation purposes of the PHMM, as will be seen in the applications of the next section.

10.3.2 Parameter Estimation under PHMM

Assume that the data consist of possibly right-censored observations from n clusters, with n_i subjects each, $i = 1, \ldots, n$. Let $N = \sum_i n_i$. Within a cluster the observations can be dependent, but conditional on the cluster-specific $d \times 1$ vector of random effects b_i, the observations are independent. We are mostly interested in clustered data where at least one of the clusters has more than one observation. For the *i.i.d.* case (i.e., all $n_i = 1$), recent work of Kosorok et al. (2001) provides conditions for identifiability under the frailty model (10.1). Identifiability and asymptotics of parameter estimate (see below) under the more general model (10.7) have been established in Xu et al. (2006).

The observed data from subject j in cluster i can generally be written as $y_{ij} = (X_{ij}, \delta_{ij}, Z_{ij}, W_{ij})$. Let $\mathbf{y}_i = (y_{i1}, \ldots, y_{in_i})'$ be the observed data for cluster i. Also, denote $\mathbf{y} = (\mathbf{y}_1', \ldots, \mathbf{y}_n')'$, and $\mathbf{b} = (b_1', \ldots, b_n')'$. Under model (10.7), for cluster i, conditional on the random effects, the log-likelihood is

$$l_i = l_i(\beta, \lambda; \mathbf{y}_i | b_i) = \sum_{j=1}^{n_i} \{\delta_{ij} \log \lambda_0(X_{ij}) + \delta_{ij}(\beta' Z_{ij} + b_i' W_{ij}) - \Lambda_0(X_{ij}) e^{\beta' Z_{ij} + b_i' W_{ij}}\}, \tag{10.10}$$

where $\Lambda_0(t) = \int_0^t \lambda_0(s)\, ds$. Vaida and Xu (2000) used a nonparametric maximum likelihood approach, with the infinite dimensional parameter $\lambda_0(\cdot)$ replaced by the vector $\lambda = (\lambda_1, \ldots, \lambda_s)'$, where $\lambda_i = \lambda_0(t_i)$ and $t_1, \ldots, t_s$ are the distinct uncensored failure times. The need to discretize the baseline hazard function lies in the fact that the likelihood function is otherwise unbounded; see also Johansen (1993) and Murphy (1994). The likelihood of the observed data is then

$$L(\theta) = \prod_{i=1}^{n} \int \exp(l_i) p(b_i; \Sigma)\, \mathrm{d}b_i, \tag{10.11}$$

where $\theta = (\beta, \Sigma, \lambda)$, and $p(\cdot; \Sigma)$ is the multivariate normal density function. Usually no closed-form expression is available for $L(\theta)$ and its calculation involves d-dimensional integration.

The EM algorithm (Dempster et al., 1997) can be used to compute the maximum likelihood estimate, and its standard convergence properties were shown to hold in Vaida and Xu (2000). This is mainly because the parameter vector θ in the nonparametric likelihood is in fact finite dimensional, once the data are given. Specifically, in the E-step the following conditional expectation of the joint log-likelihood of $(\mathbf{y}, \mathbf{b})$ given the data is computed:

$$Q(\theta) = \sum_i E\{l_i(\beta, \lambda; \mathbf{y}_i | b_i) | \mathbf{y}_i\} + \sum_i E\{\log p(b_i; \Sigma) | \mathbf{y}_i\}, \tag{10.12}$$

where the expectations are taken under the current parameter values. The E-step reduces to computing conditional expectations of the type $E\{h(b_i)|\mathbf{y}_i\} = \int h(b_i) p(b_i|\mathbf{y}_i)\, db_i$. Although Gaussian quadrature may be used when the dimension of the random effects $d \leq 2$, a Gibbs sampler was used in Vaida and Xu (2000) for general purposes to approximate these integrals. The M-step involves maximizing the two terms on the right-hand side of Equation 10.12 with respect to (β, λ) and Σ separately

$$\begin{aligned} Q_1(\beta, \lambda) &= \sum_i E\{l_i(\beta, \lambda; \mathbf{y}_i | b_i) | \mathbf{y}_i\} \\ &= \sum_{i,j} \{\delta_{ij} \log \lambda_0(X_{ij}) + \delta_{ij}(\beta' Z_{ij} + W'_{ij} E[b_i]) - \Lambda_0(X_{ij}) e^{\beta' Z_{ij} + \log E[b'_i W_{ij}]}\} \end{aligned} \tag{10.13}$$

where $E[h(b_i)] = E\{h(b_i)|\mathbf{y}_i\}$, and

$$Q_2(\Sigma) = \sum_i E\{\log p(b_i; \Sigma) | \mathbf{y}_i\} \tag{10.14}$$

with the exact expression for Q_2 depending on the parametrization of Σ. Note that Q_1 has the same form as the log likelihood in a classic Cox regression model with known offsets $\log E[b_i' W_{ij}]$. Therefore, the maximization turns out to be equivalent to using the partial likelihood with the offsets for β and Breslow's estimate with the same offsets for λ. The maximizer of Q_2 is the usual maximum likelihood estimator (MLE) for Σ, with the sufficient statistics replaced by their respective conditional expectations given the observed data; see Vaida and Xu (2000) for more details.

Following the estimation of θ, the estimated variance of $\hat{\theta}$ can be obtained using Louis' (1982) formula. The computation is similar to the E-step, and formulas are given in Vaida and Xu (2000). In addition, empirical Bayes estimate of the random effects $\hat{b}_i = E(b_i|\mathbf{y}_i, \hat{\theta})$ can be obtained as a by-product of the algorithm, since at the last step of EM a Monte Carlo sample from $p(b_i|\mathbf{y}_i, \hat{\theta})$ has already been obtained. The use of the predicted random effects in data analysis and clinical research will be illustrated in the next section.

We note that most publications on frailty models used nonparametric likelihood, although other approaches have also been suggested that include the residual maximum likelihood (REML) based on the partial likelihood (McGilchrist, 1993; Yau, 2001), hierarchical likelihood (Ha et al., 2001), and estimating equations for the mean treatment effect (Cai et al., 1999). We also note that for the more general random effects model (10.7), Ripatti and Palmgren (2000) used penalized partial likelihood, Ripatti et al. (2002) implemented the Monte Carlo EM (MCEM) algorithm with an automated stopping rule, and Abrahantes and Burzykowski (2005) implemented an EM algorithm with Laplace approximation in the E-steps. The Laplace approximation is good only when the cluster sizes n_i's are large. In Section 10.3.4, we show simulation results comparing the nonparametric maximum likelihood estimators (NPMLE) and the MCEM algorithm of Vaida and Xu (2000) with those of Ripatti and Palmgren (2000) and Abrahantes and Burzykowski (2005). The NPMLE with MCEM generally provides more accurate estimates than the other two, especially for the variance components where the other methods can lead to substantial bias in the estimates. In addition, the maximum likelihood estimate obtained through the MCEM algorithm has been stable in all our experience with both simulated and real data sets. Figure 10.4 shows the convergence of the EM sequence for the lung cancer data within 30 steps.

10.3.3 Computation: Gibbs-EM Implementation

Following the previous notation, for any $k = 1, \ldots, d$, write

$$\mathbf{b}_{i(-k)} = (b_{i1}, \ldots, b_{i,k-1}, b_{i,k+1}, \ldots, b_{id}).$$

As can be verified (see Vaida and Xu, 2000), $p(b_i|\mathbf{y}_i)$ and the univariate full-conditional densities $p(b_{ik}|\mathbf{b}_{i(-k)}, \mathbf{y}_i)$, $k = 1, \ldots, d$, are both log-concave. Therefore, the Gibbs sampler can be used, which proceeds by successively sampling from $p(b_{ik}|\mathbf{b}_{i(-k)}, \mathbf{y}_i)$ for $k = 1, \ldots, d$. Here we use the adaptive rejection sampling algorithm of Gilks et al. (1995).

We can generally start the EM algorithm with $M = 100$ Gibbs samples per EM step, and increase this number to $M = 1000$ after an "MCEM burn-in" period. The "burn-in" period is typically 30–50 iterations, except for the cases when the *MLE* for (some of) the variance components converge to zero, which can be much slower. It is known that when the true underlying variance component is zero, there is a positive probability that the *MLE* of it is also zero (Crainiceanu and Ruppert, 2004). At each EM step the Gibbs sampler for b_i starts with the last value of $\mathbf{b}_i$ from the previous EM step. All samples generated at the E-step are used within the subsequent M-step maximization, since the Gibbs "burn-in" is essentially realized by the previous EM steps. We stop the algorithm when the relative variation of all parameter values is less than 1%, or after a fixed number of steps (e.g., at most 1000 steps). Convergence is then assessed by visual inspection of the estimates; see examples below. For a fixed Gibbs sample size M, the EM sequence of estimates is approximately an AR(1) process with the MLE as the mean (Chan and Ledolter, 1995). Therefore, the final MLE is computed as the average of the EM sequence over the convergent portion of the chain.

We implemented the algorithm in the *C* programming language, using the *arms* program written by Gilks and Wild. The compiled program as well as the source code is available from the first author. A version of the program with *R* interface should soon be available at http://www.math.ucsd.edu/~rxu.

10.3.4 Simulation

Here we carry out some simulation using the MCEM algorithm described above. As no simulation results have so far appeared in the literature, the primary purpose is to illustrate that the algorithm performs well. In addition, we also compare the accuracy of the MCEM algorithm to the Laplace EM of Abrahantes and Burzykowski (2005), and the penalized partial likelihood estimation of Ripatti and Palmgren (2000). Simulation results for these latter two methods were reported in Abrahantes and Burzykowski (2005), and are given here in terms of bias and mean squared error (MSE) along with the MCEM results.

As in Abrahantes and Burzykowski (2005), the data are generated according to the following model for a pair of bivariate failure times for each subject j in cluster i, $i = 1, \ldots, n$, $j = 1, \ldots, n_i'$; note that in our previous notation $n_i = 2n_i'$.

$$\lambda_{ij1}(t) = \lambda_{01}(t)\exp(\beta Z_{ij} + b_{i1})$$
$$\lambda_{ij2}(t) = \lambda_{02}(t)\exp(\beta Z_{ij} + b_{i2}),$$

with

$$\begin{pmatrix} b_{i1} \\ b_{i2} \end{pmatrix} \sim N\left\{\begin{pmatrix} 0 \\ 0 \end{pmatrix}, \begin{pmatrix} \sigma_1^2 & \sigma_{12} \\ \sigma_{12} & \sigma_2^2 \end{pmatrix}\right\}.$$

The covariate Z_{ij}'s are *i.i.d.* Bernoulli(0.5) random variables. The censoring time process is simulated by a pair of independent uniform($0,\tau$) random variable, with τ selected so that approximately 20% of the observations are censored. For each of the parameter settings described in Abrahantes and Burzykowski (2005), we generate 250 datasets and compare the simulation results to those in their paper. For all settings, $\beta = 1$. The simulation results are given in Tables 10.1 and 10.2. In Table 10.1 $\sigma_1^2 = \sigma_2^2 = 0.2$, while in Table 10.2 $\sigma_1^2 = \sigma_2^2 = 1$. In both tables the correlation coefficient $\rho = 0.5$ and 0.9.

From the tables we see that the MCEM algorithm has the smallest MSE for majority of the cases; between the other two methods the Laplace EM tends to have smaller MSE when $\sigma_1^2 = \sigma_2^2 = 0.2$, whereas the opposite holds when $\sigma_1^2 = \sigma_2^2 = 1$. The main difficulty of the Laplace EM and the penalized partial likelihood methods lies in the estimation of the variance components (Abrahantes and Burzykowski, 2005) when the number of clusters is small ($n = 10$) and the variance components are small ($\sigma_1^2 = \sigma_2^2 = 0.2$), where the relative bias can be over 60%. The MCEM algorithm performs much better in comparison, especially for $n_i = 20$ where the biases for the other two methods are severe. The same also reflects in the estimation of the correlation coefficient ρ, where the MCEM appears much more accurate for the small sample sizes. With increasing number of clusters and number of subjects per cluster, both the bias and MSE decrease for all three methods. Note that n_i is at least 20 in these simulations, mainly owing to the fact that the Laplace approximation only works well for reasonably large n_i.

10.4 Examples

The proportional hazards mixed model, which includes the earlier frailty models as special cases, has been applied in various areas of biomedical research. Among them, Guo and Rodriguez (1992) and Sastry (1997) applied frailty models to child survival in South America, with clustering by family and by community. Li and Thompson (1997) and Siegmund et al. (1999) applied the frailty model to identify a major (Mendelian) gene from the rest shared familial risk for age at disease onset data. Moger et al. (2004) applied an individual frailty model, that is, with cluster sizes equal to one, to Danish and Norwegian cancer registry data for modelling the heterogeneity in the development of testicular cancer. Finally, Ripatti et al. (2003) applied the PHMM to twin data in a setting similar to the example given below, but with no differentiation between monozygotic and dizygotic twins.

In the following we provide two detailed examples of the application of PHMM.

TABLE 10.1

The Estimated Bias for 250 Simulated Datasets for the MCEM Algorithm (First Row for Each n), the Method of Abrahantes and Burzykowski (Second Row), and the Method of Ripatti and Palmgren (Third Row), When $\sigma_1^2 = \sigma_2^2 = 0.2$ and the Indicated Values for σ_{12}, with 20% Censoring. The Empirical Mean Squared Errors are in Parentheses

n	β	σ_1^2	σ_2^2	σ_{12}	ρ
			$\sigma_{12} = 0.1$		
			$n_i = 20$		
10	−0.015 (0.028)	−0.024 (0.019)	−0.029 (0.022)	−0.011 (0.011)	0.012
	0.069 (0.097)	0.069 (0.020)	0.063 (0.019)	0.022 (0.014)	−0.042
	0.007 (0.077)	0.058 (0.030)	0.035 (0.026)	0.034 (0.029)	0.046
50	0.006 (0.005)	−0.007 (0.005)	−0.009 (0.005)	−0.004 (0.003)	−0.001
	0.069 (0.019)	0.010 (0.003)	0.011 (0.002)	0.001 (0.003)	−0.019
	−0.010 (0.011)	0.000 (0.005)	−0.008 (0.004)	−0.003 (0.003)	−0.003
100	0.004 (0.003)	0.003 (0.003)	0.001 (0.002)	−0.001 (0.001)	−0.010
	0.074 (0.013)	0.006 (0.002)	0.008 (0.001)	0.000 (0.002)	−0.014
	−0.008 (0.005)	−0.004 (0.003)	−0.006 (0.002)	−0.003 (0.001)	−0.002
			$n_i = 100$		
10	−0.007 (0.006)	−0.015 (0.011)	−0.008 (0.010)	−0.011 (0.005)	−0.027
	0.015 (0.012)	−0.020 (0.010)	−0.018 (0.009)	−0.013 (0.006)	−0.018
	0.004 (0.013)	−0.019 (0.011)	−0.020 (0.010)	−0.006 (0.007)	0.018
50	0.000 (0.001)	−0.006 (0.002)	0.000 (0.002)	−0.003 (0.001)	−0.007
	0.010 (0.002)	−0.005 (0.002)	−0.005 (0.002)	−0.003 (0.001)	−0.005
	0.002 (0.002)	−0.003 (0.002)	−0.005 (0.002)	−0.001 (0.001)	0.004
100	0.001 (0.000)	−0.005 (0.001)	−0.002 (0.001)	−0.002 (0.001)	−0.002
	0.012 (0.001)	−0.001 (0.001)	−0.002 (0.001)	−0.001 (0.001)	−0.001
	−0.001 (0.001)	−0.001 (0.001)	0.000 (0.001)	0.000 (0.000)	0.001
			$\sigma_{12} = 0.18$		
			$n_i = 20$		
10	0.010 (0.035)	0.007 (0.027)	−0.033 (0.015)	−0.013 (0.014)	0.001
	0.080 (0.076)	0.122 (0.044)	0.101 (0.035)	0.083 (0.016)	−0.055
	0.020 (0.073)	0.118 (0.046)	0.090 (0.038)	0.083 (0.019)	−0.033
50	−0.004 (0.006)	−0.004 (0.005)	−0.002 (0.004)	−0.003 (0.003)	−0.002
	0.071 (0.019)	0.062 (0.007)	0.055 (0.006)	0.046 (0.004)	−0.023
	0.000 (0.011)	0.056 (0.016)	0.041 (0.012)	0.041 (0.009)	−0.009
100	0.000 (0.003)	−0.002 (0.002)	−0.003 (0.002)	−0.003 (0.001)	−0.004
	0.076 (0.013)	0.029 (0.002)	0.022 (0.002)	0.019 (0.001)	−0.015
	−0.003 (0.006)	0.024 (0.007)	0.018 (0.006)	0.018 (0.003)	−0.005
			$n_i = 100$		
10	0.001 (0.008)	−0.015 (0.011)	−0.006 (0.011)	−0.014 (0.009)	−0.024
	0.015 (0.012)	−0.009 (0.011)	−0.009 (0.010)	−0.010 (0.009)	−0.011
	0.007 (0.013)	−0.006 (0.008)	−0.005 (0.007)	−0.004 (0.007)	0.003
50	−0.003 (0.001)	−0.003 (0.002)	−0.005 (0.002)	−0.004 (0.002)	−0.003
	0.010 (0.002)	−0.004 (0.002)	−0.005 (0.002)	−0.005 (0.002)	−0.004
	0.002 (0.002)	0.000 (0.004)	−0.003 (0.003)	−0.001 (0.003)	0.002
100	0.002 (0.001)	−0.003 (0.001)	−0.003 (0.001)	−0.003 (0.001)	0.000
	0.012 (0.001)	0.000 (0.001)	−0.001 (0.001)	−0.001 (0.001)	−0.001
	−0.001 (0.001)	−0.001 (0.001)	−0.001 (0.001)	0.000 (0.001)	0.003

TABLE 10.2

The Estimated Bias for 250 Simulated Datasets for the MCEM Algorithm (First Row for Each n), the Method of Abrahantes and Burzykowski (Second Row), and the Method of Ripatti and Palmgren (Third Row), When $\sigma_1^2 = \sigma_2^2 = 1$ and the Indicated Values for σ_{12}, with 20% Censoring. The Empirical Mean Squared Errors are in Parentheses

n	β	σ_1^2	σ_2^2	σ_{12}	ρ
			$\sigma_{12} = 0.5$		
			$n_i = 20$		
10	−0.028 (0.035)	−0.099 (0.276)	−0.104 (0.276)	−0.084 (0.140)	−0.037
	0.078 (0.082)	−0.100 (0.482)	−0.105 (0.392)	−0.086 (0.157)	−0.039
	−0.006 (0.081)	−0.075 (0.252)	−0.114 (0.242)	−0.056 (0.136)	−0.009
50	0.002 (0.007)	−0.032 (0.051)	−0.032 (0.061)	−0.021 (0.029)	−0.005
	0.068 (0.020)	−0.026 (0.190)	−0.026 (0.082)	−0.028 (0.038)	−0.015
	−0.005 (0.013)	−0.018 (0.033)	−0.029 (0.035)	−0.010 (0.013)	0.002
100	0.000 (0.004)	0.011 (0.029)	0.000 (0.035)	0.004 (0.019)	0.001
	0.075 (0.013)	−0.018 (0.080)	−0.016 (0.043)	−0.019 (0.022)	−0.011
	−0.005 (0.006)	−0.017 (0.033)	−0.021 (0.027)	−0.007 (0.018)	0.003
			$n_i = 100$		
10	0.005 (0.006)	−0.036 (0.210)	−0.076 (0.237)	−0.042 (0.134)	−0.014
	0.011 (0.012)	−0.107 (0.249)	−0.099 (0.216)	−0.061 (0.128)	−0.011
	0.007 (0.013)	−0.101 (0.241)	−0.089 (0.204)	−0.034 (0.119)	−0.004
50	0.000 (0.001)	−0.005 (0.052)	−0.008 (0.044)	−0.007 (0.027)	−0.004
	0.009 (0.002)	−0.024 (0.081)	−0.021 (0.041)	−0.014 (0.029)	−0.002
	0.003 (0.002)	−0.021 (0.051)	−0.024 (0.039)	−0.009 (0.027)	−0.001
100	0.002 (0.000)	0.020 (0.027)	0.001 (0.022)	0.003 (0.013)	−0.002
	0.011 (0.001)	−0.009 (0.026)	0.003 (0.021)	−0.003 (0.015)	−0.002
	−0.001 (0.001)	−0.010 (0.022)	0.004 (0.019)	0.004 (0.014)	0.004
			$\sigma_{12} = 0.9$		
			$n_i = 20$		
10	0.005 (0.036)	−0.084 (0.343)	−0.072 (0.343)	−0.085 (0.255)	−0.016
	0.081 (0.084)	−0.119 (0.269)	−0.124 (0.264)	−0.120 (0.168)	−0.013
	−0.005 (0.076)	−0.114 (0.182)	−0.140 (0.180)	−0.121 (0.075)	−0.007
50	−0.002 (0.006)	−0.042 (0.052)	−0.036 (0.053)	−0.039 (0.041)	−0.004
	0.069 (0.020)	−0.023 (0.073)	−0.031 (0.060)	−0.031 (0.044)	−0.007
	−0.008 (0.012)	−0.021 (0.051)	−0.029 (0.052)	−0.022 (0.016)	0.000
100	0.000 (0.004)	−0.005 (0.027)	−0.015 (0.033)	−0.012 (0.022)	−0.003
	0.076 (0.013)	−0.020 (0.035)	−0.029 (0.027)	−0.027 (0.021)	−0.005
	−0.007 (0.006)	−0.015 (0.031)	−0.024 (0.029)	−0.016 (0.025)	0.001
			$n_i = 100$		
10	0.008 (0.007)	−0.097 (0.181)	−0.111 (0.179)	−0.098 (0.157)	−0.005
	0.013 (0.012)	−0.107 (0.266)	−0.098 (0.236)	−0.102 (0.194)	−0.011
	0.005 (0.012)	−0.099 (0.237)	−0.096 (0.210)	−0.085 (0.197)	0.003
50	−0.004 (0.001)	−0.012 (0.042)	−0.009 (0.038)	−0.010 (0.035)	−0.001
	0.010 (0.002)	−0.020 (0.053)	−0.026 (0.047)	−0.024 (0.043)	−0.004
	−0.001 (0.002)	−0.020 (0.054)	−0.024 (0.046)	−0.025 (0.050)	−0.005
100	0.000 (0.001)	0.001 (0.018)	−0.008 (0.018)	−0.003 (0.016)	0.000
	0.012 (0.001)	−0.007 (0.030)	−0.003 (0.023)	−0.006 (0.023)	−0.002
	−0.002 (0.001)	−0.006 (0.025)	−0.001 (0.021)	−0.002 (0.021)	0.001

TABLE 10.3

Estimates from the Lung Cancer Data

		$d = 0$	$d = 1$	$d = 2$
Treatment	β	−0.25 (0.09)	−0.25 (0.10)	−0.25 (0.12)
Bone		0.22 (0.09)	0.21 (0.10)	0.23 (0.14)
Liver		0.43 (0.09)	0.42 (0.09)	0.39 (0.09)
ps		−0.60 (0.10)	−0.64 (0.11)	−0.65 (0.13)
Weight loss		0.20 (0.09)	0.22 (0.09)	0.21 (0.09)
Treatment	σ	—	0.27 (0.13)	0.21 (0.43)
Bone		—	—	0.36 (0.12)

10.4.1 Multicenter Clinical Trial

As the first example we consider the multicenter non-small cell lung cancer trial mentioned earlier, which was also used as an example in Vaida and Xu (2000). The trial enrolled 579 patients from 31 institutions. The primary end point was patient death. There were two randomized treatment arms, and the other important covariates that affected patient survival were: presence or absence of bone metastases, presence or absence of liver metastases, performance status at study entry, and whether there was weight loss prior to entry. Gray (1995) used a score test for the existence of random treatment effect, and found it to be significant.

Vaida and Xu considered several different models for the data. The first column ($d = 0$) of Table 10.3 is the classic Cox model fit with the five covariates but no random effects. The second column ($d = 1$) is PHMM (Equation 10.7) with fixed effects for the five covariates and a random treatment effect, that is, random treatment by institution interaction. The estimated standard deviation $\hat{\sigma} = 0.27$ of the random treatment effect is comparable in magnitude to the fixed treatment effect of −0.25 ($Z_1 = 1$ for the CAV-HEM arm, 0 for the CAV arm). It is therefore not surprsing in Figure 10.3 that some institutions have estimated treatment effects close to zero, whereas others have approximately double the fixed treatment effect. Note that Figure 10.3 is the plot of $\hat{\beta}_1 + \hat{b}_i$, where $\hat{\beta}_1$ is the estimated fixed treatment effect, and $\hat{b}_i$ is the empirical Bayes estimate of the random treatment effect for institution i. Vaida and Xu (2000) also provided the emprical Bayes credibility intervals for the estimated treatment effect of each institution. Figure 10.4, taken from Vaida and Xu (2000), plots the EM sequences for each of the six finite dimensional parameters. Plots like these are a good way for monitoring EM convergence, especially in the absence of an automated stopping rule. Automated stopping rules, on the other hand, also need to be validated through comparison to the empirical convergent sequences.

We may also attempt to fit model (10.1) to the data, that is, a random effect on the baseline hazard. It was shown in Gray (1995) that there was not a significant variation of patient survival in the CAV arm across all institutions; note the coding $Z_1 = 0$ for CAV arm implies that this corresponds to nonsignificant

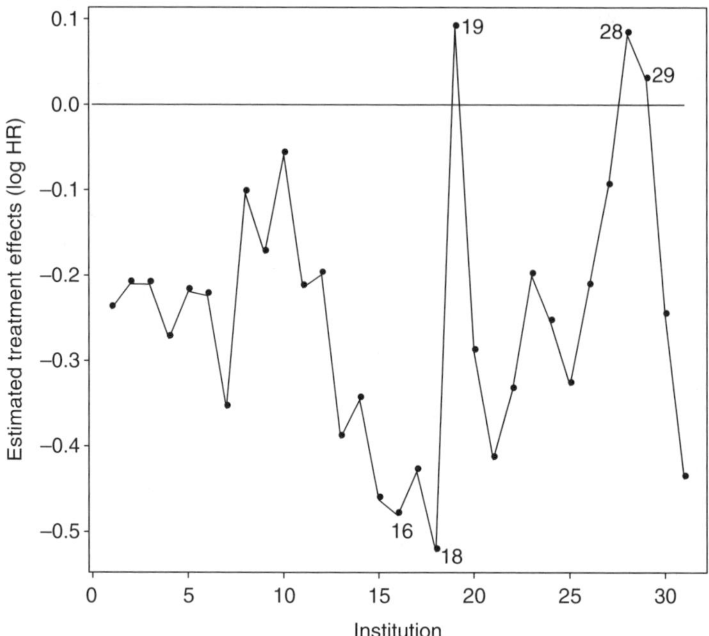

FIGURE 10.3
Estimated treatment effects by institution for the lung cancer data.

variance component under model (10.1). When we tried to fit model (10.1) to the data, we found that the variance parameter of b_i converged to zero during the EM algorithm. The result of the above two model fits means that there was significant institutional variation in patient survival in the CAV-HEM arm, but not in the CAV arm. This was indeed the case due to the complexity of the medical procedure in the CAV-HEM arm, as compared to the standard treatment CAV arm. Further investigation into the procedures carried out at institutions 16 and 18, especially in comparison to institutions 19, 28, and 29, may shed light on the ways to achieve maximum benefit of the regimen should further research be desired.

Since institutional differences might also be caused by differences in patient population, Vaida and Xu (2000) tried to fit all five covariates with random effects, and with a diagonal Σ matrix for the variances. As a result the variance parameters of the random effects for liver metastases, performance status, and weight loss converged to zero during the EM algorithm. As mentioned before if the true value of a variance parameter is zero, there is a positive probability that its MLE is zero. This has been established in theory under the linear mixed models both asymptotically and in finite samples, and is closely related to the fact that the likelihood ratio statistic for testing zero variance components has a mixed chi-squared distribution under the null hypothesis (Crainiceanu and Ruppert, 2004).

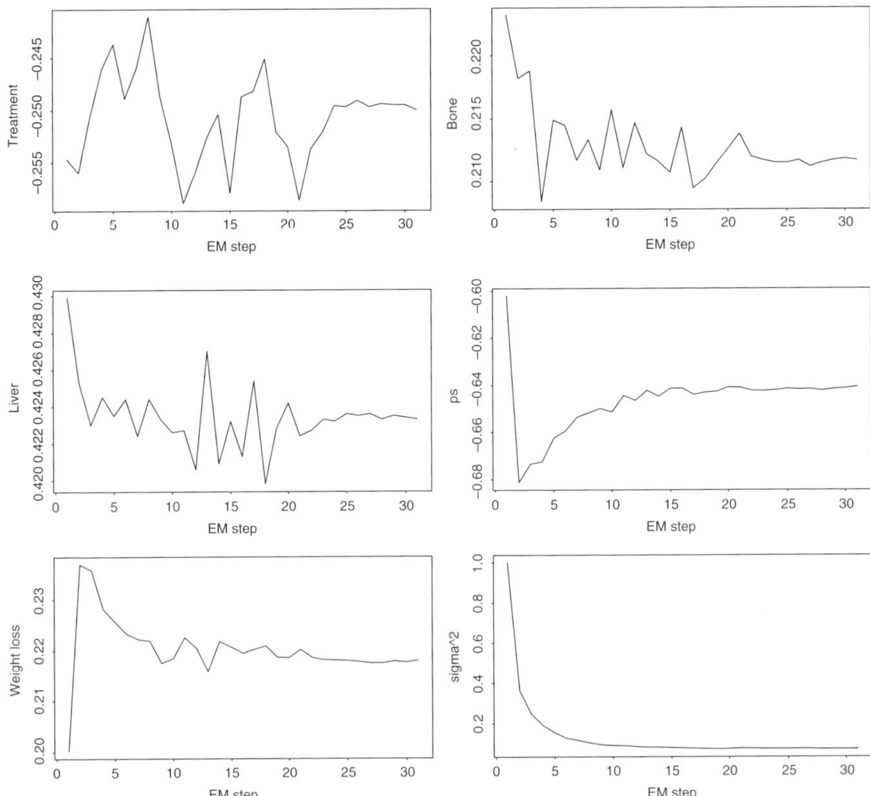

FIGURE 10.4
EM sequence of parameter estimates for the lung cancer data.

Finally, a model with the remaining two random effects terms, one for treatment and one for bone metastases, was fit. The results are in the last column of Table 10.3 ($d = 2$). The bone metastases turned out to have a fairly strong random effect, with estimated standard deviation of 0.36. Note that this is also a binary variable coded with 0 and 1. Although standard errors are given for all the parameter estimates, those for the variance components should not be used directly to test against zero by assuming normality, since the null hypothesis lies on the boundary of the parameter space. See the discussion section for testing zero variance components. Note also that for this data set the estimates of the fixed regression effects did not change substantially after inclusion of the random effects. The emphasis of the random effects model here is the investigation of institutional variations in the covariate effects.

10.4.2 Twin Data

As mentioned earlier, one of the applications of PHMM is genetic data. In particular, much work has been done in genetic epidemiology to make use of twin data, as they provide a way to examine the relative contributions of genetic and environmental factors to disease onset. The assumption here is that monozygotic (MZ) twins have the exact same genes, and any difference in disease within twin pairs should come from environmental factors. In contrast, dizygotic (DZ) twins only share half of their genes, and differences in disease could be the result of environment and genetics. More specifically, the total variance of twin resemblance on traits is partitioned into three parts: additive genetic (G), common environment (C), and unique environment (E). The additive genetic factor represents twin similarity arising from addictive effects of alleles; the common environmental factor includes all nongenetic factors shared within a twin pair that leads to twin similarity; unique environmental factor includes all nongenetic factors unshared within a twin pair and leads to twin dissimilarity.

When the end point of a twin study is time to event with possible right-censoring, such as age at onset of a disease, the PHMM can be used. The more classic linear mixed model handles continuous outcomes with no censoring, and the generalized linear mixed model handles dichotomized or categorical outcomes. Another method that is sometimes used in genetic epidemiology is structural equation modelling, which also requires dichotomizing age at onset on divided time intervals, and is unable to handle some of the bivariate censored cases in twin data (Liu et al., 2004a). To model the dependence structure of the MZ and DZ twins using PHMM, we need a vector of six random effects: $b = (b_1, b_2, \ldots, b_6)'$. For an MZ twin pair i, let b_{1i} denote the contribution from the common genetic G and the common environmental C factors, and b_{2i} and b_{3i} denote the unique environmental E factors for twin 1 and twin 2, respectively, so that

$$\lambda_{i1}(t) = \lambda_0(t)\exp(b_{1i} + b_{2i}), \tag{10.15}$$

$$\lambda_{i2}(t) = \lambda_0(t)\exp(b_{1i} + b_{3i}), \tag{10.16}$$

when no covariates are included. We can write

$$\mathrm{Var}(b_{1i}) = \sigma_G^2 + \sigma_C^2, \quad \mathrm{Var}(b_{2i}) = \mathrm{Var}(b_{3i}) = \sigma_E^2. \tag{10.17}$$

For a DZ twin pair i, let b_{4i} denote the common genetic (1/2) and the common environmental factors, and b_{5i} and b_{6i} denote the unique genetic (1/2) and the unique environmental factors for twin 1 and twin 2, respectively, so that

$$\lambda_{i1}(t) = \lambda_0(t)\exp(b_{4i} + b_{5i}), \tag{10.18}$$

$$\lambda_{i2}(t) = \lambda_0(t)\exp(b_{4i} + b_{6i}). \tag{10.19}$$

We can then write

$$\mathrm{Var}(b_{4i}) = \frac{\sigma_G^2}{2} + \sigma_C^2, \quad \mathrm{Var}(b_{5i}) = \mathrm{Var}(b_{6i}) = \frac{\sigma_G^2}{2} + \sigma_E^2. \tag{10.20}$$

Owing to the construction of these random effects, that is, to decompose the total variance, it is necessary to assume that they are independent of each other. Therefore, the variance matrix for b is

$$\Sigma = \begin{pmatrix} \sigma_G^2 + \sigma_C^2 & & & & & \\ & \sigma_E^2 & & & & \\ & & \sigma_E^2 & & & \\ & & & \frac{\sigma_G^2}{2} + \sigma_C^2 & & \\ & & & & \frac{\sigma_G^2}{2} + \sigma_E^2 & \\ & & & & & \frac{\sigma_G^2}{2} + \sigma_E^2 \end{pmatrix}. \tag{10.21}$$

The above twin model was applied to the VET Registry data mentioned earlier. The telephone-administrated psychiatric interviews were based on the Diagnostic and Statistical Manual of Mental Disorder Version III Revised (DSM-III-R, Robins et al., 1989). After giving verbal informed consent, the twins were asked questions about several aspects of alcohol use, including frequency, amount, age at onset of various symptoms, duration of each symptom, and mood or perception changes associated with alcohol drinking. Of the 10,253 eligible individuals, 8,169 (80%) were successfully interviewed. Zygosity was determined by responses to questions relating to the similarity of physical appearance, along with blood group typing method. The 3372 complete twin pairs of known zygosity (1874 MZ pairs and 1498 DZ pairs) were included in the analysis described below.

By applying standard DIS-III-R computing algorithms, individual symptoms and the diagnoses of alcohol abuse were derived from the responses to the questions. Using the age at onset of alcohol abuse as outcome, we fit the above twin model to the VET Registry data, first with no covariates. The estimates are given in Table 10.4, where in the parentheses are the standard errors. In order to interpret the results as genetic and environmental contributions to the actual age at onset of alcohol abuse, as opposed to the hazard of it, we make use of the transformation model Equation 10.9. Here the total variance is $\sigma_G^2 + \sigma_C^2 + (\sigma_E^2 + 1.645)$, and the estimated genetic contribution σ_G^2 as a percentage of it is 36%, whereas the common environmental contribution is estimated to be 9%. Note that without taking into account the extra individual (residual) variance of 1.645, we would have had a genetic contribution

TABLE 10.4

Estimates from the VET Registry Data

Contribution	Estimates (SE)	Proportion of Total* (SE)
σ_G^2	1.08 (0.15)	36% (5%)
σ_C^2	0.27 (0.05)	9% (2%)
σ_E^2	0.02 (0.01)	55% (4%)

* Total is $\sigma_G^2 + \sigma_C^2 + \sigma_E^2 + 1.645$.

of 79%. This 79% is in fact the genetic contribution to the variation in the log hazard of age at onset, but not to the variation of the random variable age at onset itself. The 36% genetic contribution turned out to be very comparable to what had been previously reported in the literature on genetic contributions to the onset of alcohol abuse.

The twin model described above can be easily extended to incorporate covariates as in Equation 10.7. It is also possible to incorporate covariate by gene interactions. Liu et al. (2004a) considered the possibly different genetic and common environment contributions to the transition period from regular alcohol use to alcohol dependence, among early and late regular alcohol users. The early users were defined as those who started regular alcohol use before the age of 40, while the late users after age 40. For such a dichotomized covariate, a vector of 12 random effects was used to model the covariate interaction with the genetic and environmental factors. The variance matrix is similar to Equation 10.21, albeit with two sets of the variance parameters, one for the early users and one for the late users. The authors found substantially larger genetic contributions among the early users (45%) than among the late users (16%), while the common environmental contribution was much larger among the late users. Such findings may have potential prevention or intervention implications, since for late users appropriate strategies might be considered to intervene on environmental determinants so as to reduce the risk of developing alcohol dependence.

Finally, the above twin model can be extended to handle multivariate outcomes in twin data, such as the ages at onset of multiple categories of alcohol dependence symtoms (see Robins et al., 1989, for definition), or joint modelling of ages at onset of alcohol and tobacco abuses. In the case of bivariate outcomes, such as alcohol and tobacco abuses, again a vector of 12 random effects is used. But unlike the case with a binary covariate above, the variance matrix is no longer diagonal, since we cannot assume independence between the genetic contributions to alcohol and to tobacco abuses. More specifically, denote $(b_{1A}, b_{2A}, \ldots, b_{6A})$ the random effects corresponding to alcohol abuse, and $(b_{1T}, b_{2T}, \ldots, b_{6T})$ the random effects corresponding to tobacco abuse. Then the variance matrix Σ for the vector of random effects

$b = (b_{1A}, b_{1T}, b_{2A}, b_{2T}, \ldots, b_{6A}, b_{6T})'$ is block-diagonal:

$$\mathrm{diag}\left[\begin{pmatrix} \sigma^2_{GA} + \sigma^2_{CA} & \tau_1 \\ \tau_1 & \sigma^2_{GT} + \sigma^2_{CT} \end{pmatrix}, \begin{pmatrix} \sigma^2_{EA} & \tau_2 \\ \tau_2 & \sigma^2_{ET} \end{pmatrix}, \begin{pmatrix} \sigma^2_{EA} & \tau_2 \\ \tau_2 & \sigma^2_{ET} \end{pmatrix},\right.$$

$$\begin{pmatrix} \frac{\sigma^2_{GA}}{2} + \sigma^2_{CA} & \tau_3 \\ \tau_3 & \frac{\sigma^2_{GT}}{2} + \sigma^2_{CT} \end{pmatrix}, \begin{pmatrix} \frac{\sigma^2_{GA}}{2} + \sigma^2_{EA} & \tau_4 \\ \tau_4 & \frac{\sigma^2_{GT}}{2} + \sigma^2_{ET} \end{pmatrix},$$

$$\left.\begin{pmatrix} \frac{\sigma^2_{GA}}{2} + \sigma^2_{EA} & \tau_4 \\ \tau_4 & \frac{\sigma^2_{GT}}{2} + \sigma^2_{ET} \end{pmatrix}\right],$$

where $\tau_1, \ldots, \tau_4$ are the covariances of the pairwise random effects, and σ^2_{GA}, σ^2_{GT}, and so forth are the variance decompositions corresponding to alcohol and tobacco abuse, respectively.

10.5 Discussion

In this chapter we reviewed the extension of the most commonly used proportional hazards regression model in failure time data analysis to applications involving correlated failure times. Alternatives to the proportional hazards model have also been extended in this context. Most of these works were under the linear mixed model, or equivalently, the accelerated failure time (AFT) model with random effects. The linear mixed models are attractive especially for interpretation purposes, since the variation in the outcome variable is directly decomposed into different attributes. In addition, the linear marginal and conditional models are nested, whereas this is not the case for the proportional hazards regression (see also Hougaard et al., 1994; Keiding et al., 1997). Hughes (1999) extended the linear mixed models with normal errors by treating the right-censored data as missing in the EM algorithm. Pan and Louis (2000) used a slightly different algorithm that required the normality working assumption of the error terms only in the sampling steps. Ha et al. (2002) applied hierarchical likelihood to models with only random intercept, also under the normal error assumption.

Relatively few works have been done on model diagnostics under either model (10.1) or model (10.7). Xu and Gamst (2007) studied the effect of non-proportional hazards on the parameter estimates and proposed methods to check the proportional hazards assumption under model (10.7). Glidden (1999), Viswanathan and Manatunga (2001) and Economou and Caroni (2005) proposed methods for checking the frailty distribution under Equation 10.1. O'Quigley and Stare (2002) and Duchateau et al. (2002), on the other hand, showed that the estimate of the fixed effects and the variance components can be robust to the misspecification of the frailty distribution. O'Quigley and Stare (2002) also compared the efficiency of estimation under model (10.1) relative to the stratified proportional hazards model.

Another aspect under PHMM is hypothesis testing, especially testing whether the variances of the random effects are significantly different from zero; that is, whether the random effects should be included in the model. As mentioned before, the null hypothesis in this case lies on the boundary of the parameter space. Gray (1995) and Commenges and Andersen (1995) proposed a score test for no random effects in the model. Andersen et al. (1999) showed that the score test is more efficient than modelling the cluster effects as fixed effects for a categorical covariate that indicates the clusters. The score test, however, cannot test for a subset of the random effects where the null model also includes some random effects. Xu et al. (2006) as well as forthcoming work by one of the authors and colleague (presented at 2005 *WNAR* in Fairbanks, Alaska) developed methods for model selection and for checking the proportional hazards assumption under the PHMM.

Future work is needed to create a professional software, so that the computational methods described above can be more accessible to applied users. In creating such a software, although automated stopping rules should be incorporated, it is perhaps also important to have visual monitoring of the EM sequence. In our experience the convergence pattern can be different for different data structure and different models fitted to the same data. A noteworthy case is the convergence of some of the parameter values to zero in the EM sequence, which can be slow. A possible improvement is through the parameter expansion of Liu et al. (1998). Another issue is the computation of the variance of the estimators. The observed information matrix includes that for the discrete baseline hazard function, and is therefore of the dimension greater than the number of failures in the dataset. Inversion of such a matrix can occasionally be a problem for large registry data. An alternative method is to use the profile likelihood with the baseline hazard as a nuisance parameter, when of interest are the fixed and random effects. Computation of the profile likelihood is also discussed in Xu et al. (2006).

References

Abrahantes, J. C. and Burzykowski, T. (2005). A version of the EM algorithm for proportional hazard model with random effects. *Biometrical Journal* **6**, 847–862.

Andersen, P. K., Klein, J. P., and Palacios, R. (1997). Estimation for variance in Cox regression model with shared gamma frailties. *Biometrics* **53**, 1475–1484.

Andersen, P. K., Klein, J. P., and Zhang, M. (1999). Testing for centre effects in multicenter survival studies: A Monte Carlo comparison of fixed and random effects tests. *Statistics in Medicine* **18**, 1489–1500.

Anderson, G. L. and Fleming, T. R. (1995). Model misspecification in proportional hazards regression. *Biometrika* **82**, 527–541.

Bretagnolle, J. and Huber-Carol, C. (1988). Effects of omitting covariates in Cox's model for survival data. *Scandinavian Journal of Statistics* **15**, 125–138.

Cai, J., Sen, P. K., and Zhou, H. (1999). A random effects model for multivariate failure time data from multicenter clinical trials. *Biometrics* **55**, 182–189.

Carlin, B. P. and Hodges, J. S. (1999). Hierarchical proportional hazards regression models for highly stratified data. *Biometrics* **55**, 4.

Chan, K.-S. and Ledolter, J. (1995). Monte Carlo EM estimation of time series models involving counts. *Journal of the American Statistical Association* **90**, 429, 242–252.

Clayton, D. and Cuzick, J. (1985). Multivariate generalizations of the proportional hazards model (with discussion). *Journal of the Royal Statistical Society, Series A* **148**, 82–117.

Commenges, D. and Andersen, P. K. (1995). Score test of homogeneity for survival data. *Lifetime Data Analysis* **1**, 1145–1156.

Cox, D. (1972). Regression models and life-tables (with discussion). *Journal of the Royal Statistical Society, Series B* **34**, 187–220.

Crainiceanu, C. and Ruppert, D. (2004). Likelihood ratio tests in linear mixed models with one variance component. *Journal of the Royal Statistical Society, Series B* **66**, 165–185.

Dempster, A. P., Laird, N. M., and Rubin, D. B. (1977). Maximum likelihood from incomplete data via the EM algorithm (with discussion). *Journal of the Royal Statistical Society, Series B* **39**, 1, 1–38.

Duchateau, L., Janssen, P., Lindsey, P., Legrand, C., Nguti, R., and Sylvester, R. (2002). The shared frailty model and the power for heterogeneity tests in multicenter trials. *Computational Statisitics and Data Analysis* **40**, 603–620.

Economou, P. and Caroni, C. (2005). Graphical tests for the assumption of gamma and inverse gaussian frailty. *Lifetime Data Analysis* **11**, 565–582.

Ford, I., Norrie, J., and Ahmadi, S. (1995). Model inconsistency, illustrated by the Cox proportional hazards model. *Statistics in Medicine* **14**, 735–746.

Gail, M. H., Wieand, S., and Piantadosi, S. (1984). Biased estimation of treatment effect in randomized experiment with nonlinear regressions and omitted covariates. *Biometrika* **71**, 431–444.

Gilks, W. R., Best, N. G., and Tan, K. K. C. (1995) Adaptive Metropolis sampling within Gibbs sampling. *Applied Statistics* **44**, 455–472.

Glidden, D. (1999). Checking the adequacy of the gamma frailty model for multivariate failure time. *Biometrika* **86**, 381–393.

Glidden, D. and Vittinghoff, E. (2004). Modelling clustered survival data from multicenter clinical trials. *Statistics in Medicine* **23**, 369–388.

Gray, R. (1994). A Bayesian analysis of institutional effects in a multicenter cancer clinical trial. *Biometrics* **50**, 244–253.

Gray, R. (1995). Tests for variation over groups in survival data. *Journal of the American Statistical Association* **90**, 198–203.

Guo, G. and Rodriguez, G. (1992). Estimating a multivariate proportional hazards model for clustered data using the EM algorithm, with an application to child survival in guatemala. *Journal of the American Statistical Association* **87**, 969–976.

Gustafson, P. (1997). Large hierarchical bayesian analysis of multivariate survival data. *Biometrics* **53**, 230–242.

Ha, I., Lee, Y., and Song, J. (2001). Hierarchical likelihood approach for frailty models. *Biometrika* **88**, 233–243.

Ha, I., Lee, Y., and Song, J. (2002). Hierarchical-likelihood approach for mixed linear models with censored data. *Lifetime Data Analysis* **8**, 163–176.

Hougaard, P. (2000). *Analysis of Multivariate Survival Data*. Springer, New York.

Hougaard, P., Myglegaard, P., and Borch-Johnsen, K. (1994). Heterogeneity models of disease susceptibility, with application to diabetic nephropathy. *Biometrics* **50**, 1178–1188.

Hughes, J. P. (1999). Mixed effects models with censored data with applications to HIV RNA levels. *Biometrics* **55**, 625–629.

Johansen, S. (1993). An extension of cox's regression model. *International Statisitical Review*, **51**, 258–262.

Keiding, N., Andersen, P. K., and Klein, J. P. (1997). The role of frailty models and accelerated failure time models in describing heterogeneity due to omitted covariates. *Statistics in Medicine* **16**, 215–224.

Klein, J. (1992). Semiparametric estimation of random effects using the Cox model based on the EM algorithm. *Biometrics* **48**, 795–806.

Kosorok, M. R., Lee, B. L., and Fine, J. P. (2001). Semiparametric inference for proportional hazards frailty regresssion models. Technical report, Department of Biostatistics, University of Wisconsin.

Lancaster, T. and Nickell, S. (1980). The analysis of re-employment probabilities for the unemployed. *Journal of the Royal Statistical Society, Series A* **143**, 141–165.

Li, H. and Thompson, E. (1997). Semiparametric estimation of major gene and family-specific random effects for age of onset. *Biometrics* **53**, 282–293.

Li, Y. and Lin, X. (2000). Covariate measurement errors in frailty models for clustered survival data. *Biometrika* **87**, 849–866.

Liu, C., Rubin, D. B., and Wu, Y. N. (1998). Parameter expansion to accelerate EM: The PX-EM algorithm. *Biometrika* **85**, 755–770.

Liu, I., Blacker, D., Xu, R., Fitzmaurice, G., Lyons, M., and Tsuang, M. (2004a). Genetic and environmental contributions to the development of alcohol dependence in male twins. *Archives of General Psychiatry* **61**, 897–903.

Liu, I., Blacker, D., Xu, R., Fitzmaurice, G., Tsuang, M., and Lyons, M. J. (2004b). Genetic and environmental contributions to age of onset of alcohol dependence symptoms in male twins. *Addiction* **99**, 1403–1409.

Louis, T. A. (1982). Finding the observed information matrix when using the EM algorithm. *Journal of the Royal Statistical Society, Series B* **44**, 2, 190–200.

Ma, R., Krewski, D., and Burnett, R. (2003). Random effects cox model: A poisson modelling approach. *Biometrika* **90**, 157–169.

McGilchrist, C. (1993). REML estimation for survival models with fraility. *Biometrics* **49**, 221–225.

Moger, T. A., Aalen, O. O., Halvorsen, T. O., Tretti, S., and Storm, H. H. (2004). Frailty modelling of testicular cancer incidence using scandinavian data. *Biostatistics* **5**, 1–14.

Murphy, S. (1994). Consistency in a proportional hazards model incorporating a random effect. *Annals of Statistics* **22**, 712–731.

Murphy, S. (1995). Asymptotic theory for the frailty model. *The Annals of Statistics*, **23**, 182–198.

Murray, D., Varnell, S., and Blitstein, J. (2004). Design and analysis of group-randomized trials: A review of recent methodological developments. *American Journal of Public Health* **94**, 423–432.

Nielsen, G., Gill, R., Anderson, P., and Sørensen, T. (1992). A counting process approach to maximum likelihood estimation in frailty models. *Scandinavian Journal of Statistics* **19**, 25–44.

Oakes, D. (1992). Frailty models for multiple event times. In *Survival Analysis: State of the Art*, 371–379, Ed. J. P. Klein and P. K. Goel. Kluwer Academic, Dordrecht; Norwell, MA.

Omori, Y. and Johnson, R. A. (1993). The influence of random effects on the unconditional hazard rate and survival functions. *Biometrika* **80**, 910–914.

O'Quigley, J. and Stare, J. (2002). Proportional hazards models with frailties and random effects. *Statistics in Medicine* **21**, 3219–3233.

Pan, W. and Louis, T. (2000). A linear mixed-effects model for multivariate censored data. *Biometrics* **56**, 1, 160–166.

Parner, E. (1998). Asymptotic theory for the correlated gamma-frailty model. *Annals of Statistics* **26**, 183–214.

Petersen, J. H. (1998). An additive frailty model for correlated life times. *Biometrics* **54**, 646–661.

Petersen, J. H., Andersen, P. K., and Gill, R. D. (1996). Variance components models for survival data. *Statistica Neerlandica* **50**, 1, 193–211.

Ripatti, S., Gatz, M., Pedersen, N. L., and Palmgren, J. (2003). Three-state frailty model for age at onset of dementia and death in swedish twins. *Genetic Epidemiology* **24**, 139–149.

Ripatti, S., Larsen, K., and Palmgren, J. (2002). Maximum likelihood inference for multivariate frailty models using an automated Monte Carlo EM algorithm. *Lifetime Data Analysis* **8**, 349–360.

Ripatti, S. and Palmgren, J. (2000). Estimation of multivariate frailty models using penalized partial likelihood. *Biometrics* **56**, 1016–1022.

Robins, L., Helzer, J., Cottler, L., and Goldring, E. (1989). *Diagnostic Interview Schedule Version III Revised*. National Institute of Mental Health, Rockville, MD.

Sargent, D. J. (1998). A general framework for random effects survival analysis in the Cox proportional hazards setting. *Biometrics* **54**, 1486–1497.

Sastry, N. (1997). A nested frailty model for survival data, with an application to the study of child survival in northeast Brazil. *Journal of the American Statistical Association* **92**, 426–435.

Siegmund, K. D., Todorov, A. A., and Province, M. A. (1999). A frailty approach for modelling diseases with variable age of onset in families: The nhlbi family heart study. *Statistics in Medicine* **18**, 1517–1528.

Struthers, C. A. and Kalbfleisch, J. D. (1986). Misspecified proportional hazards models. *Biometrika* **73**, 363–369.

Sylvester, R., van Glabbeke, M., Collette, L., Suciu, S., Baron, B., Legrand, C., Gorlia, T., et al. (2002). Statistical methodology of phase iii cancer clinical trials: Advances and future perspectives. *European Journal of Cancer* **38**, S162–S168.

Therneau, T. and Grambsch, P. (2000). *Modeling Survival Data*. Springer, New York.

Vaida, F. and Xu, R. (2000). Proportional hazards model with random effects. *Statistics in Medicine* **19**, 3309–3324.

Vaupel, J. W., Manton, K. G., and Stallard, E. (1979). The impact of heterogeneity in individual frailty on the dynamics of mortality. *Demography* **16**, 439–454.

Viswanathan, B. and Manatunga, A. (2001). Diagnostic plots for assessing the frailty distribution in multivariate survival data. *Lifetime Data Analysis* **7**, 143–155.

Xu, R. (1996). *Inference for the Proportional Hazards Model*. Ph.D. thesis, University of California, San Diego.

Xu, R. and Gamst, A. (2007). On proportional hazards assumption under the random effects models. *Lifetime Data Analysis*, in print.

Xu, R., Gamst, A., Donohue, M., Vaida, F., and Harrington, D. (2006). Using profile likelihood for semiparametric model selection with application to proportional hazards mixed models. *Harvard University Biostatistics Working Paper Series*. http://www.bepress.com/harvardbiostat/paper43.

Xue, X. and Brookmeyer, R. (1996). Bivariate frailty model for the analysis of multivariate survival time. *Lifetime Data Analysis* **2**, 277–289.

Yau, K. K. W. (2001). Multilevel models for survival analysis with random effects. *Biometrics* **57**, 96–102.

11

Classification Rules for Repeated Measures Data from Biomedical Research

Anuradha Roy and Ravindra Khattree

CONTENTS

11.1 Introduction

Innovative statistical methods are needed to take advantage of recent advances in biological, biomedical, and medical research, especially in biotechnology and the human genome project for disease diagnosis and prognosis. Some of these methods include: linear discriminant analysis, quadratic discriminant analysis, logistic discrimination, k-nearest neighbor classifier, support vector machine, classification trees, and random forests. These methods play important role in the classificatory problems. Improving the accuracy of the classification of diseased versus nondiseased samples is one of the ongoing challenges in the world of biological research. This can be done only by developing an appropriate method that is suitable for the data at hand.

A conventional classification problem usually considers observations on more than one response variable or characteristic taken at a given time point. However, in medical research it is often found that observations on one variable or characteristic are taken repeatedly over time. These are called univariate repeated measures data. However, in many medical survival studies, patients with diseases such as AIDS or cancer are followed up at successive intervals of time, and these clinical trials often report repeated outcome measures on more than one variable. Such data are often referred to as the multivariate repeated measures data or doubly multivariate data. Also, the number of such repeated measurements rarely form a long series, hence sophisticated time series modeling may not be applicable. Nonetheless, these repeated measurements on the same unit or subject are stochastically dependant or correlated with each other, and these correlations should be taken into account for developing the classification rules. For continuous data this dependence is often modeled through some covariance structure. Also, the assumption of multivariate normality is usually made.

In this chapter, we will review the extensions of the classical discriminant analysis methods to consider the classification for continuous univariate and multivariate repeated measures data. For repeated measures, a covariance structure will often be assumed. Some of the possibilities for the covariance structure are compound symmetry (CS), autoregressive of order one (AR(1)), moving average, and circulant. A CS structure specifies that correlation is same among all repeated measures regardless of occasion of measurement. An AR(1) structure specifies that the correlations are larger for time points close to each other than for those that are far away. A circulant or circular covariance structure arises naturally in many situation (Khattree and Naik, 1994). In this structure the observations in the immediate neighbors are more strongly correlated. Fuller (1976) has shown that a circular covariance can be used as an approximation for any stationary absolutely summable covariance structure, and thus can be used in a generic way.

Classification problems using repeated measures on a single variable were first addressed by Choi (1972). Choi developed a mixed effects model for classification using repeated measurements on one variable by assuming that the repeated measurements retain the same mean over time and are equicorrelated in a given group. Gupta (1980, 1986) extended the method to the multivariate scenario, where a number of variables are measured repeatedly over time. Gupta and Logan (1990) examined the classical approach to this problem, as well as Bayesian predictive discrimination (Logan and Gupta, 1993) that uses a diffuse prior for the mean vectors and the variance–covariance matrices. Nevertheless, none of these studies addressed the problem of small sample case. The issue of small sample case was finally addressed by Roy (2002), and Roy and Khattree (2005a,b, 2006). They considered the problem in small sample situation by assuming Kronecker product structure on variance–covariance matrix. Later Roy (2006a) extended the problem of classification of multivariate repeated measures data with missing values in a mixed model set up.

This article provides a review of classification rules under the assumption of multivariate normality for populations with certain structured and unstructured mean vectors and under certain covariance structures. These rules are especially applicable and have been shown to be effective when the number of observations is not large enough to estimate the variance–covariance matrix unstructurally, and thus the traditional classification rules fail. We will consider the classification in the context of two populations. The general case of k populations follows in a straightforward manner. The classification rules mentioned in this article are under the assumption of equal prior probabilities and equal costs. Changes where they vary can be incorporated in a routine fashion as indicated in Johnson and Wichern (2002). Classification rules for univariate repeated measures data will be discussed in Section 11.2 and classification rules for multivariate repeated measures data with complete data set will be discussed in Section 11.3. Testing for various covariance structures is considered in Section 11.4 and two real life data examples are discussed in Section 11.5. Furthermore, classification rules for multivariate repeated measures data with missing values will be discussed in Section 11.6.

11.2 Classification in Case of Univariate Repeated Measures Data

Let $\boldsymbol{y}_{p\times 1}$ represent the repeated measures on the subject taken over p time points. Additional multiple subscripts describing population and subject within population will be used on $\boldsymbol{y}$ for further description. We assume that for each of the two populations j, the distribution of $\boldsymbol{y}$ is described by p-variate normal distribution with mean $\boldsymbol{\mu}_j$ and variance–covariance matrix $\boldsymbol{\Omega}_j, j = 1, 2$.

Let y_{jit} be the measurement on the ith individual in the jth population at the -th time point, $j = 1, 2; i = 1, 2, \ldots, n_j; t = 1, 2, \ldots, p$. Thus, $\boldsymbol{y}_{ji} = (y_{ji1}, y_{ji2}, \ldots, y_{jip})'$ is a $p \times 1$ random vector corresponding to the ith individual in the jth population. The matrix $\boldsymbol{\Omega}_j$ is assumed to be $p \times p$ positive definite. With p repeated measurements per patient, $p(p+1)/2$ number of unknown parameters must be estimated in an unstructured covariance matrix. Often n, the number of observations is less than p. In such cases the traditional classification methods (Kshirsagar, 1972; McLachlan, 1992; Khattree and Naik, 2000) cannot be applied. This is another practical reason that we must impose certain covariance structures on $\boldsymbol{\Omega}_j$. Even when $n > p$, one may like to have a more parsimonious model with as few parameters as possible. We will adopt the CS covariance structure in Section 11.2.1 and the AR(1) covariance structure in Section 11.2.2.

We assume some structures on mean vectors as well. Examples of constant mean vector structure can be found in various genetic studies, where it is often found that mutations do not occur in genes over time; they just stay dormant. We assume that the means of various repeated measures remain constant over time. Thus, in the population j, $\boldsymbol{\mu}_j = \mu_j \mathbf{1}_p$, $j = 1, 2$, where $\mathbf{1}_p$ is a $p \times 1$ vector containing all elements as unity.

11.2.1 Repeated Measures with Compound Symmetric Covariance Structure

Let the variance–covariance matrix $\boldsymbol{\Omega}_j$ be CS. Then,

$$\boldsymbol{\Omega}_j = \sigma_j^2[(1-\rho_j)\boldsymbol{I}_p + \rho_j \mathbf{1}_p \mathbf{1}_p'],$$

where $\boldsymbol{I}_p$ represents the $p \times p$ identity matrix. Since $\boldsymbol{\Omega}_j$ must be positive definite, we must require that $-1/(p-1) < \rho_j < 1$. We further assume that the intraclass correlation ρ_j must satisfy the condition $0 < \rho_j < 1$. It is well known that the quantity $|\boldsymbol{\Omega}_j|$ and the matrix $\boldsymbol{\Omega}_j^{-1}$ are as follows:

$$|\boldsymbol{\Omega}_j| = \sigma_j^{2p}[1 + (p-1)\rho_j](1-\rho_j)^{p-1},$$

and

$$\boldsymbol{\Omega}_j^{-1} = \frac{1}{\sigma_j^2(1-\rho_j)}\left[\boldsymbol{I}_p - \frac{\rho_j}{1+(p-1)\rho_j}\mathbf{1}_p\mathbf{1}_p'\right].$$

We have the following four possibilities under this structure:

Case 1: $\boldsymbol{\Omega}_1 = \boldsymbol{\Omega}_2$ $(\sigma_1^2 = \sigma_2^2,\ \rho_1 = \rho_2)$,
Case 2: $\boldsymbol{\Omega}_1 \neq \boldsymbol{\Omega}_2$ $(\sigma_1^2 = \sigma_2^2, \rho_1 \neq \rho_2)$,
Case 3: $\boldsymbol{\Omega}_1 \neq \boldsymbol{\Omega}_2$ $(\sigma_1^2 \neq \sigma_2^2, \rho_1 = \rho_2)$,
Case 4: $\boldsymbol{\Omega}_1 \neq \boldsymbol{\Omega}_2$ $(\sigma_1^2 \neq \sigma_2^2, \rho_1 \neq \rho_2)$.

Case 1: $\mathbf{\Omega}_1 = \mathbf{\Omega}_2 = \mathbf{\Omega}$ $(\sigma_1^2 = \sigma_2^2 = \sigma^2,\ \rho_1 = \rho_2 = \rho)$

Andrews et al. (1986) have considered this case in the spirit of logistic discrimination. As they have shown, the linear classification rule based on $\boldsymbol{y}$ simplifies to,

> Allocate an individual with response $\boldsymbol{y}$ to Population 1 if
>
> $$\sum_{l=1}^{p} y_l \geq \frac{\mu_1 + \mu_2}{2} p,$$
>
> and to Population 2 otherwise.

Clearly this classification rule is independent of $\mathbf{\Omega}$.

Given $n_1 + n_2 = n$ (say) random observations $\boldsymbol{Y}_1 = (\boldsymbol{y}_{11}, \boldsymbol{y}_{12}, \ldots, \boldsymbol{y}_{1n_1})$ from Population 1 and $\boldsymbol{Y}_2 = (\boldsymbol{y}_{21}, \boldsymbol{y}_{22}, \ldots, \boldsymbol{y}_{2n_2})$ from Population 2, define $\boldsymbol{W}_j = \sum_{i=1}^{n_j} (\boldsymbol{y}_{ji} - \bar{\boldsymbol{y}}_j)(\boldsymbol{y}_{ji} - \bar{\boldsymbol{y}}_j)'$, for $j = 1, 2$, as the within sum of squares and cross products matrix for the sample from the jth population, where $\bar{\boldsymbol{y}}_1$ and $\bar{\boldsymbol{y}}_2$ are two sample mean vectors for the two populations.

The maximum likelihood estimates of μ_1 and μ_2 are given by:

$$\hat{\mu}_j = \frac{\mathbf{1}_p' \mathbf{\Omega}_j^{-1} \bar{\boldsymbol{y}}_j}{\mathbf{1}_p' \mathbf{\Omega}_j^{-1} \mathbf{1}_p} = \frac{\mathbf{1}_p' \bar{\boldsymbol{y}}_j}{p} \ (= m_{j1}, \text{say}), \quad j = 1, 2. \tag{11.1}$$

Case 2: $\mathbf{\Omega}_1 \neq \mathbf{\Omega}_2$ $(\sigma_1^2 = \sigma_2^2 = \sigma^2, \rho_1 \neq \rho_2)$

The quadratic classification rule based on $\boldsymbol{y}$ in this case, simplifies to,

> Allocate an individual with response $\boldsymbol{y}$ to Population 1 if
>
> $$-\frac{1}{2\sigma^2}\left[\left(\frac{1}{1-\rho_1} - \frac{1}{1-\rho_2}\right)\sum_{l=1}^{p} y_l^2 - \left(\frac{\rho_1}{(1-\rho_1)(1+(p-1)\rho_1)} - \frac{\rho_2}{(1-\rho_2)(1+(p-1)\rho_2)}\right)\left(\sum_{l=1}^{p} y_l\right)^2\right]$$
> $$+\frac{1}{\sigma^2}\left(\frac{\mu_1}{1+(p-1)\rho_1} - \frac{\mu_2}{1+(p-1)\rho_2}\right)\sum_{l=1}^{p} y_l$$
> $$\geq \frac{p}{2\sigma^2}\left(\frac{\mu_1^2}{1+(p-1)\rho_1} - \frac{\mu_2^2}{1+(p-1)\rho_2}\right) - \frac{p-1}{2}\ln\left(\frac{1-\rho_2}{1-\rho_1}\right)$$
> $$-\frac{1}{2}\ln\left(\frac{1+(p-1)\rho_2}{1+(p-1)\rho_1}\right),$$
>
> and to Population 2 otherwise.

The maximum likelihood estimates μ_1 and μ_2 are the same as in Equation 11.1. The maximum likelihood estimates of ρ_1, ρ_2 and σ^2 are obtained by simultaneously and iteratively solving the following three equations:

$$\begin{aligned}
&-np(1-\rho_1)\{1+(p-1)\rho_1\}(1-\rho_2)\{1+(p-1)\rho_2\}\sigma^2 \\
&\quad+\{1+(p-1)\rho_1\}(1-\rho_2)\{1+(p-1)\rho_2\}a_8 \\
&\quad-(1-\rho_2)\{1+(p-1)\rho_2\}\rho_1 b_8 \\
&\quad+(1-\rho_1)\{1+(p-1)\rho_1\}\{1+(p-1)\rho_2\}a_9 \\
&\quad-(1-\rho_1)\{1+(p-1)\rho_1\}\rho_2 b_9 = 0,
\end{aligned}$$

$$\begin{aligned}
&n_1 p(p-1)\rho_1(1-\rho_1)\{1+(p-1)\rho_1\}\sigma^2 \\
&\qquad-\{1+(p-1)\rho_1\}^2 a_8+\{1+(p-1)\rho_1^2\}b_8 = 0,
\end{aligned}$$

and

$$\begin{aligned}
&n_2 p(p-1)\rho_2(1-\rho_2)\{1+(p-1)\rho_2\}\sigma^2 \\
&\qquad-\{1+(p-1)\rho_2\}^2 a_9+\{1+(p-1)\rho_2^2\}b_9 = 0,
\end{aligned}$$

where $a_8 = \mathrm{tr}\boldsymbol{W}_{11}, a_9 = \mathrm{tr}\boldsymbol{W}_{21}, b_8 = \mathrm{tr}(\boldsymbol{J}_p\boldsymbol{W}_{11}), b_9 = \mathrm{tr}(\boldsymbol{J}_p\boldsymbol{W}_{21})$ and $\boldsymbol{W}_{j1} = \boldsymbol{W}_j + n_j(\bar{\boldsymbol{y}}_j - m_{j1}\mathbf{1}_p)(\bar{\boldsymbol{y}}_j - m_{j1}\mathbf{1}_p)', j = 1, 2$, and $\boldsymbol{J}_p$ is a $p \times p$ matrix containing all elements as unity.

This can be achieved by using the Newton–Raphson method.

Case 3: $\boldsymbol{\Omega}_1 \neq \boldsymbol{\Omega}_2$ $(\sigma_1^2 \neq \sigma_2^2, \rho_1 = \rho_2 = \rho)$

In this case the quadratic classification rule based on $\boldsymbol{y}$ is given by:

Allocate an individual with response $\boldsymbol{y}$ to Population 1 if

$$\begin{aligned}
&-\frac{1}{2(1-\rho)}\left(\frac{1}{\sigma_1^2}-\frac{1}{\sigma_2^2}\right)\left[\sum_{l=1}^{p} y_l^2 - \frac{\rho}{1+(p-1)\rho}\left(\sum_{l=1}^{p} y_l\right)^2\right] \\
&+\frac{1}{(1+(p-1)\rho)}\left(\frac{\mu_1}{\sigma_1^2}-\frac{\mu_2}{\sigma_2^2}\right)\sum_{l=1}^{p} y_l \\
&\geq \frac{p}{2(1+(p-1)\rho)}\left(\frac{\mu_1^2}{\sigma_1^2}-\frac{\mu_2^2}{\sigma_2^2}\right)-\frac{p}{2}\ln\frac{\sigma_2^2}{\sigma_1^2},
\end{aligned}$$

and to Population 2 otherwise.

It can be seen that the means μ_j can be estimated by formula given in Equation 11.1. The maximum likelihood estimates of ρ, σ_1^2 and σ_2^2 are obtained by simultaneously and iteratively solving,

$$-n_1p(1-\rho)\{1+(p-1)\rho\}\sigma_1^2+\{1+(p-1)\rho\}a_8-\rho b_8=0,$$
$$-n_2p(1-\rho)\{1+(p-1)\rho\}\sigma_2^2+\{1+(p-1)\rho\}a_9-\rho b_9=0,$$

and

$$n(p-1)(1-\rho)\{1+(p-1)\rho\}p\rho\sigma_1^2\sigma_2^2-(a_8\sigma_2^2+a_9\sigma_1^2)\{1+(p-1)\rho\}^2 + (b_8\sigma_2^2+b_9\sigma_1^2)\{1+(p-1)\rho^2\}=0,$$

where a_8, b_8, a_9 and b_9's have been already defined earlier.

Case 4: $\mathbf{\Omega}_1 \neq \mathbf{\Omega}_2$ ($\sigma_1^2 \neq \sigma_2^2, \rho_1 \neq \rho_2$)

In this case the quadratic classification rule based on $\boldsymbol{y}$ is given by:

Allocate an individual with response $\boldsymbol{y}$ to Population 1 is

$$-\frac{1}{2}\left(\frac{1}{\sigma_1^2(1-\rho_1)}-\frac{1}{\sigma_2^2(1-\rho_2)}\right)\sum_{l=1}^{p}y_l^2$$
$$+\frac{1}{2}\left(\frac{\rho_1}{\sigma_1^2(1-\rho_1)(1+(p-1)\rho_1)}\right.$$
$$\left.-\frac{\rho_2}{\sigma_2^2(1-\rho_2)(1+(p-1)\rho_2)}\right)\left(\sum_{l=1}^{p}y_l\right)^2$$
$$+\left(\frac{\mu_1}{\sigma_1^2(1+(p-1)\rho_1)}-\frac{\mu_2}{\sigma_2^2(1+(p-1)\rho_2)}\right)\sum_{l=1}^{p}y_l$$
$$\geq\frac{p}{2}\left(\frac{\mu_1^2}{\sigma_1^2(1+(p-1)\rho_1)}-\frac{\mu_2^2}{\sigma_2^2(1+(p-1)\rho_2)}\right)$$
$$-\frac{p}{2}\ln\frac{\sigma_2^2}{\sigma_1^2}-\frac{p-1}{2}\ln\left(\frac{1-\rho_2}{1-\rho_1}\right)-\frac{1}{2}\ln\left(\frac{1+(p-1)\rho_2}{1+(p-1)\rho_1}\right),$$

and to Population 2 otherwise.

Means μ_j are estimated as earlier. However, the maximum likelihood estimates of ρ_1 and σ_1^2, ρ_2 and σ_2^2 are obtained by solving,

$$-n_1p(1-\rho_1)\{1+(p-1)\rho_1\}\sigma_1^2+\{1+(p-1)\rho_1\}a_8-\rho_1b_8=0,$$
$$-n_2p(1-\rho_2)\{1+(p-1)\rho_2\}\sigma_2^2+\{1+(p-1)\rho_2\}a_9-\rho_2b_9=0,$$

$$n_1p(p-1)\rho_1\{1+(p-1)\rho_1\}(1-\rho_1)\sigma_1^2 - \{1+(p-1)\rho_1\}^2a_8+\{1+(p-1)\rho_1^2\}b_8=0,$$

and

$$n_2 p(p-1)\rho_2\{1+(p-1)\rho_2\}(1-\rho_2)\sigma_2^2 \\ - \{1+(p-1)\rho_2\}^2 a_9 + \{1+(p-1)\rho_2^2\} b_9 = 0.$$

See Roy and Khattree (2003a) for more details on all four cases discussed above.

11.2.2 Discrimination with Repeated Measures with AR(1) Covariance Structure

Suppose the repeated measures are modeled using the first order autoregressive (AR(1)) covariance structure given by the process

$$y_t - \mu = \rho(y_{t-1} - \mu) + \epsilon_t,$$

where $\{\epsilon_t\}$ is a sequence of independent and identically distributed (i.i.d) normal random variables with mean 0 and variance σ^2, and ρ is an unknown parameter satisfying the condition $|\rho| < 1$. Thus, for $j = 1, 2$,

$$\mathbf{\Omega}_j = \sigma_j^2 \begin{bmatrix} 1 & \rho_j & \rho_j^2 & \cdots & \rho_j^{p-1} \\ \rho_j & 1 & \rho_j & \cdots & \rho_j^{p-2} \\ \rho_j^2 & \rho_j & 1 & \cdots & \rho_j^{p-3} \\ \cdot & \cdot & \cdot & \cdots & \cdot \\ \cdot & \cdot & \cdot & \cdots & \cdot \\ \cdot & \cdot & \cdot & \cdots & \cdot \\ \rho_j^{p-1} & \rho_j^{p-2} & \rho_j^{p-3} & \cdots & 1 \end{bmatrix},$$

and $\mathbf{\Omega}_j^{-1}$ is a tridiagonal matrix given by,

$$\mathbf{\Omega}_j^{-1} = \frac{1}{\sigma_j^2(1-\rho_j^2)} \mathbf{\Omega}_{0j},$$

where

$$\mathbf{\Omega}_{0j} = \begin{bmatrix} 1 & -\rho_j & 0 & \cdots & 0 & 0 \\ -\rho_j & 1+\rho_j^2 & -\rho_j & \cdots & 0 & 0 \\ 0 & -\rho_j & 1+\rho_j^2 & \cdots & 0 & 0 \\ \cdot & \cdot & \cdot & \cdots & \cdot & \\ \cdot & \cdot & \cdot & \cdots & \cdot & \\ \cdot & \cdot & \cdot & \cdots & \cdot & \\ 0 & 0 & 0 & \cdots & 1+\rho_j^2 & -\rho_j \\ 0 & 0 & 0 & \cdots & -\rho_j & 1 \end{bmatrix},$$

and

$$|\mathbf{\Omega}_j| = (\sigma_j^2)^p\,(1-\rho_j^2)^{p-1}.$$

Similar to what was done in Section 11.2.1, we will consider various cases.

Case 1: $\mathbf{\Omega}_1 = \mathbf{\Omega}_2 = \mathbf{\Omega}$ $(\sigma_1^2 = \sigma_2^2 = \sigma^2, \rho_1 = \rho_2 = \rho)$
The linear classification rule based on response $\boldsymbol{y}$ is,

> Allocate an individual with response $\boldsymbol{y}$ to Population 1 if
> $$[p\bar{y}_{1,p} - \rho(p-2)\bar{y}_{2,p-1}] \geq \frac{1}{2}[p-\rho(p-2)](\mu_1+\mu_2),$$
> and to Population 2 otherwise.

Where $\bar{y}_{l,m} = (y_l + y_{l+1} + \cdots + y_m)/(m-l+1)$; $l, m = 1, 2, \ldots, p$. It may be noted that the classification rule is independent of σ^2.

The maximum likelihood estimates of μ_1, μ_2, ρ, and σ^2 are obtained by simultaneously and iteratively solving the following four equations:

$$(p-2)\rho\mu_1 - p\mu_1 + pm_{11} - (p-2)\rho\, m_{12} = 0,$$
$$(p-2)\rho\mu_2 - p\mu_2 + pm_{21} - (p-2)\rho\, m_{22} = 0,$$

$$\begin{aligned}
&n(p-1)\sigma^2\rho - n(p-1)\sigma^2\rho^3 - \{\rho(\alpha_1+\beta_1) - \gamma_1\rho^2 - \gamma_1\}\\
&\quad + n_1\mu_1\{\rho(\alpha_2+\beta_2) - \gamma_2\rho^2 - \gamma_2\}\\
&\quad + n_2\mu_2\{\rho(\alpha_3+\beta_3) - \gamma_3\rho^2 - \gamma_3\}\\
&\quad - (n_1\mu_1^2 + n_2\mu_2^2)\{\rho(2p-2) - (p-1)\rho^2 - (p-1)\} = 0,
\end{aligned}$$

and

$$\begin{aligned}
\sigma^2 = {}& \frac{1}{np(1-\rho^2)}[(\beta_1\rho^2 - 2\gamma_1\rho + \alpha_1)\\
&- n_1\mu_1(\beta_2\rho^2 - 2\gamma_2\rho + \alpha_2) - n_2\mu_2(\beta_3\rho^2 - 2\gamma_3\rho + \alpha_3)\\
&+ (n_1\mu_1^2 + n_2\mu_2^2)((p-2)\rho^2 - 2(p-1)\rho + p)],
\end{aligned}$$

where $m_{j1} = \mathbf{1}_p'\bar{\mathbf{y}}_j/p$, $m_{j2} = (\mathbf{1}_p'\bar{\mathbf{y}}_j - \bar{y}_{j1} - \bar{y}_{jp})/(p-2)$, $j = 1, 2$ and $\bar{y}_{j1}$ and $\bar{y}_{jp}$ are respectively the first and pth elements of the vector $\bar{\mathbf{y}}_j$. The vectors $\bar{\mathbf{y}}_1$

and $\bar{\boldsymbol{y}}_2$ are two sample mean vectors for the two populations and $\boldsymbol{W}_j, j = 1, 2$ have been defined in the previous section.

$$
\begin{aligned}
\alpha_1 &= \operatorname{tr} \boldsymbol{W}_0, \quad \beta_1 = \operatorname{tr} \boldsymbol{W}_0 - w_{0,11} - w_{0,pp}, \\
\gamma_1 &= w_{0,12} + w_{0,23} + \cdots + w_{0,p-1\,p}, \\
\alpha_2 &= \operatorname{tr} \boldsymbol{W}_5, \quad \beta_2 = \operatorname{tr} \boldsymbol{W}_5 - w_{5,11} - w_{5,pp}, \\
\gamma_2 &= w_{5,12} + w_{5,23} + \cdots + w_{5,p-1\,p} \quad \text{and} \\
\alpha_3 &= \operatorname{tr} \boldsymbol{W}_6, \quad \beta_3 = \operatorname{tr} \boldsymbol{W}_6 - w_{6,11} - w_{6,pp}, \\
\gamma_3 &= w_{6,12} + w_{6,23} + \cdots + w_{6,p-1\,p}.
\end{aligned}
$$

The quantity $w_{l,ij}$ here denotes the (i, j)th element of the matrix $\boldsymbol{W}_l$. $\boldsymbol{W}_0 = \boldsymbol{W}_1 + \boldsymbol{W}_2 + n_1 \boldsymbol{W}_3 + n_2 \boldsymbol{W}_4$, $\boldsymbol{W}_3 = \bar{\boldsymbol{y}}_1 \bar{\boldsymbol{y}}_1'$, $\boldsymbol{W}_4 = \bar{\boldsymbol{y}}_2 \bar{\boldsymbol{y}}_2'$, $\boldsymbol{W}_5 = (\mathbf{1}_p \bar{\boldsymbol{y}}_1' + \bar{\boldsymbol{y}}_1 \mathbf{1}_p')$, $\boldsymbol{W}_6 = (\mathbf{1}_p \bar{\boldsymbol{y}}_2' + \bar{\boldsymbol{y}}_2 \mathbf{1}_p')$.

Case 2: $\boldsymbol{\Omega}_1 \neq \boldsymbol{\Omega}_2$ $(\sigma_1^2 = \sigma_2^2 = \sigma^2, \rho_1 \neq \rho_2)$
The quadratic classification rule based on $\boldsymbol{y}$ is given by:

Allocate an individual with response $\boldsymbol{y}$ to Population 1 if

$$
\begin{aligned}
&- \frac{1}{2\sigma^2}\left(\frac{1}{1-\rho_1^2} - \frac{1}{1-\rho_2^2}\right)\sum_{l=1}^{p} y_l^2 \\
&- \frac{1}{2\sigma^2}\left(\frac{\rho_1^2}{1-\rho_1^2} - \frac{\rho_2^2}{1-\rho_2^2}\right)\sum_{l=2}^{p-1} y_l^2 \\
&+ \frac{1}{\sigma^2}\left(\frac{\rho_1}{1-\rho_1^2} - \frac{\rho_2}{1-\rho_2^2}\right)\sum_{l=1}^{p-1} y_l y_{l+1} \\
&+ \frac{p}{\sigma^2}\left(\frac{\mu_1}{1+\rho_1} - \frac{\mu_2}{1+\rho_2}\right)\bar{y}_{1,p} \\
&- \frac{p-2}{\sigma^2}\left(\frac{\mu_1\rho_1}{1+\rho_1} - \frac{\mu_2\rho_2}{1+\rho_2}\right)\bar{y}_{2,p-1} \\
&\geq \frac{1}{2\sigma^2}\left[\frac{\mu_1^2}{1+\rho_1}(p-(p-2)\rho_1) - \frac{\mu_2^2}{1+\rho_2}(p-(p-2)\rho_2)\right] \\
&- \frac{p-1}{2}\ln\left(\frac{1-\rho_2^2}{1-\rho_1^2}\right),
\end{aligned}
$$

and to Population 2 otherwise.

The maximum likelihood estimates of $\mu_1, \mu_2, \rho_1, \rho_2$, and σ^2 are obtained by solving the following equations. Details can be found in Roy (2002).

$$(p-2)\rho_1\mu_1 - p\mu_1 + pm_{11} - (p-2)\rho_1 m_{12} = 0,$$
$$(p-2)\rho_2\mu_2 - p\mu_2 + pm_{21} - (p-2)\rho_2 m_{22} = 0,$$

$$\begin{aligned}\sigma^2 = &\frac{1}{np(1-\rho_1^2)}[(\beta_8\rho_1^2 - 2\gamma_8\rho_1 + \alpha_8) - n_1\mu_1(\beta_2\rho_1^2 - 2\gamma_2\rho_1 + \alpha_2)\\ &+ n_1\mu_1^2((p-2)\rho_1^2 - 2(p-1)\rho_1 + p)]\\ &+ \frac{1}{np(1-\rho_2^2)}[(\beta_9\rho_2^2 - 2\gamma_9\rho_2 + \alpha_9) - n_2\mu_2(\beta_3\rho_2^2 - 2\gamma_3\rho_2 + \alpha_3)\\ &+ n_2\mu_2^2((p-2)\rho_2^2 - 2(p-1)\rho_2 + p)],\end{aligned}$$

$$\begin{aligned}&n_1(p-1)\sigma^2\rho_1 - n_1(p-1)\sigma^2\rho_1^3 - (\rho_1(\alpha_8+\beta_8) - \gamma_8\rho_1^2 - \gamma_8)\\ &+ n_1\mu_1(\rho_1(\alpha_2+\beta_2) - \gamma_2\rho_1^2 - \gamma_2)\\ &- n_1\mu_1^2(\rho_1(2p-2) - (p-1)\rho_1^2 - (p-1)) = 0,\end{aligned}$$

and

$$\begin{aligned}&n_2(p-1)\sigma^2\rho_2 - n_2(p-1)\sigma^2\rho_2^3 - (\rho_2(\alpha_9+\beta_9) - \gamma_9\rho_2^2 - \gamma_9)\\ &+ n_2\mu_2(\rho_2(\alpha_3+\beta_3) - \gamma_3\rho_2^2 - \gamma_3)\\ &- n_2\mu_2^2(\rho_2(2p-2) - (p-1)\rho_2^2 - (p-1)) = 0,\end{aligned}$$

where

$$\begin{aligned}&\alpha_8 = \text{tr}\,\boldsymbol{W}_8, \quad \beta_8 = \text{tr}\boldsymbol{W}_8 - w_{8,11} - w_{8,pp},\\ &\gamma_8 = w_{8,12} + w_{8,23} + \cdots + w_{8,p-1p},\\ &\alpha_9 = \text{tr}\,\boldsymbol{W}_9, \quad \beta_9 = \text{tr}\,\boldsymbol{W}_9 - w_{9,11} - w_{9,pp},\\ &\gamma_9 = w_{9,12} + w_{9,23} + \cdots + w_{9,p-1\,p},\\ &\boldsymbol{W}_8 = \boldsymbol{W}_1 + n_1\boldsymbol{W}_3,\end{aligned}$$

and

$$\boldsymbol{W}_9 = \boldsymbol{W}_2 + n_2\boldsymbol{W}_4.$$

Case 3: $\boldsymbol{\Omega}_1 \neq \boldsymbol{\Omega}_2$ $(\sigma_1^2 \neq \sigma_2^2, \rho_1 = \rho_2 = \rho)$

The quadratic classification rule based on $\boldsymbol{y}$ is given by:

Allocate an individual with response $\boldsymbol{y}$ to Population 1 if

$$
\begin{aligned}
&-\frac{1}{2(1-\rho^2)}\left(\frac{1}{\sigma_1^2}-\frac{1}{\sigma_2^2}\right)\left[\sum_{l=1}^{p} y_l^2 + \rho^2 \sum_{l=2}^{p-1} y_l^2 - 2\rho \sum_{l=1}^{p-1} y_l y_{l+1}\right] \\
&+\frac{1}{(1+\rho)}\left(\frac{\mu_1}{\sigma_1^2}-\frac{\mu_2}{\sigma_2^2}\right)[p\bar{y}_{1,p} - \rho(p-2)\bar{y}_{2,p-1}] \\
&\geq \frac{1}{2(1+\rho)}\left(\frac{\mu_1^2}{\sigma_1^2}-\frac{\mu_2^2}{\sigma_2^2}\right)[p-\rho(p-2)] - \frac{p}{2}\ln\frac{\sigma_2^2}{\sigma_1^2},
\end{aligned}
$$

and to Population 2 otherwise.

The maximum likelihood estimates of $\mu_1, \mu_2, \rho, \sigma_1^2$, and σ_2^2 are obtained by simultaneously and iteratively solving the following equations:

$$
\begin{aligned}
(p-2)\rho\mu_1 - p\mu_1 + pm_{11} - (p-2)\rho\, m_{12} = 0, \\
(p-2)\rho\mu_2 - p\mu_2 + pm_{21} - (p-2)\rho\, m_{22} = 0,
\end{aligned}
$$

$$
\begin{aligned}
\sigma_1^2 = &\frac{1}{n_1 p(1-\rho^2)}[\beta_8\rho^2 - 2\gamma_8\rho + \alpha_8 - n_1\mu_1(\beta_2\rho^2 - 2\gamma_2\rho + \alpha_2) \\
&+ n_1\mu_1^2((p-2)\rho^2 - 2(p-1)\rho + p)],
\end{aligned}
$$

$$
\begin{aligned}
\sigma_2^2 = &\frac{1}{n_2 p(1-\rho^2)}[\beta_9\rho^2 - 2\gamma_9\rho + \alpha_9 - n_2\mu_2(\beta_3\rho^2 - 2\gamma_3\rho + \alpha_3) \\
&+ n_2\mu_2^2((p-2)\rho^2 - 2(p-1)\rho + p)],
\end{aligned}
$$

and

$$
\begin{aligned}
&n(p-1)(\rho-\rho^3)\sigma_1^2\sigma_2^2 - \sigma_2^2((\alpha_8+\beta_8)\rho - \gamma_8\rho^2 - \gamma_8) \\
&\quad + n_1\mu_1\sigma_2^2((\alpha_2+\beta_2)\rho - \gamma_2\rho^2 - \gamma_2) \\
&\quad - n_1\mu_1^2\sigma_2^2((2p-2)\rho - (p-1)\rho^2 - (p-1)) \\
&\quad - \sigma_1^2((\alpha_9+\beta_9)\rho - \gamma_9\rho^2 - \gamma_9) \\
&\quad + n_2\mu_2\sigma_1^2((\alpha_3+\beta_3)\rho - \gamma_3\rho^2 - \gamma_3) \\
&\quad - n_2\mu_2^2\sigma_1^2((2p-2)\rho - (p-1)\rho^2 - (p-1)) = 0.
\end{aligned}
$$

Case 4: $\mathbf{\Omega}_1 \neq \mathbf{\Omega}_2$ $(\sigma_1^2 \neq \sigma_2^2, \rho_1 \neq \rho_2)$

The quadratic classification rule based on $\boldsymbol{y}$ is given by:

Allocate an individual to Population 1 if

$$
\begin{aligned}
&- \frac{1}{2}\left(\frac{1}{\sigma_1^2(1-\rho_1^2)} - \frac{1}{\sigma_2^2(1-\rho_2^2)}\right)\sum_{l=1}^{p} y_l^2 \\
&- \frac{1}{2}\left(\frac{\rho_1^2}{\sigma_1^2(1-\rho_1^2)} - \frac{\rho_2^2}{\sigma_2^2(1-\rho_2^2)}\right)\sum_{l=2}^{p-1} y_l^2 \\
&+ \left(\frac{\rho_1}{\sigma_1^2(1-\rho_1^2)} - \frac{\rho_2}{\sigma_2^2(1-\rho_2^2)}\right)\sum_{l=1}^{p-1} y_l y_{l+1} \\
&+ \left(\frac{\mu_1}{\sigma_1^2(1+\rho_1)} - \frac{\mu_2}{\sigma_2^2(1+\rho_2)}\right) p\bar{y}_{1,p} \\
&- (p-2)\left(\frac{\mu_1\rho_1}{\sigma_1^2(1+\rho_1)} - \frac{\mu_2\rho_2}{\sigma_2^2(1+\rho_2)}\right)\bar{y}_{2,p-1} \\
&\geq \frac{1}{2}\left(\frac{\mu_1^2}{\sigma_1^2(1+\rho_1)}(p-(p-2)\rho_1) - \frac{\mu_2^2}{\sigma_2^2(1+\rho_2)}(p-(p-2)\rho_2)\right) \\
&- \frac{p}{2}\ln\frac{\sigma_2^2}{\sigma_1^2} - \frac{p-1}{2}\ln\left(\frac{1-\rho_2^2}{1-\rho_1^2}\right),
\end{aligned}
$$

and to Population 2 otherwise.

The maximum likelihood estimates of $\mu_1, \rho_1, \sigma_1^2$ and $\mu_2, \rho_2, \sigma_2^2$ are obtained by simultaneously and iteratively solving the following equations:

$$(p-2)\rho_1\mu_1 - p\mu_1 + pm_{11} - (p-2)\rho_1 m_{12} = 0,$$

$$(p-2)\rho_2\mu_2 - p\mu_2 + pm_{21} - (p-2)\rho_2 m_{22} = 0,$$

$$
\begin{aligned}
\sigma_1^2 = {} & \frac{1}{n_1 p(1-\rho_1^2)}[\beta_8\rho_1^2 - 2\gamma_8\rho_1 + \alpha_8 - n_1\mu_1(\beta_2\rho_1^2 - 2\gamma_2\rho_1 + \alpha_2) \\
& + n_1\mu_1^2((p-2)\rho_1^2 - 2(p-1)\rho_1 + p)],
\end{aligned}
$$

$$
\begin{aligned}
\sigma_2^2 = {} & \frac{1}{n_2 p(1-\rho_2^2)}[\beta_9\rho_2^2 - 2\gamma_9\rho_2 + \alpha_9 - n_2\mu_2(\beta_3\rho_2^2 - 2\gamma_3\rho_2 + \alpha_3) \\
& + n_2\mu_2^2((p-2)\rho_2^2 - 2(p-1)\rho_2 + p)],
\end{aligned}
$$

$$n_1(p-1)(\rho_1-\rho_1^3)\sigma_1^2-(\rho_1(\alpha_8+\beta_8)-\gamma_8\rho_1^2-\gamma_8)$$
$$+n_1\mu_1(\rho_1(\alpha_2+\beta_2)-\gamma_2\rho_1^2-\gamma_2)$$
$$-n_1\mu_1^2(\rho_1(2p-2)-(p-1)\rho_1^2-(p-1))=0,$$

and

$$n_2(p-1)(\rho_2-\rho_2^3)\sigma_2^2-(\rho_2(\alpha_9+\beta_9)-\gamma_9\rho_2^2-\gamma_9)$$
$$+n_2\mu_2(\rho_2(\alpha_3+\beta_3)-\gamma_3\rho_2^2-\gamma_3)$$
$$-n_2\mu_2^2(\rho_2(2p-2)-(p-1)\rho_2^2-(p-1))=0.$$

See Roy and Khattree (2003a) for details.

11.2.3 Some Simulation Results

We apply simulation to study the impact of the different parameters on misclassification error rate (MER) when the true variance–covariance matrices were CS/AR(1). Comparisons are made of the situations when maximum likelihood estimate was performed under these covariance structures as opposed to ignoring the structure and taking the covariance matrices as unstructured. Traditional discriminant function both linear and quadratic (McLachlan, 1992) are used to calculate the MER for the case of unstructured variance covariance matrix.

Data are randomly generated from the independent p-variate normal distributions with respective means $\boldsymbol{\mu}_j$ $(=\mu_j\mathbf{1}_p)$ and variance–covariance matrices $\boldsymbol{\Omega}_j$ (σ_j^2, ρ_j) from the populations $j, j = 1, 2$, both for moderate (25, 25) and large (50, 50) samples. The discriminant function is calculated on the basis of moderate and large training samples, and then actual MER is calculated based on the 2000 test samples drawn from each population.

In order to see the effects of various parameters, a number of combinations of $\mu_1, \mu_2, \sigma_1^2, \sigma_2^2, \rho_1$, and ρ_2 are taken. The values of ρs are chosen as 0.1, 0.3, 0.5, 0.7, and 0.9 and the number of repeated measures p are chosen as 3, 5, 8, and 10. Numerous simulations were performed to see the effect of these parameters on MER. Complete details of the simulations and the results are given in Roy and Khattree (2003a).

As mentioned earlier, Case 1 for CS covariance structure has been studied by Andrews et al. (1986) in the logistic discrimination set up. They have explored the effect of the value of ρ on the a posteriori odds ratio. We will, therefore, not consider that case here. We study the Case 1 for AR(1) covariance structure. We observe that for fixed parameters μ_1, μ_2, σ, and ρ the MER decreases with p (Table 11.1). In other cases also we observe that for fixed parameters $\mu_1, \mu_2, \sigma_1^2, \sigma_2^2, \rho_1$, and ρ_2 the MER decreases with p. It is evident from the Tables 11.2 through 11.4. In all these tables for each value of ρ, the second row

TABLE 11.1

Covariance Structure AR(1)
$(\sigma_1^2 = \sigma_2^2 = 1, \mu_1 = 2, \mu_2 = 0)$

(n_1,n_2)	$p \rightarrow$ $\rho \downarrow$	3	5	8	10
(50,50)	0.1	5.700	1.875	0.350	0.100
		6.150	2.700	0.375	0.225
	0.3	8.025	3.875	1.100	0.625
		8.450	4.800	1.400	1.000
	0.5	10.675	6.800	3.325	2.050
		10.900	8.100	3.225	2.775
	0.7	13.200	9.975	6.850	5.375
		13.025	11.975	7.175	6.125
	0.9	15.000	14.375	11.850	11.050
		15.225	15.900	12.625	11.925
(25,25)	0.1	5.375	1.950	0.375	0.100
		5.950	3.075	0.650	0.450
	0.3	7.750	3.800	1.375	0.675
		8.475	5.150	2.025	1.275
	0.5	10.475	6.600	3.250	2.050
		10.850	7.525	4.325	3.525
	0.7	12.625	9.875	7.025	5.275
		12.800	11.250	7.750	6.825
	0.9	14.400	14.225	12.175	11.350
		14.975	15.275	12.875	12.700

represents the MER corresponding to the unstructured variance–covariance matrix.

The MER decreases with p owing to the fact that additional repeated measures provide additional information. We also expect that after certain point additional repeated measures do not provide any significant extra information, and also that we can see in our simulation result. Decrease in MER is quite significant when p increases from 3 to 5 and for smaller value of ρ. Decrease in MER is not significant when p increases from 8 to 10. In fact, sometimes when p increases from 8 to 10, MER increases little bit for higher values of ρ. Table 11.4 clearly exhibit this phenomenon. This increase depends on the combinations of μ_1, μ_2, σ_1^2, and σ_2^2. This pattern is not strictly followed in the case of unstructured covariance matrix.

In Cases 2 and 4, MER increases with either ρ_1 or ρ_2 when the other parameters are fixed (Tables 11.2, 11.3, and 11.4). In other cases, when $\rho_1 = \rho_2 = \rho$ (say), similar things happen, that is, MER increases with ρ. One would expect this to happen since an increase in ρ means repeated measurements are less informative and this leads to increase in MER.

In Cases 3 and 4, MER increases with either σ_1^2 or σ_2^2 when the other parameters are fixed. In Cases 1 and 2, when $\sigma_1^2 = \sigma_2^2 = \sigma^2$ (say), similar things

TABLE 11.2

Covariance Structure: Compound Symmetric
$(\rho_1 = 0.9, \sigma_1^2 = \sigma_2^2 = 2, \mu_1 = 4, \mu_2 = 0)$

(n_1,n_2)	$p \rightarrow$ $\rho_2 \downarrow$	3	5	8	10
(50,50)	0.1	2.975	1.250	0.550	0.200
		3.250	1.700	0.750	0.400
	0.3	3.725	2.050	0.975	0.475
		4.200	2.575	1.300	0.775
	0.5	4.600	3.175	1.525	1.025
		5.050	3.925	2.325	1.625
	0.7	5.850	5.025	2.925	2.450
		6.025	6.200	4.650	4.500
(25,25)	0.1	2.625	1.100	0.750	0.250
		3.100	1.600	0.875	0.725
	0.3	3.575	1.950	1.125	0.600
		4.075	2.725	1.400	1.575
	0.5	4.650	2.825	1.575	1.050
		4.725	4.675	2.200	3.450
	0.7	5.550	4.800	3.150	2.500
		6.150	8.150	4.450	6.875

TABLE 11.3

Covariance Structure Compound Symmetric
$(\rho_1 = 0.9, \sigma_1^2 = 1, \sigma_2^2 = 2, \mu_1 = 4, \mu_2 = 0)$

(n_1,n_2)	$p \rightarrow$ $\rho_2 \downarrow$	3	5	8	10
(50,50)	0.1	1.025	0.325	0.050	0.025
		0.975	0.250	0.150	0.025
	0.3	1.600	0.525	0.100	0.075
		1.650	0.625	0.225	0.175
	0.5	2.175	1.025	0.350	0.175
		2.375	0.925	0.500	0.275
	0.7	2.800	1.700	0.750	0.375
		2.925	2.125	1.350	0.950
(25,25)	0.1	0.725	0.225	0.075	0.075
		1.125	0.400	0.200	0.075
	0.3	1.350	0.275	0.150	0.075
		1.550	0.800	0.375	0.400
	0.5	1.950	0.550	0.400	0.175
		2.300	1.600	0.676	1.050
	0.7	2.725	1.300	0.875	0.475
		3.075	3.175	1.400	2.600

TABLE 11.4

Covariance Structure Compound Symmetric
$(\rho_1 = 0.9, \sigma_1^2 = 2, \sigma_2^2 = 1, \mu_1 = 4, \mu_2 = 0)$

(n_1,n_2)	$p \rightarrow$ $\rho_2 \downarrow$	3	5	8	10
(50,50)	0.1	2.300	1.600	0.675	0.525
		2.525	1.750	0.900	0.750
	0.3	3.075	2.225	1.350	1.100
		3.225	2.550	1.450	1.450
	0.5	3.750	3.075	1.875	1.950
		3.925	3.975	2.450	2.875
	0.7	4.650	4.550	2.950	3.575
		4.550	5.700	3.700	5.225
(25,25)	0.1	1.850	1.125	0.750	0.675
		2.375	1.975	1.075	1.175
	0.3	2.650	1.775	1.400	1.250
		3.000	2.600	1.650	2.300
	0.5	3.275	2.450	1.925	2.000
		3.650	5.875	2.150	4.275
	0.7	3.975	3.650	2.875	3.600
		4.375	9.550	4.025	8.075

happen, that is, MER increases with σ^2. MER is not a function of σ_1^2/σ_2^2. When $\rho_1 > \rho_2$, MER is more if σ_1^2 is larger than σ_2^2, as it is evident from Tables 11.3 and 11.4. This is also expected here because with less information, large σ^2 leads to increase in MER. For unstructured covariance matrix case also, MER increases with either σ_1^2 or σ_2^2 when the other parameters are fixed.

In all the above cases, MER is consistently larger in the case of unstructured covariance matrix than the CS/AR(1) structured covariance matrix. This is more evident (roughly 1.5–3 times) in the case of moderate sample and when the value of p is 5 or more.

11.3 Classification in Case of Multivariate Repeated Measures Data

Let $\boldsymbol{y}$ represent the vector containing the relevant information of a subject. We assume that there are q response variables and on each variable observations are taken over p equidistant time points. Thus, $\boldsymbol{y}$ is a $pq \times 1$ vector with elements arranged as follows. At first time point all q responses are arranged one below the other. This is followed by q responses at second time point and so on. As earlier additional multiple subscripts will be used on $\boldsymbol{y}$ for further description.

Let $\boldsymbol{y}_{jit}$ be a $q \times 1$ vector of measurements on the ith individual in the jth population at the tth time point, $j = 1,2$; $i = 1,2,\ldots,n_j$; $t = 1,2,\ldots,p$. Thus, $\boldsymbol{y}_{ji} = (\boldsymbol{y}'_{ji1}, \boldsymbol{y}'_{ji2}, \ldots, \boldsymbol{y}'_{jip})'$ is a $pq \times 1$ random vector corresponding to the ith individual in the jth population. We assume $\boldsymbol{y}_{ji} \sim N_{pq}(\boldsymbol{\mu}_j, \boldsymbol{\Omega}_j)$. The matrix $\boldsymbol{\Omega}_j$ is assumed to be $pq \times pq$ positive definite. However, as pq can be large, thereby resulting in a large number of unknown parameters, we will impose some structure on $\boldsymbol{\Omega}_j$. We thus assume $\boldsymbol{\Omega}_j = \boldsymbol{V}_j \otimes \boldsymbol{\Sigma}_j$, where $\boldsymbol{V}_j$ and $\boldsymbol{\Sigma}_j$, respectively, are $p \times p$ and $q \times q$ positive definite matrices and $\otimes$ denotes the Kroneker product defined by $\boldsymbol{A} \otimes \boldsymbol{B} = (a_{ij}\boldsymbol{B})$. The matrix $\boldsymbol{\Sigma}_j$, $j = 1,2$ represents the variance–covariance matrix between the measurements on all response variables at a given time point. It is assumed that it does not depend on a particular time point and is same for all time points. The matrix $\boldsymbol{V}_j$, $j = 1,2$ is the correlation matrix of the repeated measures on a given response variable and it is assumed to be the same for all response variables. Thus, $\text{cov}(\boldsymbol{y}_{jit}) = v_{jtt}\boldsymbol{\Sigma}_j$ and $\text{cov}(\boldsymbol{y}_{jil}, \boldsymbol{y}_{jim}) = v_{jlm}\boldsymbol{\Sigma}_j$, where $\boldsymbol{V}_j = (v_{jlm})$, $l = 1,\ldots,p$; $m = 1,\ldots,p$.

We will discuss CS correlation structure on $\boldsymbol{V}_j$ in Section 11.3.1 and AR(1) correlation structure on $\boldsymbol{V}_j$ in Section 11.3.2. We will develop classification rules under the following four cases:

Case 1: $\boldsymbol{\Omega}_1 = \boldsymbol{\Omega}_2$ $(\boldsymbol{V}_1 = \boldsymbol{V}_2, \boldsymbol{\Sigma}_1 = \boldsymbol{\Sigma}_2)$,
Case 2: $\boldsymbol{\Omega}_1 \neq \boldsymbol{\Omega}_2$ $(\boldsymbol{V}_1 \neq \boldsymbol{V}_2, \boldsymbol{\Sigma}_1 = \boldsymbol{\Sigma}_2)$,
Case 3: $\boldsymbol{\Omega}_1 \neq \boldsymbol{\Omega}_2$ $(\boldsymbol{V}_1 = \boldsymbol{V}_2, \boldsymbol{\Sigma}_1 \neq \boldsymbol{\Sigma}_2)$,
Case 4: $\boldsymbol{\Omega}_1 \neq \boldsymbol{\Omega}_2$ $(\boldsymbol{V}_1 \neq \boldsymbol{V}_2, \boldsymbol{\Sigma}_1 \neq \boldsymbol{\Sigma}_2)$.

Although it is common to assign a covariance structure for repeated measures, there is no obvious justification to impose a structure on $\boldsymbol{\Sigma}_j$. Therefore, no structure whatsoever on $\boldsymbol{\Sigma}_j$ will be assumed except that it is positive definite.

As in the univariate case here also possibly there can be some structure on mean vector. The structured mean vector case, when the mean remains unchanged over time as well as the unstructured mean vector will be discussed in each of the following subsections.

11.3.1 Classification Rules with CS Correlation Structure on *V*

Here we assume $\boldsymbol{V}_j$ to have the CS correlation structure. Thus, $\boldsymbol{V}_j = (1-\rho_j)\boldsymbol{I}_p + \rho_j \boldsymbol{1}_p\boldsymbol{1}'_p$. Since $\boldsymbol{V}_j$ must be positive definite, we also require that $-(1/(p-1)) < \rho_j < 1$. We will further assume that $0 < \rho_j < 1$.

The spectral decomposition of $\boldsymbol{V}_j$ yields $\boldsymbol{V}_j = \boldsymbol{P}\boldsymbol{\Lambda}_j\boldsymbol{P}'$, where $\boldsymbol{P}$ is an orthogonal matrix with first column as $(1/\sqrt{p})\boldsymbol{1}_p$ and $\boldsymbol{\Lambda}_j$ is a diagonal matrix given by

$$\boldsymbol{\Lambda}_j = \begin{bmatrix} \lambda_{j1} & \boldsymbol{0} \\ \boldsymbol{0} & \lambda_{j2}\boldsymbol{I}_{p-1} \end{bmatrix}.$$

Here λ_{j1} and λ_{j2} ($\lambda_{j1} > \lambda_{j2}$) are two distinct eigenvalues of $\boldsymbol{V}_j$. Since the elements of $\boldsymbol{P}$ do not depend on $\boldsymbol{V}_j$, and $\lambda_{j1} = 1 + (p-1)\rho_j$ and $\lambda_{j2} = 1 - \rho_j$, we make the canonical transformation of the data as,

$$\boldsymbol{w}_{ji} = (\boldsymbol{P}' \otimes \boldsymbol{I}_q)\boldsymbol{y}_{ji}.$$

Then $\boldsymbol{w}_{ji}$ is distributed as pq-dimensional normal with

$$\mathrm{E}(\boldsymbol{w}_{ji}) = (\boldsymbol{P}' \otimes \boldsymbol{I}_q)\mathrm{E}(\boldsymbol{y}_{ji}) = (\boldsymbol{P}' \otimes \boldsymbol{I}_q)\boldsymbol{\mu}_j,$$

and

$$\mathrm{Cov}(\boldsymbol{w}_{ji}) = \begin{bmatrix} \lambda_{j1}\boldsymbol{\Sigma}_j & \mathbf{0} & \cdots & \mathbf{0} \\ \mathbf{0} & \lambda_{j2}\boldsymbol{\Sigma}_j & \cdots & \mathbf{0} \\ \cdot & \cdot & \cdots & \cdot \\ \cdot & \cdot & \cdots & \cdot \\ \mathbf{0} & \mathbf{0} & \cdots & \lambda_{j2}\boldsymbol{\Sigma}_j \end{bmatrix},$$

which is a block diagonal matrix. This transformation simplifies the classification problem. We present here the two cases.

11.3.1.1 Classification Rules with Structured Mean Vectors

Here we assume that the mean for a given characteristic remains constant over time. We assume in population j, for the characteristic c, $\boldsymbol{\mu}_{jc} = \delta_{jc}\mathbf{1}_p$, $c = 1, \ldots, q$. So, $\boldsymbol{\mu}_{jc}$s are all $p \times 1$ vectors. Therefore, $\boldsymbol{\mu}_j = \mathbf{1}_p \otimes \boldsymbol{\delta}_j$, where $\boldsymbol{\delta}_j = (\delta_{j1}, \delta_{j2}, \ldots, \delta_{jq})'$. Accordingly,

$$\mathrm{E}(\boldsymbol{w}_{ji}) = (\boldsymbol{P}' \otimes \boldsymbol{I}_q)\boldsymbol{\mu}_j = (\boldsymbol{P}' \otimes \boldsymbol{I}_q)(\mathbf{1}_p \otimes \boldsymbol{\delta}_j) = \boldsymbol{P}'\mathbf{1}_p \otimes \boldsymbol{\delta}_j = [\sqrt{p}\boldsymbol{\delta}_j', \mathbf{0}, \ldots, \mathbf{0}]'.$$

Partition the $pq \times 1$ vector $\boldsymbol{w}_{ji}$ into p independent blocks of $q \times 1$ vectors as

$$\boldsymbol{w}_{ji} = [\boldsymbol{w}_{ji1}', \boldsymbol{w}_{ji2}', \ldots, \boldsymbol{w}_{jip}']',$$

where $\boldsymbol{w}_{ji1}, \ldots, \boldsymbol{w}_{jip}$ are all independent and the first component $\boldsymbol{w}_{ji1}$ of $\boldsymbol{w}_{ji}$ is distributed as q-variate normal with mean vector $\sqrt{p}\,\boldsymbol{\delta}_j$ and the variance–covariance matrix $\lambda_{j1}\boldsymbol{\Sigma}_j$. Further, $\boldsymbol{w}_{ji2}, \ldots, \boldsymbol{w}_{jip}$ are identically distributed as q-variate normal with zero mean vector and with the same variance–covariance matrix $\lambda_{j2}\boldsymbol{\Sigma}_j$, $j = 1, 2$. Therefore, except for the first component $\boldsymbol{w}_{ji1}$, the problem reduces to the problem of discriminating between populations when the mean vectors for the two populations are same and are known. Bartlett and Please (1963) have studied this problem when the covariance matrices have CS structure with the same correlation coefficient for the populations. Geisser (1964), Geisser and Desu (1967, 1968), Enis and Geisser (1970) consider this problem in Bayesian framework with common mean vector (not necessarily the zero vector) and with unstructured covariance matrix.

We will have two approaches. In Approach 1, to be discussed in Section 11.3.1.1.1, we will confine and discuss the classification rule based only on the first component $\boldsymbol{w}_{j1}$. Section 11.3.1.1.2 discusses Approach 2, where we will consider all the independent components $\boldsymbol{w}_{j1}, \boldsymbol{w}_{j2}, \ldots, \boldsymbol{w}_{jp}$ of $\boldsymbol{w}_j$ to classify the individuals.

11.3.1.1.1 Classification Rules Based on $\boldsymbol{w}_{j1}$

Case 1: $\boldsymbol{\Omega}_1 = \boldsymbol{\Omega}_2(\boldsymbol{V}_1 = \boldsymbol{V}_2 = \boldsymbol{V}, \boldsymbol{\Sigma}_1 = \boldsymbol{\Sigma}_2 = \boldsymbol{\Sigma})$

Let $\lambda_1 = 1 + (p-1)\rho$ and $\lambda_2 = 1 - \rho$ are two distinct eigenvalues of $\boldsymbol{V}$ with $\lambda_1 > \lambda_2$. Then, $\boldsymbol{w}_{j1} \sim N_q(\sqrt{p}\boldsymbol{\delta}_j, \lambda_1 \boldsymbol{\Sigma})$ in Population j, $j = 1, 2$. The linear discriminant rule based on $\boldsymbol{w}_1$ simplifies to,

> Allocate an individual with response $\boldsymbol{y}$ to Population 1 if
>
> $$(\boldsymbol{\delta}_1' - \boldsymbol{\delta}_2')\boldsymbol{\Sigma}^{-1}\boldsymbol{w}_1 \geq \frac{\sqrt{p}}{2}(\boldsymbol{\delta}_1'\boldsymbol{\delta}_1 - \boldsymbol{\delta}_2'\boldsymbol{\delta}_2),$$
>
> and to Population 2 otherwise.

The population parameters $\boldsymbol{\delta}_1, \boldsymbol{\delta}_2$, and $\boldsymbol{\Sigma}$ are unknown and should be estimated on the basis of two training samples of sizes n_1 and n_2 from the respective populations. We consider maximum likelihood approach for the estimation.

Maximum Likelihood Estimation of $\boldsymbol{\delta}_1$, $\boldsymbol{\delta}_2$, and $\boldsymbol{\Sigma}$: As in the univariate case here also, given $n_1 + n_2 = n$ (say) random observations $\boldsymbol{Y}_1 = (\boldsymbol{y}_{11}, \boldsymbol{y}_{12}, \ldots, \boldsymbol{y}_{1n_1})$ from Population 1 and $\boldsymbol{Y}_2 = (\boldsymbol{y}_{21}, \boldsymbol{y}_{22}, \ldots, \boldsymbol{y}_{2n_2})$ from Population 2, we define $\boldsymbol{W}_j = \sum_{i=1}^{n_j}(\boldsymbol{y}_{ji} - \bar{\boldsymbol{y}}_j)(\boldsymbol{y}_{ji} - \bar{\boldsymbol{y}}_j)'$, for $j = 1, 2$, as the within sum of squares and cross products matrix for the sample from the jth population, where $\bar{\boldsymbol{y}}_1$ and $\bar{\boldsymbol{y}}_2$ are two sample mean vectors for the two samples, and $\bar{\boldsymbol{y}}_j = (\bar{\boldsymbol{y}}_{j1}', \bar{\boldsymbol{y}}_{j2}', \ldots, \bar{\boldsymbol{y}}_{jp}')'$ is a $pq \times 1$ vector, $\bar{\boldsymbol{y}}_{jt}$ represents the sample mean of q responses at the tth time point in the jth population. The maximum likelihood estimates of $\boldsymbol{\delta}_1$ and $\boldsymbol{\delta}_2$ are obtained in closed forms as

$$\hat{\boldsymbol{\delta}}_1 = \frac{1}{p}(\bar{\boldsymbol{y}}_{11} + \bar{\boldsymbol{y}}_{12} + \cdots + \bar{\boldsymbol{y}}_{1p}), \tag{11.2}$$

and

$$\hat{\boldsymbol{\delta}}_2 = \frac{1}{p}(\bar{\boldsymbol{y}}_{21} + \bar{\boldsymbol{y}}_{22} + \cdots + \bar{\boldsymbol{y}}_{2p}). \tag{11.3}$$

We note that the expressions for $\hat{\boldsymbol{\delta}}_1$ and $\hat{\boldsymbol{\delta}}_2$ do not involve the unknown parameters $\boldsymbol{V}$ and $\boldsymbol{\Sigma}$ on their estimates. The maximum likelihood estimates of ρ (and hence $\boldsymbol{V}$) and $\boldsymbol{\Sigma}$ are obtained by simultaneously and iteratively solving Equations 11.4 and 11.5.

$$(p-1)k_0\rho^3 + \{k_0 - (p-1)k_0 + (p-1)^2 k_3 - (p-1)k_4\}\rho^2 \\ + \{2(p-1)k_3 - k_0\}\rho + (k_3 - k_4) = 0, \tag{11.4}$$

$$\hat{\boldsymbol{\Sigma}} = \frac{1}{np}\sum_{j=1}^{2}\sum_{i=1}^{n_j}\sum_{l=1}^{p}\sum_{m=1}^{p} v^{lm}\,(\boldsymbol{y}_{jim} - \hat{\boldsymbol{\delta}}_j)(\boldsymbol{y}_{jil} - \hat{\boldsymbol{\delta}}_j)', \tag{11.5}$$

where $k_0 = nq(p-1)p$, $\boldsymbol{k}_1 = (\bar{\boldsymbol{y}}_1 - \boldsymbol{1}_p \otimes \hat{\boldsymbol{\delta}}_1)(\bar{\boldsymbol{y}}_1 - \boldsymbol{1}_p \otimes \hat{\boldsymbol{\delta}}_1)'$, $\boldsymbol{k}_2 = (\bar{\boldsymbol{y}}_2 - \boldsymbol{1}_p \otimes \hat{\boldsymbol{\delta}}_2)(\bar{\boldsymbol{y}}_2 - \boldsymbol{1}_p \otimes \hat{\boldsymbol{\delta}}_2)'$ and $\boldsymbol{W} = \boldsymbol{W}_1 + \boldsymbol{W}_2 + n_1\boldsymbol{k}_1 + n_2\boldsymbol{k}_2$. We define $k_3 = \text{tr}(\boldsymbol{I}_p \otimes \boldsymbol{\Sigma}^{-1})\boldsymbol{W}$ and $k_4 = \text{tr}(\boldsymbol{J}_p \otimes \boldsymbol{\Sigma}^{-1})\boldsymbol{W}$. The algorithm for solving these equations is as follows:

Algorithm 11.3.1

Step 1: Get the pooled sample variance–covariance matrix for repeated measures. Say, it is $\boldsymbol{G}$. Then obtain an initial estimate of ρ as $\hat{\rho}_o = (\boldsymbol{1}_p'\boldsymbol{G}\boldsymbol{1}_p - \text{tr}\,\boldsymbol{G})/p(p-1)$, and ensure that $0 < \hat{\rho}_o < 1$. Take $\hat{\boldsymbol{V}}_o = (1-\hat{\rho}_o)\boldsymbol{I}_p + \hat{\rho}_o\boldsymbol{1}_p\boldsymbol{1}_p'$ as an initial estimate of $\boldsymbol{V}$.

Step 2: Compute $\hat{\boldsymbol{\Sigma}}$ from Equation 11.5.

Step 3: Compute k_3 and k_4 using $\hat{\boldsymbol{\Sigma}}$ obtained in Step 2.

Step 4: Compute the value of $\hat{\rho}$ by solving the cubic Equation 11.4. Ensure that $0 < \hat{\rho} < 1$. Truncate $\hat{\rho}$ to 0 or 1, if it is outside this range.

Step 5: Compute the revised estimate $\hat{\boldsymbol{V}}$ from $\hat{\rho}$.

Step 6: Compute the revised estimate $\hat{\boldsymbol{\Sigma}}$ from Equation 11.5 using $\hat{\boldsymbol{V}}$ obtained in Step 5.

Step 7: Repeat Steps 3 to 6 until convergence is attained. This is ensured by verifying if the maximum of the absolute difference between two successive values of $\hat{\rho}$ and the absolute difference between two successive values of trace of $\hat{\boldsymbol{\Sigma}}$ is less than a predetermined number ϵ.

Case 2: $\boldsymbol{\Omega}_1 \neq \boldsymbol{\Omega}_2$ ($\boldsymbol{V}_1 \neq \boldsymbol{V}_2, \boldsymbol{\Sigma}_1 = \boldsymbol{\Sigma}_2 = \boldsymbol{\Sigma}$)

Let $\lambda_{j1} = 1 + (p-1)\rho_j$ and $\lambda_{j2} = 1 - \rho_j$ are the two distinct eigenvalues of $\boldsymbol{V}_j$ with $\lambda_{j1} > \lambda_{j2}$, for $j = 1, 2$. Then, $\boldsymbol{w}_{j1} \sim N_q(\sqrt{p}\,\boldsymbol{\delta}_j, \lambda_{j1}\boldsymbol{\Sigma})$ in Population j,

$j = 1, 2$. Therefore, the quadratic discriminant rule based on $\boldsymbol{w}_1$ is given by:

Allocate an individual with response $\boldsymbol{y}$ to Population 1 if

$$-\frac{1}{2}\left(\frac{1}{\lambda_{11}} - \frac{1}{\lambda_{21}}\right)(\boldsymbol{w}_1'\boldsymbol{\Sigma}^{-1}\boldsymbol{w}_1) + \sqrt{p}\left(\frac{\boldsymbol{\delta}_1'}{\lambda_{11}} - \frac{\boldsymbol{\delta}_2'}{\lambda_{21}}\right)\boldsymbol{\Sigma}^{-1}\boldsymbol{w}_1$$

$$\geq \frac{p}{2}\left(\frac{\boldsymbol{\delta}_1'\boldsymbol{\Sigma}^{-1}\boldsymbol{\delta}_1}{\lambda_{11}} - \frac{\boldsymbol{\delta}_2'\boldsymbol{\Sigma}^{-1}\boldsymbol{\delta}_2}{\lambda_{21}}\right) - \frac{1}{2}\ln\left(\frac{\lambda_{21}}{\lambda_{11}}\right),$$

and to Population 2 otherwise.

The maximum likelihood estimates of $\boldsymbol{\delta}_1$ and $\boldsymbol{\delta}_2$ are identical to Equations 11.2 and 11.3. The maximum likelihood estimates of ρ_1, ρ_2, and $\boldsymbol{\Sigma}$ are obtained by simultaneously and iteratively solving the following three equations:

$$(p-1)k_{01}\rho_1^3 + \{k_{01} - (p-1)k_{01} + (p-1)^2k_8 - (p-1)l_8\}\rho_1^2$$
$$+ \{2(p-1)k_8 - k_{01}\}\rho_1 + (k_8 - l_8) = 0,$$

$$(p-1)k_{02}\rho_2^3 + \{k_{02} - (p-1)k_{02} + (p-1)^2k_9 - (p-1)l_9\}\rho_2^2$$
$$+ \{2(p-1)k_9 - k_{02}\}\rho_2 + (k_9 - l_9) = 0,$$

and

$$\hat{\boldsymbol{\Sigma}} = \frac{1}{np}\sum_{j=1}^{2}\sum_{i=1}^{n_j}\sum_{l=1}^{p}\sum_{m=1}^{p} v_j^{lm}(\boldsymbol{y}_{jim} - \hat{\boldsymbol{\delta}}_j)(\boldsymbol{y}_{jil} - \hat{\boldsymbol{\delta}}_j)',$$

where $k_8 = \text{tr}(\boldsymbol{I}_p \otimes \boldsymbol{\Sigma}^{-1})\boldsymbol{S}_1, l_8 = \text{tr}(\boldsymbol{J}_p \otimes \boldsymbol{\Sigma}^{-1})\boldsymbol{S}_1, k_9 = \text{tr}(\boldsymbol{I}_p \otimes \boldsymbol{\Sigma}^{-1})\boldsymbol{S}_2, l_9 = \text{tr}(\boldsymbol{J}_p \otimes \boldsymbol{\Sigma}^{-1})\boldsymbol{S}_2, \boldsymbol{S}_1 = \boldsymbol{W}_1 + n_1\boldsymbol{k}_1, \boldsymbol{S}_2 = \boldsymbol{W}_2 + n_2\boldsymbol{k}_2, k_{01} = n_1q(p-1)p$ and $k_{02} = n_2q(p-1)p$.

The computation algorithm similar to that described in the previous case can be devised here and for the later cases in this section as well.

Case 3: $\boldsymbol{\Omega}_1 \neq \boldsymbol{\Omega}_2$ $(\boldsymbol{V}_1 = \boldsymbol{V}_2 = \boldsymbol{V},\ \boldsymbol{\Sigma}_1 \neq \boldsymbol{\Sigma}_2)$

Here $\boldsymbol{w}_{j1} \sim N_q(\sqrt{p}\,\boldsymbol{\delta}_j, \lambda_1\boldsymbol{\Sigma}_j)$ in Population j, $j = 1, 2$. Therefore, the quadratic discriminant rule based on $\boldsymbol{w}_1$ is given by:

Allocate an individual with response $\boldsymbol{y}$ to Population 1 if

$$-\frac{1}{2\lambda_1}\boldsymbol{w}_1'(\boldsymbol{\Sigma}_1^{-1} - \boldsymbol{\Sigma}_2^{-1})\boldsymbol{w}_1 + \frac{\sqrt{p}}{\lambda_1}(\boldsymbol{\delta}_1'\boldsymbol{\Sigma}_1^{-1} - \boldsymbol{\delta}_2'\boldsymbol{\Sigma}_2^{-1})\boldsymbol{w}_1$$

$$\geq \frac{p}{2\lambda_1}(\boldsymbol{\delta}_1'\boldsymbol{\Sigma}_1^{-1}\boldsymbol{\delta}_1 - \boldsymbol{\delta}_2'\boldsymbol{\Sigma}_2^{-1}\boldsymbol{\delta}_2) - \frac{1}{2}\ln\left(\frac{|\boldsymbol{\Sigma}_2|}{|\boldsymbol{\Sigma}_1|}\right),$$

and to Population 2 otherwise.

The maximum likelihood estimates of $\boldsymbol{\delta}_1$ and $\boldsymbol{\delta}_2$ are identical to Equations 11.2 and 11.3. We solve the following equations to obtain the maximum likelihood estimates of $\rho, \boldsymbol{\Sigma}_1$ and $\boldsymbol{\Sigma}_2$.

$$(p-1)k_0\rho^3 + \{k_0 - (p-1)k_0 + (p-1)^2(k_5+k_6) - (p-1)(k_5+k_6)\}\rho^2 \\ + \{2(p-1)(k_5+k_6) - k_0\}\rho + (k_5+k_6-l_5-l_6) = 0,$$

and

$$\hat{\boldsymbol{\Sigma}}_j = \frac{1}{n_j p}\sum_{i=1}^{n_j}\sum_{l=1}^{p}\sum_{m=1}^{p} v^{lm}(\boldsymbol{y}_{jim} - \hat{\boldsymbol{\delta}}_j)(\boldsymbol{y}_{jil} - \hat{\boldsymbol{\delta}}_j)', \quad \text{for } j = 1, 2,$$

where $k_5 = \text{tr}(\boldsymbol{I}_p \otimes \boldsymbol{\Sigma}_1^{-1})\boldsymbol{S}_1, l_5 = \text{tr}(\boldsymbol{J}_p \otimes \boldsymbol{\Sigma}_1^{-1})\boldsymbol{S}_1, k_6 = \text{tr}(\boldsymbol{I}_p \otimes \boldsymbol{\Sigma}_2^{-1})\boldsymbol{S}_2$, and $l_6 = \text{tr}(\boldsymbol{J}_p \otimes \boldsymbol{\Sigma}_2^{-1})\boldsymbol{S}_2$.

Case 4: $\boldsymbol{\Omega}_1 \neq \boldsymbol{\Omega}_2$ $(\boldsymbol{V}_1 \neq \boldsymbol{V}_2, \boldsymbol{\Sigma}_1 \neq \boldsymbol{\Sigma}_2)$

Here $\boldsymbol{w}_{j1} \sim N_q(\sqrt{p}\boldsymbol{\delta}_j, \lambda_{j1}\boldsymbol{\Sigma}_j)$ in Population $j, j = 1, 2$. Therefore, the quadratic discriminant rule based on $\boldsymbol{w}_1$ is given by:

Allocate an individual with response $\boldsymbol{y}$ to Population 1 if

$$-\frac{1}{2}\boldsymbol{w}_1'\left(\frac{\boldsymbol{\Sigma}_1^{-1}}{\lambda_{11}} - \frac{\boldsymbol{\Sigma}_2^{-1}}{\lambda_{21}}\right)\boldsymbol{w}_1 + \sqrt{p}\left(\frac{\boldsymbol{\delta}_1'\boldsymbol{\Sigma}_1^{-1}}{\lambda_{11}} - \frac{\boldsymbol{\delta}_2'\boldsymbol{\Sigma}_2^{-1}}{\lambda_{21}}\right)\boldsymbol{w}_1 \\ \geq \frac{p}{2}\left(\frac{\boldsymbol{\delta}_1'\boldsymbol{\Sigma}_1^{-1}\boldsymbol{\delta}_1}{\lambda_{11}} - \frac{\boldsymbol{\delta}_2'\boldsymbol{\Sigma}_2^{-1}\boldsymbol{\delta}_2}{\lambda_{21}}\right) - \frac{1}{2}\ln\left(\frac{|\boldsymbol{\Sigma}_2|}{|\boldsymbol{\Sigma}_1|}\right) - \frac{1}{2}\ln\left(\frac{\lambda_{21}}{\lambda_{11}}\right),$$

and to Population 2 otherwise.

Again the maximum likelihood estimates of $\boldsymbol{\delta}_1$ and $\boldsymbol{\delta}_2$ are identical to Equations 11.2 and 11.3. The maximum likelihood estimates of $\rho_1, \rho_2, \boldsymbol{\Sigma}_1$, and $\boldsymbol{\Sigma}_2$ are obtained by solving

$$(p-1)k_{01}\rho_1^3 + \{k_{01} - (p-1)k_{01} + (p-1)^2k_5 - (p-1)l_5\}\rho_1^2 \\ + \{2(p-1)k_5 - k_{01}\}\rho_1 + (k_5 - l_5) = 0,$$

$$(p-1)k_{02}\rho_2^3 + \{k_{02} - (p-1)k_{02} + (p-1)^2k_6 - (p-1)l_6\}\rho_2^2 \\ + \{2(p-1)k_6 - k_{02}\}\rho_2 + (k_6 - l_6) = 0,$$

and

$$\hat{\boldsymbol{\Sigma}}_j = \frac{1}{n_j p}\sum_{i=1}^{n_j}\sum_{l=1}^{p}\sum_{m=1}^{p} v_j^{lm}(\boldsymbol{y}_{jim} - \hat{\boldsymbol{\delta}}_j)(\boldsymbol{y}_{jil} - \hat{\boldsymbol{\delta}}_j)', \quad \text{for } j = 1, 2.$$

11.3.1.1.2 Classification Rules Based on $\boldsymbol{w}_{j1}, \boldsymbol{w}_{j2}, \ldots, \boldsymbol{w}_{jp}$

In this section classification rules are developed based on all the independent components $\boldsymbol{w}_{j1}, \boldsymbol{w}_{j2}, \ldots, \boldsymbol{w}_{jp}$. We can perhaps allocate an individual to the Population j based on the posterior probability $P(j \mid \boldsymbol{w}_1)$ or $P(j \mid \boldsymbol{w}_2), \ldots,$ or $P(j \mid \boldsymbol{w}_p)$ (McLachlan, 1992). However, each posterior probability may not classify an individual in the same way. One way to combine all the information into a single index is by computing the average posterior probability (assuming equal priors) based on all the posterior probabilities $P(j \mid \boldsymbol{w}_1), P(j \mid \boldsymbol{w}_2), \ldots,$ and $P(j \mid \boldsymbol{w}_p)$ namely,

$$P(j \mid \boldsymbol{w}) = \frac{P(j \mid \boldsymbol{w}_1) + P(j \mid \boldsymbol{w}_2) + \cdots + P(j \mid \boldsymbol{w}_p)}{p}, \quad j = 1, 2.$$

Then the classification rule is

Allocate an individual to Population 1 if
the average posterior probability $P(1 \mid \boldsymbol{w}) \geq 0.5$,
and to Population 2 otherwise.

To estimate the average posterior probability $P(1 \mid \boldsymbol{w})$ we will use the expressions for the estimation of the population parameters $\boldsymbol{\delta}_1, \boldsymbol{\delta}_2, \lambda_{11}, \lambda_{12}, \lambda_{21}, \lambda_{22}, \boldsymbol{\Sigma}_1, \boldsymbol{\Sigma}_2$. These have already been discussed in the previous section.

11.3.1.2 Classification Rules with Unstructured Mean Vectors

In this case we do not assume any structure on $\boldsymbol{\mu}_j$. However, as earlier the CS structure on $\boldsymbol{V}_j$ is assumed. Let $\boldsymbol{P}$ be the matrix as defined in Section 11.3.1 and also $\boldsymbol{w}_{ji}$ as in Section 11.3.1. Since there is no structure on the mean vector we have

$$\mathrm{E}(\boldsymbol{w}_{ji}) = (\boldsymbol{P}' \otimes \boldsymbol{I}_q)\mathrm{E}(\boldsymbol{y}_{ji}) = (\boldsymbol{P}' \otimes \boldsymbol{I}_q)\boldsymbol{\mu}_j = \boldsymbol{e}_j = [\boldsymbol{e}'_{j1}, \boldsymbol{e}'_{j2}, \ldots, \boldsymbol{e}'_{jp}]', \tag{11.6}$$

and $\mathrm{Cov}(\boldsymbol{w}_{ji})$ is same as in Section 11.3.1. Partitioning the $pq \times 1$ vector $\boldsymbol{w}_{ji}$ as in Section 11.3.1.1, we observe that $\boldsymbol{w}_{ji1}$ is distributed as q-variate normal with mean vector $\boldsymbol{e}_{j1}$ and the variance–covariance matrix $\lambda_{j1}\boldsymbol{\Sigma}_j$. Each of the other components $\boldsymbol{w}_{ji2}, \boldsymbol{w}_{ji3}, \ldots, \boldsymbol{w}_{jip}$ is also q-variate normal with respective mean vectors $\boldsymbol{e}_{j2}, \boldsymbol{e}_{j3}, \ldots, \boldsymbol{e}_{jp}$, but with the same variance–covariance matrix $\lambda_{j2}\boldsymbol{\Sigma}_j$ for $j = 1, 2$. Furthermore, $\boldsymbol{w}_{ji1}, \boldsymbol{w}_{ji2}, \ldots, \boldsymbol{w}_{jip}$ are all independent.

As in Section 11.3.1.1.2, we calculate the average posterior probability on the basis of all the independent components $\boldsymbol{w}_1, \boldsymbol{w}_2, \ldots, \boldsymbol{w}_p$ of $\boldsymbol{w}$ to classify an individual. The classification rule will be same as in Section 11.3.1.1.2, and the unknown parameters will be replaced by their MLEs, which will differ in this case.

Case 1: $\boldsymbol{\Omega}_1 = \boldsymbol{\Omega}_2 (\boldsymbol{V}_1 = \boldsymbol{V}_2 = \boldsymbol{V}, \boldsymbol{\Sigma}_1 = \boldsymbol{\Sigma}_2 = \boldsymbol{\Sigma})$

The MLEs of $\boldsymbol{\mu}_j$'s are $\hat{\boldsymbol{\mu}}_j = \bar{\boldsymbol{y}}_j, j = 1, 2$. To get the MLEs of $\boldsymbol{e}_{j1}, \boldsymbol{e}_{j2}, \ldots, \boldsymbol{e}_{jp}$ we observe that $\hat{\boldsymbol{e}}_j = (\boldsymbol{P}' \otimes \boldsymbol{I}_q)\hat{\boldsymbol{\mu}}_j = [\hat{\boldsymbol{\xi}}'_{j1}, \hat{\boldsymbol{\xi}}'_{j2}, \ldots, \hat{\boldsymbol{\xi}}'_{jp}]'$ (say). Therefore, the MLE of $\boldsymbol{e}_{jl}$ is $\hat{\boldsymbol{e}}_{jl} = \hat{\boldsymbol{\xi}}_{jl}$ for $j = 1, 2;\ \ l = 1, 2, \ldots, p$.
The maximum likelihood estimates of ρ and $\boldsymbol{\Sigma}$ are obtained by simultaneously and iteratively solving the following equations:

$$\begin{aligned}(p-1)k_0\rho^3 &+ \{k_0 - (p-1)k_0 + (p-1)^2 k'_3 - (p-1)k'_4\}\rho^2 \\ &+ \{2(p-1)k'_3 - k_0\}\rho + (k'_3 - k'_4) = 0, \end{aligned} \tag{11.7}$$

and

$$\hat{\boldsymbol{\Sigma}} = \frac{1}{np}\sum_{j=1}^{2}\sum_{i=1}^{n_j}\sum_{l=1}^{p}\sum_{m=1}^{p} v^{lm}(\boldsymbol{y}_{jim} - \bar{\boldsymbol{y}}_{jm})(\boldsymbol{y}_{jil} - \bar{\boldsymbol{y}}_{jl})', \tag{11.8}$$

where the $pq \times 1$ vector $(\boldsymbol{y}_{ji} - \bar{\boldsymbol{y}}_j)$ is partitioned into p blocks of $q \times 1$ vectors as

$$(\boldsymbol{y}_{ji} - \bar{\boldsymbol{y}}_j) = \left((\boldsymbol{y}_{ji1} - \bar{\boldsymbol{y}}_{j1})', \ldots, (\boldsymbol{y}_{jip} - \bar{\boldsymbol{y}}_{jp})'\right)'.$$

The quantity k_0 is defined in Section 11.3.1.1.1 and $\boldsymbol{W}' = \boldsymbol{W}_1 + \boldsymbol{W}_2, k'_3 = \mathrm{tr}(\boldsymbol{I}_p \otimes \boldsymbol{\Sigma}^{-1})\boldsymbol{W}'$ and $k'_4 = \mathrm{tr}(\boldsymbol{J}_p \otimes \boldsymbol{\Sigma}^{-1})\boldsymbol{W}'$.

Case 2: $\boldsymbol{\Omega}_1 \neq \boldsymbol{\Omega}_2\ (\boldsymbol{V}_1 \neq \boldsymbol{V}_2, \boldsymbol{\Sigma}_1 = \boldsymbol{\Sigma}_2 = \boldsymbol{\Sigma})$

As before, the MLEs of $\boldsymbol{\mu}_j$'s are $\hat{\boldsymbol{\mu}}_j = \bar{\boldsymbol{y}}_j$ and the MLE of $\boldsymbol{e}_{jl}$ is $\hat{\boldsymbol{\xi}}_{jl}$ for $j = 1, 2$ and $l = 1, 2, \ldots, p$. The MLEs of ρ_1, ρ_2, and $\boldsymbol{\Sigma}$ are obtained by simultaneously and iteratively solving the following equations:

$$\begin{aligned}(p-1)k_{01}\rho_1^3 &+ \{k_{01} - (p-1)k_{01} + (p-1)^2 k'_8 - (p-1)l'_8\}\rho_1^2 \\ &+ \{2(p-1)k'_8 - k_{01}\}\rho_1 + (k'_8 - l'_8) = 0, \end{aligned} \tag{11.9}$$

$$\begin{aligned}(p-1)k_{02}\rho_2^3 &+ \{k_{02} - (p-1)k_{02} + (p-1)^2 k'_9 - (p-1)l'_9\}\rho_2^2 \\ &+ \{2(p-1)k'_9 - k_{02}\}\rho_2 + (k'_9 - l'_9) = 0, \end{aligned} \tag{11.10}$$

and

$$\hat{\boldsymbol{\Sigma}} = \frac{1}{np}\sum_{j=1}^{2}\sum_{i=1}^{n_j}\sum_{l=1}^{p}\sum_{m=1}^{p} v_j^{lm}\,(\boldsymbol{y}_{jim} - \bar{\boldsymbol{y}}_{jm})(\boldsymbol{y}_{jil} - \bar{\boldsymbol{y}}_{jl})', \tag{11.11}$$

where $k'_8 = \mathrm{tr}(\boldsymbol{I}_p \otimes \boldsymbol{\Sigma}^{-1})\boldsymbol{W}_1$, $l'_8 = \mathrm{tr}(\boldsymbol{J}_p \otimes \boldsymbol{\Sigma}^{-1})\boldsymbol{W}_1$, $k'_9 = \mathrm{tr}(\boldsymbol{I}_p \otimes \boldsymbol{\Sigma}^{-1})\boldsymbol{W}_2$, and $l'_9 = \mathrm{tr}(\boldsymbol{J}_p \otimes \boldsymbol{\Sigma}^{-1})\boldsymbol{W}_2$. The quantities k_{01} and k_{02} are defined in Section 11.3.1.1.1.

Case 3: $\boldsymbol{\Omega}_1 \neq \boldsymbol{\Omega}_2$ $(\boldsymbol{V}_1 = \boldsymbol{V}_2 = \boldsymbol{V},\ \boldsymbol{\Sigma}_1 \neq \boldsymbol{\Sigma}_2)$

Again the MLEs of $\boldsymbol{\mu}_j$'s are $\hat{\boldsymbol{\mu}}_j = \bar{\boldsymbol{y}}_j$ and the MLEs of $\boldsymbol{e}_{jl}$ is $\hat{\boldsymbol{\zeta}}_{jl}$ for $j = 1,2$ and $l = 1,2,\ldots,p$. The maximum likelihood estimates of ρ, $\boldsymbol{\Sigma}_1$, and $\boldsymbol{\Sigma}_2$ are obtained by solving the following equations:

$$\begin{aligned}(p-1)k_0\rho^3 &+ \{k_0 - (p-1)k_0 + (p-1)^2(k_5' + k_6') - (p-1)(k_5' + k_6')\}\rho^2 \\ &+ \{2(p-1)(k_5' + k_6') - k_0\}\rho + (k_5' + k_6' - l_5' - l_6') = 0, \end{aligned} \tag{11.12}$$

and

$$\hat{\boldsymbol{\Sigma}}_j = \frac{1}{n_j p}\sum_{i=1}^{n_j}\sum_{l=1}^{p}\sum_{m=1}^{p} v^{lm}\,(\boldsymbol{y}_{jim} - \bar{\boldsymbol{y}}_{jm})(\boldsymbol{y}_{jil} - \bar{\boldsymbol{y}}_{jl})', \quad \text{for } j = 1,2, \tag{11.13}$$

where $k_5' = \text{tr}(\boldsymbol{I}_p \otimes \boldsymbol{\Sigma}_1^{-1})\boldsymbol{W}_1$, $l_5' = \text{tr}(\boldsymbol{J}_p \otimes \boldsymbol{\Sigma}_1^{-1})\boldsymbol{W}_1$, $k_6' = \text{tr}(\boldsymbol{I}_p \otimes \boldsymbol{\Sigma}_2^{-1})\boldsymbol{W}_2$, and $l_6' = \text{tr}(\boldsymbol{J}_p \otimes \boldsymbol{\Sigma}_2^{-1})\boldsymbol{W}_2$.

Case 4: $\boldsymbol{\Omega}_1 \neq \boldsymbol{\Omega}_2(\boldsymbol{V}_1 \neq \boldsymbol{V}_2, \boldsymbol{\Sigma}_1 \neq \boldsymbol{\Sigma}_2)$

With $\hat{\boldsymbol{\mu}}_j = \bar{\boldsymbol{y}}_j$ and $\hat{\boldsymbol{e}}_{jl} = \hat{\boldsymbol{\zeta}}_{jl}$ for $j = 1,2$ and $l = 1,2,\ldots,p$, the maximum likelihood estimates of $\rho_1, \rho_2, \boldsymbol{\Sigma}_1$, and $\boldsymbol{\Sigma}_2$ are obtained by solving

$$\begin{aligned}(p-1)k_{01}\rho_1^3 &+ \{k_{01} - (p-1)k_{01} + (p-1)^2k_5' - (p-1)l_5'\}\rho_1^2 \\ &+ \{2(p-1)k_5' - k_{01}\}\rho_1 + (k_5' - l_5') = 0, \end{aligned} \tag{11.14}$$

$$\begin{aligned}(p-1)k_{02}\rho_2^3 &+ \{k_{02} - (p-1)k_{02} + (p-1)^2k_6' - (p-1)l_6'\}\rho_2^2 \\ &+ \{2(p-1)k_6' - k_{02}\}\rho_2 + (k_6' - l_6') = 0, \end{aligned} \tag{11.15}$$

and

$$\hat{\boldsymbol{\Sigma}}_j = \frac{1}{n_j p}\sum_{i=1}^{n_j}\sum_{l=1}^{p}\sum_{m=1}^{p} v_j^{lm}\,(\boldsymbol{y}_{jim} - \bar{\boldsymbol{y}}_{jm})(\boldsymbol{y}_{jil} - \bar{\boldsymbol{y}}_{jl})', \quad \text{for } j = 1,2. \tag{11.16}$$

11.3.2 Classification Rules with AR(1) Correlation Structure on V

Often the correlation matrix of the repeated measures $\boldsymbol{V}_j$, $j = 1,2$ may have AR(1) structure. That is, for $j = 1,2$,

$$\boldsymbol{V}_j = \begin{bmatrix} 1 & \rho_j & \rho_j^2 & \cdots & \rho_j^{p-1} \\ \rho_j & 1 & \rho_j & \cdots & \rho_j^{p-2} \\ \rho_j^2 & \rho_j & 1 & \cdots & \rho_j^{p-3} \\ \cdot & \cdot & \cdot & \cdots & \cdot \\ \cdot & \cdot & \cdot & \cdots & \cdot \\ \rho_j^{p-1} & \rho_j^{p-2} & \rho_j^{p-3} & \cdots & 1 \end{bmatrix}.$$

The determinant of this matrix is given by $|\boldsymbol{V}_j| = \left(1-\rho_j^2\right)^{p-1}$, and its inverse is given by, $\boldsymbol{V}_j^{-1} = (1-\rho_j^2)^{-1}[\boldsymbol{I}_p + \rho_j^2 \boldsymbol{C}_1 + \rho_j \boldsymbol{C}_2]$, where $\boldsymbol{C}_1 = \text{diag}(0, 1, \ldots, 1, 0)$ and $\boldsymbol{C}_2$ is a tridiagonal matrix with 0 on the diagonal and 1 on the first superdiagonal and on the first subdiagonal.

11.3.2.1 Classification Rules with Structured Mean Vectors

Case 1: $\boldsymbol{\Omega}_1 = \boldsymbol{\Omega}_2$ ($\boldsymbol{V}_1 = \boldsymbol{V}_2 = \boldsymbol{V}, \boldsymbol{\Sigma}_1 = \boldsymbol{\Sigma}_2 = \boldsymbol{\Sigma}$)

The linear classification rule in this case is given by:

Allocate an individual with response $\boldsymbol{y}$ to Population 1 if

$$(\boldsymbol{1}_p' \boldsymbol{V}^{-1} \otimes (\boldsymbol{\delta}_1 - \boldsymbol{\delta}_2)' \boldsymbol{\Sigma}^{-1}) \boldsymbol{y} \geq \frac{1}{2} (\boldsymbol{1}_p' \boldsymbol{V}^{-1} \boldsymbol{1}_p) [(\boldsymbol{\delta}_1 - \boldsymbol{\delta}_2)' \boldsymbol{\Sigma}^{-1} (\boldsymbol{\delta}_1 + \boldsymbol{\delta}_2)],$$

and to Population 2 otherwise.

Maximum Likelihood Estimation of $\boldsymbol{\delta}_1, \boldsymbol{\delta}_2, \boldsymbol{V}$, and $\boldsymbol{\Sigma}$: In this case,

$$\hat{\boldsymbol{\delta}}_j = \frac{\left(\sum_{m=1}^{p} v^{1m}\right) \bar{\boldsymbol{y}}_{j1} + \left(\sum_{m=1}^{p} v^{2m}\right) \bar{\boldsymbol{y}}_{j2} + \cdots + \left(\sum_{m=1}^{p} v^{pm}\right) \bar{\boldsymbol{y}}_{jp}}{\left(\sum_{l=1}^{p} \sum_{m=1}^{p} v^{lm}\right)}, \quad j = 1, 2, \tag{11.17}$$

where $\boldsymbol{V}^{-1} = (v^{lm})$. It may be observed that $\hat{\boldsymbol{\delta}}_1$ and $\hat{\boldsymbol{\delta}}_2$ are both free from the unknown parameter $\boldsymbol{\Sigma}$. The maximum likelihood estimates $\hat{\boldsymbol{\delta}}_1, \hat{\boldsymbol{\delta}}_2, \hat{\rho}$, and $\hat{\boldsymbol{\Sigma}}$ are obtained by simultaneously and iteratively solving Equations 11.16 through 11.8.

$$-2n(p-1)q\rho^3 + c_3\rho^2 + [2n(p-1)q - 2c_1 - 2c_2]\rho + c_3 = 0, \tag{11.18}$$

$$\hat{\boldsymbol{\Sigma}} = \frac{1}{np} \sum_{j=1}^{2} \sum_{i=1}^{n_j} \sum_{l=1}^{p} \sum_{m=1}^{p} v^{lm} (\boldsymbol{y}_{jim} - \boldsymbol{\delta}_j)(\boldsymbol{y}_{jil} - \boldsymbol{\delta}_j)', \tag{11.19}$$

where $c_1 = \text{tr}(\boldsymbol{I}_p \otimes \boldsymbol{\Sigma}^{-1})\boldsymbol{W}$, $c_2 = \text{tr}(\boldsymbol{C}_1 \otimes \boldsymbol{\Sigma}^{-1})\boldsymbol{W}$, $c_3 = \text{tr}(\boldsymbol{C}_2 \otimes \boldsymbol{\Sigma}^{-1})\boldsymbol{W}$ and $\boldsymbol{W} = \boldsymbol{W}_1 + \boldsymbol{W}_2 + n_1\boldsymbol{k}_1 + n_2\boldsymbol{k}_2$, $\boldsymbol{k}_1 = (\bar{\boldsymbol{y}}_1 - \boldsymbol{1}_p \otimes \hat{\boldsymbol{\delta}}_1)(\bar{\boldsymbol{y}}_1 - \boldsymbol{1}_p \otimes \hat{\boldsymbol{\delta}}_1)'$, and $\boldsymbol{k}_2 = (\bar{\boldsymbol{y}}_2 - \boldsymbol{1}_p \otimes \hat{\boldsymbol{\delta}}_2)(\bar{\boldsymbol{y}}_2 - \boldsymbol{1}_p \otimes \hat{\boldsymbol{\delta}}_2)'$.

The computations can be carried out by the following algorithm.

Algorithm 11.3.2

Step 1: Get the initial pooled sample variance–covariance matrix for repeated measures. Say it is $\boldsymbol{G}^{\star}$. Get the average of the first superdiagonal elements of $\boldsymbol{G}^{\star}$, say, $\rho_{1\star}$. Then get the average of the second superdiagonal elements of $\boldsymbol{G}^{\star}$, say, $\rho_{2\star}$ and so on. The initial estimate of ρ is obtained as

$$\hat{\rho}_o = \left(\frac{(\rho_{1\star})^{p-1} + (\rho_{2\star})^{(p-1)/2} + (\rho_{3\star})^{(p-1)/3} + \cdots + \rho_{p-1\star}}{p-1} \right)^{1/p-1},$$

and thus $\hat{\boldsymbol{V}}_o$, an initial estimate of $\boldsymbol{V}$ is obtained by replacing ρ by $\hat{\rho}_o$.

Step 2: Compute $\hat{\boldsymbol{\delta}}_1$ and $\hat{\boldsymbol{\delta}}_2$ from Equation 11.16.

Step 3: Compute $\boldsymbol{k}_1$ and $\boldsymbol{k}_2$.

Step 4: Compute $\hat{\boldsymbol{\Sigma}}$ from Equation 11.18.

Step 5: Compute c_1, c_2, and c_3.

Step 6: Compute the value of $\hat{\rho}$ by solving the cubic Equation 11.17. Ensure that $0 < \hat{\rho} < 1$. Truncate $\hat{\rho}$ to 0 or 1, if it is outside this range.

Step 7: Compute the revised estimate $\hat{\boldsymbol{V}}$ from $\hat{\rho}$.

Step 8: Repeat steps 2 through 7 until convergence is attained. This is ensured by verifying if the maximum of the absolute difference between two successive values of $\hat{\rho}$, and the L_1 distance between two successive values of $\hat{\boldsymbol{\delta}}_1$ and $\hat{\boldsymbol{\delta}}_2$, and the absolute difference between two successive values of trace of $\hat{\boldsymbol{\Sigma}}$ is less than a predetermined number ϵ.

Case 2: $\boldsymbol{\Omega}_1 \neq \boldsymbol{\Omega}_2$ $(\boldsymbol{V}_1 \neq \boldsymbol{V}_2, \boldsymbol{\Sigma}_1 = \boldsymbol{\Sigma}_2 = \boldsymbol{\Sigma})$

The quadratic classification rule is given by

Allocate an individual with response $\boldsymbol{y}$ to Population 1 if

$$-\frac{1}{2}\boldsymbol{y}'[(\boldsymbol{V}_1^{-1} - \boldsymbol{V}_2^{-1}) \otimes \boldsymbol{\Sigma}^{-1}]\boldsymbol{y} + (\boldsymbol{1}_p'\boldsymbol{V}_1^{-1} \otimes \boldsymbol{\delta}_1'\boldsymbol{\Sigma}^{-1} - \boldsymbol{1}_p'\boldsymbol{V}_2^{-1} \otimes \boldsymbol{\delta}_2'\boldsymbol{\Sigma}^{-1})\boldsymbol{y}$$
$$\geq \frac{1}{2}[(\boldsymbol{1}_p'\boldsymbol{V}_1^{-1}\boldsymbol{1}_p)(\boldsymbol{\delta}_1'\boldsymbol{\Sigma}^{-1}\boldsymbol{\delta}_1) - (\boldsymbol{1}_p'\boldsymbol{V}_2^{-1}\boldsymbol{1}_p)(\boldsymbol{\delta}_2'\boldsymbol{\Sigma}^{-1}\boldsymbol{\delta}_2)] - \frac{q}{2}\ln\left(\frac{|\boldsymbol{V}_2|}{|\boldsymbol{V}_1|}\right),$$

and to Population 2 otherwise.

The maximum likelihood estimates of $\boldsymbol{\delta}_1, \boldsymbol{\delta}_2, \rho_1, \rho_2$, and $\boldsymbol{\Sigma}$ are obtained by simultaneously and iteratively solving the following equations:

$$\hat{\boldsymbol{\delta}}_j = \frac{\left(\sum_{m=1}^{p} v_j^{1m}\right)\bar{\boldsymbol{y}}_{j1} + \left(\sum_{m=1}^{p} v_j^{2m}\right)\bar{\boldsymbol{y}}_{j2} + \cdots + \left(\sum_{m=1}^{p} v_j^{pm}\right)\bar{\boldsymbol{y}}_{jp}}{\left(\sum_{l=1}^{p}\sum_{m=1}^{p} v_j^{lm}\right)}, \quad j = 1, 2, \tag{11.20}$$

$$-2n_1(p-1)q\rho_1^3 + d_3\rho_1^2 + [2n_1(p-1)q - 2d_1 - 2d_2]\rho_1 + d_3 = 0, \tag{11.21}$$

$$-2n_2(p-1)q\rho_2^3 + e_3\rho_2^2 + [2n_2(p-1)q - 2e_1 - 2e_2]\rho_2 + e_3 = 0, \tag{11.22}$$

and

$$\hat{\boldsymbol{\Sigma}} = \frac{1}{np}\sum_{j=1}^{2}\sum_{i=1}^{n_j}\sum_{l=1}^{p}\sum_{m=1}^{p} v_j^{lm}(\boldsymbol{y}_{jim} - \boldsymbol{\delta}_j)(\boldsymbol{y}_{jil} - \boldsymbol{\delta}_j)',$$

where $\boldsymbol{S}_1 = \boldsymbol{W}_1 + n_1\boldsymbol{k}_1$, $\boldsymbol{S}_2 = \boldsymbol{W}_2 + n_2\boldsymbol{k}_2$, $d_1 = \text{tr}(\boldsymbol{I}_p \otimes \boldsymbol{\Sigma}^{-1})\boldsymbol{S}_1$, $e_1 = \text{tr}(\boldsymbol{I}_p \otimes \boldsymbol{\Sigma}^{-1})\boldsymbol{S}_2$, $d_2 = \text{tr}(\boldsymbol{C}_1 \otimes \boldsymbol{\Sigma}^{-1})\boldsymbol{S}_1$, $d_3 = \text{tr}(\boldsymbol{C}_2 \otimes \boldsymbol{\Sigma}^{-1})\boldsymbol{S}_1$, $e_2 = \text{tr}(\boldsymbol{C}_1 \otimes \boldsymbol{\Sigma}^{-1})\boldsymbol{S}_2$, and $e_3 = \text{tr}(\boldsymbol{C}_2 \otimes \boldsymbol{\Sigma}^{-1})\boldsymbol{S}_2$.

The computation algorithm similar to that described in the previous case can be devised here and for the later cases in this section as well.

Case 3: $\boldsymbol{\Omega}_1 \neq \boldsymbol{\Omega}_2$ ($\boldsymbol{V}_1 = \boldsymbol{V}_2 = \boldsymbol{V}, \boldsymbol{\Sigma}_1 \neq \boldsymbol{\Sigma}_2$)

The quadratic classification rule is given by

Allocate an individual with response $\boldsymbol{y}$ to Population 1 if

$$-\frac{1}{2}\boldsymbol{y}'[\boldsymbol{V}^{-1} \otimes (\boldsymbol{\Sigma}_1^{-1} - \boldsymbol{\Sigma}_2^{-1})]\boldsymbol{y} + [\boldsymbol{1}_p'\boldsymbol{V}^{-1} \otimes (\boldsymbol{\delta}_1'\boldsymbol{\Sigma}_1^{-1} - \boldsymbol{\delta}_2'\boldsymbol{\Sigma}_2^{-1})]\boldsymbol{y}$$
$$\geq \frac{1}{2}(\boldsymbol{1}_p'\boldsymbol{V}^{-1}\boldsymbol{1}_p)(\boldsymbol{\delta}_1'\boldsymbol{\Sigma}_1^{-1}\boldsymbol{\delta}_1 - \boldsymbol{\delta}_2'\boldsymbol{\Sigma}_2^{-1}\boldsymbol{\delta}_2) - \frac{p}{2}\ln\left(\frac{|\boldsymbol{\Sigma}_2|}{|\boldsymbol{\Sigma}_1|}\right),$$

and to Population 2 otherwise.

The maximum likelihood estimates of $\boldsymbol{\delta}_j$, $j = 1, 2$ are identical to Equation 11.16. Thus, the maximum likelihood estimates of $\boldsymbol{\delta}_1, \boldsymbol{\delta}_2, \rho, \boldsymbol{\Sigma}_1$, and $\boldsymbol{\Sigma}_2$ are obtained by solving the following equations, along with Equation 11.16.

$$-2n(p-1)q\rho^3 + (f_3 + g_3)\rho^2 + [2n(p-1)q - 2(f_1 + g_1) - 2(f_2 + g_2)]\rho + (f_3 + g_3) = 0, \tag{11.23}$$

and

$$\hat{\boldsymbol{\Sigma}}_j = \frac{1}{n_j p}\sum_{i=1}^{n_j}\sum_{l=1}^{p}\sum_{m=1}^{p} v^{lm}(\boldsymbol{y}_{jim} - \boldsymbol{\delta}_j)(\boldsymbol{y}_{jil} - \boldsymbol{\delta}_j)', \quad j = 1, 2,$$

where $f_1 = \text{tr}(\boldsymbol{I}_p \otimes \boldsymbol{\Sigma}_1^{-1})\boldsymbol{S}_1$, $g_1 = \text{tr}(\boldsymbol{I}_p \otimes \boldsymbol{\Sigma}_2^{-1})\boldsymbol{S}_2$, $f_2 = \text{tr}(\boldsymbol{C}_1 \otimes \boldsymbol{\Sigma}_1^{-1})\boldsymbol{S}_1$, $f_3 = \text{tr}(\boldsymbol{C}_2 \otimes \boldsymbol{\Sigma}_1^{-1})\boldsymbol{S}_1$, $g_2 = \text{tr}(\boldsymbol{C}_1 \otimes \boldsymbol{\Sigma}_2^{-1})\boldsymbol{S}_2$, and $g_3 = \text{tr}(\boldsymbol{C}_2 \otimes \boldsymbol{\Sigma}_2^{-1})\boldsymbol{S}_2$.

Case 4: $\boldsymbol{\Omega}_1 \neq \boldsymbol{\Omega}_2$ $(\boldsymbol{V}_1 \neq \boldsymbol{V}_2, \boldsymbol{\Sigma}_1 \neq \boldsymbol{\Sigma}_2)$

The quadratic classification rule is given by

Allocate an individual with response $\boldsymbol{y}$ to Population 1 if

$$-\frac{1}{2}\boldsymbol{y}'(\boldsymbol{V}_1^{-1} \otimes \boldsymbol{\Sigma}_1^{-1} - \boldsymbol{V}_2^{-1} \otimes \boldsymbol{\Sigma}_2^{-1})\boldsymbol{y} + (\mathbf{1}_p'\boldsymbol{V}_1^{-1} \otimes \boldsymbol{\delta}_1'\boldsymbol{\Sigma}_1^{-1} - \mathbf{1}_p'\boldsymbol{V}_2^{-1} \otimes \boldsymbol{\delta}_2'\boldsymbol{\Sigma}_2^{-1})\boldsymbol{y}$$
$$\geq \frac{1}{2}[(\mathbf{1}_p'\boldsymbol{V}_1^{-1}\mathbf{1}_p)(\boldsymbol{\delta}_1'\boldsymbol{\Sigma}_1^{-1}\boldsymbol{\delta}_1) - (\mathbf{1}_p'\boldsymbol{V}_2^{-1}\mathbf{1}_p)(\boldsymbol{\delta}_2'\boldsymbol{\Sigma}_2^{-1}\boldsymbol{\delta}_2)]$$
$$-\frac{p}{2}\ln\left(\frac{|\boldsymbol{\Sigma}_2|}{|\boldsymbol{\Sigma}_1|}\right) - \frac{q}{2}\ln\left(\frac{|\boldsymbol{V}_2|}{|\boldsymbol{V}_1|}\right),$$

and to Population 2 otherwise.

The maximum likelihood estimates of $\boldsymbol{\delta}_j$, $j = 1,2$ are identical to Equation 11.19. The maximum likelihood estimates of $\boldsymbol{\delta}_1, \boldsymbol{\delta}_2, \rho_1, \rho_2, \boldsymbol{\Sigma}_1$, and $\boldsymbol{\Sigma}_2$ are obtained by simultaneously and iteratively solving the following equations along with Equation 11.19.

$$-2n_1(p-1)q\rho_1^3 + f_3\rho_1^2 + [2n_1(p-1)q - 2f_1 - 2f_2]\rho_1 + f_3 = 0, \qquad (11.24)$$
$$-2n_2(p-1)q\rho_2^3 + g_3\rho_2^2 + [2n_2(p-1)q - 2g_1 - 2g_2]\rho_2 + g_3 = 0, \qquad (11.25)$$

and

$$\hat{\boldsymbol{\Sigma}}_j = \frac{1}{n_j p}\sum_{i=1}^{n_j}\sum_{l=1}^{p}\sum_{m=1}^{p} v_j^{lm}(\boldsymbol{y}_{jim} - \boldsymbol{\delta}_j)(\boldsymbol{y}_{jil} - \boldsymbol{\delta}_j)', \quad j = 1,2.$$

11.3.2.2 Classification Rules with Unstructured Mean Vectors

The classification rules for each of the four cases mentioned in the introduction and the corresponding computational schemes for the maximum likelihood estimation of the unknown parameters are given below.

Case 1: $\boldsymbol{\Omega}_1 = \boldsymbol{\Omega}_2(\boldsymbol{V}_1 = \boldsymbol{V}_2 = \boldsymbol{V}, \boldsymbol{\Sigma}_1 = \boldsymbol{\Sigma}_2 = \boldsymbol{\Sigma})$

The linear classification rule is given by:

Allocate an individual with response $\boldsymbol{y}$ to Population 1 if

$$(\boldsymbol{\mu}_1 - \boldsymbol{\mu}_2)'(\boldsymbol{V} \otimes \boldsymbol{\Sigma})^{-1}\boldsymbol{y} \geq \frac{1}{2}(\boldsymbol{\mu}_1 - \boldsymbol{\mu}_2)'(\boldsymbol{V} \otimes \boldsymbol{\Sigma})^{-1}(\boldsymbol{\mu}_1 + \boldsymbol{\mu}_2),$$

and to Population 2 otherwise.

The MLEs of $\boldsymbol{\mu}_j$ are $\hat{\boldsymbol{\mu}}_j = \bar{\boldsymbol{y}}_j$, $j = 1,2$. The maximum likelihood estimates of ρ and $\boldsymbol{\Sigma}$ are obtained by solving the following two equations:

$$-2n(p-1)q\rho^3 + c_3'\rho^2 + [2n(p-1)q - 2c_1' - 2c_2']\rho + c_3' = 0,$$

and

$$\hat{\boldsymbol{\Sigma}} = \frac{1}{np}\sum_{j=1}^{2}\sum_{i=1}^{n_j}\sum_{l=1}^{p}\sum_{m=1}^{p} v^{lm}(\boldsymbol{y}_{jim} - \bar{\boldsymbol{y}}_{jm})(\boldsymbol{y}_{jil} - \bar{\boldsymbol{y}}_{jl})',$$

where $c_1' = \text{tr}(\boldsymbol{I}_p \otimes \boldsymbol{\Sigma}^{-1})\boldsymbol{W}'$, $c_2' = \text{tr}(\boldsymbol{C}_1 \otimes \boldsymbol{\Sigma}^{-1})\boldsymbol{W}'$, $c_3' = \text{tr}(\boldsymbol{C}_2 \otimes \boldsymbol{\Sigma}^{-1})\boldsymbol{W}'$, $\boldsymbol{W}' = \boldsymbol{W}_1 + \boldsymbol{W}_2$, and $\boldsymbol{W}_j, j = 1, 2$ are defined before.

Case 2: $\boldsymbol{\Omega}_1 \neq \boldsymbol{\Omega}_2$ ($\boldsymbol{V}_1 \neq \boldsymbol{V}_2, \boldsymbol{\Sigma}_1 = \boldsymbol{\Sigma}_2 = \boldsymbol{\Sigma}$)
The quadratic classification rule is given by

Allocate an individual with response $\boldsymbol{y}$ to Population 1 if

$$-\frac{1}{2}\boldsymbol{y}'[(\boldsymbol{V}_1^{-1} - \boldsymbol{V}_2^{-1}) \otimes \boldsymbol{\Sigma}^{-1}]\boldsymbol{y} + [\boldsymbol{\mu}_1'(\boldsymbol{V}_1 \otimes \boldsymbol{\Sigma})^{-1} - \boldsymbol{\mu}_2'(\boldsymbol{V}_2 \otimes \boldsymbol{\Sigma})^{-1}]\boldsymbol{y}$$
$$\geq \frac{1}{2}[\boldsymbol{\mu}_1'(\boldsymbol{V}_1 \otimes \boldsymbol{\Sigma})^{-1}\boldsymbol{\mu}_1 - \boldsymbol{\mu}_2'(\boldsymbol{V}_2 \otimes \boldsymbol{\Sigma})^{-1}\boldsymbol{\mu}_2] + \frac{q}{2}\ln\left(\frac{|\boldsymbol{V}_1|}{|\boldsymbol{V}_2|}\right),$$

and to Population 2 otherwise.

As in the previous case $\hat{\boldsymbol{\mu}}_j = \bar{\boldsymbol{y}}_j, j = 1, 2$. The maximum likelihood estimates of ρ_1, ρ_2, and $\boldsymbol{\Sigma}$ are obtained by solving Equations 11.20 and 11.21 after changing d_1, e_1, d_2, d_3, e_2, and e_3 by $d_1', e_1', d_2', d_3', e_2'$, and e_3', respectively, and the following equation:

$$\hat{\boldsymbol{\Sigma}} = \frac{1}{np}\sum_{j=1}^{2}\sum_{i=1}^{n_j}\sum_{l=1}^{p}\sum_{m=1}^{p} v_j^{lm}(\boldsymbol{y}_{jim} - \bar{\boldsymbol{y}}_{jm})(\boldsymbol{y}_{jil} - \bar{\boldsymbol{y}}_{jl})',$$

where $d_1' = \text{tr}(\boldsymbol{I}_p \otimes \boldsymbol{\Sigma}^{-1})\boldsymbol{W}_1, e_1' = \text{tr}(\boldsymbol{I}_p \otimes \boldsymbol{\Sigma}^{-1})\boldsymbol{W}_2, d_2' = \text{tr}(\boldsymbol{C}_1 \otimes \boldsymbol{\Sigma}^{-1})\boldsymbol{W}_1, d_3' = \text{tr}(\boldsymbol{C}_2 \otimes \boldsymbol{\Sigma}^{-1})\boldsymbol{W}_1, e_2' = \text{tr}(\boldsymbol{C}_1 \otimes \boldsymbol{\Sigma}^{-1})\boldsymbol{W}_2$, and $e_3' = \text{tr}(\boldsymbol{C}_2 \otimes \boldsymbol{\Sigma}^{-1})\boldsymbol{W}_2$.

Case 3: $\boldsymbol{\Omega}_1 \neq \boldsymbol{\Omega}_2$ ($\boldsymbol{V}_1 = \boldsymbol{V}_2 = \boldsymbol{V}, \boldsymbol{\Sigma}_1 \neq \boldsymbol{\Sigma}_2$)
The quadratic classification rule is given by

Allocate an individual with response $\boldsymbol{y}$ to Population 1 if

$$-\frac{1}{2}\boldsymbol{y}'[\boldsymbol{V}^{-1} \otimes (\boldsymbol{\Sigma}_1^{-1} - \boldsymbol{\Sigma}_2^{-1})]\boldsymbol{y} + [\boldsymbol{\mu}_1'(\boldsymbol{V} \otimes \boldsymbol{\Sigma}_1)^{-1} - \boldsymbol{\mu}_2'(\boldsymbol{V} \otimes \boldsymbol{\Sigma}_2)^{-1}]\boldsymbol{y}$$
$$\geq \frac{1}{2}[\boldsymbol{\mu}_1'(\boldsymbol{V} \otimes \boldsymbol{\Sigma}_1)^{-1}\boldsymbol{\mu}_1 - \boldsymbol{\mu}_2'(\boldsymbol{V} \otimes \boldsymbol{\Sigma}_2)^{-1}\boldsymbol{\mu}_2] + \frac{p}{2}\ln\left(\frac{|\boldsymbol{\Sigma}_1)|}{|\boldsymbol{\Sigma}_2)|}\right),$$

and to Population 2 otherwise.

Again $\hat{\boldsymbol{\mu}}_j = \bar{\boldsymbol{y}}_j, j = 1, 2$. The maximum likelihood estimates of ρ, $\boldsymbol{\Sigma}_1$, and $\boldsymbol{\Sigma}_2$ are obtained by solving Equation 11.22 after changing f_1, g_1, f_2, f_3, g_2, and g_3

by $f_1', g_1', f_2', f_3', g_2'$, and g_3' and the following two equations:

$$\hat{\Sigma}_j = \frac{1}{n_j p} \sum_{i=1}^{n_j} \sum_{l=1}^{p} \sum_{m=1}^{p} v^{lm} (y_{jim} - \bar{y}_{jm})(y_{jil} - \bar{y}_{jl})', \quad j = 1, 2.$$

Here $f_1' = \text{tr}(I_p \otimes \Sigma_1^{-1})W_1$, $g_1' = \text{tr}(I_p \otimes \Sigma_2^{-1})W_2$, $f_2' = \text{tr}(C_1 \otimes \Sigma_1^{-1})W_1$, $f_3' = \text{tr}(C_2 \otimes \Sigma_1^{-1})W_1$, $g_2' = \text{tr}(C_1 \otimes \Sigma_2^{-1})W_2$, and $g_3' = \text{tr}(C_2 \otimes \Sigma_2^{-1})W_2$.

Case 4: $\Omega_1 \neq \Omega_2$ ($V_1 \neq V_2, \Sigma_1 \neq \Sigma_2$)

The quadratic classification rule is given by

Allocate an individual with response y to Population 1 if

$$-\frac{1}{2}y'(V_1^{-1} \otimes \Sigma_1^{-1} - V_2^{-1} \otimes \Sigma_2^{-1})y + [\mu_1'(V_1 \otimes \Sigma_1)^{-1} - \mu_2'(V_2 \otimes \Sigma_2)^{-1}]y$$

$$\geq \frac{1}{2}[\mu_1'(V_1 \otimes \Sigma_1)^{-1}\mu_1 - \mu_2'(V_2 \otimes \Sigma_2)^{-1}\mu_2] + \frac{p}{2}\ln\left(\frac{|\Sigma_1)|}{|\Sigma_2)|}\right) + \frac{q}{2}\ln\left(\frac{|V_1|}{|V_2|}\right),$$

and to Population 2 otherwise.

As in the previous cases $\hat{\mu}_j = \bar{y}_j, j = 1, 2$. The maximum likelihood equations of ρ_1, ρ_2, Σ_1, and Σ_2 are obtained by simultaneously and iteratively solving Equations 11.23 and 11.24 after changing f_1, g_1, f_2, f_3, g_2, and g_3 by f_1', g_1', f_2', f_3', g_2', and g_3' and the following two equations:

$$\hat{\Sigma}_j = \frac{1}{n_j p} \sum_{i=1}^{n_j} \sum_{l=1}^{p} \sum_{m=1}^{p} v_j^{lm} (y_{jim} - \bar{y}_{jm})(y_{jil} - \bar{y}_{jl})', \quad j = 1, 2.$$

11.3.3 A Simulation Study

To study the impact of the incorrect correlation structure on the MER when the actual correlation structure is AR(1) and estimating and performing classification by assuming some other simple correlation structure, such as CS, we use simulated data sets. To get a comparative picture we also study the result parallelly when the actual correlation structure is AR(1) and estimating and performing classification by assuming it as AR(1). We assume the structure of the mean vectors to be $\mu_j = \mathbf{1}_p \otimes \delta_j$, $j = 1, 2$, where $\delta_1 = (2, 1, 1)'$, and $\delta_2 = (0, 1, 0)'$. The variance–covariance matrices to be $\Omega_j = V_j \otimes \Sigma_j$, $j = 1, 2$, where $V_j = V_j(\rho_j)$, has AR(1) structure. We study the impact of the correlation coefficient ρ_j's by assuming its values as 0.1, 0.3, 0.5, 0.7, and 0.9. The values of p are chosen as 3 and 5. The variance–covariance matrices Σ_1 and Σ_2 are taken as

$$\Sigma_1 = \begin{bmatrix} 2 & 1 & 2 \\ 1 & 4 & 3 \\ 2 & 3 & 5 \end{bmatrix}, \quad \text{and} \quad \Sigma_2 = \begin{bmatrix} 1 & 0 & 0 \\ 0 & 1 & 0 \\ 0 & 0 & 1 \end{bmatrix}.$$

TABLE 11.5

MERs for the Simulated Data in Case 1

$\delta_1 = (2,1,1)', \delta_2 = (0,1,0)'$

$(n_1, n_2) \rightarrow$		(20,20)		(50,50)	
$\rho \downarrow p \rightarrow$	**Estimated Covariance Structure**	3	5	3	5
0.1	AR(1)	10.075	5.925	10.125	6.200
	CS	10.175	5.875	10.175	6.075
0.3	AR(1)	13.275	9.200	13.200	9.375
	CS	13.225	9.100	13.275	9.375
0.5	AR(1)	15.975	12.475	16.050	13.025
	CS	16.425	12.925	16.575	13.225
0.7	AR(1)	18.075	16.250	18.425	17.075
	CS	18.575	17.075	18.950	17.350
0.9	AR(1)	20.825	20.150	20.825	20.425
	CS	20.875	20.425	21.075	21.000

Training samples of sizes $(n_1, n_2) = (20, 20)$ and $(50, 50)$ and test samples of sizes (2000, 2000) are generated from pq-variate normal populations $N_{pq}(\boldsymbol{\mu}_j, \boldsymbol{\Omega}_j)$, where $\boldsymbol{\Omega}_j$ is defined in Section 11.3. On the basis of these samples, we estimate $(\boldsymbol{\delta}_1, \boldsymbol{\delta}_2), (\boldsymbol{\Sigma}_1, \boldsymbol{\Sigma}_2)$, and (ρ_1, ρ_2) using the maximum likelihood method under CS correlation structure and also under AR(1) correlation structure as discussed in Sections 11.3.1 and 11.3.2. Using these estimates, the classification is performed separately on the test samples of size 2000 from each of the two populations and the MERs are calculated. Tables 11.5 through 11.8 show the MERs when the actual correlation structure is AR(1) and we estimate it as AR(1) and as well as CS. The tables reveal that the MER increases if the actual correlation structure is AR(1) and we estimate them as CS, and if the value of $\rho \geq 0.5$, in particular. That is, with higher correlation coefficient the MER increases under wrong assumption of correlation structure. Increase in MER under wrong assumption of correlation structure is very prominent for the cases when $\boldsymbol{\Sigma}_1 \neq \boldsymbol{\Sigma}_2$ (Tables 11.7 and 11.8). We anticipate this, since in this case a large number of parameters are estimated.

We notice that for both small and large sample sizes, the larger value of p (=5) corresponds to smaller MER. This is expected as more repeated measures would provide more information. Further as ρ increases MER increases. As mentioned in Roy and Khattree, (2005b) the classification rule for Case 2 under CS covariance structure is not balanced. Therefore, Table 11.6 only displays the MER under AR(1) correlation structure.

For $p = 3$, we notice that there is not much change in the MERs for incorrect assumption of the correlation structures for Case 1, but not in Cases 3 and 4. We see that MER for incorrect classification increases with ρ, especially for large p. This fact is evident from Tables 11.7 and 11.8. So, we feel that for large number of repeated (p) measures, checking the structure of correlation

TABLE 11.6

MERs for the Simulated Data in Case 2

$\rho_2 = 0.9, \delta_1 = (2,1,1)', \delta_2 = (0,1,0)'$

$(n_1, n_2) \rightarrow$		(20,20)		(50,50)	
$\rho \downarrow p \rightarrow$	**Estimated Cov Structure**	**3**	**5**	**3**	**5**
0.1	AR(1)	2.325	0.600	2.400	0.350
0.3	AR(1)	3.800	1.350	4.050	0.900
0.5	AR(1)	5.900	2.750	6.175	1.825
0.7	AR(1)	10.600	7.875	10.900	5.950

TABLE 11.7

MERs for the Simulated Data in Case 3

$\delta_1 = (2,1,1)', \delta_2 = (0,1,0)'$

$(n_1, n_2) \rightarrow$		(20,20)		(50,50)	
$\rho_1 \downarrow p \rightarrow$	**Estimated Cov Structure**	**3**	**5**	**3**	**5**
0.1	AR(1)	3.125	2.125	3.300	1.100
	CS	4.050	3.050	4.150	2.950
0.3	AR(1)	4.150	2.925	4.225	1.850
	CS	5.150	4.100	5.375	3.800
0.5	AR(1)	4.900	3.900	5.075	2.400
	CS	6.175	5.600	6.225	4.975
0.7	AR(1)	5.800	5.125	5.875	3.000
	CS	7.075	7.300	7.300	6.400
0.9	AR(1)	6.975	5.900	6.650	3.575
	CS	8.375	9.475	8.175	8.275

structure on repeated measures before applying the classification rules is of crucial importance, especially when the autoregressive correlation coefficient ρ is large.

With the help of simulation study, we have shown that the MER decreases with the number of repeated measures p, for both univariate and multivariate repeated measures data. This is due to the fact that additional repeated measures provide more information. We also expect that after certain point additional repeated measures do not provide any significant extra information and this was observed in our simulation results. We have also shown that MER increases with ρ. One would expect this to happen since an increase in ρ means repeated measurements are less informative.

TABLE 11.8

MERs for the Simulated Data in Case 4
$\rho_2 = 0.9, \boldsymbol{\delta}_1 = (2,1,1)', \boldsymbol{\delta}_2 = (0,1,0)'$

$(n_1, n_2) \rightarrow$		(20,20)		(50,50)	
$\rho_1 \downarrow p\rightarrow$	Estimated Cov Structure	3	5	3	5
0.1	AR(1)	0.375	0.450	0.225	0.050
	CS	2.675	3.625	2.200	2.050
0.3	AR(1)	0.650	0.550	0.375	0.125
	CS	3.400	4.325	2.825	2.200
0.5	AR(1)	1.125	0.850	0.775	0.250
	CS	4.000	4.875	3.775	2.600
0.7	AR(1)	2.400	2.050	1.925	0.550
	CS	5.050	5.375	5.300	3.300

11.4 Tests for Structures

In the previous section, we provided classification rules under various structures. How does one know that a particular structure is applicable? This calls for developing some appropriate statistical tests for the structures, which we have considered in the previous section. Tests are obtained under likelihood ratio criterion (Roy and Khattree, 2003b).

11.4.1 Test Specifying the Mean Vector

We assume that the covariance matrix $\boldsymbol{\Omega}$ has the covariance structure $\boldsymbol{\Omega} = \boldsymbol{V} \otimes \boldsymbol{\Sigma}$ where $\boldsymbol{V}$ and $\boldsymbol{\Sigma}$, respectively, are $p \times p$ and $q \times q$ positive definite matrices. We assume $\boldsymbol{V}$ to have the CS correlation structure with $0 < \rho < 1$. Under these assumptions we may want to test the null hypothesis on mean.

$$(1) \quad H_1 : \boldsymbol{\mu} = \mathbf{1}_p \otimes \boldsymbol{\delta} \quad \text{versus} \quad K_1 : \boldsymbol{\mu} \text{ unstructured},$$

where $\boldsymbol{\delta} = (\delta_1 \delta_2, \ldots, \delta_q)'$. The null hypothesis implies that mean remains constant over time for all characteristics. The likelihood ratio Λ_1 for testing H_1 is given by

$$\Lambda_1 = \frac{|\hat{\boldsymbol{V}}_o|^{-(qn/2)} |\hat{\boldsymbol{\Sigma}}_o|^{-(pn/2)} e^{-(1/2)\,\mathrm{tr}(\hat{\boldsymbol{V}}_o \otimes \hat{\boldsymbol{\Sigma}}_o)^{-1} \boldsymbol{W}_s}}{|\hat{\boldsymbol{V}}|^{-(qn/2)} |\hat{\boldsymbol{\Sigma}}|^{-(pn/2)} e^{-(1/2)\,\mathrm{tr}(\hat{\boldsymbol{V}} \otimes \hat{\boldsymbol{\Sigma}})^{-1} \boldsymbol{S}}},$$

where $\bar{\boldsymbol{y}} = (\bar{\boldsymbol{y}}_1', \bar{\boldsymbol{y}}_2', \ldots, \bar{\boldsymbol{y}}_p')'$ is the sample mean vector, $\boldsymbol{S} = \sum_{i=1}^{n}(\boldsymbol{y}_i - \bar{\boldsymbol{y}})(\boldsymbol{y}_i - \bar{\boldsymbol{y}})'$ and $\boldsymbol{W}_s = \boldsymbol{S} + n(\bar{\boldsymbol{y}} - \boldsymbol{1}_p \otimes \hat{\boldsymbol{\delta}})(\bar{\boldsymbol{y}} - \boldsymbol{1}_p \otimes \hat{\boldsymbol{\delta}})'$; $\hat{\boldsymbol{V}}_o$, $\hat{\boldsymbol{\Sigma}}_o$ are the maximum likelihood estimates of $\boldsymbol{V}$ and $\boldsymbol{\Sigma}$ under H_1 and $\hat{\boldsymbol{V}}$, $\hat{\boldsymbol{\Sigma}}$ are those under no restrictions.

When n is large the sampling distribution of $-2\ln\Lambda_1$ is well approximated by a $\chi^2_{\nu_1}$ distribution, where $\nu_1 = q(p-1)$.

The maximum likelihood estimate of $\boldsymbol{\delta}$ is given by

$$\hat{\boldsymbol{\delta}} = \frac{1}{p}(\bar{\boldsymbol{y}}_1 + \bar{\boldsymbol{y}}_2 + \cdots + \bar{\boldsymbol{y}}_p).$$

The quantities $\hat{\boldsymbol{\Sigma}}_o$ and $\hat{\rho}_o$ are obtained by solving the following equations:

$$(p-1)k_0\rho^3 + \{k_0 - (p-1)k_0 + (p-1)^2 s_1 - (p-1)s_2\}\rho^2 + \{2(p-1)s_1 - k_0\}\rho + (s_1 - s_2) = 0,$$

and

$$\hat{\boldsymbol{\Sigma}} = \frac{1}{np}\sum_{i=1}^{n}\sum_{l=1}^{p}\sum_{m=1}^{p} v^{lm}\,(\boldsymbol{y}_{im} - \hat{\boldsymbol{\delta}})(\boldsymbol{y}_{il} - \hat{\boldsymbol{\delta}})',$$

where $s_1 = \mathrm{tr}(\boldsymbol{I}_p \otimes \boldsymbol{\Sigma}^{-1})\boldsymbol{W}_s$ and $s_2 = \mathrm{tr}(\boldsymbol{J}_p \otimes \boldsymbol{\Sigma}^{-1})\boldsymbol{W}_s$.

Thus, the maximum likelihood estimates of $\boldsymbol{V}$ say $\hat{\boldsymbol{V}}_o$ is given as $\hat{\boldsymbol{V}}_o(\hat{\rho}_o) = (1-\hat{\rho}_o)\boldsymbol{I}_p + \hat{\rho}_o\boldsymbol{1}_p\boldsymbol{1}_p'$.

The maximum likelihood estimates of $\boldsymbol{\Sigma}$ and ρ say, $\hat{\boldsymbol{\Sigma}}$ and $\hat{\rho}$ are obtained by solving the equations,

$$(p-1)k_0\rho^3 + \{k_0 - (p-1)k_0 + (p-1)^2 s_3 - (p-1)s_4\}\rho^2 + \{2(p-1)s_3 - k_0\}\rho + (s_3 - s_4) = 0,$$

and

$$\hat{\boldsymbol{\Sigma}} = \frac{1}{np}\sum_{i=1}^{n}\sum_{l=1}^{p}\sum_{m=1}^{p} v^{lm}(\boldsymbol{y}_{im} - \bar{\boldsymbol{y}}_m)(\boldsymbol{y}_{il} - \bar{\boldsymbol{y}}_l)',$$

where $s_3 = \mathrm{tr}(\boldsymbol{I}_p \otimes \boldsymbol{\Sigma}^{-1})\boldsymbol{S}$ and $s_4 = \mathrm{tr}(\boldsymbol{J}_p \otimes \boldsymbol{\Sigma}^{-1})\boldsymbol{S}$.

Therefore, MLE of $\boldsymbol{V}$ is, say, $\hat{\boldsymbol{V}} = \hat{\boldsymbol{V}}(\hat{\rho}) = (1-\hat{\rho})\boldsymbol{I}_p + \hat{\rho}\boldsymbol{1}_p\boldsymbol{1}_p'$.

11.4.2 Test Specifying the Covariance Matrices

It is well known that the difference between covariance matrices, may affect the performance of the classification rule. The choice between the linear or quadratic classification rules can be made following a test for equality of the two correlation matrices $\boldsymbol{V}_1$ and $\boldsymbol{V}_2$ and the two covariance matrices $\boldsymbol{\Sigma}_1$ and $\boldsymbol{\Sigma}_2$. We assume $\boldsymbol{\Omega}_1 = \boldsymbol{V}_1 \otimes \boldsymbol{\Sigma}_1$ and $\boldsymbol{\Omega}_2 = \boldsymbol{V}_2 \otimes \boldsymbol{\Sigma}_2$. The parameters $\boldsymbol{\mu}_1, \boldsymbol{\mu}_2, \boldsymbol{\Omega}_1$, and $\boldsymbol{\Omega}_2$ are all assumed to be unknown.

We will construct the likelihood ratio tests for the following hypotheses of interest

(2) $H_2 : \boldsymbol{\Sigma}_1 = \boldsymbol{\Sigma}_2 = \boldsymbol{\Sigma}$ (say) versus $K_2 : \boldsymbol{\Sigma}_1 \neq \boldsymbol{\Sigma}_2$,

$\boldsymbol{V}_j = \boldsymbol{V}(\rho), \quad j = 1, 2$ compound symmetric matrices.

(3) $H_3 : \boldsymbol{\Sigma}_1 = \boldsymbol{\Sigma}_2 = \boldsymbol{\Sigma}$ (say) versus $K_3 : \boldsymbol{\Sigma}_1 \neq \boldsymbol{\Sigma}_2$,

$\boldsymbol{V}_j = \boldsymbol{V}_j(\rho_j), \quad j = 1, 2$ compound symmetric matrices.

(4) $H_4 : \boldsymbol{V}_1 = \boldsymbol{V}_2 = \boldsymbol{V}$ (say), $\boldsymbol{\Sigma}_1 = \boldsymbol{\Sigma}_2 = \boldsymbol{\Sigma}$ (say)

versus

$K_4 : \boldsymbol{V}_1 \neq \boldsymbol{V}_2, \ \boldsymbol{\Sigma}_1 \neq \boldsymbol{\Sigma}_2$

$\boldsymbol{V}_j = \boldsymbol{V}_j(\rho_j), \quad j = 1, 2$ are compound symmetric matrices.

The likelihood ratio Λ_2 is given by

$$\Lambda_2 = \frac{|\hat{\boldsymbol{V}}_o|^{-(qn/2)} |\hat{\boldsymbol{\Sigma}}_o|^{-(pn/2)} e^{-((1/2)\,\mathrm{tr}(\hat{\boldsymbol{V}}_o \otimes \hat{\boldsymbol{\Sigma}}_o)^{-1} \boldsymbol{W}')}}{|\hat{\boldsymbol{V}}|^{-(qn/2)} |\hat{\boldsymbol{\Sigma}}_1|^{-(pn_1/2)} |\hat{\boldsymbol{\Sigma}}_2|^{-(pn_2/2)} e^{-((1/2)\mathrm{tr}(\hat{\boldsymbol{V}} \otimes \hat{\boldsymbol{\Sigma}}_1)^{-1} \boldsymbol{W}_1)} e^{-((1/2)\,\mathrm{tr}(\hat{\boldsymbol{V}} \otimes \hat{\boldsymbol{\Sigma}}_2)^{-1} \boldsymbol{W}_2)}}.$$

The degrees of freedom ν_2 for the null distribution of $-2 \ln \Lambda_2$ is given by, $\nu_2 = q(q+1)/2$.

The maximum likelihood estimates $\hat{\rho}_o$ and $\hat{\boldsymbol{\Sigma}}_o$ under H_2 are obtained by simultaneously and iteratively solving Equations 11.6 and 11.7 and the maximum likelihood estimates of $\hat{\rho}$, $\hat{\boldsymbol{\Sigma}}_1$, and $\hat{\boldsymbol{\Sigma}}_2$ under K_2 are obtained by solving the Equations 11.11 and 11.12.

Now the likelihood ratio test statistic Λ_3 is given by

$$\begin{aligned}\Lambda_3 = \Big\{ &|\hat{\boldsymbol{V}}_{1o}|^{-(qn_1/2)} |\hat{\boldsymbol{V}}_{2o}|^{-(qn_2/2)} |\hat{\boldsymbol{\Sigma}}_o|^{-(pn/2)} e^{-((1/2)\,\mathrm{tr}(\hat{\boldsymbol{V}}_{1o} \otimes \hat{\boldsymbol{\Sigma}}_o)^{-1} \boldsymbol{W}_1)} \\ &\times e^{-((1/2)\,\mathrm{tr}(\hat{\boldsymbol{V}}_{2o} \otimes \hat{\boldsymbol{\Sigma}}_o)^{-1} \boldsymbol{W}_2)} \Big\} \Big\{ |\hat{\boldsymbol{V}}_1|^{-(qn_1/2)} |\hat{\boldsymbol{V}}_2|^{-(qn_2/2)} |\hat{\boldsymbol{\Sigma}}_1|^{-(pn_1/2)} \\ &\times |\hat{\boldsymbol{\Sigma}}_2|^{-(pn_2/2)} e^{-((1/2)\,\mathrm{tr}(\hat{\boldsymbol{V}}_1 \otimes \hat{\boldsymbol{\Sigma}}_1)^{-1} \boldsymbol{W}_1)} e^{-((1/2)\,\mathrm{tr}(\hat{\boldsymbol{V}}_2 \otimes \hat{\boldsymbol{\Sigma}}_2)^{-1} \boldsymbol{W}_2)} \Big\}^{-1}.\end{aligned}$$

The degrees of freedom ν_3 for the null distribution of $-2 \ln \Lambda_3$ is given by, $\nu_3 = q(q+1)/2$.

The maximum likelihood estimates of $\hat{\rho}_{1o}$, $\hat{\rho}_{2o}$, and $\hat{\boldsymbol{\Sigma}}_o$ are obtained by simultaneously and iteratively solving the Equations 11.8, 11.9, and 11.10.

The maximum likelihood estimates $\hat{\rho}_1$, $\hat{\rho}_2$, $\hat{\boldsymbol{\Sigma}}_1$, and $\hat{\boldsymbol{\Sigma}}_2$ are obtained by simultaneously and iteratively solving the Equations 11.13, 11.14, and 11.15.

The likelihood ratio Λ_4 is given by

$$\Lambda_4 = \left\{|\hat{V}_o|^{-(qn/2)}|\hat{\Sigma}_o|^{-(pn/2)}e^{-((1/2)\,\mathrm{tr}(\hat{V}_o\otimes\hat{\Sigma}_o)^{-1}W')}\right\}\left\{|\hat{V}_1|^{-(qn_1/2)}|\hat{V}_2|^{-(qn_2/2)}\right.$$
$$\left.\times\,|\hat{\Sigma}_1|^{-(pn_1/2)}|\hat{\Sigma}_2|^{-(pn_2/2)}e^{-((1/2)\,\mathrm{tr}(\hat{V}_1\otimes\hat{\Sigma}_1)^{-1}W_1)}e^{-((1/2)\,\mathrm{tr}(\hat{V}_2\otimes\hat{\Sigma}_2)^{-1}W_2)}\right\}^{-1}.$$

The degrees of freedom ν_4 for the null distribution of $-2\ln\Lambda_4$ is given by, $\nu_4 = (q(q+1)/2) + 1$.

The maximum likelihood estimates $\hat{\rho}_o$ and $\hat{\Sigma}_o$ are obtained from Equations 11.6 and 11.7 and those of $\hat{\rho}_1$, $\hat{\rho}_2$, $\hat{\Sigma}_1$, and $\hat{\Sigma}_2$ are obtained from Equations 11.13, 11.14, and 11.15.

See Roy (2006b), for similar hypotheses testing when both V_1 and V_2 have AR(1) correlation structures.

11.5 Two Examples

To illustrate the previous results, two real data sets are considered. The first of these data sets is of relatively smaller in size, whereas the second one is moderate in size. The data sets have very different features and as it turns out, would require different covariance structure assumptions.

EXAMPLE 1

Dental data: The first data set is from Timm (1980, Table 7.2). The data were originally collected by T. Zullo of the School of Dental Medicine at the University of Pittsburgh. There are nine subjects in each of two orthopedic adjustment group ($k = 2, n_1 = 9, n_2 = 9$). Three measurements at three different time points ($p = 3$) were made on each of $q = 3$ characteristic to assess the change in the vertical position of the mandible. The problem here is to classify an unknown subject into one of the two orthopedic populations. Clearly, appropriate assumptions on the correct structure on the mean vectors as well as the covariance structures on both the populations are very important for selecting the classification rules. Thus, we test whether the data set has a structure on mean (Hypothesis 1) as well as test Hypotheses 2, 3, and 4 for covariance structures given by

$$(V_1 = V_2, \Sigma_1 = \Sigma_2), (V_1 \neq V_2, \Sigma_1 = \Sigma_2), \quad \text{or} \quad (V_1 = V_2, \Sigma_1 \neq \Sigma_2).$$

In this data set the correlation matrices V_1 and V_2 of the repeated measures have CS structures with p values 0.3344 and 0.3068, respectively. From Table 11.9 we see that the data set probably has unstructured mean vector in each population. Also, from Table 11.10 we see that the data set possibly

TABLE 11.9

Testing Results for Hypothesis 1 for Dental Data

Hypothesis	Population	p Value
1	1	0.0004
	2	≈ 0

TABLE 11.10

Testing Results for Hypotheses 2, 3, and 4 for Dental Data

Hypothesis	p Value
2	≈ 0
3	0.0376
4	≈ 0

has the covariance structure $(\boldsymbol{V}_1 \neq \boldsymbol{V}_2, \boldsymbol{\Sigma}_1 \neq \boldsymbol{\Sigma}_2)$, because all the hypotheses stated above are rejected.

Thus, we assume the data are multivariate normal with mean vector $\boldsymbol{\mu}_j$ and covariance matrix $\boldsymbol{\Omega}_j = \boldsymbol{V}_j \otimes \boldsymbol{\Sigma}_j, j = 1, 2$. The maximum likelihood estimates of $\boldsymbol{\mu}_1$ and $\boldsymbol{\mu}_2$ in the two populations are

$$\hat{\boldsymbol{\mu}}_1 = [118.2222, 63.2222, 24.6667, 121.7222, 64.0000, 25.1667, 122.9444, 65.6111, 25.1111]',$$

and

$$\hat{\boldsymbol{\mu}}_2 = [122.7778, 64.2778, 24.0000, 125.7222, 66.0000, 24.2222, 127.2778, 67.1667, 24.2889]',$$

respectively. The maximum likelihood estimates of $\boldsymbol{\Sigma}_1$ and $\boldsymbol{\Sigma}_2$ in the two populations are

$$\hat{\boldsymbol{\Sigma}}_1 = \begin{bmatrix} 58.6154 & 16.2587 & -8.0093 \\ 16.2587 & 17.7243 & 0.6180 \\ -8.0093 & 0.6180 & 23.2864 \end{bmatrix},$$

and

$$\hat{\boldsymbol{\Sigma}}_2 = \begin{bmatrix} 51.5092 & 37.5211 & 7.4671 \\ 37.5211 & 38.8700 & 6.1830 \\ 7.4671 & 6.1830 & 14.1340 \end{bmatrix},$$

TABLE 11.11

Classification Table for Dental Data when the Mean Vector is Unstructured and Repeated Measures has CS Structure

	Case 1		Case 2		Case 3		Case 4	
	Pop1	**Pop2**	**Pop1**	**Pop2**	**Pop1**	**Pop2**	**Pop1**	**Pop2**
Pop1	6	2	7	2	3	0	7	1
Pop2	3	7	2	7	6	9	2	8

and these appear to be different. The MLEs of ρ_1 and ρ_2 are 0.9266 and 0.9931, respectively, and they also appear to be different to some extent. Therefore, $\hat{\boldsymbol{V}}_1$ and $\hat{\boldsymbol{V}}_2$ appear to be different.

The classification results of the Dental data for all the four cases for unstructured mean vectors and CS correlation structure on the repeated measures are shown in Table 11.11. The table consists of four confusion matrices corresponding to four cases mentioned in the introduction. The rows are predicted categories of the dependents and columns are true categories of the dependents. The MERs for the four cases under the CS correlation structure are 27.78%, 22.22%, 33.33%, and 16.67%, respectively. We see that MER is smallest in Case 4 where $\boldsymbol{V}_1 \neq \boldsymbol{V}_2$ and $\boldsymbol{\Sigma}_1 \neq \boldsymbol{\Sigma}_2$ (the situation we suspected on the basis of our hypothesis testing) and when the assumption of unstructured mean vector is adapted. A total of 7 out of 9 subjects (77.78%) from Population 1, and 8 out of 9 subjects (88.89%) from Population 2 were classified correctly.

EXAMPLE 2

Osteoporosis data: The data are from a clinical study in the prevention of osteoporosis. Bone mineral density (BMD) in gm/cm^2 was measured on patients at the baseline visit and then for four subsequent follow-up visits, every 6 months, for 2 years ($p = 5$). The BMD assessments are obtained at different anatomic locations namely spine, radius, femoral neck, and total hip ($q = 4$). This data set has two treatment groups consisting of 21 patients in one treatment group, and 23 patients in the other treatment group. Complete data were observed only on 32 patients; 13 patients are in one treatment group, and 19 patients are in the other treatment group. There were ten investigators for this study. In this section we will analyze only the complete data set, that is, we will consider two populations with $n_1 = 13$ and $n_2 = 19$. We will use the incomplete data set along with the investigator information as a covariate in Section 11.6. In the complete data set after testing we conclude that the correlation matrices $\boldsymbol{V}_1$ and $\boldsymbol{V}_2$ of the repeated measures have CS structures with p values 0.9697 and 0.9648, respectively. From Table 11.12 we see that the data set probably has an unstructured mean vector in each population. Also, from Table 11.13 we see that the data set is likely to have the covariance structure $(\boldsymbol{V}_1 = \boldsymbol{V}_2, \boldsymbol{\Sigma}_1 = \boldsymbol{\Sigma}_2)$.

TABLE 11.12

Testing Results for Hypothesis 1 for Osteoporosis Data

Hypothesis	Population	p Value
1	1	≈ 0
	2	0.0015

TABLE 11.13

Testing Results for Hypotheses 2, 3, and 4 for Osteoporosis Data

Hypothesis	p Value
2	0.0463
3	0.4239
4	0.0594

Assuming unstructured mean vectors the maximum likelihood estimates of $\boldsymbol{\mu}_1$ and $\boldsymbol{\mu}_2$ in the two populations are

$$\hat{\boldsymbol{\mu}}_1 = [1.0718, 0.7941, 0.8818, 0.9185, 1.0906, 0.7932, 0.8841, 0.9313, 1.1069, 0.7942, 0.8839, 0.9394, 1.1075, 0.7834, 0.8841, 0.9410, 1.0989, 0.7902, 0.8878, 0.9425]',$$

and

$$\hat{\boldsymbol{\mu}}_2 = [1.1118, 0.8112, 0.8664, 0.9541, 1.1325, 0.8149, 0.8633, 0.9605, 1.1458, 0.8116, 0.8622, 0.9657, 1.1481, 0.8220, 0.8751, 0.9673, 1.1549, 0.8184, 0.8742, 0.9691]',$$

respectively. The maximum likelihood estimates of $\boldsymbol{\Sigma}_1$ and $\boldsymbol{\Sigma}_2$ in the two populations are

$$\hat{\boldsymbol{\Sigma}}_1 = \begin{bmatrix} 0.0170 & 0.0013 & 0.0081 & 0.0057 \\ 0.0013 & 0.0051 & 0.0016 & 0.0013 \\ 0.0081 & 0.0016 & 0.0167 & 0.0050 \\ 0.0057 & 0.0013 & 0.0050 & 0.0066 \end{bmatrix},$$

and

$$\hat{\boldsymbol{\Sigma}}_2 = \begin{bmatrix} 0.0201 & 0.0015 & 0.0084 & 0.0079 \\ 0.0015 & 0.0062 & 0.0010 & 0.0010 \\ 0.0084 & 0.0010 & 0.0193 & 0.0101 \\ 0.0079 & 0.0010 & 0.0101 & 0.0112 \end{bmatrix}.$$

TABLE 11.14

Classification Table for Osteoporosis Data when the Mean Vector is Unstructured and Repeated Measures has CS Structure

	Case 1		Case 2		Case 3		Case 4	
	Pop1	Pop2	Pop1	Pop2	Pop1	Pop2	Pop1	Pop2
Pop1	12	2	6	1	10	2	9	1
Pop2	1	17	7	18	3	17	4	18

TABLE 11.15

Classification Table for Osteoporosis Data when the Mean Vector is Unstructured and Repeated Measures has AR(1) Structure

	Case 1		Case 2		Case 3		Case 4	
	Pop1	Pop2	Pop1	Pop2	Pop1	Pop2	Pop1	Pop2
Pop1	9	2	10	4	9	3	9	3
Pop2	4	17	3	15	4	16	4	16

These appear to be quite close to each other. The maximum likelihood estimates of ρ_1 and ρ_2 are 0.9711 and 0.9637, respectively, and they also appear to be close. Therefore, $\hat{\boldsymbol{V}}_1$ and $\hat{\boldsymbol{V}}_2$ appear to be close. The classification results of the complete Osteoporosis data with CS correlation structure on the repeated measures are shown in Table 11.14. The MERs for the four cases under the CS correlation structure, respectively, are 9.37%, 25.00%, 15.62%, and 15.62%. We observe that the MER is smallest in Case 1 when $\boldsymbol{V}_1 = \boldsymbol{V}_2$ and $\boldsymbol{\Sigma}_1 = \boldsymbol{\Sigma}_2$ (the situation we suspected on the basis of our hypothesis testing) and when assumption of unstructured mean vector is adapted. A total of 12 out of 13 subjects (92.31%) from Population 1 and 17 out of 19 subjects (89.47%) from Population 2 were correctly classified.

The assumption of the Kronecker product structure on the covariance matrix $\boldsymbol{\Omega}$ is necessary for both the data sets since we do not have enough observations in both the data sets to estimate the parameters in $\boldsymbol{\Omega}$. In the Dental data set the number of unknown parameters in the unstructured $\boldsymbol{\Omega}$ is 45, whereas we have only 9 observations in each population. For the Osteoporosis data set the number of unknown parameters in unstructured $\boldsymbol{\Omega}$ is 210 and we have only 13 and 19 observations in the two populations, respectively.

For comparison sake, classification results of the complete Osteoporosis data with AR(1) correlation structure on the repeated measures, and under unstructured mean vector assumption is shown in Table 11.15. The overall MERs in the four cases are 18.75%, 21.87%, 21.87%, and 21.87%, respectively. We observe that most of the times the MERs in each of the four cases are more than the corresponding MERs with the CS correlation structure on the repeated measures. This suggests that in the context of classification, assumption of incorrect structure may lead to very high MERs. This was already addressed in the previous section, through a simulation study.

11.6 Multivariate Repeated Measures Data with Missing Values

Missing observations are almost inevitable in longitudinal studies. Missing values frequently occur in biomedical and biological data as observations are lost, not recorded, or experimental units inadvertently dropout owing to a variety of reasons. It is customary to fill in the missing values in some way. Discarding data with missing components can result in appreciable information loss, and may bias the results if the subjects, who provide complete data, are not representative of the population. Early approaches to missing values problems simply ignored them or replaced them with some plausible values (e.g., mean values or regression predictions) calculated over the observations present. Another way to handle missing values is to develop some learning algorithms that can deal with the missing values in some meaningful way. A third approach is to impute all missing values in the first phase. Imputation although solves or appears to solve the missing data problem, it may have an undesirable consequence in that it can dampen the relationships among variables or can artificially inflate the correlations between the variables.

Multiple imputation provides a useful strategy for dealing with missing values. Instead of filling in a single value for each missing value, Rubin's multiple imputation procedure (Rubin, 1987, 1996) replaces each missing value with a set of plausible values that represent the uncertainty about the right value to impute. These multiple imputed data sets are then analyzed by using standard procedures for complete data and results from these analyses are then combined. Roy (2006a) has shown, in the case of multivariate repeated measures data, that multiple imputation method is not the best choice of analysis for all missing data problems as it introduces noise; especially when there is some special structure in the data. Roy has studied the classification problem with special reference to multivariate repeated measures data with missing values, as well as covariates in a mixed effects model setup under the mechanism of missing at random (MAR), that is, when the probability that a value is missing depends only on the observed variable, and time point values of the individual, but neither on the missing variable nor on the missing time point values of the individual. Covariate information was first used in the discrimination function in 1948 by Cochran and Bliss. They found that by using the covariates the MER was improved in their examples. However, Cochran and Bliss did not handle missing values. Later, Lachenburch (1977) worked on covariance adjusted discriminant functions. However, his work did not result in a satisfactory performance in case of incomplete data. Discriminant analysis on incomplete data on a single variable with covariates and in the mixed model setup for a repeated measures scenario, was first studied by Tomasco et al. (1999). Here we briefly present a solution to this problem using an approach that builds on the ideas promoted by Tomasco et al. (1999), and Laird and Ware (1982). Complete details are available in Roy (2006a).

Let the random variable $\boldsymbol{y}_i$ denote the response of interest for the ith subject, $i = 1, 2, \ldots, n$. As before suppose each individual is observed on q response variables and each variable is observed over p time points. Thus, $\boldsymbol{y}_i$ is a vector as described in Section 11.3 with many patients having missing values. Suppose the ith patient has $(pq - r_i)$ number of missing values; then $\boldsymbol{y}_i$ is a r_i-dimensional vector, $1 \leq r_i \leq pq$.

Consider a linear mixed-effects model as described by Laird and Ware (1982)

$$\begin{aligned}
\boldsymbol{y}_i &= \boldsymbol{X}_i\boldsymbol{\beta} + \boldsymbol{Z}_i\boldsymbol{b}_i + \boldsymbol{\epsilon}_i,\\
\boldsymbol{b}_i &\sim N_m(\boldsymbol{0}, \boldsymbol{D}),\\
\boldsymbol{\epsilon}_i &\sim N_{r_i}(\boldsymbol{0}, \boldsymbol{\Omega}_i),
\end{aligned}$$

where $\boldsymbol{b}_1, \boldsymbol{b}_2, \ldots, \boldsymbol{b}_n, \boldsymbol{\epsilon}_1, \boldsymbol{\epsilon}_2, \ldots, \boldsymbol{\epsilon}_n$ are all independent, and $\boldsymbol{y}_1, \boldsymbol{y}_2, \ldots, \boldsymbol{y}_n$ are also independent. The matrices $\boldsymbol{X}_i$ and $\boldsymbol{Z}_i$ are $r_i \times l$ and $r_i \times m$ dimensional matrices of known covariates; $\boldsymbol{\beta}$ is a l-dimensional vector containing the fixed effects; $\boldsymbol{b}_i$ is the m-dimensional vector containing the random effects (RE); and $\boldsymbol{\epsilon}_i$ is an r_i-dimensional vector of residual components. The variance–covariance matrix $\boldsymbol{D}$ is a general $(m \times m)$ dimensional matrix, and $\boldsymbol{\Omega}_i$ is a $(r_i \times r_i)$ covariance matrix that depends on i only through its dimension r_i.

It is clear that $\boldsymbol{y}_i \sim N_{r_i}(\boldsymbol{X}_i\boldsymbol{\beta}, \boldsymbol{Z}_i\boldsymbol{D}\boldsymbol{Z}_i' + \boldsymbol{\Omega}_i)$. For complete data, $r_i = pq$ and $\boldsymbol{\Omega}_i = \boldsymbol{V} \otimes \boldsymbol{\Sigma}$. For the ith subject with $pq - r_i$ missing values we define

$$\boldsymbol{\Omega}_i = \dim_{r_i}(\boldsymbol{V} \otimes \boldsymbol{\Sigma}),$$

which is an $r_i \times r_i$ submatrix obtained from $\boldsymbol{V} \otimes \boldsymbol{\Sigma}$ by retaining only the rows and columns corresponding to nonmissing observations. The number of RE and the form of $\boldsymbol{Z}_i$ can be chosen to fit the observed covariance matrix for the ith individual as:

$$\mathrm{Cov}(\boldsymbol{y}_i) = \boldsymbol{Z}_i\boldsymbol{D}\boldsymbol{Z}_i' + \dim_{r_i}(\boldsymbol{V} \otimes \boldsymbol{\Sigma}).$$

The new linear classification rule in this case is,

Allocate the ith individual with response $\boldsymbol{y}_i$ to Population 1 if

$$(\boldsymbol{\mu}_{1i} - \boldsymbol{\mu}_{2i})'(\boldsymbol{Z}_i\boldsymbol{D}\boldsymbol{Z}_i' + \dim_{r_i}(\boldsymbol{V} \otimes \boldsymbol{\Sigma}))^{-1}\left(\boldsymbol{y}_i - \frac{1}{2}(\boldsymbol{\mu}_{1i} + \boldsymbol{\mu}_{2i})\right) \geq 0, \quad (11.26)$$

and to Population 2 otherwise,

where $\boldsymbol{\mu}_{1i}$ and $\boldsymbol{\mu}_{2i}$ are the subject specific ith subject means in Population 1 and in Population 2 respectively. The parameters $\boldsymbol{\mu}_{1i}$, $\boldsymbol{\mu}_{2i}$, $\boldsymbol{D}$, $\boldsymbol{V}$, and $\boldsymbol{\Sigma}$ are unknown, and should be estimated from independent training samples, n_1 and n_2, from the respective populations. The sample version of the classification rule is obtained by replacing $\boldsymbol{\mu}_{1i}$, $\boldsymbol{\mu}_{2i}$, $\boldsymbol{D}$, $\boldsymbol{V}$, and $\boldsymbol{\Sigma}$ by their maximum likelihood estimates. This can be readily done using commonly available software (such as Proc Mixed in SAS) which fit the mixed effects models. To demonstrate the potential of the new classification rule (Equation 11.25) we use the Osteoporosis data set described in the previous section. We have already tested and accepted in Section 11.5 the hypothesis that $\boldsymbol{\Omega} = \boldsymbol{V} \otimes \boldsymbol{\Sigma}$, with $\boldsymbol{V}$ as CS structure. For illustration, we will use the investigator information as the covariate in the model. The classification results with CS correlation structure on $\boldsymbol{V}$ on the repeated measures for the complete data and for the incomplete data for both without/with RE are given in Tables 11.16 and 11.17, respectively. This new classification rule is found to be as good as the classification rule that is described in Section 11.3.1.2. In fact, introduction of the random effect marginally improves the performance. The overall model can be selected by Akaike's Information Criterion (AIC) and by -2 log-likelihood (smaller is better). Tables 11.18 and 11.19 respectively give the values of the AIC and -2 log-likelihood, and the MERs for all combinations of covariates and RE of the models. The covariate information, however, does not improve the MER. This indicates that the investigator is not a very informative covariate for classification purpose, even though AIC and -2 log-likelihood criteria have been improved slightly. See Roy (2006a) for more details.

Using MI procedure of SAS (Version 9.1) for multiple imputation we create the multiply imputed data set. Median MER and its range on the five imputed

TABLE 11.16

Classification Table for Complete Data by our Proposed Method

	No RE		**RE**	
	Pop1	**Pop2**	**Pop1**	**Pop2**
Pop1	12	2	12	1
Pop2	1	17	1	18

TABLE 11.17

Classification Table for Incomplete Data by our Proposed Method

	No RE		**RE**	
	Pop1	**Pop2**	**Pop1**	**Pop2**
Pop1	16	4	16	4
Pop2	5	19	5	19

TABLE 11.18

MERs for Complete Data by our Proposed Method

Case Number	Covariates	RE	AIC	−2 log-likelihood	MER (%)
1	No	No	−2602.9	−2704.3	9.375
2	No	Yes	−2611.2	−2715.2	6.250
3	Yes	No	−2654.7	−2774.7	9.375
4	Yes	Yes	−2655.5	−2777.5	9.375

TABLE 11.19

MERs for Incomplete Data by our Proposed Method

Case Number	Covariates	RE	AIC	−2 log-likelihood	MER (%)
1	No	No	−3382.9	−3484.9	20.455
2	No	Yes	−3383.0	−3487.0	20.455
3	Yes	No	−3434.2	−3554.2	20.455
4	Yes	Yes	−3433.7	−3555.7	20.455

TABLE 11.20

Median MERs for Multiply Imputed Data by our Proposed Method

Case Number	Covariates	RE	AIC	−2 log-likelihood	MER (%)	Range
1	No	No	−3059.1	−3161.6	27.273	11.364
2	No	Yes	−3060.1	−3164.1	25.000	13.636
3	Yes	No	−3131.4	−3251.4	31.818	11.364
4	Yes	Yes	−3128.6	−3250.6	34.091	13.364

data sets are reported in Table 11.20. We see that the median MER from multiply imputed data sets is greater than the MER on the incomplete data set. This suggests us to use the new classification rule directly on the incomplete data set rather than imputed and complete data set. This is also confirmed by the increase of AIC and the −2 log-likelihood values in all four cases for all five imputed data sets. A comparison of Tables 11.19 and 11.20 suggests that multiple imputation is not, in fact, efficient in this case.

11.7 Concluding Remarks

This chapter summarizes some of the recent work on discrimination and classification in the context of univariate and multivariate repeated measures data. The work extends the classical discriminant analysis but at the expense of computational complexity that is added by the extra dimension in which extension is made. The first author has written extensive SAS codes (Roy, 2002) to solve most of the problems discussed here. However, while these codes are functional and with minor modifications can analyze a large variety of data sets, there is a genuine need for more efficient and user-friendly algorithms and corresponding computer codes to solve these problems. We hope that this review provides a platform for the development of these algorithms.

It must be added that only one specific aspect of this classification problem has been addressed here. Issues pertaining to variable selection as well as time point selection have not been considered in the present work. These are important issues as it is well known in the classical discriminant analysis that it is possible for the performance of the classification rule to go down as more variables are added. The same issue is likely to appear here with respect to variables as well as repeated measures. Work on these issues is under progress.

References

Andrews, D.F., Brant, R., Percy, M.E. (1986). Incorporation of Repeated Measurements in Logistic Discremination, *The Canadian Journal of Statistics*, 14(3), 263–266.

Bartlett, M.S., Please, N.W. (1963). Discrimination in the Case of Zero Mean Differences, *Biometrika*, 50, 17–21.

Choi, S.C. (1972). Classification of Multiply Observed Data, *Biometrical Journal*, 14, 8–11.

Cochran, W.G., Bliss, C.I. (1948). Discriminant functions with covariance. *Annals of Mathematical Statistics*, 19, 151–176.

Enis, P., Geisser, S. (1970). Sample Discriminants which Minimize Posterior Squared Error Loss. *South African Statistical Journal*, 4, 85–93.

Fuller, R.A. (1976). *Introduction to Statistical Time Series*, John Wiley & Sons, New York.

Geisser, S. (1964). Posterior Odds for Multivariate Normal Classification, *Journal of the Royal Statistical Society Series B*, 26, 69–76.

Geisser, S., Desu, M.M. (1967). Bayesian Zero-Mean Uniform Distribution, *Research Report No. 10*, Department of Statistics, State University of New York at Buffalo, 1–19.

Geisser, S., Desu, M.M. (1968). Predictive Zero-Mean Uniform Discrimination, *Biometrika*, 55, 519–524.

Gupta, A.K. (1980). On a Multivariate Statistical Classification Model. In R.P. Gupta (Ed.), *Multivariate Statistical Analysis*, Amsterdam: North-Holland, 83–93.

Gupta, A.K. (1986). On a Classification Rule for Multiple Measurements, *Computers and Mathematics with Applications*, 12A, 301–308.

Gupta, A.K., Logan, T.P. (1990). On a Multiple Observations Model in Discriminant Analysis, *Journal of Statistical Computation and Simulation*, 34, 119–132.

Johnson, R.A., Wichern, D.W. (2002). *Applied Multivariate Statistical Analysis*, Prentice-Hall, New Jersey.

Khattree, R., Naik, D.N. (1994). Estimation of Interclass Correlation under Circular Covariance, *Biometrika*, 81(3), 612–617.

Khattree, R., Naik, D.N. (2000). *Multivariate Data Reduction and Discrimination with SAS Software*, Wiley, New York.

Kshirsagar, A.M. (1972). *Multivariate Analysis*, Dekker, Inc., New York, NY.

Lachenburch, P.A. (1977). Covariance Adjusted Discriminant Functions, *Annals of the Institute of Statistics and Mathematics*, 29, Part A:247–257.

Laird, N.M., Ware, J.H. (1982). Random Effects Models for Longitudional Data, *Biometrics*, 38, 963–974.

Logan, T.P., Gupta, A.K. (1993). Bayesian Discrimination Using Multiple Observations, *Communications in Statistics—Theory and Methods*, 22(6), 1735–1754.

McLachlan, G.J. (1992). *Discriminant Analysis and Statistical Pattern Recognition*, Wiley, New York.

Roy, A. (2002). *Some Contributions to Discrimination and Classification with Repeated Measures Data with Special Emphasis on Biomedical Applications*, Ph.D. Dissertation, Department of Mathematics and Statistics, Oakland University, Rochester, MI.

Roy, A. (2006a). A New Classification Rule for Incomplete Doubly Multivariate Data Using Mixed Effects Model with Performance Comparisons on the Imputed Data, *Statistics in Medicine*, 25(10), 1715–1728.

Roy, A. (2006b). Testing of Kronecker Product Structured Mean Vectors and Covariance Matrices, *Journal of Statistical Theory and Applications*, 5(1), 53–69.

Roy, A., Khattree, R. (2003a). Discrimination with Repeated Measures Data: A Study of Certain Covariance Structures, *Technical Report*, 9, Department of Mathematics and Statistics, Oakland University, Rochester, MI.

Roy, A., Khattree, R. (2003b). Tests for Mean and Covariance Structures Relevant in Repeated Measures Based Discriminant Analysis, *Journal of Applied Statistical Science*, 12(2), 91–104.

Roy, A., Khattree, R. (2005a). Discrimination and Classification with Repeated Measures Data under Different Covariance Structures, *Communications in Statistics—Simulation and Computation*, 34(1), 167–178.

Roy, A., Khattree, R. (2005b). On Discrimination and Classification with Multivariate Repeated Measures Data, *Journal of Statistical Planning and Inference*, 134(2), 462–485.

Roy, A., Khattree, R. (2006). Classification of Multivariate Repeated Measures Data with Temporal Autocorrelation, *Journal of Applied Statistical Science*, 15(3), 41–53.

Rubin, D.B. (1987). *Multiple Imputation for Nonresponse in Surveys*, John Wiley & Sons, New York.

Rubin, D.B. (1996). Multiple Imputation after 18+ Years, *Journal of the American Statistical Association*, 91, 473–489.

SAS Institute Inc. *SAS/STAT User's Guide Version 9.1*, SAS Institute Inc., Cray, NC, 2004.

Timm, N.H. (1980). Multivariate Analysis of Variance of Repeated Measurements. In P.R. Krishnaiah (Ed.), *Handbook of Statistics, Vol. 1*, North-Holland, 41–87.

Tomasco, L., Helms, R.D., Snapian, S.M. (1999). A Discriminant Analysis Extension to Mixed Models, *Statistics in Medicine*, 18, 1249–1260.

12

Estimation Methods for Analyzing Longitudinal Data Occurring in Biomedical Research

N. Rao Chaganty and Deepak Mav

CONTENTS

12.1 Introduction

Statistical methods for analyzing longitudinal data are important tools in biomedical research. The phrase "repeated or longitudinal data" is used for data consisting of responses taken on subjects or experimental units at different time points or under multiple treatments. Such data occur commonly in many scientific disciplines especially in biomedical studies. Although longitudinal data can be viewed as multivariate data, there are some key differences between the two. Multivariate data is usually a snapshot of different variables taken at a single time point, whereas longitudinal data consists of snapshots of the same variable taken at different time points. Thus, even though there are similarities between the two types, the data analysis goals are usually not the same and each pose different challenges and require different approaches for statistical analysis. Much research has already been done on multivariate continuous response variables using linear models and the multivariate normal distribution. Modelling multivariate discrete data is difficult because there is not a single multivariate discrete distribution as prominent as the multivariate normal distribution that has nice properties. For discrete univariate outcomes, exponential dispersion families are among the most popular and widely studied probability models. These statistical models that relate the random outcomes or responses to covariates are useful to understand variation in responses as a function of the covariates. We begin this chapter with a brief discussion of these models and maximum likelihood (ML) estimation of model parameters.

12.2 Univariate Exponential Dispersion Models

Suppose we have a collection $\{y_i,\ i = 1, \ldots, n\}$ of univariate observations or responses of a dependent variable as well as vectors of observations $\{\mathbf{x}_i,\ i = 1, \ldots, n\}$ on some covariates or explanatory variables taken on n independent subjects. The simple linear model that relates the response variables with the explanatory variables is given by

$$y_i = \mathbf{x}_i' \beta + \epsilon_i, \tag{12.1}$$

where β is the vector of unknown regression coefficients and the ϵ_i's are assumed to be independent, normal random variables with zero mean and constant variance. This traditional simple linear model has been used extensively in statistical data analysis but has several limitations. First, the response variables y_i could be proportions with range 0 to 1, yet $\mathbf{x}_i' \beta$ is free to vary and may not fall within this range. Second, the response variables y_i could be binary or counts so that the assumption of normality for ϵ_i is invalid. Third, it may not be realistic to assume that the variance of y_i is a constant. For example, if y_i represents Poisson counts then the variance and the mean are equal.

To overcome the aforementioned limitations, Nelder and Wedderburn (1972) proposed a class of exponential dispersion families. A probability mass function or probability density function of a random variable y_i is said to be a member of the exponential dispersion family if it can be written in the form

$$f(y_i; \theta_i, \phi) = \exp\left[\frac{y_i\theta_i - b(\theta_i)}{\phi} + c(y_i, \phi)\right], \tag{12.2}$$

TABLE 12.1

Canonical Link and Variance Functions

Distribution	Canonical Link Function		Variance Function $\tau(\cdot)$
Poisson	Log	$\eta_i = \log(\mu_i)$	μ_i
Binomial	Logit	$\eta_i = \log\left[\frac{\mu_i}{n_i - \mu_i}\right]$	$\mu_i\left[1 - \frac{\mu_i}{n_i}\right]$
Normal	Identity	$\eta_i = \mu_i$	1
Gamma	Reciprocal	$n_i = \frac{1}{\mu_i}$	μ_i^2
Inverse Gaussian	Reciprocal^2	$\eta_i = \frac{1}{\mu_i^2}$	μ_i^3

where θ_i is the canonical form of the location parameter and ϕ is the scale parameter of the family. For the probability distribution (Equation 12.2) we can verify that the mean $E(y_i) = \mu_i = b'(\theta_i)$ and $Var(y_i) = \phi\, b''(\theta_i)$. The function $\tau(\mu_i) = b''(\theta_i)$ is known as the variance function. In the generalized linear model we assume that the distribution of the ith sample outcome y_i is a member of the exponential dispersion family. Further, the relationship between the mean μ_i and the linear predictor $\eta_i = \mathbf{x}_i'\beta$ is given by $\mu_i = g^{-1}(\eta_i)$, where function g is known as the link function. The link function is a monotonic and differentiable function. If $\theta_i = \eta_i$, then the generalized linear model is called the canonical model, and the corresponding link function g defined by $g^{-1}(\theta_i) = b'(\theta_i)$ is called the canonical link function. Table 12.1 has the canonical link and variance functions for some of the standard distributions.

12.2.1 Parameter Estimation

The popular and efficient method of estimating the unknown regression parameter β and the scale parameter ϕ in the generalized linear model is the method of ML. The ML estimate need not be unique in general. However, for the models with canonical link functions the ML estimate of the location parameter is unique, which is obtained maximizing the log likelihood function

$$\log\left[L(\theta_i, \phi; y_i)\right] = \sum_{i=1}^{n}\left[\frac{[y_i\theta_i - b(\theta_i)]}{\phi} + c\,(y_i, \phi)\right]. \tag{12.3}$$

Recall that $\mu_i = b'(\theta_i)$ and $g(\mu_i) = \eta_i = \mathbf{x}_i'\beta$. A common procedure to get the ML estimate of β is to use the Newton–Raphson iterative algorithm. Starting with a trial value β_0 for β, at the rth step of the iterative algorithm we compute

$$\widehat{\beta}_r = \widehat{\beta}_{r-1} - H^{-1}S.$$

The gradient vector S, also known as the score function, is given by

$$S = \frac{\partial \log[L(\theta_i, \phi; y_i)]}{\partial \beta} = \sum_{i=1}^{n} \left[\frac{y_i - \mu_i}{\phi}\right] \frac{\partial \theta_i}{\partial \mu_i} \frac{\partial \mu_i}{\partial \eta_i} \mathbf{x}_i$$

$$= \sum_{i=1}^{n} \left[\frac{y_i - \mu_i}{\phi}\right] \frac{1}{\tau(\mu_i)} \frac{1}{g'(\mu_i)} \mathbf{x}_i.$$

The Hessian matrix H consists of the second order partial derivatives of Equation 12.3 with respect to β. In matrix notation $H = \mathbf{X}W\mathbf{X}'$, where $\mathbf{X} = (\mathbf{x}_1, \ldots, \mathbf{x}_n)$ and W is the diagonal matrix with ith diagonal

$$w_i = \left[\frac{y_i - \mu_i}{\phi}\right] \frac{\partial^2 \theta_i}{\partial \eta_i^2} - \frac{1}{\phi\, \tau(\mu_i)} \left(\frac{1}{g'(\mu_i)}\right)^2.$$

The asymptotic covariance matrix of the ML estimate of β is the inverse of the information matrix $\mathcal{I} = -E(H) = \mathbf{X}\,[-E(W)]\,\mathbf{X}'$.

12.3 Extensions to Longitudinal Data

Our discussion so far has been focused on data that consists of one response or one observation on each subject. We now turn our attention to the analysis of data that consists of several measurements taken on each subject. Such data occur in longitudinal studies where individuals are measured repeatedly through time or in cross-sectional studies where measurements are taken on each subject under several treatment plans. Many cross-sectional studies arise from cross-over trials where the covariates tend to be fixed and not time varying. Unlike cross-sectional studies, in longitudinal studies the covariates tend to change over time within individuals. The primary interest in these studies is to distinguish changes over time within individuals.

During the last two decades several methods based on extensions of univariate generalized linear models were developed for the analysis of longitudinal measurements. We will discuss some of these in the next few sections. These methods do not specify the joint probability model, but instead assume only a functional form for the marginal distributions for the repeated measurements and specify a covariance structure for the within subject measurements. The covariance structure across time is usually treated as a nuisance parameter, though it can be important in some situations. The estimation methods exploit the independence across subjects to obtain consistent estimates of the parameters and their asymptotic standard errors. The earliest and the most popular method for analyzing longitudinal data is the method of generalized estimating equations (GEE) the details of which are presented in the next section.

12.4 Generalized Estimating Equations

Suppose that instead of y_i we have a vector $Y_i = (y_{i1}, \ldots, y_{it_i})'$ consisting of t_i repeated measurements taken on the ith subject. Associated with each measurement y_{ij} we also have a vector of covariates $x_{ij} = (x_{ij1}, \ldots, x_{ijp})'$, $1 \leq j \leq t_i$, $1 \leq i \leq n$. Here, we assume that the marginal distribution of y_{ij} belongs to the exponential dispersion family so that $E(y_{ij}) = \mu_{ij}$ and $\text{Var}\,(y_{ij}) = \phi\, \tau(\mu_{ij})$, where $\phi > 0$ may be a known constant or an unknown scale parameter. And as before $g(\mu_{ij}) = x'_{ij}\beta$, where $\beta = (\beta_1, \ldots, \beta_p)'$ is a p-dimensional vector of regression coefficients. Because the repeated measurements on each subject are correlated, Liang and Zeger (1986) proposed the GEEs, which adjust for the within subject correlation. This method does not specify a multivariate distribution for Y_i but makes an assumption concerning the within subject correlation, in the form of a "working correlation" matrix $R_i(\alpha)$, which is parametrized by a vector $\alpha = (\alpha_1, \ldots, \alpha_q)'$ of dimension q, and does not depend on the marginal means. It is treated as a nuisance parameter, but is crucial to efficient estimation of the regression parameter. We thus have $E(Y_i) = \mu_i(\beta)$ and $\text{Cov}\,(Y_i) = \phi\, \Sigma_i(\beta, \alpha)$, where $\mu_i(\beta) = (\mu_{i1}, \ldots, \mu_{it_i})'$. The covariance matrix $\Sigma_i(\beta, \alpha) = A_i^{1/2}(\beta)\, R_i(\alpha) A_i^{1/2}(\beta)$, where $A_i(\beta) = \text{diag}\,(\tau(\mu_{i1}), \tau(\mu_{i2}), \ldots, \tau(\mu_{it_i}))$ is the diagonal matrix of variances of the y_{ij}'s. In the GEE method the parameters β and α are obtained by an iterative procedure starting with a trial value β_0 for β and estimating α by the method of moments using the residuals $Z_i(\beta_0)$ where $Z_i(\beta) = A_i^{-1/2}(\beta)\,(Y_i - \mu_i(\beta))$, $1 \leq i \leq n$. Next an updated estimate of β is obtained solving the generalized estimating equation

$$\sum_{i=1}^{n} D_i'(\beta)\, A_i^{-1/2}(\beta)\, R_i^{-1}(\alpha)\, Z_i = 0, \tag{12.4}$$

where $D_i(\beta) = \partial \mu_i(\beta)/\partial \beta'$. The iterative procedure is continued until the estimates converge. Software for this estimation procedure is now available in popular statistical packages including SAS, Splus, STATA, and SPSS.

Since the introduction of the GEE method by Liang and Zeger (1986), numerous authors have extended and suggested different versions of the method. Most of these versions use the same estimating Equation 12.4 but differ in estimation of the correlation parameter. Noteworthy to mention is the method of Prentice (1988) for analyzing correlated binary data, who suggested replacing the moment estimate of α by another estimating equation, known as GEE1. Prentice and Zhao (1991) gave a single GEE type estimating equation treating both β and α as a single parameter, known as the GEE2 method, which includes GEE and GEE1 as special cases. Finally, Hall and Severini (1998) suggested a unified approach for simultaneously estimating all the three parameters β, α, and ϕ. Their approach uses ideas from extended quasi-likelihood and is known as the Extended Generalized Estimating

Equation method (EGEE). However, Hall (2001) has shown that the EGEE approach is a special case of GEE1. In particular, EGEE amounts to estimating the correlation parameter by maximizing the Gaussian likelihood function. We will study the Gaussian method of estimation in detail in Section 12.5. See Sutradhar (2003) for an extensive list of GEE-related acronyms with and without subscripts.

12.4.1 Shortcomings of the GEE Method

The GEE method has been popular for analyzing longitudinal data because it bypasses the difficult problem of specifying the full likelihood and the computationally challenging problem of obtaining the ML estimates. However, despite its popularity the GEE method has several shortcomings. It falls short of the purpose it was introduced—that is, to handle correlated data efficiently. The GEE method has pitfalls in theoretical and in software implementations, particularly in the estimation of the correlation parameters.

Crowder (1995) was the first to identify these pitfalls in the GEE method. He argued with simple examples that the working correlation, when misspecified, lacks a proper definition and thus causes a breakdown of the asymptotic properties of the estimation procedure. See the discussion on page 57 in Crowder (2001). Further, even if the working correlation is correctly specified there is no guarantee that the moment estimate of α will fall within the set of feasible values, that is, α may not fall in the range where the correlation matrix is positive definite (Shults and Chaganty, 1998). Lindsey and Lambert (1998, pp. 465–467) outlined a long list of drawbacks with the GEE procedure. Lee and Nelder (2004, pp. 224–225) have argued that for correlated errors there is no general quasi-likelihood such that GEEs are score equations. Because of the lack of a likelihood basis, they do not regard GEEs as being a proper extension of Wedderburn's (1974) quasi-likelihood approach to analyze models with correlated observations.

There are additional problems with the GEE method when applied to binary data. From Godambe's (1960, 1991) optimal estimating equation theory, it follows that the estimate of the regression parameter β is efficient only if the working correlation is the true correlation of the repeated measurements. However, for multivariate binary distributions the range of the correlations depends on the marginal means, so that the working correlation cannot in general be the true correlation of the data. See Chaganty and Joe (2004, 2006). Furthermore, there is no parametric multivariate binary distribution where the means and variances are functions of the covariates but the correlations are not. Therefore, the GEE framework may be inappropriate for analyzing correlated binary data.

Another major problem with the GEE method is the absence of an objective function that is being minimized (maximized). Such an objective function, if it exists, is useful in testing adequacy or goodness of fit. A solution to this problem was given by Crowder (2001). He suggested the use of the Gaussian likelihood function as an objective function to estimate the correlation parameters. This method is known as Gaussian Estimation (GE). We will discuss this method in the next section.

12.5 Gaussian Estimation

The Gaussian method of estimation was originally introduced by Whittle (1961) as a general method for estimating the parameters in time series data. It was brought into the limelight by Crowder (1995) for the analysis of correlated binomial data, and more recently by Crowder (2001) as a general procedure and an alternative to the GEE method. Here the parameters α and ϕ are estimated by maximizing the Gaussian (normal) log likelihood. Fixing β, this amounts to minimizing the objective function

$$\sum_{i=1}^{n} \left\{ \log |\phi\, R_i(\alpha)| + \frac{1}{\phi}\, Z_i(\beta)'\, R_i^{-1}(\alpha)\, Z_i(\beta) \right\}.$$

Differentiation leads to the estimating equations

$$\sum_{i=1}^{n} \operatorname{tr}\left[\frac{\partial R_i^{-1}(\alpha)}{\partial \alpha_j} (Z_i(\beta)\, Z_i(\beta)' - \phi\, R_i(\alpha)) \right] = 0 \quad \text{for} \quad j = 1, \ldots, q, \tag{12.5}$$

and

$$\widehat{\phi} = \frac{\sum_{i=1}^{n} Z_i(\beta)'\, R_i^{-1}(\alpha)\, Z_i(\beta)}{\sum_{i=1}^{n} t_i}. \tag{12.6}$$

Thus, the Gaussian method of estimation involves solving Equations 12.4 and 12.5 recursively using $\widehat{\phi}$, starting with a trial value for β.

12.6 Quasi-Least Squares Estimation

An alternative method of estimating the correlation parameters is the quasi-least squares (QLS) that was developed in the three papers Chaganty (1997), Shults and Chaganty (1998), and Chaganty and Shults (1999). Unlike the method of moments, for continuous data this method yields an estimate of α such that $R_i(\alpha)$ is positive definite. The method is motivated by the principle of generalized least squares and the theory of unbiased estimating equations. Here starting with a trial value for β we first solve the equations

$$\sum_{i=1}^{n} Z_i(\beta)' \frac{\partial R_i^{-1}(\alpha)}{\partial \alpha_j} Z_i(\beta) = 0, \quad 1 \le j \le q, \tag{12.7}$$

and get an estimate $\widetilde{\alpha}$ of α. Next we get the QLS estimate $\widehat{\alpha}$ of α solving the equation

$$\sum_{i=1}^{n} b_i(\widetilde{\alpha}, \alpha) = 0, \tag{12.8}$$

where $b_i(\widetilde{\alpha}, \alpha) = (b_{i1}(\widetilde{\alpha}, \alpha), \ldots, b_{iq}(\widetilde{\alpha}, \alpha))'$ and $b_{ij}(\widetilde{\alpha}, \alpha) = \mathrm{tr}((\partial R_i^{-1}(\alpha)/\partial \alpha_j)|_{\alpha=\widetilde{\alpha}} R_i(\alpha))$, for $j = 1, \ldots, q$. And as before we update the value of β solving Equation 12.4 using $\widehat{\alpha}$. The procedure is repeated until convergence. Finally, the scale parameter could be estimated as in the GE method.

12.7 Asymptotic Distributions

In this section we will obtain the asymptotic distribution of the Gaussian and the QLS estimates, as $n \to \infty$. The following theorem due to Yuan and Jennrich (1998) is fundamental for establishing the asymptotic distributions for any unbiased estimating equation approach.

THEOREM 12.7.1

(Yuan and Jennrich). Let Z_i, $1 \leq i \leq n$, be independent random vectors of dimensions t_i generated from distributions $f_i(Z_i, \theta_0)$, $1 \leq i \leq n$. Assume that $t_i \leq t$ for all i and $\theta_0 \in \Theta$, which is a subset of $\mathbb{R}^k$. Let the multivariate functions $h_i(Z_i, \theta)$, $1 \leq i \leq n$, taking values in $\mathbb{R}^k$, satisfy the six regularity conditions stated in the Appendix. If $\widehat{\theta}$ is the solution of the unbiased estimating equation

$$\frac{1}{n} \sum_{i=1}^{n} h_i(Z_i, \theta) = 0, \tag{12.9}$$

then we have

$$(\widehat{\theta} - \theta_0) \text{ is } AMVN\left(0, \frac{[I(\theta_0)]^{-1} M(\theta_0) [I(\theta_0)]^{-1}}{n}\right). \tag{12.10}$$

The matrices $I(\cdot)$ and $M(\cdot)$ are defined in the Appendix. In practice $I(\theta_0)$ and $M(\theta_0)$ are unknown and we can estimate them as

$$I(\widehat{\theta}) = \frac{1}{n} \sum_{i=1}^{n} \mathrm{E}\left[\frac{\partial h_i(Z_i, \theta)}{\partial \theta'}\right]\bigg|_{\theta=\widehat{\theta}} \quad \text{and} \quad M(\widehat{\theta}) = \frac{1}{n} \sum_{i=1}^{n} M_i(\widehat{\theta}). \tag{12.11}$$

To establish the asymptotic normality of the Gaussian estimates, we first note that the three estimating Equations 12.4, 12.5, and 12.6 used in that method can be rewritten as $\sum_{i=1}^{n} h_{mi}(Z_i, \theta) = 0$, where $h_{mi}(Z_i, \theta) = K'_{mi}(\theta)\, \xi_i(\theta)$ and

$$K_{mi}(\theta) = \begin{bmatrix} R_i^{-1}(\alpha) A_i^{-1/2}(\beta) D_i(\beta) & & & & & \mathbf{0} \\ \mathbf{0} & b_{i1}(\alpha) & b_{i2}(\alpha) & \ldots & b_{iq}(\alpha) & \mathbf{vec}(R_i^{-1}(\alpha)) \end{bmatrix},$$

$$b_{ij}(\alpha) = \mathbf{vec}\left(\frac{\partial\, R_i^{-1}(\alpha)}{\partial \alpha_j} \right), \quad 1 \le j \le q,$$

$$\xi_i(\theta) = \begin{pmatrix} Z_i \\ \mathbf{vec}\left(Z_i Z_i' - \phi\, R_i(\alpha)\right) \end{pmatrix}.$$

The asymptotic distribution of the Gaussian estimates follows as a consequence of Theorem 12.7.1 and it is given below.

THEOREM 12.7.2
Let $\theta = (\beta, \alpha, \phi)$ and $\widehat{\theta}_m = (\widehat{\beta}_m, \widehat{\alpha}_m, \widehat{\phi}_m)$ be the Gaussian estimates. If $h_{mi}(Z_i, \theta)$'s satisfy the six regularity conditions then we have $\sqrt{n}(\widehat{\theta}_m - \theta)$ is asymptotically multivariate normal with mean 0 and covariance matrix $\Sigma_m(\theta) = I_m^{-1}(\theta) M_m(\theta) I_m^{-1}(\theta)$, where

$$I_m(\theta) = \frac{1}{n} \sum_{i=1}^{n} K'_{mi}(\theta) \nabla_i(\theta),$$

$$M_m(\theta) = \frac{1}{n} \sum_{i=1}^{n} K'_{mi}(\theta) \Psi_i(\theta) K_{mi}(\theta),$$

$$\nabla_i(\theta) = \mathrm{E}\left[-\frac{\partial \xi_i(\theta)}{\partial \theta} \right] = \begin{bmatrix} A_i^{-1/2}(\beta) D_i(\beta) & & & & & \mathbf{0} \\ \mathbf{0} & \phi c_{i1}(\alpha) & \phi c_{i2}(\alpha) & \ldots & \phi c_{iq}(\alpha) & \mathbf{vec}(R_i(\alpha)) \end{bmatrix},$$

$$c_{ij}(\alpha) = \mathbf{vec}\left(\frac{\partial R_i(\alpha)}{\partial \alpha_j} \right), \quad 1 \le j \le q.$$

and $\Psi_i(\theta) = \mathrm{Cov}\,(\xi_i(\theta))$.

To derive the asymptotic distribution of the QLS estimates we first note that the estimating equations 12.4, 12.7, 12.8, and 12.6 can be rewritten as $\sum_{i=1}^{n} h_{qi}(Z_i, \theta) = 0$, where $h_{qi}(Z_i, \theta) = K'_{qi}(\theta) \xi_i(\theta)$ and

$$K_{qi}(\theta) = \begin{bmatrix} R_i^{-1}(\alpha) A_i^{-1/2}(\beta) D_i(\beta) & & & & & \mathbf{0} \\ \mathbf{0} & b_{1i}(\widetilde{\alpha}) & b_{2i}(\widetilde{\alpha}) & \ldots & b_{qi}(\widetilde{\alpha}) & \mathbf{vec}(R_i^{-1}(\alpha)) \end{bmatrix}.$$

Theorem 12.7.3 establishes the asymptotic distribution of the QLS estimates.

THEOREM 12.7.3
Let $\theta = (\beta, \alpha, \phi)$ and let $\widehat{\theta}_q = (\widehat{\beta}_q, \widehat{\alpha}_q, \widehat{\phi}_q)$ be the QLS estimates. Assume that for each α, there exists $\widetilde{\alpha}$ such that

$$\sum_{i=1}^{n} \operatorname{tr}\left(\left. \frac{\partial R_i^{-1}(\alpha)}{\partial \alpha} \right|_{\alpha=\widetilde{\alpha}} R_i(\alpha) \right) = 0.$$

Under the regularity conditions 1–6 of Theorem 12.7.1 we have $\sqrt{n}(\widehat{\theta}_q - \theta)$ is asymptotically multivariate normal with mean 0 and covariance matrix $\Sigma_q(\theta) = I_q^{-1}(\theta) M_q(\theta) I_q^{-1}(\theta)$, where

$$I_q(\theta) = \frac{1}{n} \sum_{i=1}^{n} K'_{q_i}(\theta) \nabla_i(\theta)$$

$$M_q(\theta) = \frac{1}{n} \sum_{i=1}^{n} K'_{q_i}(\theta) \Psi_i(\theta) K_{q_i}(\theta),$$

and $\nabla_i(\theta)$, $\Psi_i(\theta)$ are as in Theorem 12.7.2.

In general, the Gaussian estimate $\widehat{\alpha}_g$ and the QLS estimate $\widehat{\alpha}_q$ are not the same. For example the following results can be found in Chaganty (2003). When $t_i = t$ for all i, Z_i's are multivariate normal with mean 0 and $R_i(\alpha)$ has an AR(1) structure then

$$\sqrt{n}(\widehat{\alpha}_g - \alpha) \xrightarrow{\text{d}} N\left(0, \frac{t(1-\alpha^2)^2}{(t-1)\,(2\alpha^2 + t(1-\alpha^2))}\right),$$

$$\sqrt{n}(\widehat{\alpha}_q - \alpha) \xrightarrow{\text{d}} N\left(0, \frac{t(1-\alpha^2) - (1-\alpha^{2t})}{(t-1)^2}\right). \tag{12.12}$$

12.8 Analysis of Longitudinal Binary Data

The two estimation methods for the correlation parameter described in Sections 12.5 and 12.6 are mainly applicable for analyzing continuous longitudinal data. It is well known that for binary variables the range of the correlation depends on the marginal means. Both estimating equation methods do not take into account these bounds and can result in invalid estimates. For analyzing longitudinal binary data we recommend using the multivariate probit model originally proposed by Ashford and Sowden (1970). This method is widely used in econometrics (Chib and Greenberg (1996), Chib and Greenberg (1998)), genetics (Mendell and Elston (1974)), and psychometrics (Maydeu-Olivares (2001), Muthén (1978)), but neglected in biostatistics and medical statistics (Lesaffre and Molenberghs (1991)). The basic idea behind the multivariate probit model is that there exists a latent random vector L_i distributed as multivariate normal with mean 0 and correlation matrix $\Omega(\gamma)$ such that $Y_i = I(L_i \le \mathbf{x}_i' \beta)$, where $I(\cdot)$ is the indicator function. The parameter γ is known as the latent correlation. Lesaffre and Kaufmann (1992) have established existence and uniqueness of the ML estimates for the multivariate probit model. Recently, Chaganty and Joe (2004) have shown that the ML estimate of β for this model is more efficient than the GEE estimate.

As an example we consider the data studied in Preisser and Qaqish (1999) and made available publicly by the authors at http://www.bios.unc.edu/~jpreisse/ for further analysis by the readers. See page 578, Preisser and Qaqish (1999). The data is from a randomized clinical trial conducted to assess the Guidelines for Urinary Incontinence (UI) Discussion and Evaluation as adopted by primary care providers. The goal was to identify factors among urinary incontinent men and women age 76 and older that could predict their dichotomous response to the question: "Do you consider this accidental loss of urine a problem that interferes with your day to day activities or bothers you in other ways?" This question was put to a total of 137 patients from 38 practices. In this example, the practices are independent and treated as subjects or clusters. The observations within practices are treated as equicorrelated measurements. The binary response y_{ij} of the jth patient from the ith practice, equals 1 if the patient is bothered and 0 if not. There are six covariates: (1) age; (2) gender (male or female); (3) the number of leaking accidents in a day (dayacc); (4) severity of the leak, takes 1 = the leak creates a moisture, 2 = wet their underwear, 3 = trickle down the thigh, 4 = wet the floor, (severe); (5) number of times they go to the toilet to urinate (toilet); and (6) an indicator variable for female. The age variable is centered at 76 and scaled down by 0.1.

Table 12.2 contains the parameter estimates obtained using logit link function and equicorrelated structure for the three methods of estimation: GEE, GE, and QLS. The last two columns contain the estimates and standard

TABLE 12.2

UI Data: Parameter Estimates

	GEE		GE		QLS		Probit	
Parameter	**Estimates**	**S.E.**	**Estimates**	**S.E.**	**Estimates**	**S.E.**	**Estimates**	**S.E.**
Intercept	−3.054	0.959	−2.929	0.959	−2.996	0.959	−1.759	0.683
Female	−0.745	0.600	−0.782	0.591	−0.762	0.596	−0.425	0.341
Age	−0.676	0.561	−0.694	0.556	−0.684	0.558	−0.383	0.340
Dayacc	0.392	0.093	0.381	0.092	0.386	0.092	0.228	0.053
Severe	0.812	0.359	0.802	0.357	0.808	0.358	0.485	0.197
Toilet	0.108	0.099	0.106	0.098	0.107	0.098	0.054	0.047
α	0.093	—	0.153	0.074	0.120	0.076	0.288*	0.215

* Latent correlation.

errors for the multivariate probit model. We have used PROC GENMOD to obtain the estimates for the GEE method, which does not give the standard error for the estimate of α. The GEE estimates can be found also in Table 12.2, page 576 in Preisser and Qaqish (1999). Note that the estimates of the regression parameter obtained by the first three methods are similar. Using the estimates of the regression parameter we calculated the upper bound for the correlation parameter as given on page 204 for the equicorrelated structure in Chaganty and Joe (2006). The upper bound for α is 0.0175 for GEE, 0.0195 for GE, and 0.0185 for the QLS. Clearly, the estimate of α is out of bounds for the three methods and thus the validity of the estimates is in doubt. On the other hand the multivariate probit model, which has a sound theoretical basis, yields regression parameter estimates that are different from the other methods. The standard errors are also uniformly smaller for the multivariate probit model, and hence it is preferable. A software package "mprobit" for fitting the multivariate probit model is available at http://cran.r-project.org/src/contrib/PACKAGES.html.

12.9 Analysis of Longitudinal Poisson Counts

The Poisson distribution has been a common marginal model for longitudinal count data. Similar to binary variables, the range of the correlation between two dependent Poisson variables depends on the marginal means through Fréchet bounds. These bounds are not in closed form, but they can be obtained numerically. We will see that the bounds are a little more flexible than they are for binary variables, since the support of Poisson variables is not a set of cardinality two, but it is countably infinite.

In recent years a method due to Sim (1993) for simulating a multivariate Poisson distribution with given means and correlations has gained popularity. Here we will discuss a slight generalization of Sim (1993)'s algorithm. This algorithm, when successful, leads to a complicated multivariate Poisson distribution. ML estimation is intractable for this complex multivariate distribution and there may be a need to use methods based on estimating equations, like Gaussian and the QLS estimation. We will use simulated data generated using Sim's algorithm to compare performance of the two estimating methods.

12.9.1 Correlation Bounds for Poisson Variables

Let Y_1 and Y_2 be two Poisson random variables with means λ_1 and λ_2. Suppose that $F_i(y_i)$ denotes the marginal cumulative distribution function of Y_i and the joint cumulative distribution function of (Y_1, Y_2) is denoted by $F(y_1, y_2)$. The nonparametric range of the correlation between Y_1 and Y_2 comes from the Fréchet bounds

$$F_L(y_1, y_2) \leq F(y_1, y_2) \leq F_U(y_1, y_2), \tag{12.13}$$

where $F_L(y_1, y_2) = \max(0, F_1(y_1) + F_2(y_2) - 1)$ and $F_U(y_1, y_2) = \min(F_1(y_1), F_2(y_2))$. The correlation coefficient corresponding to the joint distribution $F_U(y_1, y_2)$ is given by

$$\rho_U(\lambda_1, \lambda_2) = \frac{1}{\lambda_1\lambda_2} \sum_{y_1=1}^{\infty} \sum_{y_2=1}^{\infty} \min[F_1(y_1)(1 - F_2(y_2)), F_2(y_2)(1 - F_1(y_1))], \tag{12.14}$$

and the correlation coefficient corresponding to the joint distribution $F_L(y_1, y_2)$ is

$$\rho_L(\lambda_1, \lambda_2) = \frac{1}{\lambda_1\lambda_2} \sum_{y_1=1}^{\infty} \sum_{y_2=1}^{\infty} -\min[(1 - F_1(y_1))(1 - F_2(y_2)), F_1(y_1)F_2(y_2)]. \tag{12.15}$$

The range of the correlation between Y_1 and Y_2 is given by the interval $[\rho_L(\lambda_1, \lambda_2), \rho_U(\lambda_1, \lambda_2)]$. Figure 12.1 shows the complex behavior of this range as a function of $\log(\lambda_2/\lambda_1)$ for various values of λ_1. The curves below the x-axis are the graphs of $\rho_L(\lambda_1, \lambda_2)$ and the corresponding graph of $\rho_U(\lambda_1, \lambda_2)$ is above the x-axis. The complex behavior of these graphs poses difficulty in

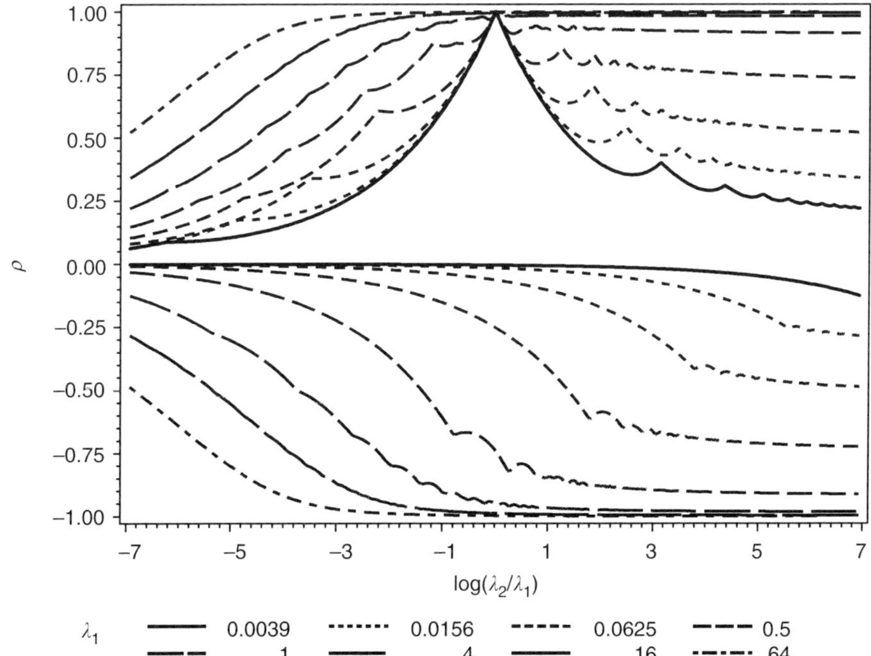

FIGURE 12.1
Bivariate Poisson: Feasible region for the correlation.

incorporating the bounds on the correlation in estimation procedures. The next lemma gives some properties of the functions $\rho_U(\lambda_1, \lambda_2)$ and $\rho_L(\lambda_1, \lambda_2)$.

LEMMA 12.9.1
Let Y_1 and Y_2 be two Poisson random variables with means λ_1 and λ_2 respectively. Assume without loss of generality $\lambda_1 \leq \lambda_2$. Let $\rho_L(\lambda_1, \lambda_2)$ and $\rho_U(\lambda_1, \lambda_2)$ be the correlations corresponding to the distribution functions $F_L(y_1, y_2)$ and $F_U(y_1, y_2)$, respectively. Then we have

$$(a)\ \ \rho_U(\lambda_1, \lambda_2) \leq \frac{1}{\sqrt{\lambda_1\lambda_2}}\left(\frac{(\lambda_2 - \lambda_1)^2}{2} + \lambda_1\right),$$

$$(b)\ \ \rho_L(\lambda_1, \lambda_2) = -\sqrt{\lambda_1\lambda_2},\ if\ \lambda_2 \leq \log(2).$$

PROOF If (Y_1, Y_2) is distributed as $F_U(y_1, y_2)$ then,

$$E(Y_1 Y_2) = \sum_{y_1=1}^{\infty}\sum_{y_2=1}^{\infty} \min[\overline{F_1}(y_1), \overline{F_2}(y_2)],$$

where $\overline{F_i}(y_i) = P(Y_i \geq y_i)$ for $i = 1, 2$. Since, $\lambda_2 \geq \lambda_1$, and the Poisson family possesses TP_2 property we have $\overline{F_2}(y_2) \geq \overline{F_1}(y_2) \geq \overline{F_1}(y_1)$ for $y_1 \geq y_2$. Now

$$\begin{aligned}
E(Y_1 Y_2) &= \sum_{y_1=1}^{\infty}\sum_{y_2=1}^{y_1} \min[\overline{F_1}(y_1), \overline{F_2}(y_2)] + \sum_{y_1=1}^{\infty}\sum_{y_2=y_1+1}^{\infty} \min[\overline{F_1}(y_1), \overline{F_2}(y_2)] \\
&= \sum_{y_1=1}^{\infty} y_1\overline{F_1}(y_1) + \sum_{y_2=2}^{\infty}\sum_{y_1=1}^{y_2-1} \min[\overline{F_1}(y_1), \overline{F_2}(y_2)] \\
&\leq \sum_{y_1=1}^{\infty} y_1\overline{F_1}(y_1) + \sum_{y_2=2}^{\infty} (y_2 - 1)\overline{F_2}(y_2).
\end{aligned} \tag{12.16}$$

The first term on the right-hand side of Equation 12.16 is

$$\begin{aligned}
\sum_{y_1=1}^{\infty} y_1\overline{F_1}(y_1) &= \sum_{y_1=1}^{\infty} y_1 P(Y_1 \geq y_1) = \sum_{y_1=1}^{\infty}\sum_{y=y_1}^{\infty} y_1 P(Y_1 = y) \\
&= \sum_{y=1}^{\infty}\sum_{y_1=1}^{y} y_1 P(Y_1 = y) = \sum_{y=1}^{\infty} \frac{y(y+1)}{2} P(Y_1 = y) \\
&= \frac{(E(Y_1^2) + E(Y_1))}{2} = \frac{\lambda_1^2}{2} + \lambda_1.
\end{aligned} \tag{12.17}$$

Now the second term on the right-hand side of Equation 12.16 is

$$\sum_{y_2=2}^{\infty}(y_2-1)\,\overline{F_2}(y_2) = \sum_{y_2=1}^{\infty} y_2\overline{F_2}(y_2) - \sum_{y_2=1}^{\infty} y_2 P(Y_2=y_2)$$
$$= \frac{\lambda_2^2}{2} + \lambda_2 - \lambda_2 = \frac{\lambda_2^2}{2}. \tag{12.18}$$

Substituting Equations 12.17 and 12.18 in Equation 12.16 we get

$$E(Y_1Y_2) \le \frac{\lambda_1^2+\lambda_2^2}{2} + \lambda_1. \tag{12.19}$$

It is easy to verify inequality (a) using Equation 12.19. The proof of (b) is simple. When $0 \le \lambda_1 \le \lambda_2 \le \log(2)$, we have $P(Y_i = 0) = e^{-\lambda_i} \ge 0.5$ for $i = 1, 2$. Hence $2\,F_i(y) \ge 1$ or equivalently $F_i(y) \ge (1 - F_i(y))$ for all $y \ge 1$ and $i = 1, 2$. Therefore,

$$F_1(y_1)F_2(y_2) \ge (1 - F_1(y_1))(1 - F_2(y_2))$$

and Equation 12.14 simplifies to

$$\rho_L(\lambda_1, \lambda_2) = \frac{-1}{\sqrt{\lambda_1\lambda_2}}\sum_{y_1=1}^{\infty}\sum_{y_2=1}^{\infty}(1 - F_1(y_1))(1 - F_2(y_2)) = \frac{-\lambda_1\lambda_2}{\sqrt{\lambda_1,\lambda_2}} = -\sqrt{\lambda_1,\lambda_2}.$$

This completes the proof of the lemma. ∎

There are situations where the range of the correlation between Y_1 and Y_2 could be very narrow. For example, when $\lambda_1 = \lambda_2 = \lambda$ converges to zero, the random variables Y_1 and Y_2 converge to zero in probability and thus the range of the correlation between Y_1 and Y_2 converges to the singleton set $\{1\}$. On the other hand for fixed λ_1, we have $\lim_{\lambda_2\to\infty}\rho_U(\lambda_1, \lambda_2) = 0$ and $\lim_{\lambda_2\to 0}\rho_L(\lambda_1, \lambda_2) = 0$. Therefore, for fixed λ_1 when λ_2 is too large or too small the range of the correlation becomes a singleton set containing zero.

12.9.2 Multivariate Poisson

Unlike the multivariate Gaussian distribution, which is uniquely determined by the marginal means and covariance matrix, there can be several multivariate distributions with marginals as Poisson and a specified covariance matrix. In the bivariate case we can easily construct a joint distribution with specified correlation in the feasible region by taking a linear combination of the Fréchet lower and upper bounds given in Equation 12.13. For higher dimensions, Bernoulli random variables play an important role in the construction of multivariate Poisson distributions with given means and correlations. In particular, the binomial thinning operator, which has been extensively used in modelling reliability data involving imperfect repairs [see Barlow and Proschan (1975)], provides the foundation for simulating multivariate Poisson distributions.

Let M be a nonnegative integer valued random variable and $0 < \theta < 1$ be fixed. The binomial thinning of M with θ, is a random variable denoted by $\theta \star M$ that equals in distribution to $\sum_{i=0}^{M} I_i$, where $I_0 = 0$ and $I_i, i = 1, 2, \ldots, M$ are i.i.d. Bernoulli (θ) random variables independent of M. The binomial thinning is closed within the class of Poisson distributions as shown in the next lemma. Joe (1996) and Zhu and Joe (2003) have developed approaches to construct nonstationary Poisson time series models by exploiting this closure property of the binomial thinning operator. See also Joe (1997). It is easy to establish the following lemma.

LEMMA 12.9.2
Let M be a Poisson random variable with mean λ. Then,

1. *$\theta \star M$ is distributed as Poisson random variable with mean $\theta\lambda$.*
2. $\text{Cov}\,(\theta_1 \star M, \theta_2 \star M) = \theta_1\theta_2\lambda$.

for all θ, θ_1 and $\theta_2 \in (0, 1)$.

Using the binomial thinning operator, we now introduce a constructive definition for the multivariate Poisson distribution. This is a slight generalization of Sim's (1993) algorithm. Let $Z_1, Z_2, \ldots, Z_q$ be independently distributed random variables such that Z_k is Poisson with mean λ_k, $1 \le k \le q$. Let $\Theta = [\theta_{jk}]$ be a $(p \times q)$ matrix of constants, $0 < \theta_{jk} < 1$. Define

$$Y_j = \sum_{k=1}^{q} \theta_{jk} \star Z_k \quad \text{for } j = 1, 2, \ldots, p. \tag{12.20}$$

Then $\mathbf{Y} = (Y_1, Y_2, \ldots, Y_p)$ is distributed as multivariate Poisson with mean vector $\boldsymbol{\mu} = (\mu_1, \mu_2, \ldots, \mu_p)'$. It is clear from the definition, the conditional probability mass function of Y_j given $(z_1, \ldots, z_q)$ is

$$P(y_j; \mathbf{z}) = \sum_{W \in A_{\mathbf{z}}(y_j)} \prod_{k=1}^{q} \binom{z_k}{w_k} \theta_{jk}^{w_k} (1 - \theta_{jk})^{z_k - w_k}, \tag{12.21}$$

where $A_{\mathbf{z}}(y) = \{W = (w_1, w_2, \ldots, w_q) : w_k \in \mathbb{N}, w_k \le z_k \text{ and } \sum_{k=1}^{q} w_k = y\}$. Therefore, the unconditional joint probability mass function of $\mathbf{Y}$ is

$$P(\mathbf{y}) = \sum_{\mathbf{z}} \left\{ \prod_{i=1}^{p} P(y_j; \mathbf{z}) \right\} \left\{ \prod_{k=1}^{q} \frac{e^{-\lambda_k} \lambda_k^{z_k}}{z_k!} \right\}. \tag{12.22}$$

Using Lemma 12.9.2, we can check that the moment generating function of $\mathbf{Y}$ is given by

$$\begin{aligned} M(\mathbf{t}) &= E\left[\prod_{j=1}^{p} e^{t_j Y_j}\right] \\ &= E_{\mathbf{Z}}\left[\prod_{j=1}^{p}\prod_{k=1}^{q} \mathbf{e}^{t_j(\theta_{jk}\star Z_k)}\right] \\ &= E_{\mathbf{Z}}\left\{\prod_{k=1}^{q}\prod_{j=1}^{p}\left[\theta_{jk}e^{t_j} + (1-\theta_{jk})\right]^{Z_k}\right\}. \end{aligned} \tag{12.23}$$

Further, the moment generating function (12.23) can be written as

$$M(\mathbf{t}) = \prod_{k=1}^{q} \exp\{\lambda_k(e^{S_k(\mathbf{t})} - 1)\} \tag{12.24}$$

where $S_k(\mathbf{t}) = \sum_{j=1}^{p} \log[\theta_{jk}e^{t_j} + (1-\theta_{jk})]$. The representation (12.24) shows that the moment generating function of $\mathbf{Y}$ is equal to the joint moment generating function of q independently distributed Poisson random variables centered at $(S_1(\mathbf{t}), \ldots, S_q(\mathbf{t}))$. Also, $S_k(\mathbf{t})$ can be viewed as the joint cumulant generating function of p independently distributed Bernoulli random variables. The central moment generating function of $\mathbf{Y}$ is given by

$$\begin{aligned} K(\mathbf{t}) &= E[\exp(\mathbf{t}'(\mathbf{Y} - \boldsymbol{\mu}))] \\ &= \prod_{k=1}^{q} \exp\{\lambda_k(U_k(\mathbf{t}) - 1)\}, \end{aligned} \tag{12.25}$$

where $U_k(\mathbf{t}) = e^{S_k(\mathbf{t})} - \sum_{j=1}^{p} \theta_{jk}t_j$. Moments up to order four of the multivariate Poisson mass function (12.22) are needed to calculate the asymptotic relative efficiencies of the estimating methods GE and QLS. These can be obtained by differentiating the central moment generating function (12.25). We will need the following derivatives to obtain simplified expressions for the higher order moments. Note that $U_k(\mathbf{t})$ can also be written as the following power-series:

$$U_k(\mathbf{t}) = \sum_{d_1=0}^{\infty} \cdots \sum_{d_p=0}^{\infty} \prod_{j=1}^{p}\left[\frac{t_j^{d_j}}{d_j!}\pi_{jk}^{(d_j)}\right] - \sum_{j=1}^{p} \theta_{jk}t_j, \tag{12.26}$$

where

$$\pi_{jk}^{(d)} = \begin{cases} 1 & \text{if } d = 0 \\ \theta_{jk} & \text{otherwise.} \end{cases}$$

Therefore, for all $d_j = 0, 1, \ldots; j = 1, 2, \ldots, p$ such that $\sum_{j=1}^{p} d_j \geq 1$ we have

$$\begin{aligned} U_k^{(\mathbf{d})}(\mathbf{t}) &= \frac{\partial^{\sum_{j=1}^{p} d_j} U_k(\mathbf{t})}{\partial t_1^{d_1} \ldots \partial t_p^{d_p}} \\ &= \left\{ \sum_{r_1=0}^{\infty} \cdots \sum_{r_p=0}^{\infty} \prod_{j=1}^{p} \left[\frac{t_j^{r_j}}{r_j!} \pi_{jk}^{(r_j+d_j)} \right] \right\} - \delta_k^{(\mathbf{d})}(\mathbf{t}), \end{aligned} \tag{12.27}$$

where

$$\delta_k^{(\mathbf{d})}(\mathbf{t}) = \begin{cases} \theta_{jk} & \text{if } d_j = 1 \text{ and } d_{j'} = 0 \text{ for all } j' \neq j \\ 0 & \text{otherwise.} \end{cases}$$

Differentiating Equation 12.25 with respect to t_{j_1} we get

$$\frac{\partial K(\mathbf{t})}{\partial t_{j_1}} = \omega_{j_1}(\mathbf{t}) K(\mathbf{t}) \tag{12.28}$$

where

$$\omega_{j_1}(\mathbf{t}) = \sum_{k=1}^{q} U_k^{(\mathbf{d}_1)}(\mathbf{t}) \lambda_k$$

and $\mathbf{d}_1 = e_{j_1}$ (j_1th unit vector in $\mathbb{R}^p$). The covariance matrix of $\mathbf{Y}$ can be obtained by differentiating Equation 12.28 with respect to t_{j_2} and substituting $\mathbf{t}$ with $\mathbf{0}$ as

$$\frac{\partial^2 K(\mathbf{t})}{\partial t_{j_1} \partial t_{j_2}} = \left\{ \omega_{j_1}(\mathbf{t}) \omega_{j_2}(\mathbf{t}) + \omega_{j_1 j_2}^{(2)}(\mathbf{t}) \right\} K(\mathbf{t}), \tag{12.29}$$

where

$$\omega_{j_1 j_2}^{(2)}(\mathbf{t}) = \sum_{k=1}^{q} U_k^{(\mathbf{d}_2)}(\mathbf{t}) \lambda_k$$

and $\mathbf{d}_2 = (e_{j_1} + e_{j_2})$. Therefore, the elements of $\text{Cov}(\mathbf{Y}) = \Sigma$ are given by

$$\begin{aligned}\sigma_{j_1j_2} &= \text{Cov}(Y_{j_1}, Y_{j_2}) \\ &= \omega^{(2)}_{j_1j_2}(\mathbf{0}) \\ &= \sum_{k=1}^{q}\prod_{l=1}^{p}\pi_{lk}^{(d_{2l})}\lambda_k \quad \text{for all } 1 \le j_1, j_2 \le p. \end{aligned} \tag{12.30}$$

Differentiating Equation 12.29 with respect to t_{j_3} we get

$$\begin{aligned}\frac{\partial^3 K(\mathbf{t})}{\partial t_{j_1}\partial t_{j_2}\partial t_{j_3}} &= \Big\{\omega_{j_1}(\mathbf{t})\omega_{j_2}(\mathbf{t})\omega_{j_3}(\mathbf{t}) + \omega_{j_1}(\mathbf{t})\omega^{(2)}_{j_2j_3}(\mathbf{t}) + \omega_{j_2}(\mathbf{t})\omega^{(2)}_{j_1j_3}(\mathbf{t}) \\ &\quad + \omega_{j_3}(\mathbf{t})\omega^{(2)}_{j_1j_2}(\mathbf{t}) + \omega^{(3)}_{j_1j_2j_3}(\mathbf{t})\Big\} K(\mathbf{t}),\end{aligned} \tag{12.31}$$

where

$$\omega^{(3)}_{j_1j_2j_3}(\mathbf{t}) = \sum_{k=1}^{q} U_k^{(\mathbf{d}_3)}(\mathbf{t})\lambda_k$$

is defined using Equation 12.27 and $\mathbf{d}_3 = (e_{j_1} + e_{j_2} + e_{j_3})$. Let $\zeta = \text{vec}((\mathbf{Y} - \boldsymbol{\mu}) \times (\mathbf{Y} - \boldsymbol{\mu})')$. A typical element of $\text{Cov}(\zeta, (\mathbf{Y} - \boldsymbol{\mu})) = \boldsymbol{\tau}$ is given by

$$\begin{aligned}\tau_{j_1j_2j_3} &= \text{Cov}((Y_{j_1} - \mu_{j_1})(Y_{j_2} - \mu_{j_2}), (Y_{j_3} - \mu_{j_3})) \\ &= \omega^{(3)}_{j_1j_2j_3}(\mathbf{0}) \\ &= \sum_{k=1}^{q}\prod_{l=1}^{p}\pi_{lk}^{(d_{3l})}\lambda_k.\end{aligned} \tag{12.32}$$

Similarly, differentiating Equation 12.31 with respect to t_{j_4} we get

$$\begin{aligned}\frac{\partial^4 K(\mathbf{t})}{\partial t_{j_1}\ldots\partial t_{j_4}} &= \Big\{\omega_{j_1}(\mathbf{t})\omega_{j_2}(\mathbf{t})\omega_{j_3}(\mathbf{t})\omega_{j_4}(\mathbf{t}) + \omega^{(2)}_{j_1j_2}(\mathbf{t})\left[\omega^{(2)}_{j_3j_4}(\mathbf{t}) + \omega_{j_3}(\mathbf{t})\omega_{j_4}(\mathbf{t})\right] \\ &\quad + \omega^{(2)}_{j_1j_3}(\mathbf{t})\left[\omega^{(2)}_{j_2j_4}(\mathbf{t}) + \omega_{j_2}(\mathbf{t})\omega_{j_4}(\mathbf{t})\right] + \omega^{(2)}_{j_2j_3}(\mathbf{t})\left[\omega^{(2)}_{j_1j_4}(\mathbf{t}) + \omega_{j_1}(\mathbf{t})\omega_{j_4}(\mathbf{t})\right] \\ &\quad + \omega_{j_1}(\mathbf{t})\omega^{(3)}_{j_2j_3j_4}(\mathbf{t}) + \omega_{j_2}(\mathbf{t})\omega^{(3)}_{j_1j_3j_4}(\mathbf{t}) + \omega_{j_3}(\mathbf{t})\omega^{(3)}_{j_1j_2j_4}(\mathbf{t}) + \omega_{j_4}(\mathbf{t})\omega^{(3)}_{j_1j_2j_3}(\mathbf{t}) \\ &\quad + \omega^{(4)}_{j_1j_2j_3j_4}(\mathbf{t})\Big\} K(\mathbf{t}),\end{aligned}$$

where

$$\omega^{(4)}_{j_1j_2j_3j_4}(\mathbf{t}) = \sum_{k=1}^{q} U_k^{(\mathbf{d}_4)}(\mathbf{t})\lambda_k$$

is defined using Equation 12.27 and $\mathbf{d}_4 = (e_{j_1} + e_{j_2} + e_{j_3} + e_{j_4})$. A typical element of the matrix $\text{Cov}(\zeta) = \boldsymbol{K}$ is given by the following equations:

$$\begin{aligned}\kappa_{j_1j_2j_3j_4} &= \text{Cov}((Y_{j_1} - \mu_{j_1})(Y_{j_2} - \mu_{j_2}), (Y_{j_3} - \mu_{j_3})(Y_{j_4} - \mu_{j_4})) \\ &= \omega^{(2)}_{j_1j_3}(\mathbf{0})\omega^{(2)}_{j_2j_4}(\mathbf{0}) + \omega^{(2)}_{j_2j_3}(\mathbf{0})\omega^{(2)}_{j_1j_4}(\mathbf{0}) + \omega^{(4)}_{j_1j_2j_3j_4}(\mathbf{0}) \\ &= \sigma_{j_1j_3}\sigma_{j_2j_4} + \sigma_{j_1j_4}\sigma_{j_2j_3} + \sum_{k=1}^{q}\prod_{l=1}^{p}\pi_{lk}^{(d_{l4})}\lambda_k. \end{aligned} \tag{12.33}$$

Finally, the covariance matrix of first and second order standardized deviations is

$$\Psi = \text{Cov}(\xi) = D^{-1/2}(\boldsymbol{\sigma})\,\mathbf{V}\,D^{-1/2}(\boldsymbol{\sigma}), \tag{12.34}$$

where

$$\xi = D^{-1/2}(\boldsymbol{\sigma})\,\xi^*,$$

$$\xi^* = \begin{pmatrix} (\mathbf{Y} - \boldsymbol{\mu}) \\ \zeta - \text{vec}(\Sigma) \end{pmatrix}$$

$$\mathbf{V} = \begin{pmatrix} \Sigma & \boldsymbol{\tau}' \\ \boldsymbol{\tau} & \boldsymbol{K} \end{pmatrix},$$

$D(\boldsymbol{\sigma}) = \text{diag}\,(\sigma_{11}, \ldots, \sigma_{pp}, \sigma_{11}^2, \sigma_{11}\sigma_{22}, \ldots, \sigma_{pp}\sigma_{(p-1)(p-1)}, \sigma_{pp}^2)$. Note that $\mathbf{V}$ is the covariance matrix of ξ^*.

We can use the constructive definition to simulate multivariate Poisson vectors with a fixed covariance matrix Σ. The first step in this approach is to find a q dimensional vector $\boldsymbol{\lambda}$ and $(p \times q)$ matrix Θ as a function of Σ satisfying Equation 12.30. In general there could be more than one solution to Equation 12.30. This is often the case when the covariance matrix Σ is structured or the targeted value of q is greater than p. For example, when $\Sigma = \phi R(\alpha)$ and $R(\alpha) = (1 - \alpha)\mathbf{I}_p + \alpha\mathbf{J}_p$, we can check that the choices

$$\{\boldsymbol{\lambda}_1 = \phi[(1 - \alpha)\mathbf{1}_p' \;\; \alpha]', \;\; \Theta_1 = [\mathbf{I}_p \;\; \mathbf{1}_p]\},$$

and

$$\{\boldsymbol{\lambda}_2 = \phi\mathbf{1}_{(p+1)}, \;\; \Theta_2 = [(1 - \sqrt{\alpha})\mathbf{I}_p \;\; \sqrt{\alpha}\mathbf{1}_p]\}$$

satisfy Equation 12.30. In the above $\mathbf{I}_p$ is the identity matrix of order p, $\mathbf{1}_p$ is a column vector of ones of order p, and $\mathbf{J}_p = \mathbf{1}_p\mathbf{1}_p'$. The selection of pair $(\boldsymbol{\lambda}, \boldsymbol{\Theta})$ introduces additional constraints on feasibility of the Σ matrix. The study of the mappings of Σ to $(\boldsymbol{\lambda}, \boldsymbol{\Theta})$ is an important problem, whose solution will not be pursued here.

12.10 Epileptic Seizure Data

In this section we study the performance of the GE and QLS estimation methods for the multivariate Poisson model described in Section 12.9.2. For the simulations we will use a model fitted for a real life data using some well established methods. Leppik et al. (1985) conducted a 2×2 randomized double-blinded crossover clinical trial to study the effectiveness of anti-epileptic progabide drug on 59 patients suffering from simple or complex seizures. At each of the four successive postrandomization visits the number of seizures occurring during the previous 2 weeks were reported. The four pre-crossover responses are in Thall and Vail (1990, Table 2, p. 664).

Inspection of the data clearly indicates that the patient with id #207 is an outlier, since the baseline and the first visit seizure count is more than 100. As a preliminary analysis we have fitted the log-linear model that was given by Thall and Vail (1990) for the mean responses

$$\log(\mu_{ij}) = \beta_0 + \text{base}_i\beta_1 + \text{trt}_i\beta_2 + (\text{base}\times\ \text{trt})_i\beta_3 + \text{age}_i\beta_5 + \text{visit4}_j\beta_6, \tag{12.35}$$

where base $=$ log(0.25 $\times$ baseline seizure count), trt is a binary indicator for inclusion in treatment group, age $=$ log(age of patient) and visit4 is a binary indicator for the fourth visit. The Gaussian and QLS estimates with and without patient #207 are given in Tables 12.3 and 12.4. The results with patient #207 included are different from the results without that patient. For example, we can see from the tables, the treatment and the baseline-treatment interaction become insignificant if we exclude patient #207.

TABLE 12.3
Seizure Data: Gaussian Estimates

	With Patient #207			Without Patient #207		
Parameter	**Estimate**	**S.E.**	**p Value**	**Estimate**	**S.E.**	**p Value**
Intercept	−2.7729	0.9489	0.0035	−2.3407	0.8766	0.0076
Base	0.9499	0.0974	0.0000	0.9505	0.0973	0.0000
trt	−1.3401	0.4272	0.0017	−0.5206	0.4164	0.2112
Base × trt	0.5627	0.1742	0.0012	0.1383	0.1941	0.4763
Age	0.9011	0.2756	0.0011	0.7722	0.2550	0.0025
Visit4	−0.1611	0.0656	0.0140	−0.1479	0.0763	0.0527
α	0.1906	0.2731	0.4853	0.1819	0.2765	0.5106

TABLE 12.4

Seizure Data: QLS Estimates

	With Patient #207			Without Patient #207		
Parameter	**Estimate**	**S.E.**	***p* Value**	**Estimate**	**S.E.**	***p* Value**
Intercept	−2.7939	0.9561	0.0035	−2.3579	0.8838	0.0076
BASE	0.9504	0.0987	0.0000	0.9509	0.0983	0.0000
trt	−1.3386	0.4296	0.0018	−0.5196	0.4185	0.2145
Base × trt	0.5633	0.1749	0.0013	0.1388	0.1947	0.4758
Age	0.9066	0.2772	0.0011	0.7768	0.2567	0.0025
Visit4	−0.1611	0.0656	0.0140	−0.1479	0.0763	0.0527
α	0.3582	0.2547	0.1596	0.3393	0.2621	0.1955

For our simulations we have used Equation 12.35. To avoid additional variability of estimates, patient #207 was excluded from our simulations. Details of the simulations are described in the following steps:

1. We fixed the true parameter $\boldsymbol{\beta} = \boldsymbol{\beta}_0$, where

$$\boldsymbol{\beta}_0 = (-2.3579, 0.9509, -0.5196, 0.1388, 0.7768, -0.1479)'$$

   is the QLS estimate for the epileptic seizure data obtained using an exchangeable correlation structure and excluding patient with id #207.

2. For the true correlation matrix we used the exchangeable structure and later repeated the whole simulation procedure for both the autoregressive and moving average correlation structures of the first order.
3. We varied the true correlation parameter α from 0 to $\alpha_{\max}$ in increments of 0.0125, where $\alpha_{\max}$ is the upper extreme of feasible region.
4. For each value of α, $N = 5000$ datasets consisting of 59 multivariate Poisson random vectors were simulated using the algorithm described in Section 12.10.1.
5. For the ith simulated data set we calculated $\widehat{\theta}_{m_i}$ and $\widehat{\theta}_{q_i}$, the Gaussian and QLS estimates of θ, respectively. We also computed the covariance matrices $\widehat{\text{Cov}}(\widehat{\theta}_{m_i}) = \Sigma_m(\widehat{\theta}_{m_i})$ and $\widehat{\text{Cov}}(\widehat{\theta}_{q_i}) = \Sigma_q(\widehat{\theta}_{q_i})$. In calculating these matrices we have used formulas given by Equation 12.34 for the Ψ matrix.
6. The infeasibility or divergent solution probability is estimated for each of the estimates $\widehat{\theta}_m$ and $\widehat{\theta}_q$ as the proportion of times the estimate did not converge or the estimate is deemed to be inconsistent. The infeasible/divergent cases were discarded from further analysis.

7. The joint asymptotic relative efficiency of $\widehat{\theta}_q$ with respect to $\widehat{\theta}_m$ is computed using the trace and determinant of the matrix

$$\Gamma = \left\{ \frac{1}{N_c} \sum_{i=1}^{N_c} \Sigma_q(\widehat{\theta}_{qi}) \right\}^{-1} \left\{ \frac{1}{N_c} \sum_{i=1}^{N_c} \Sigma_m(\widehat{\theta}_{mi}) \right\}, \tag{12.36}$$

where N_c is the number of simulated data sets that yielded convergent estimates for both Gaussian and the QLS methods.

8. We calculated the coverage probability for simultaneous confidence region of the Gaussian method as

$$\frac{1}{N_c} \sum_{i=1}^{N_c} I\left((\widehat{\theta}_{mi} - \theta_0)' \widehat{\Sigma}_{mi}^{-1} (\widehat{\theta}_{mi} - \theta_0) \le \chi^2_{K+q,0.05} \right), \tag{12.37}$$

where $I(\cdot)$ is the indicator function. The coverage probability for QLS method is calculated similarly.

12.10.1 Simulation Results

In this section, we report the findings of our simulations. Consider first the case where the true correlation structure is equicorrelated (EQC). Figure 12.2 contains the plots of estimated probabilities (proportions of simulated data sets) as a function of α, where the Gaussian and QLS estimates of $\theta = (\beta, \alpha, \phi)$ did not converge or the final values of the estimates are deemed to be invalid. It is clear from the plots, both the Gaussian and QLS methods have similar infeasibility problems. The estimated probabilities are low when α is in the interior of the feasible range. On the other hand, the proportions of invalid estimates are high when α is close to zero; the reason being the estimate of α could be negative in this case and Sim's algorithm is valid only for positively correlated Poisson variables and negative estimates of α are automatically discarded.

Figure 12.3 contains plot of simultaneous coverage probability of the 95% confidence ellipsoids as defined in Equation 12.37 for both the Gaussian and QLS estimates. The coverage probabilities for the QLS method are closer to the nominal level compared to the coverage probabilities of the Gaussian method over a wide range of α. Thus, when the true structure is exchangeable, confidence ellipsoids constructed using QLS are preferable than those constructed using the Gaussian method. Table 12.5 contains the asymptotic relative efficiencies of QLS estimates with respect to the Gaussian estimates. We have presented efficiencies for two regression coefficients; $\widehat{\beta}_0$ is the coefficient of a time independent covariate, and $\widehat{\beta}_5$ is the coefficient of a time dependent covariate. Table 12.5 shows that the QLS estimate of the regression coefficient for time independent covariate is more efficient, whereas the opposite is true for the regression coefficient of time dependent covariate.

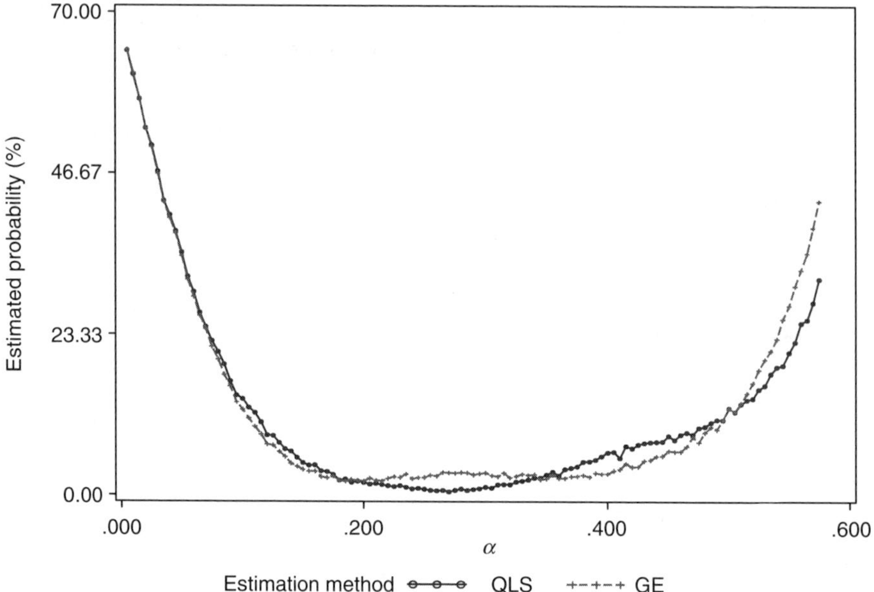

FIGURE 12.2
EQC: Infeasibility/divergent solutions probabilities.

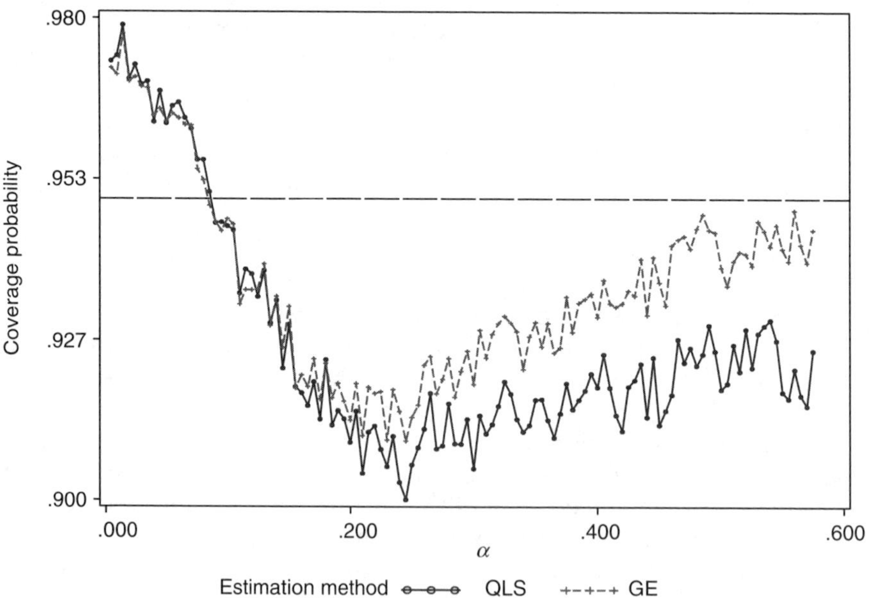

FIGURE 12.3
EQC: Coverage probability of simultaneous confidence region.

TABLE 12.5

EQC: ARE of QLS versus GE

α_0	N_c	$\widehat{\beta_0}$	$\widehat{\beta_5}$	$\widehat{\alpha}$	Λ_1	ν_1	Λ_2	ν_2
0.02	2294	1.0059	0.99765	0.99474	1.0277	6.0274	1.0226	7.0234
0.12	4516	1.0119	0.99421	0.96431	1.0548	6.0538	1.0173	7.0237
0.22	4826	1.0205	0.98673	0.89914	1.0922	6.0894	0.9782	7.0026
0.32	4801	1.0300	0.97402	0.80735	1.1295	6.1243	0.8981	6.9547
0.42	4488	1.0344	0.96027	0.70581	1.1375	6.1326	0.7771	6.8680
0.52	3842	1.0356	0.94502	0.60452	1.1258	6.1232	0.6453	6.7612
0.56	2991	1.0374	0.93672	0.57174	1.1258	6.1241	0.6063	6.7300

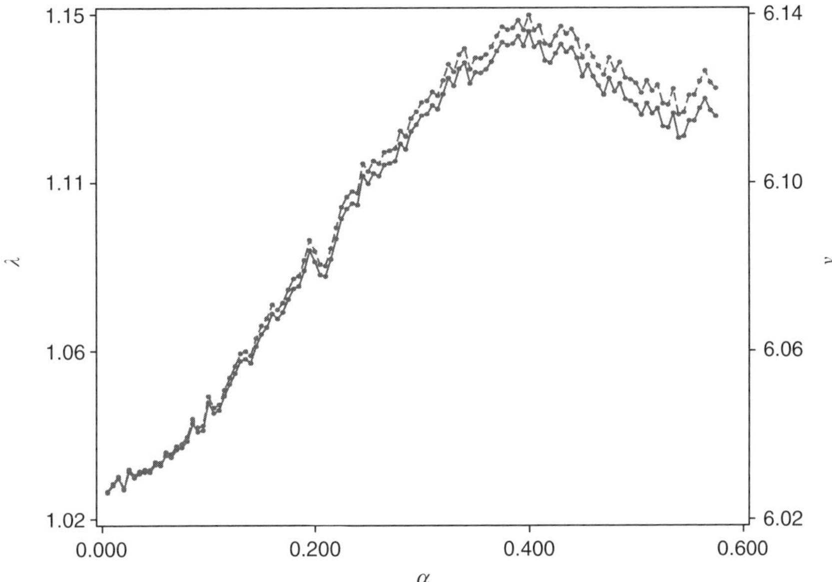

FIGURE 12.4
EQC: ARE of $\widehat{\beta_q}$ vs. $\widehat{\beta_m}$.

Our simulations also show that the QLS estimate of α is less efficient than the Gaussian estimate, and furthermore, the efficiency is decreasing as α increases. However, for the regression parameter, the efficiency of the QLS method as measured by the determinant (Λ_1) or the trace (ν_1) criteria of the submatrix of Γ defined in Equation 12.36, is better than the Gaussian estimate over the entire feasible range of the correlation parameter α. Plots of these efficiencies are in Figure 12.4. Note that the efficiencies are increasing functions of α. Hence, when the regression parameter is of primary interest, the covariates are time independent or mixed, and the correlation is exchangeable but the parameter α is treated as a nuisance parameter, then QLS is preferable over the Gaussian method.

Table 12.5 also shows the overall efficiency of the QLS estimates of the regression and correlation parameters with respect to the Gaussian estimates, as measured by the determinant (Λ_2) or the trace criteria (ν_2) of the matrix Γ defined in Equation 12.36. For small values of α, the overall performance of the QLS is better than the Gaussian estimates. But when there is a strong correlation, Gaussian estimates tend to be more efficient than the QLS estimates.

Simulation results concerning infeasibility, coverage probabilities of the confidence ellipsoids, and the asymptotic relative efficiencies of the regression parameter when the true correlation structure is first order autoregressive (AR(1)) are presented in Figures 12.5 through 12.7, respectively. Table 12.6 contains asymptotic relative efficiencies for the AR(1) structure, similar to the ones that we presented for the exchangeable case in Table 12.5. An examination of the figures and the table of efficiencies shows that the behavior of the QLS method with respect to Gaussian method is very similar to that when the true correlation is exchangeable. Though not reported here, we also performed simulations when the true correlation is moving average of order one (MA(1)). The behavior of the efficiency of the QLS estimate of the regression parameter in this case is different from the other structures. The efficiency as a function of α is approximately concave with a maximum in a neighborhood

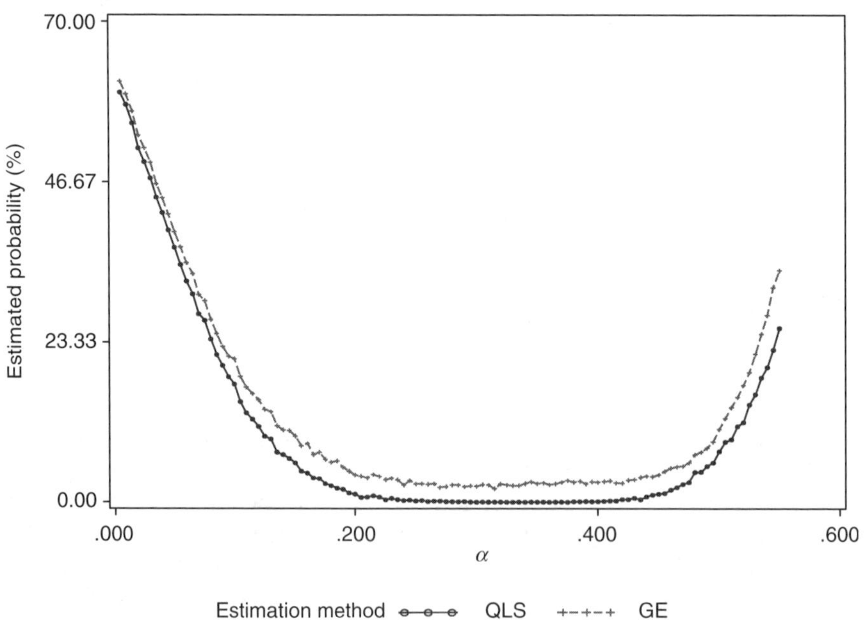

FIGURE 12.5
AR(1): Infeasibility/divergent solutions probabilities.

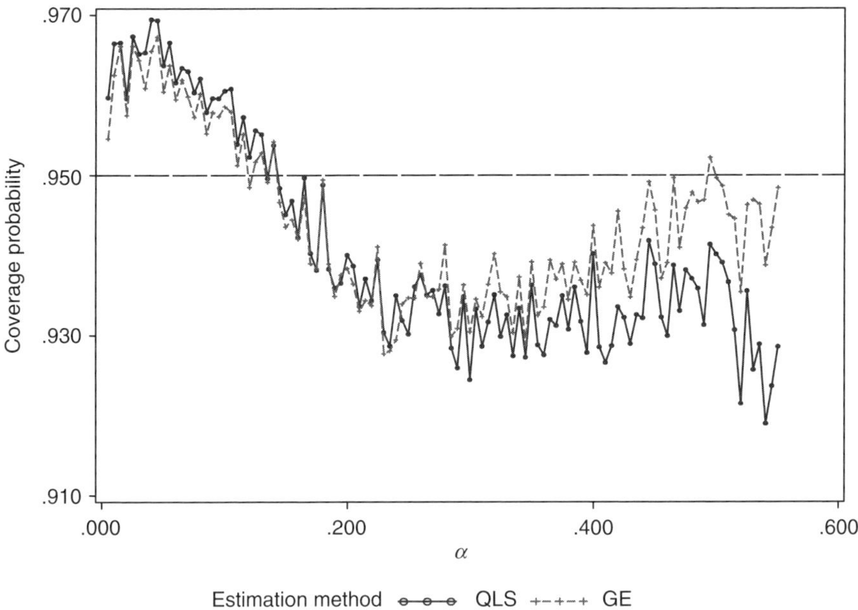

FIGURE 12.6
AR(1): Coverage probability of simultaneous confidence region.

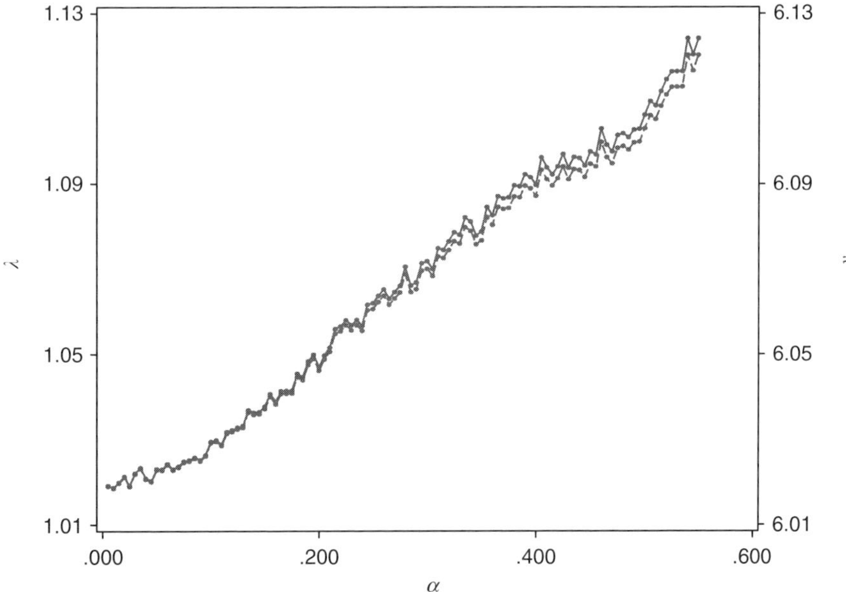

FIGURE 12.7
AR(1): ARE of $\widehat{\beta}_q$ vs. $\widehat{\beta}_m$.

TABLE 12.6
AR(1): ARE of QLS versus GE

α_0	N_c	$\widehat{\beta_0}$	$\widehat{\beta_5}$	$\widehat{\alpha}$	Λ_1	ν_1	Λ_2	ν_2
0.02	2328	1.0045	0.99881	0.98929	1.0213	6.0212	1.0107	7.0115
0.12	4254	1.0068	0.9977	0.97061	1.0323	6.0319	1.0025	7.0057
0.22	4812	1.0122	0.99443	0.92653	1.0565	6.0554	0.97865	6.99
0.32	4867	1.0171	0.98917	0.85847	1.0765	6.0745	0.91908	6.9467
0.42	4847	1.0221	0.98087	0.77332	1.0941	6.0913	0.83091	6.8822
0.52	3948	1.0286	0.96797	0.68886	1.1145	6.111	0.74001	6.8185
0.54	3381	1.0313	0.96383	0.67498	1.1243	6.1202	0.72904	6.8138

of $\alpha = 0.3$, and the efficiency is more than over the entire range of α. When the true structure is MA(1), the coverage probabilities of the Gaussian estimates are closer to the nominal level than the coverage probabilities of the QLS estimates.

In summary, the simulation results show that for commonly used correlation structures, when the covariates are time independent or mixed the QLS estimate of the regression parameter as a whole is more efficient than the Gaussian estimate. But if all the covariates are time varying or if the correlation parameter is not a nuisance parameter, the Gaussian method is preferable over QLS.

12.11 Appendix

The following are the regularity conditions needed for Yuan and Jennrich (1998) theorem:

1. The multivariate distributions $f_i(Z_i, \theta_0)$ exist for $1 \leq i \leq n$.
2. For each i, $\mathrm{E}\,(h_i(Z_i, \theta_0)) = 0$ and $\mathrm{Var}\,(h_i(Z_i, \theta_0)) = M_i(\theta_0)$; and let $1/n \sum_{i=1}^{n} M_i(\theta_0) \to M(\theta_0)$, as $n \to \infty$. Assume that $M(\theta_0)$ is positive definite.
3. For all $\lambda \in \mathbb{R}^k$ of length one there exists positive numbers B and δ such that for all i $\mathrm{E}\,[(\lambda' h_i(Z_i, \theta_0))^2/(1 + \lambda' M_i(\theta_0)\lambda)]^{1+\delta} \leq B$.
4. For each i, $h_i(Z_i, \theta)$ is twice differentiable almost surely on Θ.
5. For each $\theta \in \Theta$, $1/n \sum_{i=1}^{n} \mathrm{E}\,[\partial h_i(Z_i, \theta)/\partial\theta'] \to I(\theta)$ as $n \to \infty$. Assume that $I(\theta_0)$ is nonsingular.
6. Suppose that $|\partial^2(\lambda' h_i(Z_i, \theta))/\partial\theta\partial\theta'| \leq T$ for all i and for all $\lambda \in \mathbb{R}^k$ of length one. Here $|\cdot|$ denotes the determinant.

Acknowledgments

We thank Profs. Dayanand Naik and Ravi Khattree for inviting us to write this chapter. Thanks are also due to Roy Sabo for proof-reading this article.

References

Ashford, J. R. and Sowden, R. R. (1970), "Multivariate probit analysis," *Biometrics*, 26, 535–546.

Barlow, R. E. and Proschan, F. (1975), *Statistical Theory of Reliability and Life Testing: Probability Models*, Holt, Rinehart and Winston, Inc., New York.

Chaganty, N. R. (1997), "An alternative approach to the analysis of longitudinal data via generalized estimating equations," *Journal of Statistical Planning and Inference*, 63, 39–54.

Chaganty, N. R. (2003), "Analysis of growth curves with patterned correlation matrices using quasi-least squares," *Journal of Statistical Planning and Inference*, 117, 123–139.

Chaganty, N. R. and Joe, H. (2004), "Efficiency of generalized estimating equations for binary responses," *Journal of Royal Statistical Society, Ser. B*, 66, 851–860.

Chaganty, N. R. and Joe, H. (2006), "Range of correlation matrices for dependent Bernoulli random variables," *Biometrika*, 93, 197–206.

Chaganty, N. R. and Shults, J. (1999), "On eliminating the asymptotic bias in the quasi-least squares estimate of the correlation parameter," *Journal of Statistical Planning and Inference*, 76, 145–161.

Chib, S. and Greenberg, E. (1996), "Bayesian analysis of multivariate probit models," Econometrics 9608002, Economics Working Paper Archive at WUSTL, available at http://ideas.repec.org/p/wpa/wuwpem/9608002.html.

Chib, S. and Greenberg, E. (1998), "Analysis of multivariate probit models," *Biometrika*, 85, 347–361.

Crowder, M. (1995), "On the use of working correlation matrix in using generalised linear models for repeated measures," *Biometrika*, 82, 407–410.

Crowder, M. (2001), "On repeated measures analysis with misspecified covariance structure," *Journal of Royal Statistical Society*, 63, 55–62.

Godambe, V. P. (1960), "An optimum property of regular maximum likelihood estimation," *Annals of Mathematical Statistics*, 1208–1212.

Godambe, V. P. (ed.) (1991), *Estimating Functions*, Oxford: Oxford University Press.

Hall, D. B. (2001), "On the application of extended quasi-likelihood to the clustered data case," *Candadian Journal of Statistics*, 29, 77–97.

Hall, D. B. and Severini, T. A. (1998), "Extended generalized estimating equations for clustered data," *Journal of the American Statistical Association*, 93, 1365–1375.

Joe, H. (1996), "Time series models with univariate margins in the convolution-closed infinitely divisible class," *Journal of Applied Probability*, 33, 664–677.

Joe, H. (1997), *Multivariate Models and Dependence Concepts*, Chapman & Hall, New York.

Lee, Y. and Nelder, J. A. (2004), "Conditional and marginal models: Another view," *Statistical Science*, 19, 219–238.

Leppik, I. E., et al. (1985), "A double-blind crossover evaluation of progabide in partial seizures," *Neurology*, 35, 285.

Lesaffre, E. and Kaufmann, H. (1992), "Existence and uniqueness of the maximum likelihood estimator for multivariate probit model," *Journal of the American Statistical Association*, 87, 805–811.

Lesaffre, E. and Molenberghs, G. (1991), "Multivariate probit analysis: A neglected procedure in medical statistics," *Statistics in Medicine*, 10, 1391–1403.

Liang, K.-Y. and Zeger, S. L. (1986), "Longitudinal data analysis using generalized linear models," *Biometrika*, 73, 13–22.

Lindsey, J. K. and Lambert, P. (1998), "On the appropriate of marginal models for repeated measurements in clinical trials," *Statistics in Medicine*, 17, 447–469.

Maydeu-Olivares, A. (2001), "Multidimensional item response theory modeling of binary data: Large sample properties of NOHARM estimates," *Journal of Educational and Behavioral Statistics*, 26, 49–69.

Mendell, N. and Elston, R. (1974), "Multifactorial qualitative traits: Genetic analysis and prediction of recurrence risks," *Biometrics*, 30, 41–57.

Muthén, B. (1978), "Contributions to factor analysis of dichotomous variables," *Psychometrika*, 43, 551–560.

Nelder, J. A. and Wedderburn, R. (1972), "Generalized linear models," *Journal of Royal Statistical Society A*, 135, 370–384.

Preisser, J. S. and Qaqish, B. F. (1999), "Robust regression for clustered data with applcation to binary responses," *Biometrics*, 55, 574–579.

Prentice, R. L. (1988), "Correlated binary regression with covariates specific to each binary observation," *Biometrics*, 44, 1033–1048.

Prentice, R. L. and Zhao, L. P. (1991), "Estimating equations for parameters in the means and covariances of multivariate discrete and continuous responses," *Biometrika*, 47, 825–839.

Shults, J. and Chaganty, N. R. (1998), "Analysis of serially correlated data using quasi-least squares," *Biometrics*, 54, 1622–1630.

Sim, C. H. (1993), "Generation of Poisson and gamma random vectors with given marginals and covariance matrix," *Journal of Statistical Computation and Simulation*, 47, 1–10.

Sutradhar, B. C. (2003), "An overview on regression models for discrete longitudinal responses," *Statistical Science*, 18, 377–393.

Thall, P. F. and Vail, S. C. (1990), "Some covariance models for longitudinal count data with overdispersion," *Biometrics*, 46, 657–671.

Wedderburn, R. W. M. (1974). "Quasilikelihood functions, generalized linear models and the Gauss-Newton method," *Biometrika*, 61, 439–447.

Whittle, P. (1961), "Gaussian estimation in stationary time series," *Bulletin of the International Statistical Institute*, 39, 1–26.

Yuan, K.-H. and Jennrich, R. I. (1998), "Asymptotics of estimating equations under natural conditions," *Journal of Multivariate Analysis*, 65, 245–260.

Zhu, R. and Joe, H. (2003), "A new type of discrete self-decomposability and its application to continuous-time Markov processes for modeling count data time series," *Stochastic Models*, 19, 235–254.

Index

D